Analysis
of
Trace Organics
in the
Aquatic Environment

Editors

B. K. Afghan, Ph.D.
Chief
Water Quality National Laboratory
Canada Centre for Inland Waters
Burlington, Ontario, Canada

Alfred S. Y. Chau, M.Sc.
Chief
Quality Assurance Program
National Water Research Institute
Canada Centre for Inland Waters
Burlington, Ontario, Canada

CRC Press, Inc.
Boca Raton, Florida

Library of Congress Cataloging-in-Publication Data

Analysis of trace organics in the aquatic environment / editors, B. K.
  Afghan, Alfred S. Y. Chau.
        p.   cm.
        Bibliography: p.
     Includes index.
     ISBN 0-8493-4626-6
     1. Organic water pollutants—Environmental aspects.   2. Trace
  elements in water.   3. Aquatic ecology.   4. Water quality.   5. Water
  Chemistry.   I. Afghan, B. K.   II. Chau, Alfred S. Y.
  TD427.07A529   1989
  628.1'61—dc19   88-37888
                                                                              CIP

This book represents information obtained from authentic and highly regarded sources. Reprinted material is
quoted with permission, and sources are indicated. A wide variety of references are listed. Every reasonable effort
has been made to give reliable data and information, but the authors and the publisher cannot assume responsibility
for the validity of all materials or for the consequences of their use.

Direct all inquiries to CRC Press, Inc., 2000 Corporate Blvd., N.W., Boca Raton, Florida 33431.

© 1989 by CRC Press, Inc.

International Standard Book Number 0-8493-4626-6

Library of Congress Card Number 88-37888

Printed in the United States of America   2 3 4 5 6 7 8 9 0

PREFACE

The importance of environment assessment pollution control to protect and upgrade environmental quality, has led to the necessity and demand in the study and monitoring of the pollutants in the aquatic environment. All research and surveillance activities to assess the quality of the environment depend upon the availability of analytical data, since scientific conclusions and political decisions on environmental issues are based on their interpretation. Questionable data result in questionable conclusions.

One of the key elements of effective quality assurance to ensure reliable data generation is the availability of reliable analytical methods and their appropriate application. Many books have been published which cover measurement techniques for chemical constituents. The present book is part of a multi-volume work which is designed to provide, in a single source, theories and important analytical techniques and methodologies necessary for understanding and determining the trace organic contaminants in various compartments of the aquatic environment.

The first three volumes dealt with the analysis of pesticide residues and their toxic degradation products in the aquatic environment. This final volume covers the toxaphene, organometallic compounds, humic acids, and related substances and industrial chemicals such as PCBs, phenols, chlorinated dioxins, and volatile organics.

The book is intended to serve as a general reference for university and college students as well as a practical reference for environmental chemists and technologists. It provides sufficient theoretical and practical detail to ensure that a reader would be able to determine trace constituents in an accurate manner. Each chapter covers practical applications of techniques, and provides state-of-the-art methodologies in current use with special reference to sampling, concentration, cleanup, quantitative and confirmatory analysis, etc. The detail provided is such that it should not be necessary to refer to other protocols or manuals to set up a particular method.

The authors of the chapters have attempted to emphasize the practical aspects, and with the amount of information provided, it should be possible to make critical comparisons of the available methods in order to satisfy specific needs.

The authors would like to acknowledge the assistance and support provided by Mrs. T. Searle, Miss M. Srdanovic, and Miss M. Miscione for typing various chapters and coordinating the material as well as maintaining a close liaison with CRC Press.

B. K. Afghan
Alfred S. Y. Chau

THE EDITORS

B. K. Afghan, Ph.D. is presently the Chief, National Water Quality Laboratory of the Water Quality Branch at the Canada Centre for Inland Waters in Burlington, Ontario. He has completed a career of nearly 25 years of active research and service in analytical chemistry.

Dr. Afghan has conducted independent research for over 20 years and has also directed research activities related to analytical chemistry, radiochemistry, and application of electron microscopy to biological studies.

He has published and/or co-authored over 50 technical papers and reports. The work has included the development of new and/or improved methods for analysis of inorganic and organic trace constituents in environmental samples. The techniques utilized included polarographic and related electroanalytical techniques, spectrophotometry, spectrofluorometry, and luminescence, high pressure liquid chromatography, gas chromatography, and gas chromatography/mass spectrometry. He has also written chapters for books and has co-edited books and symposium proceedings.

Dr. Afghan has contributed significantly to the promotion of analytical chemistry and international cooperation in scientific matters. He has served as a member and/or chairperson of various working groups devoted to standardization of chemical and biochemical water quality parameters. He has served in an official capacity to professional societies such as the Chemical Institute of Canada, the American Society for Testing and Materials, and the International Standards Organization.

Dr. Afghan has lectured widely in Canada, the U.S., Europe, and Australia on advances in analytical chemistry.

Alfred S. Y. Chau, M.Sc., is Chief, Quality Assurance Program at the Research and Application Branch National Water Research Institute, Canada Centre for Inland Waters. He obtained his B.Sc. degree from the University of British Columbia in 1961 and his M.Sc. degree from Carleton University, Ottawa, Ontario, Canada.

From 1965 to 1970 he held the position of pesticide analyst in the Department of Agriculture. At the Department of the Environment he was Head of the Organic Laboratories Section and then Head of the Special Services Section. He has held his current position since 1987.

Mr. Chau was the General Referee and is a member of the Association of Official Analytical Chemists, a Fellow of the Chemical Institute of Canada, a Life Fellow of the International Association, U.K., and Fellow of the American Biographical Institute, and is currently on the Advisory Board, National Division of the American Biographical Institute, Chairman of several binational and multiagency quality assurance committees and a member of the Canadian Advisory Committee on Quality Assurance Management, and International Standards Organizations. Mr. Chau is listed in 23 international books of recognition including *Who's Who in the World, Who's Who in America, Personalities of America, International Who's Who of Contemporary Achievement, and International Who's Who of Intellectuals.* He has published over 100 scientific papers in the areas of analytical methodologies, research, and development of certified reference materials for organic and inorganic contaminants in lake sediments.

Further, Alfred Chau is well known as an accomplished nature artist having work admitted to regional and national juried exhibitions such as the Ontario Society for Artists Touring Exhibition (Collectors' Choice Awards, 1981), Ontario Jury Exhibition, 1985; Etobicoke Civic Centre Art Competition (CCAC Award, 1984), World Wildlife Fund Art Auction, 1982, Canada Nature Art National Tour 1982-84. His works are in many private and permanent collections including the Dofasco Canadian Art Collection, Hiram-Walker Art Collection, the Beckett Collection, and IBM.

CONTRIBUTORS

B. K. Afghan, Ph.D.
Chief
National Water Quality Laboratory
Canada Centre for Inland Waters
Environment Canada
Burlington, Ontario, Canada

R. M. Baxter, Ph.D.
Research Scientist
Environmental Contaminants Division
National Water Research Institute
Burlington, Ontario, Canada

John M. Carron, B.Sc.
National Water Quality Laboratory
Canada Centre for Inland Waters
Environment Canada
Burlington, Ontario, Canada

Alfred S. Y. Chau, M.Sc.
Chief
Quality Assurance and Methods Section
National Water Research Institute
Canada Centre for Inland Waters
Burlington, Ontario, Canada

Y. K. Chau, Ph.D.
Research Scientist
National Water Research Institute
Canada Centre for Inland Waters
Burlington, Ontario, Canada

Ray E. Clement, Ph.D.
Senior Scientist
Drinking Water Organics Section
Laboratory Services Branch
Ontario Ministry of the
  Environment
Rexdale, Ontario, Canada

Jagmohan Kohli, Ph.D.
Chemist
National Water Quality Laboratory
Canada Centre for Inland Waters
Environment Canada
Burlington, Ontario, Canada

John Lawrence, Ph.D.
Director
Research and Applications Branch
National Water Research Institute
Burlington, Ontario, Canada

Hing-Biu Lee, Ph.D.
Research Scientist
National Water Research Institute
Environment Canada
Burlington, Ontario, Canada

Barry G. Oliver, Ph.D.
Research Scientist
Canada Centre for Inland
  Waters
National Water Research Institute
Burlington, Ontario, Canada

Francis I. Onuska, Ph.D.
Scientist
Research and Applications Branch
National Water Research Institute
Burlington, Ontario, Canada

**James F. Ryan, Chemical Technology
  Diploma**
Research Technologist
Canada Centre for Inland Waters
Environment Canada
National Water Research Institute
Burlington, Ontario, Canada

David B. Sergeant, B.Sc.(Hons)
Ultratrace Laboratory
Department of Fisheries and Oceans
Bayfield Institute
Great Lakes Laboratory for Fisheries and
  Aquatic Sciences
Burlington, Ontario, Canada

Helle M. Tosine, B.Sc.
Manager
Drinking Water Organics Section
Ministry of the Environment
Rexdale, Ontario, Canada

Paul T. S. Wong, Ph.D.
Research Scientist
Great Lakes Laboratory for Fisheries and
  Aquatic Sciences
Canada Centre for Inland
  Waters
Burlington, Ontario, Canada

TABLE OF CONTENTS

Chapter 1

ANALYSIS OF VOLATILE HALOGENATED AND PURGEABLE ORGANICS

Barry G. Oliver

TABLE OF CONTENTS

I. INTRODUCTION

A. Production and Uses

Interest in the presence of volatile halogenated compounds in water has been high for several years because of the discovery of the ubiquitous presence of trihalomethanes, THMs, in chlorinated drinking water in several countries.[1-4] The THMs, chloroform, bromodichloromethane, chlorodibromomethane, and bromoform, are formed from the reaction of chlorine with naturally occurring organics such as fulvic acid,[5-8] algae,[9-11] etc., during water disinfection. There is also concern about other volatile halogenated compounds that are *not* water chlorination byproducts such as tetrachloroethylene[12] and carbon tetrachloride[4] because they have been found at high concentrations in some drinking waters from groundwater sources.

The purgeable halogenated compounds on the U.S. Environmental Protection Agency's priority pollutant list are shown in Table 1 together with some of their physical properties.[13] In general the compounds have low boiling points (<200°C), high vapor pressures (≥1 torr), low water solubilities, and fairly low octanol/water partition coefficients.

U.S. production figures for some of the compounds are shown in Table 2.[14] It can be seen that massive quantities of these materials are produced in the U.S. which is only a fraction (1/2 to 1/5) of world production. Some of the compounds are used as chemical intermediates, for example, 1,2-dichloroethane in the production of vinyl chloride which in turn is used to produce polyvinyl chloride; chloroform and carbon tetrachloride in the manufacture of fluorohydrocarbons. Many of the compounds are used as solvents in such applications as paint, extraction or reaction media solvents, and in dry cleaning, vapor degreasing, metal cleaning, and textile processing. Metal cleaning and dry cleaning are the first and second largest solvent applications.[14]

The chlorinated hydrocarbons are produced by a variety of chlorination reactions involving hydrocarbon feedstocks such as methane (natural gas), ethane, ethylene, propylene, and propane.[14] Chlorination of benzene in the presence of a catalyst such as ferric chloride is used to produce the chlorinated benzenes. Industrial solvents are to a large extent recovered and total production represents handling losses to the environment. On the other hand, the quantity of chemical intermediates that enter the environment is only a small fraction of the annual production. It should be noted that all THMs produced during water chlorination (drinking or cooling water) will be lost to the environment.[15] In addition to anthropogenic sources, compounds such as CH_3Cl and CH_3Br are thought to have large natural sources mainly in the oceans.[16]

B. Environmental Distribution

1. Ambient Water, Sediments, and Biota

The concentration of volatiles in ambient waters is usually quite low in the ng/L or ppt range. Pearson and McConnell[17] reported C_2HCl_3, C_2CL_4, and CCl_4 concentrations in the 100 to 300 ng/l range for Liverpool Bay, England seawater. Hammer et al.[18] showed that the CCl_3F concentration averaged 20 ng/l in the Gulf of Maine, U.S. Kaiser and Valdmanis[19] report concentrations from 11 to 76 ng/l for CCl_2F_2, $CHCl_3$, CCl_4, and C_2HCl_3 in Lake Erie, Canada. Concentrations of dichlorobenzenes ranged from 2 to 64 ng/l for Lakes Huron and Ontario, Canada.[20] Somewhat higher concentrations in the µg/l or ppb range are found in wastewater effluents[20,21] or in rivers close to outfalls.[22,23,24]

A mass balance study on Lake Zurich, Switzerland on the least volatile purgeable compound in Table 1, 1,4-dichlorobenzene, showed that the major environmental pathway of this compound was volatilization to the atmosphere.[25] Using laboratory studies, Dilling et al.[26] showed that the time required for 50% evaporation from water of many of the chemicals in Table 1 was less than 1 h. Mesocosm experiments[27] also showed that the major process removing these compounds from water is volatilization. The removal of these purge-

Table 1
PURGEABLE HALOCARBONS IN THE U.S. ENVIRONMENTAL PROTECTION AGENCY'S PRIORITY POLLUTANT LIST AND SOME OF THEIR PHYSICAL PROPERTIES[13]

Chemical	Formula	Boiling Point at 760 mm (°C)	Vapor pressure at 20°C (torr)	Water solubility at 20°C (mg/l)	Log octanol/water partition coefficient
Chloromethane	CH_3Cl	-24.2	3765	6450—7250	0.91
Bromomethane	CH_3Br	4.6	1420	900	1.1
Dichlorodifluoromethane	CCl_2F_2	-29.6	4310	280	2.2
Vinyl chloride	C_2H_3Cl	-13.4	2660	60	0.6
Chloroethane	C_2H_5Cl	12.3	1000	5740	1.5
Methylene chloride	CH_2Cl_2	39.8	362	17000	1.3
Trichlorofluoromethane	CCl_3F	23.8	667	1100	2.5
1,1-Dichloroethylene	$C_2H_2Cl_2$	37	591	400	1.5
1,1-Dichloroethane	$C_2H_4Cl_2$	57.3	180	5500	1.8
Trans-1,2-dichloroethylene	$C_2H_2Cl_2$	47.5	200	600	1.5
Chloroform	$CHCl_3$	61.7	151	8200	2.0
1,2-Dichloroethane	$C_2H_4Cl_2$	83.5	61	8690	1.5
1,1,1-Trichloroethane	$C_2H_3Cl_3$	74.1	96	480—4400	2.2
Carbon tetrachloride	CCl_4	76.5	90	785	2.6
Bromodichloromethane	$CHBrCl_2$	90	50	—	1.9
1,2-Dichloropropane	$C_2H_6Cl_2$	96.8	42	2700	2.3
Trans-1,3-dichloropropene	$C_3H_4Cl_2$	112	25	2800	2.0
Trichloroethylene	C_2HCl_3	87	57.9	1100	2.3
Dibromochloromethane	$CHClBr_2$	120	15	—	2.1
1,1,2-Trichloroethane	$C_2H_3Cl_3$	133.8	19	4500	2.2
Cis-1,3-dichloropropene	$C_3H_4Cl_2$	104.3	25	2700	2.0
2-Chloroethylvinyl ether	C_4H_7OCl	108	26.8	15000	1.3
Bromoform	$CHBr_3$	149.5	10	3100	2.3
1,1,2,2-Tetrachloroethane	$C_2H_2Cl_4$	146.2	5	2900	2.6
Tetrachloroethylene	C_2Cl_4	121	14	150—200	2.9
Chlorobenzene	C_6H_5Cl	132	8.8	500	2.8
1,3-Dichlorobenzene	$C_6H_4Cl_2$	173	2.3	123	3.4
1,2-Dichlorobenzene	$C_6H_4Cl_2$	180.5	1.5	145	3.4
1,4-Dichlorobenzene	$C_6H_4Cl_2$	174	0.6	79	3.4

Table 2

**U.S. PRODUCTION FOR SOME
VOLATILE CHLORINATED
HYDROCARBONS[14]**

Compound	Thousands of metric tons	
	1970	1976
Vinyl chloride	1534	2289
1,1-Dichloroethylene	54	70
1,2-Dichloroethane	3773	4773
Chloromethane	192	168
Methylene chloride	182	227
Chloroform	109	125
Carbon tetrachloride	459	385
1,1,1-trichloroethane	166	234
Trichloroethylene	274	127
Tetrachloroethylene	320	304
Chlorobenzene	—	163
1,2-Dichlorobenzene	—	68
1,4-Dichlorobenzene		

able compounds during wastewater treatment occurs mainly by volatilization; biodegradation or adsorption to solids are minor processes.[28] When effluents containing volatile chemicals are discharged at depth (below the waterbody surface) or beneath an ice cover,[29] volatilization is restricted and the compounds make excellent tracers for industrial plumes in lakes[22] and in the ocean.[30] Some exchange of these materials between water and the atmosphere does occur,[31] and some recycling takes place since measurable concentrations of these compounds are found in rainwater.[17] But the ultimate fate of these compounds appears to be destruction in the troposphere by reaction with hydroxyl radicals or photodecomposition or reaction with ozone in the stratosphere.[32] The destruction in the stratospheric ozone layer by reactions with freons and other halocarbons is still a matter of controversy.[33,34]

The concentrations of purgeable halocarbons in ambient lake and river sediments should be low because partitioning onto solids is not a favored process for these chemicals.[35,36] Sediments in Liverpool Bay were in the low $\mu g/Kg$ or ppb range for CCl_4, $CHCl_3$, C_2Cl_4, and C_2HCl_3.[17] Great Lakes sediments[20] and southern Californian coastal sediments[21] contained similar concentrations (low ppb) for dichlorobenzenes. Exceptionally high sediment perchloroethylene concentrations up to percent levels have been reported in the St. Clair River, Canada, but this was due to direct spillage of the solvent into the river.[37]

Similarly, concentrations of these compounds in biota and fish should not be excessively high because of their low octanol/water partition coefficients, K_{ow}s, and low bioconcentration factors, BCFs. Mackay[38] has fitted the equation, $\log BCF = \log K_{ow} - 1.32$, to a set of literature data. From the K_{ow} data in Table 1 (log K_{ow}, 0.6 to 3.4), the maximum bioconcentration factor in fish for these compounds should be about 100. Values in good agreement with this equation for bluegill sunfish[39] and for rainbow trout[40] have been measured in the laboratory. Oliver and Niimi[41] have reported somewhat higher BCFs, 270 to 420, for dichlorobenzenes in rainbow trout exposed to environmental concentrations of these chemicals. Since typical environmental water concentrations for these chemicals are in the low ppt range, the maximum chemical concentrations in fish should be in the low ppb range. Field studies have shown low ppb levels of dichlorobenzenes in lake trout from the Great Lakes[20] and in Dover sole from southern California.[21] Similar low concentrations of $CHCl_3$, CCl_4, C_2HCl_3, and C_2Cl_4 have been found in various species of fish and other biota from Liverpool Bay,[17] from the U.S.,[42] and from Norway.[43] Biomagnification of these types of chemicals does not occur because of the rapid establishment of equilibrium between fish and water concentrations of the chemicals.[17,41] Thus tissue residue levels are governed mainly by the chemical concentration in the water and are not significantly influenced by the fish's food.[41]

2. Drinking Water

The contamination of drinking water with purgeable organics is a problem in many countries. Table 3 shows the concentrations of trihalomethanes in several of these countries.[2,3,44-47] Some limited data are available for other countries,[4] but only fairly extensive studies are included in the table. In general, all water that is disinfected with chlorine contains THMs because of the reaction of chlorine with natural organic matter. Concentrations up to several hundred ppb have been observed in several locations (Table 3). Chlorinated groundwaters contain somewhat lower THM concentrations than chlorinated surface waters from the same region. Groundwaters usually have a much lower total organic carbon (TOC) concentration than surface waters, so usually contain less THM precursor material.[48] Also, groundwaters are usually not as contaminated with bacteria as surface waters so they are not as heavily chlorinated.

The concentration of THMs in water supplies has been related to TOC, chlorine dose, pH, temperature, and bromide concentration — the higher the value for any of these parameters the higher the yield.[8,49-52] The presence of bromide in the water not only leads to the formation of the brominated THMs but also has been shown to increase the total THM yield.[51,53] The bromide reacts with hypochlorous acid to yield hypobromous acid ($Br- + HOCl \rightarrow HOBr + Cl-$) which is a more efficient and faster reacting halogenating agent. Several water supplies have reported quite high concentrations of brominated THMs.[2,54] Chlorination of seawater yields almost exclusively $CHBr_3$[55] because of the presence of high bromide concentrations.

High concentrations of purgeable organics, which are *not* chlorination byproducts, have been found in some drinking waters which use groundwater as the source. For example, in Dubendorf, Switzerland, average tetrachloroethylene concentrations in the water supply in one part of the city were 69 ppb.[12] One well from which the city drew its water had average C_2Cl_4 concentrations of 76 ppb.[12] Similarly, C_2Cl_4 was found at concentrations up to 300 ppb in groundwater wells used by the city of Kalamazoo, Michigan for drinking water.[56] Contamination from dry cleaning establishments were the cause of the above two problems. In Nicaragua, CCl_4 concentrations of 1.1 ppb were found in a city's water supply and tests conducted on source wells close to an industrial park showed concentrations up to 2500 ppb CCl_4.[4] More and more contaminated groundwater aquifers in the vicinity of industrial dumps are being discovered, especially in the southern U.S. where surface water supplies are in short supply.[57,58,59]

Groundwater contamination is usually the result of improper disposal of industrial chemicals. Most C_1 and C_2 chlorinated hydrocarbons have little affinity for soils[34] so, when they are dumped on the soil surface, their movement through the soil with water is only minimally retarded.[60,61,62] This mobility in the soil can result in the contamination of groundwater aquifers.[4,12,56] Most of these chemicals are highly resistant to biological degradation so they pass through the soil unaltered.[62]

C. Health Effects

1. Toxicity

The toxicity of these chemical is low to moderate so direct toxic effects are not expected at environmental concentrations.[63] Occupational exposure to high concentrations or improper use of these chemicals in confined spaces can lead to toxic responses.[64]

Table 3
TRIHALOMETHANE CONCENTRATIONS (µg/l) IN WORLD DRINKING WATERS

Location	Number of cities	$CHCl_3$		$CHBrCl_2$		$CHBr_2Cl$		$CHBr_3$		Ref.
		Range	Median	Range	Median	Range	Median	Range	Median	
U.S.	80	<0.1—311	21	<0.2—116	6	<0.4—100	1.2	<1.0-92	<1.0	2
Texas, U.S. (Ground waters)	11	1.1—14	2.5	ND[a]—53	0.3	ND—173	ND	ND—242	ND	44
Texas, U.S. (Surface water)	14	20—882	190	31—89	49	ND-125	11	ND-53	ND	44
Canada	70	ND-83	9.0	ND-15	1.0	ND-6.5	ND	ND-2.1	ND	3
Ontario, Canada	22	2—121	22	ND-9	4.0	ND-5	ND	NOT MEASURED		45
West Germany	39	0.1—52	2.5	ND-3.4	ND	ND	ND	ND	ND	46
Belgium (Ground waters)	9	ND-26	2.5	0.2—5.3	0.8	0.2—5.5	1.5	ND-2.7	0.7	47
Belgium (Surface waters)	7	4.4—106	49	2.4—38	20	0.4—13	5.8	ND-0.7	ND	47

[a] Not detected.

2. Carcinogenic and Mutagenic Responses

Reviews of the carcinogenic and mutagenic properties of these chemicals have been compiled by Fishbein.[65,68] Vinyl chloride has been clearly implicated in causing angiosarcoma of the liver in occupationally exposed humans.[65] Increased incidences of cancer of the pancreas, lung, and brain also occur in occupationally exposed workers.[65] In addition to vinyl chloride, 1,1-dichloroethylene, trichloroethylene, tetrachloroethylene, chloroform, carbon tetrachloride, 1,1,2-trichloroethane, and 1,1,2,2-tetrachloroethane have been shown to cause cancer (mainly of the liver) in experimental animals.[60,61] These compounds have also been demonstrated to be mutagenic in several bacterial assays.[65,66,69] Very little data on the carcinogenic or mutagenic properties of the fluorocarbons and the chlorobenzenes is currently available.[68]

3. Epidemiologic Studies and Risk Factor Calculations

Since the discovery of the ubiquitous presence of THMs in drinking water, several studies have been conducted to see whether exposure to chlorinated drinking water (and the chlorination byproducts) poses an additional cancer risk.[70-81] Because of the difficulties in the design of epidemiological studies such as the presence of confounding factors, inaccurate records of exposure levels, etc., these studies have not provided conclusive proof of the association between chlorinated compounds in drinking water and increased cancer risk.[82] But the evidence from these studies suggests the possibility of a causal relationship for rectal cancer and, to a lesser extent, for bladder and colon cancer.[92]

Crouch et al.[83] have attempted to calculate cancer risk factors for several U.S. drinking waters using analytical data for specific organics plus carcinogenic potency data for these organics from animal studies. The total cancer risk was assumed to be the sum of the risks for the individual chemicals. A large portion of the total risk factor was found to be due to the purgeable chlorinated organics because of their presence at reasonably high (μg/l) concentrations. The major deficiency of the study was the uncertainties caused by lack of good dose-cancer response data for the brominated trihalomethanes. The investigators concluded that the consumption of certain U.S. drinking waters might pose moderately high risk.[83]

4. Exposure Routes and Doses

A person's total daily intake of a chemical will be the sum of his water, air, and food exposure. Volatile halogenated compounds have concentrations in the range μg/l for water (~2 l consumed/d), μg/m^3 for air (~22 m^3 consumed/d), and μg/kg for food (~2 kg consumed/d).[84] For a typical North American the daily intake of $CHCl_3$ is about 60 μg — 67% from water, 23% from air, and 10% from food.[84] For CCl_4 the typical daily dose is 7.7 μg — 23% from water, 62% from air, and 15% from food.[84] For people who consume chlorinated drinking water, their main source of THM exposure will be from their drinking water.[84] If a groundwater is contaminated with, for example, tetrachloroethylene, the major source of this compound will also be the drinking water.[85]

The exposure of rats to $CHCl_3$, CCl_4, and C_2HCl_3 resulted in elevated serum and adipose concentration of these chemicals.[86] The rats quickly reached a steady state concentration after which continued exposure did not lead to increases in tissue or serum concentrations. Within three to six days after exposure was terminated, most of these halogenated compounds had disappeared from the serum and adipose tissue.[86] The analysis of blood from humans exposed to chlorinated drinking water (chloroform concentration $\approx$100 μg/l) showed that they had higher serum $CHCl_3$ levels than a control group which used unchlorinated ($CHCl_3$ free) water.[87] Serum $CHCl_3$ levels were about 12 μg/l in the exposed group.[87]

From this information it is apparent that exposure to volatile halogenated compounds at doses in the 100 μg/d range leads to chemical concentrations in the body in the μg/kg range. A concentration of chemical at the 1 μg/kg level in the body corresponds roughly to 1000 molecules of chemical per cell.[88] Whether or not long-term exposure to chemicals in this concentration range can cause problems is presently a matter of controversy, but it would seem prudent to reduce our exposure to these chemicals where possible. For example, several methods have been proposed for reducing drinking water THMs: aeration after chlorination; carbon adsorption; flocculation prior to chlorination for precursor removal; ozone, chlorine dioxide, and chloramines as disinfection alternatives, etc.[89-91] The World Health Organization has established a drinking water guideline value of 30 μg/l for chloroform.[92] The Environmental Protection Agency[93] have set 100 μg/l total THMs as the maximum permissible concentration in U.S. drinking waters. A less stringent standard of 350 μg/l has been proposed by the Department of Health and Welfare for Canada.[94]

II. GENERAL ANALYSIS PROCEDURES

A. Sampling and Sample Preservation

The collection of an appropriate representative sample is a vital part of performing an environmental assessment of a chemical. For water, grab samples are most commonly used, but the preferred method for sampling is to collect a composite sample which integrates the effluent, stream, or river over a much longer time frame, e.g., days. Commercially available composite samplers have been tested for volatiles and they collect the sample with minimal losses.[95] But the sample vessels in these devices are open so sample storage over hours or days will lead to volatilization losses. Westrick and Cummins[96] have designed a composite sampler for effluents. The collection vessel in this devise is a modified 2-l graduated cylinder into which a Teflon® float is snugly fitted. A series of samples is introduced through an inlet at the base of the cylinder using a switching valve, and the float moves upward as the sample volume increases. The float seals the liquid surface inhibiting volatilization losses. More recently Tigwell et al.[97] have combined a multichannel positive displacement Teflon® and glass sampler with the cylinder-Teflon® float collection vessel above, to collect composite samples for volatile organics analyses. Unfortunately these two composite samplers are not currently commercially available.

Recently Blanchard and Hardy[98] have proposed a new method to obtain time-weighted-average concentration values for volatiles in water. The method is based on the permeation of volatile organics through a silicon polycarbonate membrane and should prove very useful with further development.

Bottles used for sample collection and storage should be filled to overflowing and sealed with Teflon®-faced silicon rubber septa. Samples stored at room temperature in this way were shown to be stable for at least 8 to 10 days.[99] Partially filled bottles can be stored at 4°C for at least 2 days without significant losses.[95] It is recommended that water samples be stored with as little headspace as possible at ≈4°C and be analyzed within 1 week.

Because these compounds have a low tendency for adsorption or bioconcentration the most important compartment in the aquatic ecosystem will be the water phase.[100] However, if it is necessary to analyze sediments, fish, or other biota for purgeable organics, the samples should be sealed in glass jars with aluminum- or Teflon®-lined caps and frozen on collection. Amin and Narang[101] have shown that sediment samples could be stored up to 90 days if methanol was added at the time of collection. Untreated samples should be analyzed within 7 days of collection to minimize volatilization losses.

B. Extraction, Concentration, and Cleanup
1. Sediments and Fish

Very few methods have been tested for determination of volatile organics in sediments and fish. As mentioned earlier, this is primarily because only a limited amount of partitioning into these phases in the aquatic environment is expected. The methods that have been used fall into two categories — headspace analysis and solvent extraction.

Speis[102] has tested the purge-and-trap technique for sediments (purge-and-trap methodology will be discussed later). A sediment slurry consisting of 15 g of sediment in 100 ml of water is heated to 80°C and purged with helium (60 ml/min) for 4 min. Recoveries from the Tenax trap for sediments spiked with chlorform, 1,1,1-trichloroethane, toluene, tetrachloroethylene, and chlorobenzene in the concentration range 7 to 200 μg/kg were fairly low, 24 to 52%, but were consistent over the concentration range examined. Minimum detectable levels were about 0.1 μg/kg. Michael et al.[103] have described a similar purge-and-trap procedure for volatiles in soils, sediments, and sludges, and reported recoveries mainly in the 60 to 100% range. Michael et al.[104] have used a similar procedure, for recovery of volatiles from human biological samples. Their recoveries from spiked urine and blood were higher and more consistent than Speis' but their recoveries from spiked adipose tissues (which should behave similarly to fish) varied from 13% for chlorobenzene to 80% for methylene chloride.

Recently, Amin and Narange[101] have described a closed-loop-stripping technique for volatiles in sediments. The method uses a Porapak N cartridge for adsorption of the organics and methanol for elution of the chemicals from the cartridge. Recoveries over 80% were obtained for many volatile priority pollutants.

The best procedure developed to date for fish appears to be the solvent extraction procedure of Ofstad et al.[43] This procedure could also be used for sediments although they did not test it for this purpose. The method consists of extracting 5 g of homogenized fish with a mixture of 5 ml pentane and 20 ml isopropanol in a 100-ml glass-stoppered centrifuge flask. After 2 hr of shaking, enough water is added to the mixture to separate the isopropanol and to bring the pentane layer into the narrow part of the flask. After the addition of a small amount of Na_2SO_4 the flask is centrifuged to separated the phases completely. The pentane layer is transferred to a glass-stoppered 10-ml centrifuge tube. The extract is cooled in ice, treated with 3 ml of concentrated H_2SO_4 to remove the lipids, and centrifuged to separate the layers. The pentane layer is then transferred to another centrifuge tube where it is washed with 3 ml of water and then centrifuged prior to gas chromatographic analysis.

Pearson and McConnell[17] used direct pentane extraction for volatiles in sediments and fish, but did not report recoveries. Ofsted's procedure[43] would appear to be preferable since wet sediments and fish must be used for analysis, so the isopropanol in the solvent mixture would tend to dissolve the water in the sample providing better contact between the solvent and the sample.

Hiatt[105] has reported a vacuum distillation procedure using capillary gas chromatography/mass spectrometry for qualitative analysis of fish tissue for volatiles. The procedure is complex and would appear to require further development and simplification to be useful for quantitative analysis.

2. Resin Adsorption

XAD resins have been used for isolating and concentrating volatile organics from water samples.[106-115] The most commonly used resins are XAD-2 and XAD-4. Various sizes of columns have been used from 1.2 mm × 25 mm[107] up to 2.3 cm × 10 cm[108] and flow rates through the columns are kept at ≈5 to 10 times the column bed volume using nitrogen or helium pressurized water reservoirs. Sized resins with finer mesh sizes (e.g., 100/120 Chromosorb 102 or 104) have been shown to give better recoveries (63 to 100%).[109] In general, a 100-fold concentration factor can be achieved using this technique for the THMs.[109] Higher concentration factors are possible for chlorobenzenes using large water volumes,[114] but breakthrough of more volatile compounds such as chloroform occurs under these conditions.[108,109] Detection limits are generally in the low ppb range. Most investigators purify the resin using sequential soxhlet extraction with methanol, acetonitrile, and diethylether,[106] and store the resin under methanol until use. James et al.[115] reported that the use of the solvent series methanol-diethyl ether-water sample, instead of the commonly used series methanol-water sample, reduced impurities arising from the XAD-2 resins. Although resin adsorption is widely used in the analysis of medium and low volatility compounds, it has never been widely adopted for volatiles because of breakthrough problems and because it is less convenient than several other procedures.

3. Direct Aqueous Injection

Nicholson et al.[116] were first to report the determination of THMs in water using the direct aqueous injection (DAI) technique. The water sample (9 μl) is injected into a gas chromatograph equipped with a packed column and a scandium tritide electron capture detector. A comparison of the THM results from the DAI method with the purge-and-trap technique showed it produced much higher results (up to a factor of two).[116] This discrepancy was thought to be due to the conversion of nonvolatile THM intermediates to THMs in the high temperature gas chromatographic injector. The DAI method was studied further by Pfaender et al.[117] They used a [63]Ni electron capture detector and found that injection of more than 1 μl of water led to loss in detector sensitivity. Therefore, they implemented a bypass valve to vent the water so that larger injection volumes (up to 10 μl) could be used. In agreement with the earlier study, Pfaender et al.[117] showed that the DAI technique produced higher results than the purge-and-trap method. But, they showed that a value close to the true THM could be obtained if the sample was reinjected after purging it for 30 min with purified nitrogen, and subtracting this value from the THM value determined before purging.

The DAI method is simple, since it does not require any sample treatment, and it can easily be automated with a GC autoinjector. Automation would be more difficult if it was nonvolatile THM intermediates. The method detection limit is about 1 μg/l and reproducibility is in the 2 to 5% range. Provided that suitable GC separation can be achieved, the method should work well for THMs and for chlorinated ethanes and ethylenes, but it may be insufficiently sensitive for mono- and dichlorobenzenes. To date, its main use has been for monitoring THM concentrations in drinking water.[8,45,118]

4. Liquid-Liquid Extraction.

Extraction of volatile organics into a water immiscible organic solvent is one of the simplest procedures for sample concentration. The organic contaminants tend to partition preferentially into the organic solvent from the water. The equilibrium between the two phases can be expressed mathematically by the Nernst distribution law:

$$K_D = C_s/C_w \tag{1}$$

where K_D is the distribution coefficient and C_s and C_w are the concentration of the chemicals in the solvent and water, respectively. The extraction efficiency of the solvent can be represented by the following equation:

$$E = \frac{100 \, K_D}{K_D + V_w/V_s} \tag{2}$$

where E is the percent extraction efficiency and V_s and V_w are the volumes of the solvent and water. If the distribution coefficient of the chemical is known, it is possible to predict the volume of solvent and the number of extractions required to achieve the desired recovery efficiency using plots such as those developed by Robbins.[119]

The most common solvent used for liquid-liquid extraction (LLE) of volatile organics is pentane,[110,120-127] although hexane,[128] hexane-diisopropylether 1:1,[129] methylcyclohexane,[130,131] and xylene,[132] have also been used. The use of pentane (boiling point 36°C) is preferred by most investigators because it elutes from the gas chromatographic column before most of the volatile halogenated compounds being measured. Also, sufficiently pure pentane is readily available commercially. However, problems associated with solvent losses, because of its high volatility, and with bubble formation in the syringes of autoinjectors[129] have prompted some investigators to use less volatile solvents. In general, a sample to solvent ratio of between 10 and 25 is employed since recoveries for most halogenated volatiles are over 80% at these water/solvent ratios.[120-123] Higher sample to solvent ratios up to 100 have been used but the recovery efficiency can be as low as 20%.[110,127] However, recoveries are concentration independent and reproducible, so high sample/solvent ratios can be used if lower detection limits are required.[127] Solvent extracts are not usually concentrated to smaller volumes by evaporation because of the volatility of the analytes.

Enhancement of LLE recoveries can be accomplished by adding electrolytes such as Na_2SO_4 to the aqueous solution. This salting-out effect has been used by Oliver[133] for dihaloacetonitriles which are poorly recovered by simple LLE.[134]

LLE can be performed by manual shaking of the sample plus solvent in a centrifuge tube or volumetric flask for 2 to 3 min. A more efficient method, when many samples are involved, is to extract the samples in septum-sealed vials on a mechanical or orbital shaker for ≈15 min. In both cases a sample of the organic layer (top) is collected with a syringe and injected into the gas chromatograph for analysis. The method has a precision of about 7 to 14% and gives detection limits in the range 0.1 to 1.0 μg/l.[126]

Recently, Folgelquist et al.[135] reported a completely automated segmented flow LLE procedure for volatile halogenated compounds. The mechanized system, which extracts the sample in a glass coil coated with a hydrophobic layer and separates the layers with the aid of a hydrophobic membrane, is coupled on-line to a gas chromatographic on-column injector. Using an EC detector and injection volumes of up to 130 μl, concentrations down to the pg/l level in seawater have been determined.[135]

5. Static Headspace

The tendency of volatiles to partition into the headspace over aqueous solutions has been used by some investigators as a method of concentration. An aqueous solution with some headspace is equilibrated in a thermostat for 20 to 30 min and a sample of headspace gas over the solution is withdrawn with a gas-tight syringe. This headspace sample is directly injected onto the gas chromatograph for analysis. A summary of the theoretical aspects of the technique have been given by Drozd and Novak,[136] and by Vitenbert.[137] The method has been used by Kaiser and Oliver[138] and by Helz and Hsu[139] for halogenated hydrocarbons using headspace injection volumes up to 1 ml on packed GC columns. For determinations on capillary columns either a small injection volume (≈20 μl)[140] or some type of precolumn cryogenic trap[141] must be used to prevent losses in column efficiency. Procedures for optimization and enhancement of the concentration of volatiles in the headspace by changing temperature, salt concentration, and pH have been discussed by Friant and Suffet.[142] Elevated temperature and salt concentrations increase the concentrations of volatiles in the headspace. Automated systems for headspace gas chromatography have been developed[143,144] which improve the feasibility of the technique.

In general, detection limits in the 0.1 to 1 μ/l range are attainable by the technique,[138] but the reproducibility is not exceptional (±10 to 20%). The procedure also requires a fairly long equilibration time (≈30 min) and is much less convenient than the LLE procedure.

The use of distillation[145,146] and cryogenic trapping of headspace volatiles[19,147] has been used to enhance the sensitivity of the method to the ng/l range. These methods may be useful for broad surveys to pinpoint sources but, to date, the accuracy and reproducibility of the procedures have not been adequately demonstrated.

6. Purge and Trap

The most common method used for quantification of volatile organics in water is the purge-and-trap (P&T) technique first developed by Bellar and Lichtenberg.[148] Briefly, the organics are purged out of solution into the vapor phase with a stream of inert gas, the vapor is swept through a sorption trap where the organics are trapped, and the trap is heated and backflushed with an inert gas to desorb the organics onto a gas chromatographic column. A block diagram of the purge and trap system is shown in Figure 1. Various shapes of stripping vessels have been tested and it was concluded that the Bellar-Lichtenberg design (Figure 1) performed the best.[149]

The sample (5 ml) is purged with helium at ambient temperature for 11 min at a flow rate of 40 ml/min. The trap normally employed is 2,6-diphenylene oxide polymer — Tenax (60/80 mesh) or a combination of Tenax (2/3) and silica gel, grade 15 (1/3) about 0.27 cm ID and 25 cm long. The organics are desorbed from the trap by rapidly heating it to 180°C and backflushing for 4 min with a helium flow of $\approx$20 ml/min. During this desorption phase the packed GC column should be kept at $\approx$30°C to trap the volatiles at the head of the column. If a capillary column is used a much lower desorption purge rate must be used ($\approx$1 to 2 ml/min) over a longer time, and some type of cryogenic trap must be used at the head of the column to prevent serious peak broadening.

The P&T technique has been used extensively in drinking water surveys[2,3,4] and detailed reports discussing methodology and recovery studies on various volatile organics are available.[148-157] Several articles have been written showing that the P&T technique compares favorably with other volatile organic methodologies.[158-160] The detection limit of the method is in the low µg/l range and the reproducibility is good, $\pm$10%.[154] The P&T system and an automated system version are available commercially from Tekmar Company, Cincinnati, OH; Envirochem Inc., Kemblesville, PA; and Chemical Data Systems, Oxford, PA.

Although some attempts have been made to lower the detection limit of the P&T method by increasing sample volumes,[161] in practice this leads to several difficulties. Larger sample volumes require excessively long purge times[162] and large purge gas volumes which could lead to breakthrough of some of the more volatile compounds from the sorption trap. If a combination of larger sample volume and higher purging temperature is used, more water accumulates in the trap. This leads to difficulties in the subsequent gas chromatographic analysis. For these reasons, there is not much flexibility available in sample volume, purging time, or purging temperature.[154] For routine monitoring of drinking water or wastewater the P&T technique has proved to be excellent.

Pankow and Rosen[163] have reported a whole column cryotrapping technique that can be used with the P&T system. It is necessary to have a gas chromatograph with a cryogenic-equipped oven so the capillary column can be cooled to -80°C. The PT̄ device must provide a trap-drying step which involves the passage of dry carrier gas through the trap at an above ambient temperature ($\approx$40°C) for a period of about 2 min. This trap drying step prevents column plugging due to ice formation and permits the injection of all the material in the sample onto the capillary column. The method thus provides excellent sensitivity and good chromatography.

7. Closed-Loop Stripping

The last commonly used procedure for volatiles is the closed-loop-stripping (CSA) procedure developed by Grob et al.[164-167] A block diagram of a recent CSA system is shown in Figure 2.[168] A 1-l sample is thermostatted at the desired temperature ($\approx$40°C) and air is pumped through the solution, then through a small 1.5-mg activated charcoal trap, and then back to the pump in a closed circuit for a period of $\approx$2 hr. In principle, all the volatile organics are purged out of the water and are adsorbed on the charcoal trap. After the stripping/adsorption is completed, the carbon trap is removed from the apparatus and the organics

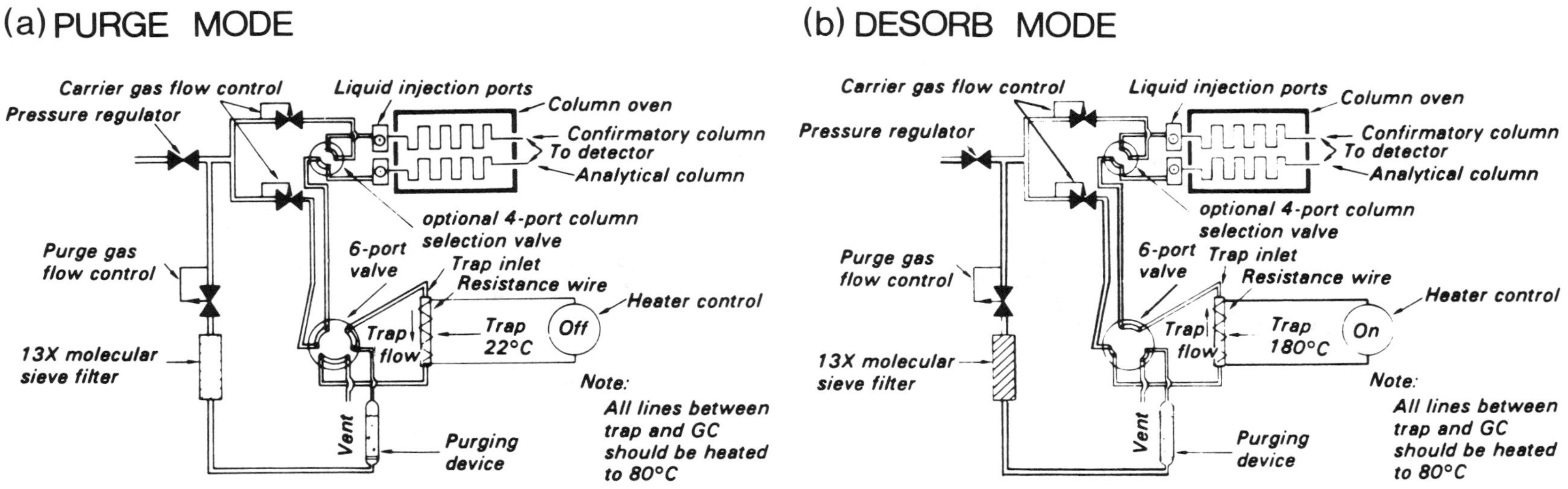

FIGURE 1.   Block diagram of the purge and trap system. (From *Test Methods. Methods for Organic Chemical Analysis of Municipal and Industrial Wastewater*, EPA-600/4-82-057, Environmental Protection Agency, Cincinnati, OH, 1982.)

(d) DESORBER

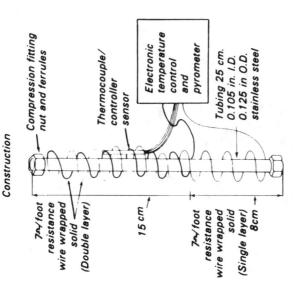

(c) PURGING DEVICE

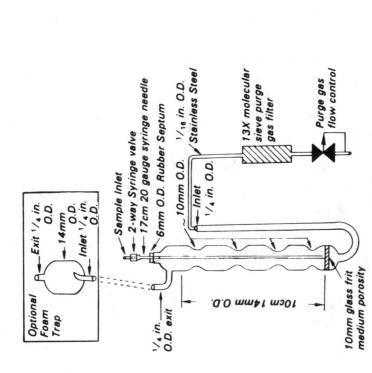

FIGURE 1 (continued)

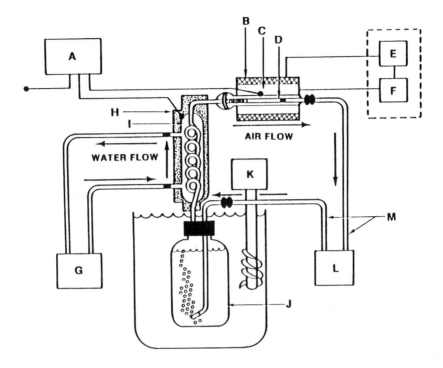

FIGURE 2.   Block diagram of the closed-loop stripping system A, temperature monitor; B, heater block insulation; C, aluminum heating block with cartridge heater; D, glass filter holder with carbon filter; E and F, temperature controller (maintains 50°C); G, thermostatically controlled water circulator for I (set at 95°C); H, foam insulation; I, glass condensing column; J, 1-L sample bottle; K, thermostatically controlled water bath circulator (set at 40°C); L, metal bellows air pump; M, 1/8-in stainless steel tubing. (From Coleman, W. E. et al., *Environ. Sci. Technol.*, 17, 571, 1983. With permission.)

are stripped off the trap with ≈ 8 μl of carbon disulfide (CS_2). About 2 μl of this CS_2 extract is then injected on the GC for analysis. Because all the volatile organics in 250 ml of water are injected into the GC, this method provides a 50-fold enhancement in sensitivity over the P&T method. Unfortunately, recoveries of the more volatile compounds such as $CHCl_3$ are low with CSA[169] so this method cannot be used for their quantification. For compounds in the volatility range between benzene and hexachlorobiphenyl, CSA recoveries are excellent.[168,169] Thus, the method is recommended for and has been successfully used for medium volatility compounds.[167-171] It has also been applied for the analysis of taste and odor producing compounds in drinking water.[172-174] The method detection limit is in the low ng/l range and its reproducibility is about ±9%;[168] the CSA system is available commercially from Brechbuhler AG, Schlieren, Switzerland and from Tekmar Company, Cincinnati, OH.

C. Separation
1. Packed Column Gas Chromatography

The dimensions of packed columns used for this analysis normally range from 2 to 3 m in length and have from 2 to 5 mm inside diameters. A wide range of packing materials, including various polar and nonpolar liquid phases bonded to supports and a variety of porous polymers, have been used for separation of the volatile organics. When the sample is introduced to the chromatograph as a discrete plug, as in the case of injection in a solvent or in headspace gas, columns with low polarity liquid phases with fairly high percentage loadings are usually used to obtain rapid separations. For example, the current recommended column for the LLE procedure is 10% Squalane or Chromosorb W AW, 80/100 mesh, and a commonly used column in headspace analysis is 10% OV-1 on Gas Chrom Q, 80%100 mesh.[138] These columns are generally operated isothermally at 67°C and 50°C, respectively. Argon/methane at 25 ml/min is used as the carrier gas and the injection port temperature is about 100°C. An added advantage of the Squalane column is that materials such as dihalo-acetonitriles, which decompose on many other liquid phases, can also be quantified.

In the P&T method a significant amount of time ($\approx$4 min) is required to desorb the contaminants from the Tenax trap onto the GC column. Because of this slow injection considerable peak broadening will occur unless the organics are trapped in a narrow band at the front of the GC column. More polar columns such as 1% SP-1000 on Carbopak B, 60/80 mesh,[150,158] 0.2% Carbowax 1500 on Carbopak C, 80/100 mesh,[158] and Porapak C[150] are used in the P&T method. To provide a better chromatographic focus it is also acceptable to pack a $\approx$5 cm section at the front on the analytical column with packing having a higher loading of liquid phase e.g., 3% SP-1000 or 3% Carbowax 1500.[158] Generally, the column temperature is kept low 45 to 60°C during the desorption phase of the P&T method and the temperature of the column is programmed at rates between 6 and 8°C/min to 160 to 220°C to elute the organics.

2. Capillary Column Gas Chromatography

Although the separations of volatile organics on packed columns is generally quite good, the use of capillary columns can improve separations and can lead to improved sensitivity because of sharper peaks.[120] A comparison of a packed vs. a capillary column chromatogram for the THMs is shown in Figure 3.[175] Special conditions must be used to attain good separations on capillary columns, since the use of normal columns with film thicknesses of 0.1 to 0.25 μm at normal operating temperature of 50° to 250°C would lead to very poor separation of the volatile organics. First of all, columns with a thicker coating of liquid phase 1 to 1.5 μ[104,123,140] are required to increase the retention time of the volatile organics in the column. Also cryogenic temperature programming of the gas chromatograph provides a further increase in retention time for the volatile organics.[147] Long capillary columns up to 100 m in length[125] and support coated open tubular columns (SCOT)[104] have also been used for volatiles.

The injection of analytes in a small volume of organic solvent or air yields an excellent capillary chromatogram provided the conditions specified above are met. But, when the P&T method is used some type of cryogenic focusing of the organics at the head of the capillary column is required. Trussel et al.[120] cooled the first loop of the capillary column in liquid nitrogen to achieve the required focusing. The organics from the P&T Tenax trap were desorbed with helium at a flow rate of 25 ml/min for 4 mn. Over this period and during the analysis the column flow rate was 1 ml/min so only a fraction of the desorbed organics were injected onto the capillary column. Trussel et al.[120] observed no difficulties with this procedure. Other investigators who have attempted to transfer to the capillary column a much larger fraction of the organics (by desorbing at flow rates comparable to capillary column flows) have run into difficulties with icing and plugging of the columns with traces of water that are also desorbed from the Tenax trap. Special arrangements to overcome this difficulty, using a section of wider bore capillary for cryogenic trapping prior to the analytical column, have been reported by Kalman et al.[176] and by Krost et al.[177] As mentioned earlier a trap-drying step can also be added to the P&T system to overcome this difficulty.[163] With the advent of gas chromatographs equipped with cryogenic ovens, whole column cryotrapping can also be used.

D. Detection

The three most commonly used detector for volatile organics are the electron capture detector (ECD), the electrolytic conductivity detector (ELCD), and the mass spectrometer

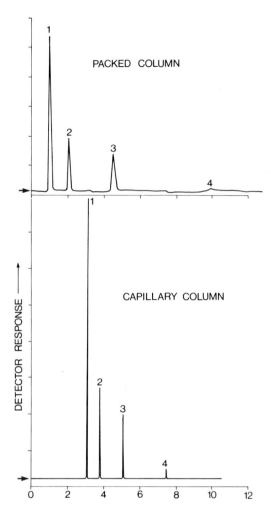

FIGURE 3. A comparison of a packed-column and capillary-column chromatogram for the trihalomethanes (CHCl$_3$, 100 µg/l; CHBrCl$_2$, 5 µg/l; CHBr$_2$ Cl, 4 µg/l; CHBr$_3$, 2 µg/l).

(MS). The ECD is usually used in the liquid-liquid extraction, the headspace, and the direct aqueous injection methods. The ELCD is used almost exclusively with the P&T method. The MS is used for all of the above methods as well as with the CSA technique. Both the ECD and ELCD are halide sensitive detectors. The ECD has an excellent and large response for compounds with three or more halide substituents but the sensitivity drops off exponentially for the di- and monohalo-substituted chemicals. The ELCD response is much lower than the ECD response but is fairly uniform for all halosubstituted compounds regardless of the number of halogens. These characteristics make the ELCD detector a better choice than the ECD for the P&T method, where a fairly large quantity of material (all the organics in the 5 ml sample) is injected into the GC.

The detection limits for the LLE/ECD, P&T/ELCD, and P&T/MS techniques are shown in Table 4. The dramatic dropoff in sensitivity between tri- and di-substituted halogenated compounds is apparent for the LLE/ECD method. A further reduction in sensitivity by a factor of between 100 and 1000 is expected between di- and monohalosubstituted compounds. This method would obviously not be the choice for compounds with few halogenated sub-

Table 4
METHOD DETECTION LIMITS (µg/l) FOR
LLE/ECD, P&T/ELCD, AND P&T/MS[178]

Compound	LLE/ECD[a]	P&T/ELCD[b]	P&T/MS[b]
Chloroform	0.5	0.05	0.1(83)[c]
1,1,1-Trichloroethane	0.07	0.03	0.2(97)
Carbon tetrachloride	0.01	0.1	0.2(117)
1,2-Dichloroethane	9.0	0.03	0.2(62)
Trichloroethylene	0.09	0.1	0.2(95)
Bromodichloromethane	0.04	0.1	0.2(83)
Tetrachloroethylene	0.5	0.03	0.2(166)
Chlorodibromomethane	0.07	0.09	0.2(129)
Bromoform	0.2	0.2	0.5(173)

[a] Effective sample volume injected 0.04 ml.
[b] Effective sample volume injected 5 ml.
[c] Quantification mass, m/z.

stituents. A much lower fraction of the total organics in the sample is injected onto the GC for the LLE procedure compared to the P&T method. In general, detection limits for all the procedures in Table 4 are more than adequate for drinking water, since THMs are usually present in the 1 to 100 µg/l range[2] and concentrations of concern for other compounds are generally in the µg/L range. If lower detection limits are required (as might be the case for ambient waters) the headspace cryogenic trapping/ECD method of Comba and Kaiser[147] could be used for the highly volatile compounds, and the CSA/MS procedure of Grob[164] could be used for the medium volatility compounds.[168] In both techniques the sensitivity enhancement is achieved by injecting onto the GC the organics present in a larger portion of the original sample 1 ml and 250 ml, respectively. If MS detection is used, enhanced sensitivity could also be achieved using selective ion monitoring.[179]

The flame ionization detector[26,128,178,180] has also been used for volatile organics even though its sensitivity, particularly for halogenated compounds, is much lower than the halide-specific detectors already discussed. It is useful for organics with few or no halogens. A microwave emission detector has recently been applied to volatile organics and is reported to have a good or better sensitivity than the ELCD and to offer element specific (bromine, chlorine, iodine) responses.[181,182]

E. Calibration

Calibration standards are prepared in methanol or other appropriate solvents by weighing. (Ligon and Grade[183] have proposed the use of the nonpurgeable poly(ethylene glycol) as a carrier solvent for the P&T technique.) A few drops of the chemical are added to a previously weighed flasks containing the solvent and the flask reweighed to obtain the concentration. For halocarbons that boil below 30°C (bromomethane, chloroethane, chloromethane, dichlorodifluoromethane, trichlorofluoromethane, and vinyl chloride) a gas sampling syringe containing the chemical is positioned just above the solvent in a preweighed solvent-containing flask and the chemicals slowly added. The heavier than air gas rapidly dissolves in the solvent, and, after the addition is completed, the flask is reweighed to determine the concentration. Samples of these individual concentrated standard solutions are pipetted into a flask containing the appropriate solvent to obtain a mixed standard solution at fairly high concentration, from which dilute standard solutions in the required concentration range can be prepared. The diluted mixed standards, which are usually freshly prepared each working day, are used to calibrate the instrumentation. Calibration must be carried out on a daily basis and may even be required periodically throughout the day if detector responses are prone to change. The concentration of chemicals in the samples being quantified must be within the concentration range of the calibration standards.

Both the external standard (ES) and internal standard (IS) approaches have been used for volatile organics. In the ES approach for the LLE procedure, a series of standards in pentane (or other solvent being used) is prepared and calibration graphs for the gas chromatographic procedure are derived for each compound to be quantified. The sample extracts are then run using the same conditions. For the headspace or P&T techniques the external standards in methanol or poly(ethylene glycol) are added directly to purified water at various concentrations and calibration graphs are obtained using the analytical protocols to be used for the samples. For example, for the analysis of chloroform by the P&T method, varying amounts of chloroform would be added to 5 ml of purified water in the P&T vessel, and each standard would be carried through the entire procedure to construct the calibration graph.

In the IS procedure a compound or series of compounds having similar properties to the analytes but not found in the sample are added to each sample or sample extract. These internal standards must also be well separated from the study compounds on the gas chromatograph. The comparative response factor (RF) of each analyte to the internal standard is first measured:

$$RF = A_s C_{Is}/A_{Is} C_s \qquad (3)$$

where A_s = response for the parameter to be measured a_{Is} = response for IS, C_{Is} = concentration of Is, and C_s = concentration of parameter to be measured. After the RFs for each compound have been determined the samples with added IS are run and the concentration of chemicals are determined using the following equation:

$$\text{Concentration } (\mu g/I) = A_s C_{Is}/A_{Is} (RF) \qquad (4)$$

Some IS that have been used are dibromoethane,[126] bromochloromethane, 2-bromo-1-chloropropane, 1,4-dichlorobutane,[150] and $\alpha\alpha\alpha$-trifluorotoluene.[151] The IS procedure should provide more consistent data than the ES procedure but there is little, if any, literature information comparing the two approaches.

III. RECOMMENDED METHODS

The LLE/ECD method is recommended for routine monitoring of drinking water for THMs and for several other halogenated solvents such as trichloro- and tetrachloroethylene because the method exhibits good sensitivity for these compounds, both the extraction and chromatographic procedures are simple, and the method is easily automated. For general volatile organics scanning of drinking water, wastewater, and ambient waters the P&T/ELCD method is recommended. This method provides good sensitivity for a broader range of compounds, is reasonably simple and can be partially automated. Detailed methods for sediments and biota are not presented because the methods are still in the developmental stage, and because the need for such methods (because of the low tendency of the compounds to partition into these phases) has yet to be demonstrated.

A. LLE/ECD Method
Add 10 ml of glass-distilled pentane to the water sample in a Teflon®-lined septum-sealed 120 ml hypovial (actual volume 160 ml). This is accomplished by injecting the pentane into the inverted bottle with a syringe, a second syringe needle (also piercing the septum) allows

10 ml of water sample to escape from the filled bottle as the pentane enters.[121] Place the samples on an orbital or wrist action shaker for 15 to 20 min. Sample the pentane (top) layer with a syringe and inject on the gas chromatograph. Alternatively remove 1 to 2 ml of the pentane layer with a syringe or pipet and place in an autosampler vial for autoinjection onto the GC.

The primary gas chromatographic column used in this analysis is a glass, 2.74 m × 2 mm ID, column-packed with 10% Squalane on 80/100 mesh Chromosorb W AW. Chromatographic conditions are: oven 68°C isothermal; injection port, 100°C; ^{63}Ni ECD, 350°C; argon:methane (90:10) carrier gas flow, 20 ml/min; injection volume, 2 μl (syringe or preferably autoinjector). A second more polar column of similar dimensions packed with a phase such as 6% OV-11 and 4% SP-2100 on 100/120 supelcoport[126] can be used for qualitative confirmation. An integrator should be coupled to the GC output since peak areas rather than peak heights are preferred.

Prepare standard solutions to the analytes in pentane to span the range of concentration required. Run the samples using the same conditions. The chemical concentrations in the sample are calculated by dividing the concentration in the sample extract, after correction for the blank, by 15. Because at the solvent to sample ratio of 1:15 recoveries for volatile organics are greater than 80%, losses due to incomplete recovery are usually neglected. The precision of the method for concentrations in the low ppb range is about ± 10 to 15%.[126]

If the gas chromatographs in your laboratory have capillary capability or if you are in the process of purchasing a gas chromatograph, you should seriously consider using capillary rather than packed columns for volatile analyses. Capillary columns are expensive but currently available bonded phases are extremely stable so columns can be used for at least six months to a year. The primary column should be either OV1 or SE54 or equivalent (1 μm film thickness, 60 m length) with the more polar OV17 as the secondary confirmation column. Split or splitless injection (2 μl) could be used depending on sensitivity requirements with helium as the carrier gas (linear velocity 20-30 cm/sec). An acceptable temperature program would be: initial temperature 70°C (initial hold 10 min), 5°C/min to 200°C (final hold 20 min).

When analyzing chloroform and less volatile compounds hexane should be substituted for pentane as the extraction solvent, since there are fewer operational problems with this less volatile solvent.[128]

B. P&T/ELCD Method

The P&T device consists of the sample purger, the trap, and the desorber (see Figure 1). The trap consists of 7.7 cm lengths of Tenax, silica gel, and activated charcoal. If it is not necessary to analyze the dichlorodifluromethane, the charcoal can be eliminated and the Tenax polymer section lengthed to 15 cm. The desorber must be capable of heating the trap from room temperature to 180°C in less than 30 sec to effect rapid desorption of the organics. P&T devices are available commercially (see Section II.B.6) and detailed design parameters have been published[150,152] so no further discussion is required here.

Inject 5 ml of sample into the purging device with a syringe. Purge sample with helium at 40 ml/min for 11 min. After purging, heat trap rapidly to 180°C while flushing with helium at 25 ml/min in the reverse direction for 4 min. After desorption, initiate temperature program on gas chromatograph. All lines between the trap and the GC must be heated to 80°C to prevent condensation.

The primary gas chromatographic column for this method is 2.44 m long × 2.7 mm I.D. stainless steel or glass packed with 1% SP-1000 on Carbopak B (60/80 mesh). Chromatographic conditions are: oven initial temperature 45°C for 3 min, program at 8°C/min to 220°C, final hold 15 min; injection port 175°C; helium carrier flow rate 40 ml/min. The secondary confirmatory column recommended has the same dimensions as the primary column but contains n-octane coated-Porasil-C (100/120 mesh). The initial column temperature for this column is 50°C (3 min hold) then the column is programmed to 170°C at 6°C/min and held for 4 min.[150] An electrolytic solvent (2-propanol/water, 50/50, v/v) flow 0.12 ml/min, a hydrogen gas flow rate of 10 ml/min, and a furnace temperature of 820°C are used for the Hall electrolytic conductivity detector.[156]

The P&T plus detection device is first calibrated by spiking 5 ml of prepurified and prepurged distilled water with the chemicals of interest and carrying them through the entire procedure or by using the internal standard approach as described in Section II.E. Blanks must also be run. The samples are then analyzed by the same procedure. The precision of the method is about 10 to 20% for concentrations in the low ppb range.

As previously recommended for the LLE method, capillary columns should be used if practicable. For the P&T method, a P&T device with an additional adsorption-column drying step and a gas chromatograph equipped with a cryogenic oven are also required if capillary columns are used.

C. Confirmation

The gas chromatographic methods with halide-sensitive detectors cannot be used to unambiguously identify volatile organics. The certainty of the identification is improved if a second column is used, but positive identification requires both qualitative and quantitative detection with a mass spectrometer. Once a positive identification of a compound in a certain vicinity has been established by MS, monitoring and surveillance activities can be carried out using LLE/ECD or the P&T/ELCD methods with the occasional confirmation by GC/MS. The coupling of the P&T device with a mass spectrometer for both qualitative and quantitative analysis has been discussed in detail by Munch et al.[178] They used the internal standards fluorobenzene for aromatics and 2-bromo-1-chloropropane for aliphatics and report good agreement with the P&T/ELCD method for specific samples. A comparative evaluation of the GC/MS and the GC/electrochemical detector approach has been made by Brass.[184] Although the capitol investment is high, the use of GC/MS must be considered a viable method for routine analysis if a mass spectrometer is available and can be devoted to this purpose. Although the cost per analysis is higher, the added certainty of the organic identifications may make this a cost-effective procedure.

D. Future Directions

The methodology for volatile organics in water is reasonably well advanced, but the methods for other matrices are only in the developmental stage. Although adsorption to sediments and bioconcentration by biota in the aquatic environment may be of minor importance as pointed out earlier, methods for matrices such as soils are needed because dumping on land can lead to groundwater contamination. Therefore, soil analysis methods should aid in the identification of pollutant sources. Also there is a complete lack of interlaboratory comparison studies for these compounds due to the difficulty in producing stable aqueous solutions for round robin studies. Research into methods of preparing samples suitable for such studies and actual interlaboratory comparisons are needed.

The improvement and increased automation of existing methodologies is also a future requirement. Automation in both the sample extraction and in the data analysis are required so that environmental agencies can keep up with increasing analytical demands with decreasing manpower resources. The trend toward capillary column chromatography to improve separations and quantitations is continuing. The direct coupling of capillary columns to mass spectrometers[185] improves the efficiency of sample transfer and also eliminates adsorption and decomposition losses at the GC/MS interface. This may be particularly important for labile compounds such as the dihaloacetonitriles. The development of mass selective detectors may lead to improved sensitivities for volatile organics if selective ion monitoring is used and will definitely lead to more certain compound identifications. These detectors are probably already inexpensive enough to be devoted to volatile organic analysis.

REFERENCES

1. **Rook, J. J.,** Formation of haloforms during the chlorination of natural waters, *Water Treat. Exam.,* 23, 234, 1974.
2. **Symons, J. M., Bellar, T. A., Carswell, J. K., DeMarco, J., Kropp, K. L., Robeck, G. G., Seeger, D. R., Slocum, C. J., Smith, B. L., and Steven, A. A.,** National organics reconnaissance survey for halogenated organics, *J. Am. Water Works Assoc.,* 67, 634, 1975.
3. National Survey for Halomethanes in Drinking Water, Report No. 77-EH-9, Department Health and Welfare Canada, Ottawa, Ontario, 1977.
4. **Trussell, A. R., Cromer, J. L., Umphres, M. D., Kelley, P. E., and Moncur, J. G.,** Monitoring of volatile halogenated organics: a survey of twelve drinking waters from various parts of the world, in *Water Chlorination: Environment Impact and Health Effects, Vol. 3,* Jolley, R. L., Brungs, W. A., and Cumming, R. B., Eds., Ann Arbor Science, Ann Arbor, MI, 1980, 39.
5. **Rook, J. J.,** Chlorination reactions of fulvic acids in natural waters, *Environ. Sci. Technol.,* 11, 478, 1977.
6. **Steven, A. A., Slocum, C. J., Seeger, D. R., and Robeck, G. G.,** Chlorination of organics in drinking water, *J. Am. Water Works Assoc.,* 68, 615, 1976.
7. **Oliver, B. G. and Lawrence, J.,** Haloforms in drinking water: a study of precursors and precursor removal, *J. Am. Water Works Assoc.,* 71, 161, 1979.
8. **Peters, C. J., Young, R. J., and Perry, R.,** Factors influencing the formation of haloforms in the chlorination of humic materials, *Environ. Sci. Technol.,* 14, 1391, 1980.
9. **Hoehn, R. C., Barnes, D. B., Thompson, B. C., Randall, C. W., Grizzard, T. J., and Shaffer, P. T. B.,** Algae as sources of trihalomethane precursors, *J. Am. Water Works Assoc.,* 72, 344, 1980.
10. **Oliver, B. G. and Shindler, D. B.,** Trihalomethanes from the chlorination of aquatic algae, *Environ. Sci. Technol.,* 14, 1502, 1980.
11. **Briley, K. F., Williams, R. F., Longley, K. E., and Sorber, C. A.,** Trihalomethane production from algal precursors, in *Water Chlorination: Environmental Impact and Health Effects, Vol. 3,* Jolley, R. L., Brungs, W. A., and Cumming, R. B., Eds., Ann Arbor Science, Ann Arbor, MI, 1980, 117.
12. **Giger, W. and Molnar-Kubica, E.,** Tetrachloroethylene in contaminated ground and drinking waters, *Bull. Environ. Contam. Toxicol.,* 19, 475, 1978.
13. *Water-Related Environmental Fate of 129 Priority Pollutants, Report EPA-440/4-79-029b,* U.S. Environmental Protection Agency, Cincinnati, OH, 1979.
14. *Kirk Othmer Encyclopedia of Chemical Technology,* Vol. 5, John Wiley & Sons, New York, 1979, 668.
15. **Jolley, R. L., Pitt, W. W., Taylor, F. G., Hartmann, S. J., Jones, G., and Thompson, J. E.,** An experimental assessment of halogenated organics in water from cooling towers and once-through systems, in *Water Chlorination: Environmental Impact and Health Effects, Vol. 2,* Jolley, R. L., Gorchev, H., and Hamilton, D. H., Eds., Ann Arbor Science, Ann Arbor, MI, 1978, 695.
16. **Lovelock, J. E.,** Natural halocarbons in the air and in the sea, *Nature,* 256, 193, 1975.
17. **Pearson, C. R. and McConnell, G.,** Chlorinated C_1 and C_2 hydrocarbons in the marine environment, *Proc. R. Soc. London Ser. B,* 189, 305, 1975.
18. **Hammer, P. M., Hayes, J. M., Jenkins, W. J., and Gagosian, R. B.,** Exploratory analyses of trichlorofluoromethane (F-11) in North Atlantic water columns, *Geophys. Res. Lett.,* 5, 645, 1978.
19. **Kaiser, K. L. E. and Valdmanis, I.,** Volatile chloro- and chlorofluorocarbons in Lake Erie — 1977 and 1978, *J. Great Lakes Res.,* 5, 160, 1979.
20. **Oliver, B. G. and Nicol, K. D.,** Chlorobenzenes in sediments, water, and selected fish from Lakes Superior, Huron, Erie, and Ontario, *Environ. Sci. Technol.,* 16, 532, 1982.
21. **Young, D. R. and Heesen, T. C.,** DDT, PCB and chlorinated benzenes in the marine ecosystem off southern California, in *Water Chlorination: Environmental Impact and Health Effects, Vol. 2,* Jolley, R. L., Gorchev, H., and Hamilton, D. H., Eds., Ann Arbor Science, Ann Arbor, MI, 1978, 267.
22. **Kaiser, K. L. E., Comba, M. E., and Huneault, H.,** Volatile halocarbon contaminants in the Niagara River and in Lake Ontario, *J. Great Lakes Res.,* 9, 212, 1983.
23. **Kaiser, K. L. E. and Comba, M. E.,** Volatile contaminants in the Welland River watershed, *J. Great Lakes Res.,* 9, 274, 1983.
24. **Correia, Y., Martens, G. J., Van Mensch, F. H., and Whim, B. P.,** The occurrence of trichloroethylene, tetrachloroethylene and 1,1,1-trichloroethane in western Europe in air and water, *Atmos. Environ.,* 11, 1113, 1977.

25. **Schwarzenbach, R. P., Molnar-Kubica, E., Giger, W., and Wakeham, S. G.,** Distribution, residence time, and fluxes of tetrachloethylene and 1,4-dichlorobenzene in Lake Zurich, Switzerland, *Environ. Sci. Technol.,* 13, 1367, 1979.
26. **Dilling, W. L., Tefertiller, N. B., and Kallos, G. J.,** Evaporation rates and reactivities of methylene chloride, chloroform, 1,1,1-trichloroethane, trichloroethylene, tetrachloroethylene, and other chlorinated compounds from dilute aqueous solutions, *Environ. Sci. Technol.,* 9, 833, 1975.
27. **Wakeham, S. G., Davis, A. C., and Karas, J. L.,** Mesocosm experiments to determine the fate and persistence of volatile organic compounds in coastal seawater, *Environ. Sci. Technol.,* 17, 611, 1983.
28. **Matter-Müller, C., Gujer, W., Giger, W., and Stumm, W.,** Nonbiological elimination mechanisms in a biological treatment plant, *Prog. Water Technol.,* 12, 299, 1980.
29. **Pecher, K. and Herrmann, R.,** Behaviour of chloroform from pulp bleaching in an ice-covered Finnish lake, *Sci. Total Environ.,* 48, 123, 1986.
30. **Fogelquist, E., Josefsson, B., and Roos, C.,** Halocarbons as tracer substances in studies of the distribution patterns of chlorinated waters in coastal areas, *Environ. Sci. Technol.,* 16, 479, 1982.
31. **Liss, P. S. and Slater, P. G.,** Flux of gases across the air-sea interface, *Nature,* 247, 181, 1974.
32. **Singh, H. B., Salas, L., Shigeishi, H., and Crawford, A.,** Urban-nonurban relationships of halocarbons, SF_6, N_2O and other atmospheric trace constituents, *Atmos. Environ.,* 11, 819, 1977.
33. **Molina, M. J. and Rowland, F. S.,** Stratospheric sink for chlorofluoromethanes: chlorine atom-catalysed destruction of ozone, *Nature,* 249, 810, 1974.
34. **STolarski, R. S. and Cicerone, R. J.,** Stratospheric chlorine: a possible sink for ozone, *Can. J. Chem.,* 52, 1610, 1974.
35. **Chiou, C. T., Peters, L. J., and Freed, V. H.,** A physical concept of soil-water equilibria for nonionic organic compounds, *Science,* 206, 831, 1979.
36. **Karickhoff, S. W.,** Semi-empirical estimation of sorption of hydrophobic pollutants on natural sediments and soils, *Chemosphere,* 10, 833, 1981.
37. **Oliver, B. G. and Pugsley, C. W.,** Chlorinated contaminants in St. Clair River sediments, *Water Pollut. Res. J. Can.,* 21, 368, 1986.
38. **Mackay, D.,** Correlation of bioconcentration factors, *Environ. Sci. Technol.,* 16, 274, 1982.
39. **Barrows, M. E., Petrocelli, S. R., Macek, K. J., and Carroll, J. J.,** Bioconcentration and elimination of selected water pollutants by bluegill sunfish *(Lepomis macrochirus)* in *Dynamics, Exposure and Hazard Assessment of Toxic Chemicals,* Hague, R., Ed., Ann Arbor Science, Ann Arbor, MI, 1980, 379.
40. **Neely, W. B., Branson, D. R., and Blau, G. E.,** Partition coefficient to measure bioconcentration potential of organic chemicals in fish, *Environ. Sci. Technol.,* 8, 1113, 1974.
41. **Oliver, B. G. and Niimi, A. J.,** Bioconcentration of chlorobenzenes from water by rainbow trout: correlations with partition coefficients and environmental residues, *Environ. Sci. Technol.,* 17, 287, 1983.
42. **Ferrario, J. B., Lawler, G. C., DeLeon, I. R., and Laseter, J. L.,** Volatile organic pollutants in biota and sediments of Lake Pontchartrain, *Bull. Environ. Contam. Toxicol.,* 34, 246, 1985.
43. **Ofstad, E. B., Drangsholt, H., and Carlberg, G. E.,** Analysis of volatile halogenated organic compounds in fish, *Sci. Total Environ.,* 20, 205, 1981.
44. **Glaze, W. H. and Rawley, R.,** A preliminary survey of trihalomethane levels in selected east Texas water supplies, *J. Am. Water Works Assoc.,* 71, 509, 1979.
45. **Smillie, R. D., Nicholson, A. A., Meresz, O., Duholke, W. K., Rees, G. A. V., Roberts, K., and Fung, C.,** *Organics in Ontario Drinking Waters, Part II, A Survey of Selected Water Treatment Plants,* Ontario Ministry of the Environment, Rexdale, Ontario, 1977.
46. **Sonneborn, M. and Bohn, B.,** Formation and occurrence of haloforms in drinking water in the Federal Republic of Germany, in *Water Chlorination: Environmental Impact and Health Effects, Vol. 2,* Jolley, R. L., Gorchev, H., and Hamilton, D. H., Eds., Ann Arbor Science, Ann Arbor, MI, 1978, 537.
47. **Quaghebeur, D. and DeWulf, E.,** Volatile halogenated hydrocarbons in Belgian drinking waters, *Sci. Total Environ.,* 14, 43, 1980.
48. **Oliver, B. G. and Thurman, E. M.,** Influence of aquatic humic substance properties on trihalomethane production, in *Water Chlorination: Environmental Impact and Health Effects, Vol. 4, Book 1,* Jolley, R. L., Brungs, W. A., Cotruvo, J. A., Cumming, R. B., Mattice, J. S., and Jacobs, V. A., Eds., Ann Arbor Science, Ann Arbor, MI, 1983, 231.
49. **Rook, J. J.,** Haloforms in drinking water, *J. Am. Water Works Assoc.,* 68, 168, 1976.
50. **Oliver, B. G.,** Effect of temperature, pH and bromide concentration on the trihalomethane reactions of chlorine with aquatic humic material, in *Water Chlorination: Environmental Impact and Health Effects, Vol. 3,* Jolley, R. L., Brungs, W. A., and Cumming, R. B., Eds., Ann Arbor Science, Ann Arbor, MI, 1980, 141.
51. **Minear, R. A. and Bird, J. C.,** Trihalomethanes: impact of bromide concentration on yield, species distribution, rate of formation and influence of other variables, in *Water Chlorination: Environmental Impact and Health Effects, Vol. 3,* Jolley, R. L., Brungs, W. A., and Cumming, R. B., Eds., Ann Arbor Science, Ann Arbor, MI, 1980, 151.

52. **Otson, R., Williams, D. T., Bothwell, P. D., and Quon, T. K.,** Comparison of trihalomethane levels and other water quality parameters for three treatment plants in the Ottawa River, *Environ. Sci. Technol.,* 15, 1075, 1981.
53. **Cooper, W. J., Zika, R. G., and Steinhauer, M. S.,** Bromide-oxidant interactions and THM formation: a literature review, *J. Am. Water Works Assoc.,* 77(4), 116, 1985.
54. **Batjer, K., Gabel, B., Koschorrek, M., Lahl, U., Lierse, K. W., Stachel, B., and Thiemann, W.,** Drinking water in Bremen: trihalomethanes and social costs. A case study of bromoform formation during the chlorination of river water highly contaminated with bromide ions, *Sci. Total Environ.,* 14, 287, 1980.
55. **Sugam, R. and Helz, G. R.,** Seawater chlorination: a description of chemical speciation in *Water Chlorination: Environmental Impact and Health Effects, Vol. 3,* Jolley, R. L., Brungs, W. A., and Cumming, R. B., Eds., Ann Arbor Science, Ann Arbor, MI, 1980, 427.
56. **Minsley, B.,** Tetrachloroethylene groundwater contamination in Kalamazoo, *J. Am. Water Works Assoc.,* 75, 272, 1983.
57. **Ballew, W. W.,** Groundwater laws: opportunities for management and protection, *J. Am. Water Works Assoc.,* 75, 280, 1983.
58. **Stover, E. L. and Kincannon, D. F.,** Contaminated groundwater treatability — a case study, *J. Am. Water Works Assoc.,* 75, 292, 1983.
59. **Westrick, J. J., Mello, J. W., and Thomas, R. F.,** The groundwater supply survey, *J. Am. Water Works Assoc.,* 76(5), 52, 1984.
60. **Greenberg, M., Anderson, R., Keene, J., Kennedy, A., Page, G. W., and Schowgurow, S.,** Empirical test of the association between gross contamination of wells with toxic substances and surrounding land use, *Environ. Sci. Technol.,* 16, 14, 1982.
61. **Schwarzenbach, R. P. and Westall, J.,** Transport of nonpolar organic compounds from surface water to groundwater. Laboratory sorption studies, *Environ. Sci. Technol.,* 15, 1360, 1981.
62. **Schwarzenbach, R. P., Giger, W., Hoehn, E., and Schneider, J. K.,** Behaviour of organic compounds during infiltration of river water to groundwater. Field Studies, *Environ. Sci. Technol.,* 17, 472, 1983.
63. **McConnell, G., Ferguson, D. M., and Pearson, C. R.,** Chlorinated hydrocarbons and the environment, *Endeavour,* 34, 13, 1975.
64. **Stecher, P. G., Ed.,** *The Merck Index, Eighth Edition,* Merck and Co., Rahway, NJ, 1968.
65. **Fishbein, L.,** Potential halogenated industrial carcinogenic and mutagenic chemicals. I. Halogenated unsaturated hydrocarbons, *Sci. Total Environ.,* 11, 111, 1979.
66. **Fishbein, L.,** Potential halogenated industrial carcinogenic and mutagenic chemicals. II. Halogenated saturated hydrocarbons, *Sci. Total Environ.,* 11, 163, 1979.
67. **Fishbein, L.,** Potential halogenated industrial carcinogenic and mutagenic chemicals. III. Alkane halides, alkanols and ethers, *Sci. Total Environ.,* 11, 223, 1979.
68. **Fishbein, L.,** Potential halogenated industrial carcinogenic and mutagenic chemicals. IV. Halogenated aryl derivatives, *Sci. Total Environ.,* 11, 259, 1979.
69. **Simmon, V. F. and Tardiff, R. G.,** The mutagenic activity of halogenated compounds found in chlorinated drinking water, in *Water Chlorination: Environmental Impact and Health Effects, Vol. 2,* Jolley, R. L., Gorchev, H., and Hamilton, D. H., Eds., Ann Arbor Science, Ann Arbor, MI, 1978, 417.
70. **Page, T., Harris, R. H., and Epstein, S. S.,** Drinking water and cancer mortality in Louisiana, *Science,* 193, 55, 1976.
71. **Tardiff, R. G.,** Health effects of organics: risk and hazard assessment of ingested chloroform, *J. Am. Water Works Assoc.,* 69, 658, 1977.
72. **Cantor, K. P. and McCabe, L. J.,** The epidemiologic approach to the evaluation of organics in drinking water, in *Water Chlorination: Environmental Impact and Health Effects, Vol. 2,* Jolley, R. L., Gorchev, H., and Hamilton, D. H., Eds., Ann Arbor Science, Ann Arbor, MI, 1978, 379.
73. **Alavanja, M., Goldstein, J.,a nd Susser, M.,** A case control study of gastrointestinal and urinary tract cancer mortality and drinking water chlorination, in *Water Chlorination: Environmental Impact and Health Effects, Vol. 2,* Jolley, R. L., Gorchev, H., and Hamilton, D. H., Eds., Ann Arbor Science, Ann Arbor, MI, 1978, 395.
74. **Brenniman, G. R., Vasilomanolakis-Lagos, J., Amsel, J., Namekata, T., and Wolff, A. H.,** Case-control study of cancer deaths in Illinois communities served by chlorinated or nonchlorinated water, in *Water Chlorination: Environmental Impact and Health Effects, Vol. 3,* Jolley, R. L., Brungs, W. A., and Cumming, R. B., Eds., Ann Arbor Science, Ann Arbor, MI, 1980, 1043.
75. **Gottlieb, M. S., Carr, J. K., and Morris, D. T.,** Cancer and drinking water in Louisiana: colon and rectum, *Int. J. Epidemiol.,* 10, 117, 1981.
76. **Williamson, S. J.,** Epidemiological studies on cancer and organic compounds in U.S. drinking water, *Sci. Total Environ.,* 18, 187, 1981.
77. **Wilkins, J. R. and Comstock, G. W.,** Source of drinking water at home and site-specific cancer incidence in Washington Country, Maryland, *Am. J. Epidemiol.,* 115, 178, 1981.

78. **Isacson, P., Bean, J. A., and Lynch, C.,** Relationship of cancer incidence rates in Iowa municipalities to chlorination status of drinking water, in *Water Chlorination: Environmental Impact and Health Effects, Vol. 4, Book 2,* Jolley, R. L., Brungs, W. A., Cotruvo, J. A., Cumming, R. B., Mattice, J. S., and Jacobs, V. A., Eds., Ann Arbor Science, Ann Arbor, MI, 1983, 1353.

79. **Young, T. B. and Kanarek, M. S.,** Matched pair case control study of drinking water chlorination and cancer mortality, in *Water Chlorination: Environmental Impact and Health Effects, Vol. 4, Book 2,* Jolley, R. L., Brungs, W. A., Cotruvo, J. A., Cumming, R. B., Mattice, J. A., and Jacobs, V. A., Eds., Ann Arbor Science, Ann Arbor, MI, 1983, 1365.

80. **Cantor, K. P.,** Epidemiologic studies of chlorination by-products in drinking water, in *Water Chlorination: Environmental Impact and Health Effects, Vol. 4, Book 2,* Jolley, R. L., Brungs, W. A., Cotruvo, J. A., Cumming, R. B., Mattice, J. S., and Jacobs, V. A., Eds., Ann Arbor Science, Ann Arbor, MI, 1983, 1381.

81. **Monarca, S., Pasquini, R., and Arcaleni, P.,** Detection of mutagens in unconcentrated and concentrated drinking water supplies before and after treatment using a microscale fluctuation test, *Chemosphere,* 14, 1069, 1985.

82. **Crump, K. S.,** Chlorinated drinking water and cancer: the strength of epidemiologic evidence, in *Water Chlorination: Environmental Impact and Health Effects, Vol. 4, Book 2,* Jolley, R. L., Brungs, W. A., Cotruvo, J. A., Cumming, R. B., Mattice, J. S., and Jacobs, V. A., Eds., Ann Arbor Science, Ann Arbor, MI, 1983, 1481.

83. **Crouch, E. A. C., Wilson, R., and Zeise, L.,** The risks of drinking water, *Water Resour. Res.,* 19, 1359, 1983.

84. *Chloroform, Carbon Tetrachloride, and Other Halomethanes: An Environmental Assessment,* National Academy of Sciences, Washington, D.C., 1978.

85. **Utzinger, R. and Schlatter, C.,** A review on the toxicity of trace amounts of tetrachloroethylene in water, *Chemosphere,* 9, 517, 1977.

86. **Pfaffenberger, C. D., Peoples, A. J., and Enos, H. F.,** Distribution of volatie halogenated organic compounds between rat blood serum and adipose tissue, *Int. J. Environ. Anal. Chem.,* 8, 55, 1980.

87. **Pfaffenberger, C. D., Cantor, K. P., Peoples, A. J., and Enos, H. F.,** Relationship between serum chloroform level and drinking water source: an interim report, in *Water Chlorination: Environmental Impact and Health Effects, Vol. 3,* Jolley, R. L., Brungs, W. A., and Cumming, R. B., Eds., Ann Arbor Science, Ann Arbor, MI, 1980, 1059.

88. **Dougherty, R. C., Whitaker, M. J, Smith, L. M., Stalling, D. L., and Kuehl, D. W.,** Negative chemical ionization studies of human and food chain contamination with xenobiotic chemicals, *Environ. Health Perspec.,* 36, 103, 1980.

89. **Roberts, P. V. and Levy, J. A.,** Energy requirements for air stripping trihalomethanes, *J. Am. Water Works Assoc.,* 77(4), 138, 1985.

90. **Suffet, I. H., Radziul, J. V., Cairo, P. R., and Coyle, J. T.,** Evaluation of the capability of granular activated carbon and resins to remove chlorinated and other trace organics from treated drinking water, in *Water Chlorination: Environmental Impact and Health Effects, Vol. 2,* Jolley, R. L., Gorchev, H., and Hamilton, D. H., Eds., Ann Arbor Science, Ann Arbor, MI, 1978, 561.

91. **Symons, J. M, Carswell, J. K., Clark, R. M., Dorsey, P., Geldreich, E. E., Heffernan, W. P., Hoff, J. C., Love, O. T., McCabe, L. J, and Steven, A. A.,** Ozone, chlorine dioxide and chloramines as alternatives to chlorine for distinction of drinking water, in *Water Chlorination: Environmental Impact and Health Effects, Vol. 2,* Jolley, R. L., Gorchev, H., and Hamilton, D. H., Eds., Ann Arbor Science, Ann Arbor, MI, 1978, 555.

92. Guidelines for Drinking Water Quality, Vol. 1, Recommendations, World Health Organization, Geneva, 1984, 77.

93. *Regulations for Control of Trihalomethanes,* Federal Register Vol. 44, No. 231, U.S. Environmental Protection Agency, Cincinnati, OH, 1979, 68624—68642.

94. *Guidelines for Canadian Drinking Water Quality 1978,* Department Health and Welfare Canada, Ottawa, Ontario, 1979.

95. **Ho, J. S. Y.,** Effect of sampling variables on recovery of volatile organics in water, *J. Am. Water Works Assoc.,* 75, 583, 1983.

96. **Westrick, J. J. and Cummins, M. D.,** Collection of automatic composite samples without atmospheric exposure, *J. Water Pollut. Control Fed.,* 51, 2948, 1979.

97. **Tigwell, D. C., Schaeffer, D. J., and Landon, L.,** Multichannel, positive displacement Teflon and glass sampler for trace organics in water, *Anal. Chem.,* 53, 1199, 1981.

98. **Blanchard, R. D. and Hardy, J. K.,** Continuous monitoring device for the collection of 23 volatile organic priority pollutants, *Anal. Chem.,* 58, 1529, 1986.

99. **Bellar, T. A. and Lichtenberg, J. J.,** Semiautomated headspace analysis of drinking waters and industrial wastes for purgeable volatile organic compounds, in *Measurement of Organic Polutants in Water and Wastewater,* Van Hall, C. E., Ed., American Society for Testing and Materials, STP686, 398, 1979.

100. **Chapman, P. M., Romberg, G. P., and Vigers, G. A.,** Design of monitoring studies for priority pollutants, *J. Water Pollut. Control Fed.,* 54, 292, 1982.

101. **Amin, T. A. and Narang, R. S.,** Determination of volatile organics in sediments at nanogram-per-gram concentrations by gas chromatography, *Anal. Chem.,* 57, 648, 1985.

102. **Speis, D. N.,** Determination of purgeable organics in sediment, in *Hydrocarbons and Halogenated Hydrocarbons in the Aquatic Environment,* Afghan, B. K. and Mackay, D., Eds., Plenum Press, New York, 1980, 201.

103. **Michael, L. C., Thomas, K. W., Sheldon, L. S., Zweidinger, R. A., and Pellizzari, E. D.,** A method for the analysis of purgeable organics in soils, sediments and sludges. Abstracts, *Am. Chem. Soc. Div. Environ. Chem.,* 23(2), 434, 1983.

104. **Michael, L. C., Erickson, M. D., Parks, S. P., and Pellizzari, E. D.,** Volatile environmental pollutants in biological matrices with a headspace purge technique, *Anal. Chem.,* 52, 1836, 1980.

105. **Hiatt, M. H.,** Determination of volatile organic compounds in fish samples by vacuum distillation and fused silica capillary gas chromatography/mass spectrometry, *Anal. Chem.,* 55, 506, 1983.

106. **Junk, G. A., Richard, J. J., Grieser, M. D., Witiak, D., Witiak, J. L., Arguello, M. D., Vich, R., Svec, H. J., Fritz, J. S., and Calder, G. V.,** Use of macroreticular resins in the analysis of water for trace organic contaminants, *J. Chromatogr.,* 99, 745, 1974.

107. **Tateda, A. and Fritz, J. S.,** Mini-column procedure for concentrating organic contaminants from water, *J. Chromatogr.,* 152, 329, 1978.

108. **VanRossum, P. and Webb, R. G.,** Isolation of organic water pollutants by XAD resins and carbon, *J. Chromatogr.,* 150, 381, 1978.

109. **Glaze, W. H., Peyton, G. R., and Raley, R.,** Total organic halogen as water quality parameter: adsorption/microcoulometric method, *Environ. Sci. Technol.,* 11, 685, 1977.

110. **Junk, G. A., Chriswell, C. D., Chang, R. C., Kissinger, L. D., Richard, J. J., Fritz, J. S., and Svec, H. J.,** Applications of resins for extracting organic components from water, *Z. Anal. Chem.,* 282, 331, 1976.

111. **Stepan, S. F. and Smith, J. F.,** Some conditions for use of macroreticular resins in the quantitative analysis of organic pollutants in water, *Water Res.,* 11, 339, 1977.

112. **Stepan, S. F., Smith, J. F., Flego, U., and Reukers, J.,** Apparatus for on-site extraction of organic compounds from water, *Water Res.,* 12, 447, 1978.

113. **Renberg, L.,** Determination of volatile halogenated hydrocarbons in water with XAD-4 resin, *Anal. Chem.,* 50, 1836, 1978.

114. **Oliver, B. G. and Bothen, K. D.,** Determination of chlorobenzenes in water by capillary gas chromatography, *Anal. Chem.,* 52, 2066, 1980.

115. **James, H. A., Steel, C. P., Wilson, I.,** Impurities arising from the use of XAD-2 resin for the extraction of organic pollutants in drinking water, *J. Chromatogr.,* 208, 89, 1981.

116. **Nicholson, A. A., Meresz, O., Lemyk, B.,** Determination of free and total potential haloforms in drinking water, *Anal. Chem.,* 49, 814, 1977.

117. **Pfaender, F. K., Jonas, R. B., Stevens, A. A., Moore, L., and Hass, J. R.,** Evaluation of the direct aqueous injection method of analysis of chloroform in drinking water, *Environ. Sci. Techol.,* 12, 438, 1978.

118. **Young, J. S. and Singer, P. C.,** Chloroform formation in public water supplies: a case study, *J. Am. Water Works Assoc.,* 71, 87, 1979.

119. **Robbins, W. K.,** Representative of extraction efficiencies, *Anal. Chem.,* 51, 1860, 1979.

120. **Trussell, A. R., Umphres, M. D., Leong, L. Y. C., and Trussell, R. R.,** Optimizing trihalomethane analysis, in *Water Chlorination: Environmental Impact and Health Effects, Vol. 2,* Jolley, R. L., Gorchev, H., and Hamilton, D. H., Eds., Ann Arbor Science, Ann Arbor, MI, 1978, 543.

121. **Henderson, J. E., Peyton, G. R., and Glaze, W. H.,** A convenient liquid-liquid extraction method for the determination of halomethanes in water at the parts-per-billion level, in *Identification and Analysis of Organic Pollutants in Water,* Keith, L. H., Ed., Ann Arbor Science, Ann Arbor, MI, 1976, 105.

122. **Richard, J. J. and Junk, G. A.,** Liquid extraction for the rapid determination of halomethanes in water, *J. Am. Water Works Assoc.,* 69, 62, 1977.

123. **Eklund, G., Josefsson, B., and Roos, C.,** Determination of volatile halogenated hydrocarbons in tap water, seawater and industrial effluents by glass capillary gas chromatography and electron capture detection, *J. High Resolut. Chromatogr. Chromatogr. Commun.,* 1, 34, 1978.

124. **Hu, H. C. and Weiner, P. H.,** Modifications to methods for volatile organics analysis at trace levels, *J. Chromatogr. Sci.,* 18, 333, 1980.

125. **Stock, W. and Alberti, J.,** Analytik von organischen Chlorverbindungen im Wasser, *Vom Wasser,* 52, 75, 1979.

126. **Glaze, W. H., Rawley, R., Burleson, J. L., Mapel, D., and Scott, D. R.**, Further optimization of the pentane liquid-liquid extraction method for the analysis of trace organic compounds in water, in *Advances in the Identification and Analysis of Organic Pollutants in Water, Vol. 1*, Keith, L. H., Ann Arbor Science, Ann Arbor, MI, 1981, 267.

127. **Junk, G. A., Ogawa, I., and Svec, H. J.**, Extraction of organic compounds from water using small amounts of solvents, in *Advances in the Identification and Analysis of Organic Pollutants in Water, Vol. 1*, Keith, L. H., Ed., Ann Arbor Science, Ann Arbor, MI, 1981, 281.

128. **Otson, R. and Williams, D. T.**, Evaluation of a liquid-liquid extraction technique for water pollutants, *J. Chromatogr.*, 212, 187, 1981.

129. **Norin, H. and Renberg, L.**, Determination of trihalomethanes (THM) in water using high efficiency solvent extraction, *Water Res.*, 14, 1397, 1980.

130. **Mieure, J. P.**, A rapid and sensitive method for determining volatile organohalides in water, *J. Am. Water Works Assoc.*, 69, 60, 1977.

131. **Nicolson, B. C., Bursill, D. B., and Couche, D. J.**, Rapid method for the analysis of trihalomethanes in water, *J. Chromatogr.*, 325, 221, 1985.

132. **Inoko, M., Tsuchiya, M., and Matsuno, T.**, Determination of lower halogenated hydrocarbons by the water-xylene extraction method, *Environ. Pollut.*, B7, 129, 1984.

133. **Oliver, B. G.**, Dihaloacetonitriles in drinking water: algae and fulvic acid as precursors, *Environ. Sci. Technol.*, 17, 80, 1983.

134. **Trehy, M. L. and Bieber, T. I.**, Detection, identification and quantitative analysis of dihaloacetonitriles in chlorinated natural waters, in *Advances in the Identification and Analysis of Organic Pollutants in Water, Vol. 2*, Keith, L. H., Ed., Ann Arbor Science, Ann Arbor, MI, 1981, 941.

135. **Fogelquist, E., Krysell, M., and Danielsson, L. G.**, On-line liquid-liquid extraction of a segmented flow directly coupled to on-column injection into a gas chromatograph, *Anal. Chem.*, 58, 1516, 1986.

136. **Drozd, J. and Novak, J.**, Headspace gas analysis by gas chromatography, *J. Chromatogr.*, 165, 141, 1979.

137. **Vitenberg, A. G.**, Theory of gas chromatographic headspace analysis with pneumatic sampling, *J. Chromatogr. Sci.*, 22, 122, 1984.

138. **Kaiser, K. L. E. and Oliver, B. G.**, Determination of volatile halogenated hydrocarbons in water by gas chromatography, *Anal. Chem.*, 48, 2207, 1976.

139. **Helz, G. R. and Hsu, R. Y.**, Volatile chloro- and bromocarbons in coastal waters, *Limnol. Oceanogr.*, 23, 858, 1978.

140. **Morton, C. E., Roberts, D. J., and Cooke, M.**, Simple gas-loop injection system for use with capillary columns, *J. Chromatogr.*, 280, 119, 1983.

141. **Drozd, J., Novak, J., and Rijks, J. A.**, Quantitative and qualitative head-space gas analysis of parts per billion amounts of hydrocarbons in water. A study of model systems by capillary-column gas chromatography with splitless sample injection, *J. Chromatogr.*, 158, 471, 1978.

142. **Friant, S. L. and Suffet, I. H.**, Interactive effects of temperature, salt concentration, and pH on headspace analysis for isolating volatile trace organics in aqueous environmental samples, *Anal. Chem.*, 51, 2167, 1979.

143. **Kolb, B., Pospisil, P., Borath, T. and Auer, M.**, Headspace gas chromatography with glass capillaries using an automatic electropneumatic dosing system, *J. High Resolut. Chromatogr. Chromatogr. Commun.*, 2, 283, 1979.

144. **Otson, R.**, Automatic liquid injector for headspace gas chromatography, *Anal. Chem.*, 53, 929, 1981.

145. **Chian, E. S. K., Kuo, P. P. K., Cooper, W. J., Cowen, W. F., and Fuentes, R. C.**, Distillation/headspace/gas chromatographic analysis of volatile polar organics at ppb level, *Environ. Sci. Technol.*, 11, 282, 1977.

146. **Kozloski, R. P.**, Simple method for concentrating volatiles in water for gas chromatographic analysis by vacuum distillation, *J. Chromatogr.*, 346, 408, 1985.

147. **Comba, M. E. and Kaiser, K. L. E.**, Determination of volatile contaminants at the ng.L^{-1} level in water by capillary gas chromatography with electron capture detector, *Intern. J. Environ. Anal. Chem.*, 16, 17, 1983.

148. **Bellar, T. A. and Lichtenberg, J. J.**, Determination of volatile organics at microgram-per litre levels by gas chromatography, *J. Am. Water Works Assoc.*, 66, 739, 1974.

149. **Kuo, P. P. K., Chian, E. S. K., DeWalle, F. B., and Kim, J. H.**, Gas stripping, sorption, and thermal desorption procedures for preconcentrating volatile polar water-soluble organics from water samples for analysis by gas chromatography, *Anal. Chem.*, 49, 1023, 1977.

150. **Longbottom, J. E. and Lichtenberg, J. J., Eds.**, Method 601, purgeable halocarbons, in *Test Methods. Methods for Organic Chemical Analysis of Municipal and Industrial Wastewater*, EPA-600/4-82-057, Environmental Protection Agency, Cincinnati, OH, 1982.

151. **Longbottom, J. E. and Lichtenberg, J. J., Eds.,** Method 602, purgeable aromatics, in *Test Methods. Methods for Organic Chemical Analysis Municipal and Industrial Wastewater,* EPA-600/4-82-057, U.S. Environmental Protection Agency, Cincinnati, OH, 1982.
152. **Bellar, T. A. and Lichtenberg, J. J.,** Semiautomated headspace analysis of drinking waters and industrial waters for purgeable volatile organic compounds, in *Measurement of Organic Pollutants in Water and Wastewater,* Van Hall, C. E., Ed., STP686, American Society for Testing and Materials, Philadelphia, PA, 1979, 108.
153. **Bellar, T. A., Lichtenberg, J. J., and Eichelberger, J. W.,** Determination of vinyl chloride at μg/l level in water by gas chromatography, *Environ. Sci. Technol.,* 10, 926, 1976.
154. **Olynyk, P., Budde, W. L., and Eichelberger, J. W.,** Simultaneous qualitative and quantitative analyses. I. Precision study of compounds amenable to the inert gas-purge-and-trap method, *J. Chromatogr. Sci.,* 19, 377, 1981.
155. **Spingarn, N. E., Northington, D. J., and Pressely, T.,** Analysis of volatile hazardous substances by GC/MS, *J. Chromatogr. Sci.,* 20, 286, 1982.
156. **Otson, R. and Williams, D. T.,** Headspace chromatographic determination of water pollutants, *Anal. Chem.,* 54, 942, 1982.
157. **Adlard, E. R. and Davenport, J. N.,** A study of some of the parameters in purge and trap gas chromatography, *Chromatographia,* 17, 421, 1983.
158. **Brass, H. J.,** The analysis of trihalomethanes in drinking water by purge and trap and liquid-liquid extraction, *Amer. Lab.,* 12(7), 23, 1980.
159. **Thomason, M. M. and Bertsch, W.,** Evaluation of sampling methods for the determination of trace organics in water, *J. Chromatogr.,* 279, 383, 1983.
160. **Nunez, A. J., Gonzalez, L. F., and Janak, J.,** Pre-concentration of headspace volatiles for trace organic analysis by gas chromatography, *J. Chromatogr.,* 300, 127, 1984.
161. **Kopfler, F. C., Melton, R. G., Lingg, R. D., and Coleman, W. E.,** GC/MS determination of volatiles for the national organics reconnaissance survey (NORS) of drinking water, in *Identification and Analysis of Organic Pollutants in Water,* Keith, L. H., Ed., Ann Arbor Science, Ann Arbor, MI, 1976, 87.
162. **Lingg, R. D., Melton, R. G., Kopfler, F. C., Coleman, W. E., and Mitchell, D. E.,** Quantitative analysis of volatile organic compounds by GC-MS, *J. Am. Water Works Assoc.,* 69, 605, 1977.
163. **Pankow, J. F. and Rosen, M. E.,** The analysis of volatile compounds by purge and trap with whole column cryotrapping (WCC) on a fused silica capillary column, *J. High Resolut. Chromatogr. Chromatogr. Commun.,* 7, 504, 1984.
164. **Grob, K.,** Organic substances in potable water and in its precursor. I. Methods for their determination by gas-liquid chromatography, *J. Chromatogr.,* 84, 255, 1973.
165. **Grob, K. and Grob, G.,** Organic substances in potable water and its precursor. II. Applications in the area of Zurich, *J. Chromatogr.,* 90, 303, 1974.
166. **Grob, K., Grob, K., Jr., and Grob, G.,** Organic substances in potable water and its precursor. III. The closed loop stripping procedure compared with rapid liquid extraction, *J. Chromatogr.,* 106, 299, 1975.
167. **Grob, K. and Zurcher, F.,** Stripping of trace organic substances from water equipment and procedure, *J. Chromatogr.,* 117, 285, 1976.
168. **Coleman, W. E., Munch, J. W., Slater, R. W., Melton, R. G., and Kopfler, F. C.,** Optimization of purging efficiency and quantification of organic contaminants from water using a 1-L closed-loop-stripping apparatus and computerized capillary column GC/MS, *Environ. Sci. Technol.,* 17, 571, 1983.
169. **Melton, R. G., Coleman, W. E., Slater, R. W., Kopfler, F. C., Allen, W. K., Aurand, T. A., Mitchell, D. E., and Voto, S. J.,** Comparison of Grob closed-loop-stripping analysis with other trace organic methods, in *Advances in the Identification and Analysis of Organic Pollutants in Water, Vol. 2,* Keith, L. H., Ed., Ann Arbor Science, Ann Arbor, MI, 1981, 597.
170. **Coleman, W. E., Melton, R. G., Slater, R. W., Kopfler, F. C., Voto, S. J., Allen, W. K., and Aurand, T. A.,** Determination of organic contaminants by Grob closed-loop-stripping technique, *J. Am. Water Works Assoc.,* 73, 119, 1981.
171. **Coleman, W. E., Allen, W. K., Slater, R. W., Voto, S. J., Melton, R. G., Kopfler, F. C., and Aurand, T. A.,** Automatic quantification and statistical evaluation of organic contaminants using computerized glass capillary gas chromatography/mass spectrometry system and Grob closed-loop-stripping, in *Advances in the Identification and Analysis of Organic Pollutants in Water, Vol. 2,* Keith, L. H., Ed., Ann Arbor Science, Ann Arbor, MI, 1981, 675.
172. **Krasner, S., W., Hwang, C. J., and McGuire, M. J.,** Development of a closed-loop-stripping technique for the analysis of taste- and odor-causing substances in drinking water, in *Advances in Identification and Analysis of Organic Pollutants in Water, Vol. 2,* Keith, L. H., Ed., Ann Arbor Science, Ann Arbor, MI, 1981, 689.
173. **Krasner, S. W., Hwang, C. J., and McGuire, M. J.,** A standard method for quantification of earthy-musty odorants in water, sediments, and algal cultures, *Wat. Sci. Technol.,* 15(6/7), 127, 1983.

174. **Savenhed, R., Boren, H., Grimvall, A., and Tjeder, A.,** Stripping techniques for the analysis of odorous compounds in drinking water, *Wat. Sci. Technol.,* 15(6/7), 139, 1983.

175. **Trussell, A. R.,** James M. Montgomery Consulting Engineers, Pasadena, CA, private communication, 1977.

176. **Kalman, D., Dills, R., Perera, C., and DeWalle, F.,** On-column cryogenic trapping of sorbed organics for determination by capillary gas chromatography, *Anal. Chem.,* 52, 1993, 1980.

177. **Krost, K. J., Pellizzari, E. D., Walburn, S. G., and Hubbard, S. G.,** Collection and analysis of hazardous organic emission, *Anal. Chem.,* 54, 810, 1982.

178. **Munch, D. J., Munch, J. W., Feige, M. A., Glick, E. M., and Brass, H. J.,** A scheme for the routine analysis of purgeable compounds by gas chromatography/mass spectrometry, in *Advances in the Identification and Analysis of Organic Pollutants in Water, Vol. 2,* Keith, L. H., Ed., Ann Arbor Science, Ann Arbor, MI, 1981, 713.

179. **Budde, W. L. and Eichelberger, J. W.,** *Organics Analysis Using Gas Chromatography/Mass Spectrometry — A Technique and Procedures Manual,* Ann Arbor Science, Ann Arbor, MI, 1979.

180. **Dowty, B. J., Antoine, S. R., and Laseter, J. L.,** Quantitative and qualitative analysis of purgeable organics by high-resolution gas chromatography and flame ionization detection, in *Measurement of Organic Pollutants in Water and Wastewater,* Hall, C. E., Ed., STP686, American Society for Testing and Materials, Philadelphia, PA, 1979, 24.

181. **Quimby, B. D., Delaney, M. F., Uden, P. C., and Barnes, R. M.,** Determination of trihalomethanes in drinking water by gas chromatography with a microwave plasma emission detector, *Anal. Chem.,* 51, 875, 179.

182. **Chiba, K. and Haraguchi, H.,** Determination of halogenated organic compounds in water by gas chromatography/atmospheric pressure helium microwave-induced plasma emission spectrometry with a heated discharge tube for pyrolysis, *Anal. Chem.,* 55, 1504, 1983.

183. **Ligon, W. V. and Grade, H.,** Poly(ethylene glycol) as a diluent for preparation of standards for volatile organics in water, *Anal. Chem.,* 53, 920, 1981.

184. **Brass, H. J.,** Procedures for analyzing organic contaminants in drinking water, *J. Am. Water Works Assoc.,* 74, 107, 1982.

185. **Jensen, T. E., Kaminsky, R., McVeety, B. D., Wozniak, T. J., and Hites, R. H.,** Coupling of fused silica capillary gas chromatographic columns to three mass spectrometers, *Anal. Chem.,* 54, 2388, 1982.

Chapter 2

POLYCHLORINATED BIPHENYLS

Barry G. Oliver, Robert M. Baxter, and Hing-Biu Lee

TABLE OF CONTENTS

I. INTRODUCTION

A. Structure, Nomenclature, Production, Properties, and Disposal

Polychlorinated biphenyls (PCBs) are produced by the reaction of biphenyl (Figure 1) with anhydrous chlorine in the presence of iron filings or ferric chloride as a catalyst[1] according to the equation

$$C_{12}H_{10} + n\ Cl_2 \rightarrow C_{12}H_{10-n}Cl_n + n\ HCl \qquad (1)$$

where n can have any value from one to ten. The term "polychlorinated biphenyl" was originally limited in its application to commercial mixtures manufactured in this way.[2] Subsequently, however, the term has, perhaps unfortunately, come to be applied to any mixture of chlorinated biphenyls and even to individual compounds.

PCBs are, or have been, manufactured in several countries. In the U.S. they were produced by Monsanto under the trade name "Aroclor®", in Germany by Bayer AG ("Clophen®"), in France by Prodelec ("Phenoclor®"), in Japan by Kanegafuchi ("Kanechlor®"), and also in the United Kingdom, Italy, Czechoslovakia, and the U.S.S.R.[3]

By controlling the quantity of chlorine used in the reaction, the extent of chlorination can be varied to produce a series of products of differing physical properties. The Aroclors®, which probably have been most studied, are generally designated by four-digit numbers in which the first two digits indicate the starting material and the last two, the percent chlorine by weight in the final product. Thus, Aroclors® 1221, 1232, 1242, 1248, 1254, 1260, 1262, and 1268 are manufactured by chlorinating biphenyl to a final chlorine content of 21 to 68%. The 25 and 44 series are manufactured from mixtures of biphenyl with 25 and 44 percent terphenyl, respectively, and the 54 series are chlorinated terphenyls.[1] Aroclor® 1016 is similar to Aroclor® 1242 in chlorine content, but contains fewer *tetra-* and *penta*-chlorobiphenyls.

The viscosities of PCBs increases with increasing chlorine content; Aroclor® 1221 is a mobile oil while Aroclor® 1260 is a fairly hard resin. They all have low solubilities in water and low vapor pressures. Their most important property from the practical point of view is their general inertness; they resist both acids and alkalis, and have thermal stability. They were, therefore, widely used wherever such properties were desirable, for example, as heat exchange and hydraulic fluids, lubricants, dielectric fluids in transformers and capacitors, and as plasticizers in applications where flame resistance is important.

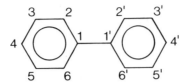

FIGURE 1. Structure and numbering system of biphenyl.

Table 1
POSSIBLE PCB CONGENERS, MOLECULAR
WEIGHTS, AND PERCENT CHLORINE OF
THE VARIOUS PCB ISOMERS

Parent biphenyl	Number of isomers	Molecular weight	Chlorine (%)
Monochloro-	3	188.7	18.8
Dichloro-	12	223.1	31.8
Trichloro-	24	257.5	41.3
Tetrachloro-	42	292.0	48.6
Pentachloro-	46	326.4	54.3
Hexachloro-	42	360.9	58.9
Heptachloro-	24	395.3	62.8
Octachloro-	12	429.8	66.0
Nonachloro	3	464.2	68.7
Decachloro-	1	498.7	71.2
Total	209		

It can be seen from the structure of the biphenyl molecule (Figure 1) that a very large number of compounds can be formed by adding chlorine atoms at one or more of the 10 possible positions; in fact, it can be shown that 209 such compounds are geometrically possible (see Table 1). In addition, it is possible that in some of these free rotation about the inter-ring bond may be hindered so that they exist in two optically active forms.[4]

In Aroclors® 1221—1260, the average number of chlorine atoms per molecule is approximately one to six respectively, but such average values give little information about the actual composition of the mixtures. Thus Aroclor® 1221 is only about 50 percent monochlorobiphenyls, the rest being unchlorinated biphenyl and dichlorobiphenyls with smaller amounts of tri-, tetra,- and even pentachlorobiphenyls.[5] Two monochlorobiphenyls and four dichlorobiphenyls are present.[6] The more highly chlorinated products are much more complex, containing scores of components.[7-9] In Table 2, a summary of the approximate compositions in weight percent of each chlorobiphenyl homologous series is given for Aroclor® preparations from 1221 to 1260.[10]

In the last 15 years, considerable effort has been expended to synthesize individual chlorobiphenyls to provide authentic standards for congener by congener analysis as well as for toxicological investigations. Two of the more useful synthesis routes, which have been reviewed by Hutzinger et al.,[11] are the Gomberg and Ullmann reactions. In the Gomberg reaction, a chloroaniline is first diazotized and the resulting salt reacts with an aromatic substrate (e.g., benzene, chlorobenzenes) in the presence of aqueous sodium hydroxide or sodium acetate to form the chlorobiphenyl (Figure 2).

In the Ullmann reaction, two molecules of chloroiodobenzene are condensed in the presence of activated copper at 220 to 280°C to yield a symmetric chlorobiphenyl (Figure 3).

If a mixture of two different chloroiodobenzenes is used as starting materials, a mixture of three chlorobiphenyls (two symmetrical and one unsymmetrical) is obtained.

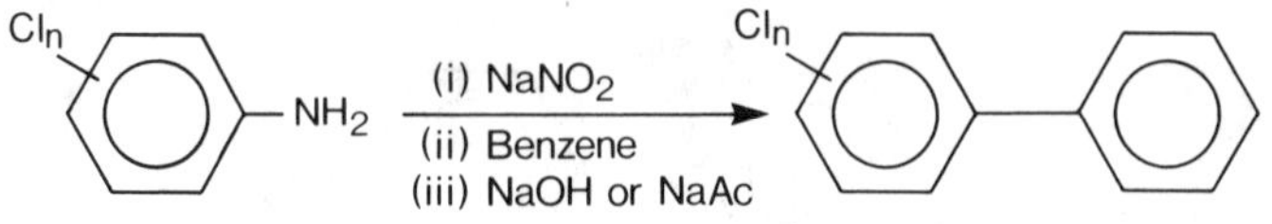

FIGURE 2.   Gomberg reaction.

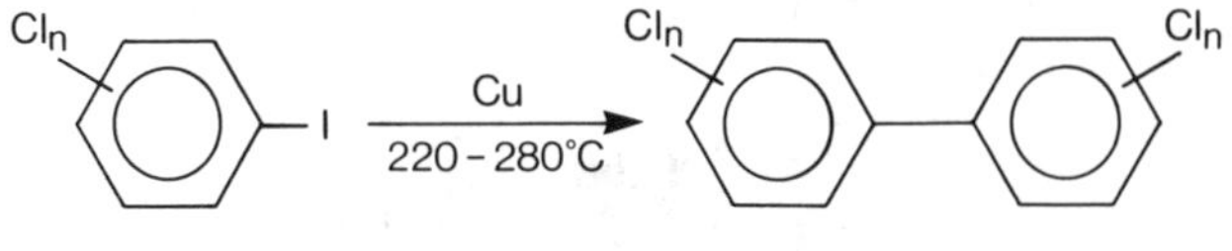

FIGURE 3.   Ullmann reaction.

Table 2.
APPROXIMATE COMPOSITION (WEIGHT PERCENT) OF AROCLOR®
PREPARATIONS[10]

Homolog # Cl/ biphenyl	1221	1232	1016	1242	1248	1254	1260
0	10						
1	50	26	2	1			
2	35	29	19	13	1		
3	4	24	57	45	21	1	
4	1	15	22	31	49	15	
5				10	27	53	12
6					2	26	42
7						4	38
8							7
9							1
10							
Av. #Cl/molecule	1.15	2	3	3	4	5	6
Approx. wt. % Cl	21	32	42	42	48	54	60

Qualitative information on individual chlorobiphenyls in Aroclors® 1242, 1254, and 1260 has been reported by Sissons and Welti[7] and in Aroclor® 1254 by Webb and McCall.[12] Also, some quantitative compositions of Aroclors® 1221, 1242, 1016, 1248, 1254, and 1260 have been provided by Albro et al.[8,9,13] Results of these studies indicated that the predominant species in Aroclor preparations are those with chlorine substitutions in the 2-, 4-, 2,3-,2,4-, 2,5-, 2,4,5-, 2,3,6-, 2,3,5-, and 2,3,4,5- positions. Some of the less favored substitution patterns are 3,4- and 3,4,5-.

The production of PCBs in the U.S. began in about 1930 and hundreds of millions of pounds were produced during the succeeding years.[14] In 1970, however, Monsanto voluntarily decreased production and imposed restrictions on their use. By 1975, production had fallen to less than 50%, and by 1978 to less than 1% of the 1971 production.[15] Subsequent government regulations have virtually eliminated the use of PCBs in the U.S. and similar restrictions have been imposed in several other countries. The main reason for these controls was the realization that certain components of PCB mixtures are among the most persistent of environmental contaminants.

Recognition of the undesirability of PCBs has lead to considerable research on methods for destroying them. This has been the subject of a book[16] and several reviews.[17,18,19] PCBs can be destroyed by burning in incinerators, diesel engines, lime kilns,[20] and high efficiency boilers.[21] Care is required because at insufficiently high temperatures the pyrolysis of PCBs may lead to the formation of polychlorinated dibenzo-*p*-dioxins (PCDDs) and polychlorinated dibenzofurans (PCDFs), some of which are much more toxic than PCBs, and to losses of some PCBs to the atmosphere. However, with proper operation, it is possible to achieve the destruction/removal efficiency (DRE) of 99.9999% required in the U.S. under the Toxic Substances Control Act,[23] and to reduce the amounts of PCDDs and PCDFs to very low concentrations.[24]

Low level wastes, such as transformer oils contaminated with relatively small amounts of PCBs may be dealt with by one of several processes involving the reaction of the chlorobiphenyls with sodium to produce sodium chloride and an inert polymer.[19] This has the advantage that the oil may be reused. Such processes are in use in Canada and the U.S.

A number of other processes are technically possible but probably not economically feasible at present. One uses an alkali metal polyethylene glycolate complex as a dehalogenating agent.[25] This method may be adapted for use in the presence of water and shows promise for the destruction of PCBs in contaminated soils.

B. Environmental Distribution

The first indication of the environmental persistence of PCB congeners was provided in the 1960s by a Swedish chemist, Sören Jensen. He was examining samples of tissues from human beings and wildlife for the presence of DDT and its metabolites. In addition to these compounds he found a number of other chlorinated hydrocarbons which eventually proved to be chlorinated biphenyls.[26] Many studies carried out during the next few years revealed that chlorobiphenyls were widely distributed throughout the environment in North America, western Europe, and Japan.[3] They have now been found in fish in many parts of the world's oceans, including the Antarctic[27] and are probably almost universally distributed through the world.

In 1968, an epidemic caused serious illness in more than 1000 people in Japan, and it was found that all those affected had used cooking oil which was grossly contaminated with Kanechlor® 400, a PCB preparation containing 48% chlorine.[28] Although this was not the result of environmental contamination, but rather of the use of a defective heat exchanger that permitted the PCB to leak into the oil during processing, it gave rise to widespread anxiety about these substances and was probably responsible for PCBs being the only class of compounds specifically mentioned in the Toxic Substances Control Act passed by the U.S. Congress in 1976.[29] Although the manufacturer of PCBs in the U.S. has now stopped production, it has been estimated that 750 million pounds are still in use, and many million pounds are buried in landfill sites or distributed throughout the environment.[30] Clearly, therefore, PCBs will be a matter of concern to environmentalists for some time.

In 1972, Nisbet and Sarofim[31] postulated four principal routes of transfer of PCBs within the environment: (1) volatilization by vaporization from plastics and inefficient burning in dumps and incinerators, followed by adsorption on particulates, transport, and eventual fall-out; (2) leaching from dumps; (3) adsorption on sediments and transport in this form in rivers; and (4) sedimentation in the sea. In 1975, Peakall[3] suggested that with the stopping of the use of PCBs in plastics, the quantity entering the atmosphere by volatilization should be much reduced. However, air analyses from many parts of the world continue to reveal the presence of chlorobiphenyls, usually at concentrations of the order of 1 ng m^{-3}.[32-36] The greater part is in the vapor phase, not bound to particulate material. Earlier analyses[36] indicated that the materials present were rather highly chlorinated, approximating the composition of Aroclor® 1254, but more recent studies have shown a preponderance of the di- and trichloro congeners, giving a composition similar to that of Aroclor® 1242.[33,34]

Chlorinated biphenyls are also widely distributed throughout the waters of the world.

Antarctic sea water was found to contain about 70 pg/l,[27] mostly di- and trichloro compounds. Samples from Lake Superior contained around 1 ng/l.[37] Here the composition corresponded to a mixture of equal amounts of Aroclor® 1242 and Aroclor® 1254. In sediments from Lake Superior a somewhat similar mixture was found,[38] whereas in sediments from Lake Geneva the compounds found were mostly penta-, hexa-, and heptachlorobiphenyls.[39] Perhaps these differences reflect differences in the sources in the two regions.

The fractionation of PCB mixtures that occurs when the mixtures enter water and sediments depends to a large extent on various physical properties of the individual congeners, such as water solubility, partition coefficient between water and nonaqueous solvents, vapor pressure, and sorption coefficient on sediments, as well as the rates of photochemical and bacterial degradation.[2] When chlorinated biphenyls are taken up by organisms, a much more drastic fractionation occurs, resulting from the factors just mentioned and also the ability of the organisms in question to degrade and modify the various components of the mixture. Thus the presence or absence of individual congeners in animal tissues provides an overall indication of the susceptibility or resistance of the various compounds to degradation by all the processes to which it may be exposed. A number of the more highly chlorinated components are found in biological samples in the same proportions as in commercial mixtures, suggesting that they are virtually totally recalcitrant.[40] Such observations have made it possible to infer certain rules governing the susceptibility of the various chlorobiphenyls to degradation. The absence of two adjacent unsubstituted carbon atoms makes a compound very resistant[41,42] especially if the 4 and 4' positions are substituted.[43] As will be shown below, these observations can be accounted for to a certain extent by the properties of the enzymes involved in the oxidation of chlorobiphenyls.

C. Bioaccumulation, Photodegradation, and Biodegradation

The movement of chlorobiphenyls from the nonliving environment to living organisms and from one organism to another has traditionally been considered to involve the phenomenon of biomagnification, i.e., transfer from prey to predator. Thus, the higher the trophic level of a particular organisms the higher would be the concentration of contaminants is its tissue. This view has, however, recently been challenged.[44,45] Although the concentrations of chlorobiphenyls increases with increasing trophic level when calculated on the basis of total weight, this increase is much less pronounced when the concentrations are calculated on the basis of lipid weight.[44] It has been suggested that the uptake of chlorobiphenyls by small planktonic organisms of high surface:volume ratio is primarily a matter of sorption to surfaces.[45] In larger organisms the uptake occurs, or at least the concentrations within the organisms are controlled, by partition between the tissues and the water.[45,46] It is only in higher organisms such as birds and mammals (including man) that the accumulation of chlorinated biphenyls in the tissues depends on the intake through food.

The photochemistry of chlorobiphenyls and the possible contribution of photochemical reactions to their degradation in the environment has been reviewed.[46,47] Although the positions of the absorption maximum vary somewhat according to the degree of chlorination, all chlorobiphenyls absorb only weakly at the wavelengths found in sunlight (300 nm). Nevertheless, some direct photolysis can occur in the environment, chiefly dechlorination. It has been estimated[48] that in a lake of the depth and latitude of Lake Erie, chlorobiphenyls would lose an average of one chlorine atom per molecule per year by photodechlorination.

A number of studies have been made of sensitized photolysis of chlorinated biphenyls using alcohols or amines as sensitizers.[46] The environmental significance of such reactions is questionable. Of greater possible importance is the photolysis of chlorobiphenyls adsorbed to semiconductors, which leads to dechlorination and apparent degradation of the biphenyl skeleton.[49]

The microbial degradation of chlorobiphenyls has been the subject of two reviews.[50,51]

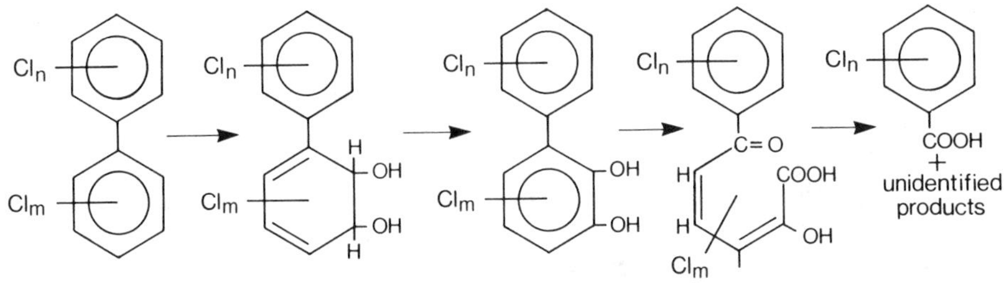

FIGURE 4. Principal pathway for the microbial degradation of chlorobiphenyls.

Several are degraded by the pathway shown in Figure 4. One ring, usually the less highly chlorinated, is oxidized, probably by way of a hypothetical dioxetane to a dihydrodiol. This is dehydrogenated to a diphenol which undergoes *meta*-cleavage to form a 2-hydroxy-6-oxo-6-phenyl-hexa-2,4-dienoic acid; these compounds are bright yellow in color. The aliphatic chain is then degraded, probably in more than one step, to yield a chlorobenzoic acid. Bacterial strains capable of oxidizing chlorobiphenyls are not usually capable of acting further on chlorobenzoic acids, so in pure cultures these compounds accumulate in the medium. Organisms capable of oxidizing chlorobenzoic acids exist, so complete mineralization of certain chlorobiphenyls to carbon dioxide, water, and chloride ion can be achieved with suitable mixed cultures.[52,53] From one such mixed culture bacterial strains have been isolated containing plasmids encoding for all the enzymes required for the total mineralization of 4-chlorobiphenyl.[54]

Different strains of bacteria differ somewhat in their ability to attack various congeners, but, in general, there is a remarkable similarity among the various organisms in their specificities, making it possible to infer certain general rules governing the susceptibility of chlorobiphenyls to degradation by this pathway.

Less highly chlorinated compounds are generally more susceptible than more highly chlorinated ones, and compounds with all their chlorine atoms on one ring tend to be more susceptible than compounds with the same number of chlorine atoms distributed over both rings. However, compounds with two or more chlorine atoms in the *ortho* position (2,6 or 2,2′) tend to be highly resistant. In such compounds free rotation about the inter-ring bond is hindered, so the molecule cannot adopt a planar configuration. Possibly, this prevents these compounds combining with the oxidizing enzyme. Compounds substituted in the 2 and 4′ positions yield more or less stable *meta*-cleavage products, as shown by the development of a permanent yellow color in the medium.

Other less well understood pathways undoubtedly exist. Thus 2,4,6-trichlorobiphenyl, which as the above rules suggest is only slowly degraded if at all by the principal pathway, is rapidly converted to an unidentified trihydroxy compound by a species of *Acinetobacter*.[50] 2,3-Substituted compounds also appear to be metabolized by a different pathway, yielding large quantities of unidentified compounds in addition to chlorobenzoic acids.[50] Tetra- and pentachlorobiphenyls having chlorine atoms in the 2- and 3- positions tend to be more readily attacked than analogous compounds without this substitution pattern.[55] A number of bacterial strains have recently been isolated that have been found capable of degrading a wide range of chlorobiphenyls, including certain penta- and hexa-chlorobiphenyls lacking adjacent un-chlorinated positions.[56] Monohydroxychlorobiphenyls have sometimes been detected among the metabolites of certain chlorobiphenyls. In many instances these have probably been artifacts produced by dehydration of dihydrodiols during isolation,[57] but in some instances they may have been true intermediates.[58] Some organisms can produce nitrochlorobiphenyl derivatives when grown with nitrate as a nitrogen source.[58]

Oxidation of the less highly chlorinated PCBs can occur in soils.[59] The extent of mineralization can be increased by enrichment of the soil with biphenyl; the rate, but not the extent, of mineralization was further increased by inoculation with a bacterial strain capable of degrading chlorobiphenyls.[60,61]

Degradation by a fundamentally different and nonspecific mechanism may also occur. A white-rot fungus, *Phanerochaete chrysosporium,* was shown to degrade Aroclor® 1254[62] and certain individual biphenyls.[63] The enzyme responsible appeared to be a lignase, which is a kind of peroxidase.[64]

Combined photochemical and biochemical degradation may occur. In one study, the rate of bacterial mineralization of 4-chlorobiphenyl increased several-fold when the reaction mixture was irradiated with simulated sunlight and it was suggested that a photodechlorination may have occurred, yielding a product more susceptible to biochemical degradation.[53] Conversely, it has been shown that the yellow *meta*-cleavage product of 2,4'-dichlorobiphenyl undergoes photochemical degradation and it has been suggested that a biochemical-photochemical sequence may assist in the degradation of compounds of this type.[65]

It has been suggested[66] that PCBs may be converted to chlorinated humus, and thereby effectively immobilized in soils or sediments. One possible mechanism for this is the spontaneous (i.e., nonenzymatic) oxidative polymerization of bacterial degradation products to high molecular weight humus-like substances.[67]

In addition to the oxidative mechanisms discussed above, there is evidence that some chlorobiphenyls may undergo reductive degradation. One of the earliest papers on bacterial degradation of PCBs[68] reported the presence of a number of nonchlorinated hydrocarbons in the culture medium after growth of bacteria in a mineral mixture with Aroclor® 1242 as a carbon source. More recent studies[69] have shown depletion of some of the more highly chlorinated PCB congeners in anoxic sediments, indicating that reductive dechlorination has occurred.

D. Metabolism, Retention, Toxicity, and Carcinogenicity in Animals and Humans

The metabolism of chlorobiphenyls in vertebrates can be studied in various ways. One that has already been alluded to is to compare the relative amounts of various chlorobiphenyls in the tissues of an organism with their relative amounts in commercial PCB mixtures. Those that are absent or depleted may have been metabolized, although it is also possible that they may have been removed or destroyed by physical, chemical or biological processes before the mixture was ingested by the animal. It can be assumed, however, that those found within an animal's tissue are relatively resistant to metabolism.

Analyses of marine fish samples revealed 83 identifiable chlorobiphenyls with three to eight chlorine atoms per molecule.[43] Fish from the Antarctic Ocean contained dichloro- to heptachloro biphenyls.[27] A related approach is to expose fish to chlorobiphenyls in water or in food, and determine the rate of their subsequent disappearance from the tissue. In the goldfish, di-,tri-, and tetrachlorobiphenyls were eliminated at rates which decreased with increasing chlorine content, and rates were correlated with water solubility.[70] In rainbow trout fed a mixture of 31 dichloro- to decachlorobiphenyls in a single dose, the rate of elimination of the various compounds similarly decreased with increasing chlorination.[71] Congeners that had no chlorine atoms in the ortho positions were eliminated more rapidly than those that did.[71]

The accumulation of chlorobiphenyls in man has been reviewed by Safe.[72] In pooled samples of adipose tissue extracts from hospital patients in Sweden[42] the chlorobiphenyls were found to be relatively depleted in the less highly chlorinated components, in components having two adjacent unsubstituted carbon atoms, and in components containing more than two *ortho* chlorine atoms. In patients with Yusho disease and in apparently healthy people in Japan[73] penta-, hexa-, and heptachlorobiphenyls were found. All of these were composed of phenyl groups substituted in the 3,4-, 2,3,4-, 2,4,5-, or 2,3,4,5- positions. They also all were substituted in the 2-, 4-, and 4′ positions. In the milk and blood of a Japanese woman who had developed PCB intoxication as a result of working in a capacitor factory several years before,[74] a similar group of chlorobiphenyls was found, along with 2,4,4′-trichlorobiphenyl. The three most abundant components in the Swedish samples also conformed to this pattern.[42]

Another approach is to attempt to identify metabolites in the excreta of organisms that have ingested chlorobiphenyls. These are principally mono- and dihydoxy derivatives; methoxy derivatives are sometimes found, and partial dechlorination occasionally occurs.[75] Phenolic derivatives are excreted not only by organisms that have been given chlorobiphenyls deliberately in laboratory experiments but also by organisms that have ingested them from the environment.[76] Most chlorobiphenyls conform to the rule that 2 adjacent unsubstituted carbon atoms are required for metabolism to occur, but 2,4,5,2′,4′,5′-hexachlorobiphenyl can be hydroxylated.[77] Another metabolic pathway leads to the formation of methylthio-, methylsulfinyl-, and methylsulfonylchlorobiphenyls, which are excreted in the feces.[78,79] Methylsulfonylchlorobiphenyls have been found in the fat of seals[80] and methylthiochlorobiphenyls have been found in lake sediment.[81]

A third approach is to investigate the characteristics of the enzymes responsible for the transformation of chlorobiphenyls. These, like foreign substances generally, are metabolized if at all by certain enzymes in the liver. This process is sometimes referred to as "detoxification" but it has been pointed out[82,83] that the metabolites may be more toxic than the parent substances. The principal biological significance of the process is that the products are usually more hydrophilic than the parent compounds and consequently more readily excreted. The enzymes which oxidize chlorobiphenyls in animals are, in general, somewhat different from those in bacteria. The bacterial enzymes are usually dioxygenases, i.e., enzymes which incorporate both atoms of an oxygen molecule into the product (Figure 4). The animal enzymes usually only incorporate one oxygen atom into the product, and hence are called monooxygenases. They are also sometimes referred to as mixed function oxidases, because a suitable reducing agent such as reduced nicotinamide-adenine dinucleotide (NADH) is simultaneously oxidized by the other oxygen atom. The first product is thought to be an epoxide, or arene oxide,[84] which can either isomerize spontaneously to a phenol or be converted by another type of enzyme, epoxide hydrase, to a dihydrodiol, which can, in turn, be oxidized to a diphenol (Figure 5).

The requirement for two adjacent unsubstituted carbon atoms for the formation of the arene oxide in the first step explains the earlier observations that chlorobiphenyls lacking this feature tend to resist degradation. The fact that some such compounds are nevertheless oxidized suggests that other metabolic processes may occur.

These monooxygenases are inducible, that is to say their concentrations in the liver are greatly increased when an animal is fed or injected with foreign substances, especially potential substrates. Unlike the enzymes involved in the usual catabolic and anabolic processes in the cell, which are usually highly specific, monooxygenases tend to display a rather broad specificity. They are complex enzyme systems composed of a number of proteins and co-factors, notably a group of hemoproteins known as cytochromes P-450 after the approximate position of the absorption maximum of the difference spectrum of their carbon monoxide addition compound. The chemistry and biology of these enzymes have been reviewed by Ullrich[85] and Paine.[86]

The various cytochromes P-450 can be classified according to their physical properties, the substrates in whose transformation they participate, or the substances that can induce them. Two forms are of particular significance in the metabolism of chlorobiphenyls. According to one classification these are designated as $P-450_{LM2}$ and $P-250_{LM4}$. "LM" indicates that they are found in the microsome fraction of liver cells, and the numbers "2" and "4" denote their mobility on electrophoresis. $P-450_{LM4}$ is also sometimes referred to as P-448

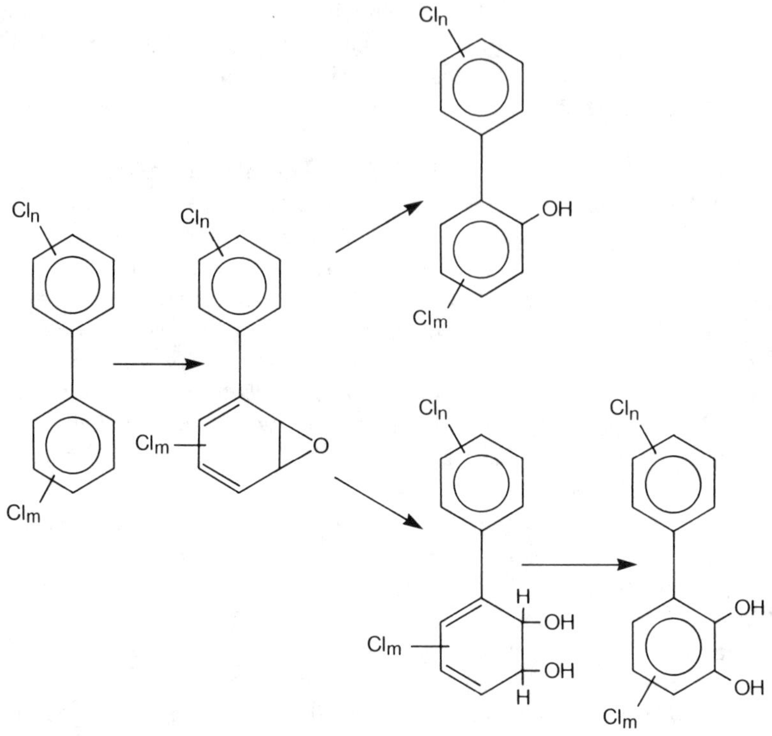

FIGURE 5. Principal steps in the biochemical transformation of chlorobiphenyls in vertebrates.

because its absorption maximum occurs at a slightly shorter wave length than that of most of the other cytochromes P-450. The P-450$_{LM4}$-dependent monooxygenase is sometimes called aryl hydrocarbon hydroxylase (AHH) or benzypyrene-3-monooxygenase because of the reactions it catalyzed. P-450$_{LM2}$ is readily induced by phenobarbitol (PB) and P-450$_{LM4}$ by 3-methylchlaonthrene (3-MC) so they are also sometimes distinguished on this basis.[87]

Commercial PCB mixtures act both as PB and as 3-MC type inducers.[88,89] A study of the activity of individual congeners indicated that most were either inactive or were PB-type inducers. A few, such as 3,4,3',4'-tetrachlorobiphenyl were fairly powerful 3-MC type inducers.[87] All of these were unsubstituted in the ortho positions and it was postulated that the molecule had to be planar in order to be active.[90] However, the congeners that proved active as 3-MC type inducers are only minor constituents of commercial PCB mixtures. It was subsequently found that a number of ortho-substituted congeners acted both as PB type inducers and as rather weak 3-MC type inducers.[91] It now appears that 3-MC type inducing activity requires substitution in both para positions, in at least two meta positions, and in not more than two ortho positions.[92]

The induction of the cytochrome P-450 enzymes is believed to involve a combination between the inducer and a certain protein in the cytoplasm of the cell (the cytosolic receptor) to form a complex which then enters the nucleus. Here it probably interacts with DNA to stimulate the production of the messenger RNA which leads to the synthesis of the proteins which make up the cytochrome P-450 system. The differences in the enzyme-inducing potencies of the various chlorobiphenyls can be accounted for to a considerable extent on the basis of their abilities to combine with the cytosolic receptor. This appears to require that the chlorobiphenyl be able to adapt a coplanar structure, which is hindered by the presence of ortho-substituents. However, certain di-ortho substituted chlorobiphenyls which have low binding affinities for the cytosolic receptor can nevertheless induce AHH.[93] The exact molecular mechanisms involved have therefore not yet been fully elucidated. Empirically, a close relationship has been found between binding affinity and electronegativity, hydrophobicity and hydrogen bonding capacity.[93]

The enzyme-inducing behavior of chlorinated biphenyls, and other compounds, has attracted a great deal of attention because there is a relationship between the ability to induce AHH and toxicity. Such compounds as certain chlorinated dibenzodioxins and dibenzofurans, which are much more toxic than chlorinated biphenyls, are also much more potent inducers of AHH.[92,94] It is not a matter of a substance being converted to a more toxic product, because some AHH-inducers are rendered less toxic by metabolism.[92] Furthermore, the toxicity of 3,4,3',4'-tetrachlorobiphenyl can be suppressed without affecting the induction of AHH,[95] and a hexabromobiphenyl, which resembles the corresponding hexachlorobiphenyl, was found to induce AHH at levels which showed little or no toxicity.[96] There is evidence that enzyme induction and toxicity may be mediated by different receptors with similar structural requirements for binding.[97]

The metabolism of chlorobiphenyls is of further concern for another reason. It was mentioned above that many, though not all, form arene oxides as intermediates. These can combine with cellular macromolecules such as proteins and nucleic acids. It has been suggested that such reactions may cause cancer under certain circumstances.[82]

The acute toxicity of PCBs appears to vary somewhat among species of animals.[98] They are quite toxic for fish, and there is some evidence that fish may be adversely affected at concentrations that may occur in the environment.[99] They are not very toxic for birds and mammals in general, but appear to be extremely toxic for mink.[100] High concentrations have been found in the blubber of Beluga whales;[101] their effects on the animals are not yet known. There is some evidence that dietary PCBs, or associated contaminants, have been responsible for reproductive failure in seals in the Baltic[102] and Wadeen[103] Seas. Chronic exposure can lead to impaired reproductive ability in various species, and long-term experimental exposure has been found to cause liver tumors in rats and mice.[98] Under some conditions, PCBs may cause impairment of the immune system.[104]

The practical significance of PCBs for human health is difficult to assess at the present time. There appears to be widespread belief that they are extremely dangerous substances, and they are often so characterized in the media, but much of the apprehension about them is probably unwarranted.[105] Some congeners appear to be fairly toxic but most are not. Even in the case of PCB poisoning in Japan, referred to earlier, and a similar episode in Taiwan, it is not certain if the symptoms were due to chlorinated biphenyls, or if they were due in whole or in part to much more toxic chlorinated dibenzofurans which were also present.[106]

The most common, and usually the only, result of accidental heavy exposure appears to be a type of dermatitis known as chloracne, which usually usually responds to treatment.[107] Studies on individuals exposed by eating fish containing PCBs or working in factories using PCBs gave no clear evidence of increased illness or deaths compared with the general population.[107]

The quantities of chlorinated biphenyls ingested by people in the ordinary course of events is likely to be small. The main source is probably fish,[108] so the actual amount will vary according to the dietary habits of the individual. The amount ingested in water is probably very much smaller. One assessment[109] suggests that representative figures would be an intake of 24 µg/d and a whole body concentration of 0.35 mg/kg. Such a concentration is two orders of magnitude below the concentration at which adverse symptoms appear. The mean residence time in the body was calculated from these results to be about 3 years. Another study reported a half-life in the body of 6 to 7 months for Aroclor® 1242 congeners and 33 to 34 months for more highly chlorinated compounds.[110]

It should be borne in mind, however, that these figures represent average behavior of all the congeners ingested, and do not take account of possible differences among them. Studies of Japanese victims of PCB poisoning,[73] and of the Japanese woman mentioned earlier,[74] seem to indicate that certain congeners may have a mean residence time of considerably more than three years. Moreover, some of these are among those considered to be relatively toxic. Particularly disturbing is the fact that these may be excreted in large quantities in milk; the Japanese woman excreted about 200 mg, and lowered the chlorobiphenyl concentration in her blood by 80 percent during the period of lactation.[74] A reconstituted mixture of the chlorobiphenyl congeners reported in human milk was found to be a potent inducer of AHH.[111] Chlorobiphenyls have been found in samples of human milk from many parts of the world.[112] Although the effects of low concentrations of chlorobiphenyls in milk on the babies ingesting it are not known, some studies have indicated that babies exposed to PCBs before birth may show delayed physical[113] and mental[114] development. Ordinary prudence would dictate that pregnant and nursing mothers should not be exposed to any significant PCB quantities through their food or otherwise.

It is obvious that problems of this nature can only be studied by the use of analytical methods that permit the separation, identification, and quantification of the various chlorobiphenyls in the samples under consideration.

II. GENERAL ANALYSIS PROCEDURES

A. Sampling and Storage

1. Water

Water samples are generally collected in glass, Teflon® or stainless steel containers. Sampling at various depths can be accomplished using varying lengths of Teflon® tubing and a pump with Teflon® or stainless steel bellows. In general, when sampling for organics the use of plastic containers or plastic tubing and connections should be avoided to minimize potential contamination of the sample. After collection, samples should be stored in clean, solvent rinsed, Pyrex® containers. The containers should be filled to the top to reduce volatilization of organics into the headspace. Teflon® or aluminum foil (dull side) faced caps or closures are recommended. The organics must be extracted from the sample as soon as possible after collection (within a few days) to minimize losses due to volatilization or adsorption to container surfaces. The samples should be kept at low temperature ($\approx$4°C) until processed. Sample extracts can be stored for long periods (months) provided they are placed in properly sealed containers and covered with aluminum foil (darkness) so that no photodegradation can occur.

The validity and usefulness of an analysis hinges largely on the representativeness of the water sample collected. For instance, when sampling a wastewater treatment plant it is of utmost importance to collect a composite sample rather than a grab sample. Flow can vary by as much as a factor of ten over the day at a typical municipal sewage treatment plant. Contaminant concentrations can vary even more, particularly in the raw sewage. Large diurnal variations in concentrations may also occur in rivers and streams subjected to direct industrial or municipal discharges, so either a large number of grab samples or a composite sampling program must be in place for the data to be meaningful. Methods for sample compositing and for using continuous flow liquid-liquid extractors have been outlined.[115] In sampling lakes one should be aware of the possibility of thermal stratification.[116] If your object is to obtain a representative whole lake sample the best time to sample would be immediately after the spring or fall turnover.

The design of specialized samplers used for rainwater, a Teflon®-lined funnel with an XAD-2 resin cartridge,[117] and for large volume (several liters) surface microlayer samples, a rotating ceramic drum,[118,119] have been described.

2. Sediments

Wet sediment samples, in clean solvent-rinsed jars, should be frozen until analysis. Freeze drying of the sediments is probably also acceptable but further studies are required to see whether this leads to any PCB losses. Again, for sediments the representativeness of the sample is crucial. Improper sampling procedures can lead to major losses of the surficial recent sediments, which could result in complete misinterpretation of the data.[120] A good review of sediment sampling procedures for surficial sediments and sediment cores has been presented by Bouma.[121]

3. Fish and Other Biota

Fish and other biota are usually wrapped in solvent-cleaned aluminum foil or placed in glass jars and frozen until analysis. Samples should be stored whole since maceration releases enzymes capable of changing sample composition. Again, sample representativeness is important, since the biological variability between individual organisms is large. In general, it is best to analyze a reasonably large number of individuals or composite a large number into a single sample to obtain useful data. Fish are generally caught using nets or by electrofishing.[122] Plankton and algae are filtered from the water using various mesh plankton nets.[123,124] Small invertebrates such as mysids or shrimps which live just above the bottom sediments can be sampled by dragging a 1-mm mesh metal screen on a sled.[125] Sediment-dwelling invertebrates such as oligochaete worms are collected from bottom sediments recovered with a dredge or other samplers by sieving the sediments (250 μm sieve). Where populations of sediment-dwelling organisms are small, a large number of dredged sediment samples can be prescreened through a 500 μm plankton net in the field and the invertebrates plus debris returned to the laboratory for sorting. The animals should be sorted and frozen as soon as possible after collection (within 24 h). If it is desired to measure the tissue PCB concentration of sediment-ingesting organisms such as oligochaetes, it is necessary to either depurate the organism before analysis[126] or preferable to determine the ash content of a portion of the organism sample (muffle at 500°C for 3 h) and correct for the PCB content of the internal sediment.

B. Reagents

1. Water and Solvents

All water used in the PCB procedures must be free from organic interferences. Double distillation in an all-glass still or distillation over chromic acid or potassium permanganate in a glass still usually produces water of sufficient purity. Commercially available water purification systems consisting of various combinations of activated carbon adsorption, ion exchange and reverse osmosis may also be used. Should further water cleanup be required, passage through a column $\approx$400 mm $\times$ 25 mm I.D. containing 20 to 30 g of XAD-2 resin effectively reduces residual organic contaminants. The column has sufficient capacity for organics to purify several gallons of distilled water and may be regenerated by repurifying the XAD-2 (see adsorbents reaction). Smaller volumes of pure water may be made by liquid-liquid extraction with solvents such as hexane. Whatever method is chosen the water purity should be checked periodically. This can be done by extracting the water with the appropriate solvent such as hexane, concentrating and cleaning up the hexane extract by procedures in the analytical method, and running the concentrated extract by the chosen gas chromatographic method.

In most cases, pesticide grade or distilled-in-glass solvents can be used without further purification. Since the quality of commercial grade solvents can vary from batch to batch, solvent concentrates should be checked prior to use, and, in all cases, solvent and complete method blanks should be run for each solvent batch. It is better to store solvents away from areas where chemicals are stored or standards are prepared, because of potential solvent contamination even into tightly closed solvent bottles.

2. Chemicals

Reagent grade anhydrous sodium sulfate should be muffled overnight at 650°C prior to use. The Na_2SO_4 should be removed from the muffle furnace when it has cooled to ≈100°C and placed in a dessicator to cool to room temperature. It can then be stored in tightly stoppered glass containers until use.

Celite, siliceous, or diatomaceous earth, which is used as a support to keep sintered glass thimbles in soxhlet extractors from becoming clogged or contaminated, can be used as received but should be soxhlet-extracted for 2 to 3 h with solvent being used prior to sample addition. The solvent from this preliminary apparatus cleanup stage is discarded and fresh solvent added for the sample extraction.

Sulfuric acid is commercially available in highly purified form but it is fairly expensive. Reagent grade hydrochloric acid must be extracted three times with hexane prior to use because it sometimes contains interfering contaminants. Triply distilled mercury can be used as received for sulfur removal,[127] but copper dust must be treated with 2 *M* HCl to remove surface oxides before it can be used for this purpose.[128]

3. Adsorbents

Amberlite XAD-2 resin is cleaned by sequential soxhlet extraction with methanol, acetonitrile and ether as described by Junk et al.[129] and stored in methanol. Florisil (synthetic magnesium silicate) is activated and dehydrated by heating for 24 h at 130°C. The Florisil can then be deactivated to the desired activity by adding the appropriate volume of water (0 to 6% water) and shaking vigorously until no clumping is observed. Florisil may also be muffled at 650°C for 2 h or at 450°C for 18 h[130] for activation and for removal of potential organic impurities. Because of variable Na_2SO_4 impurities in Florisil batches the lauric acid method[131] may be used to measure the approximate adsorption capacity of a specific Florisil batch. But before use the retention characteristics of each Florisil lot will have to be measured with the appropriate PCB and organochlorine standards. Prior to use alumina is usually heat treated at 800°C for 4 h, cooled to room temperature, and deactivated with five percent water.[132] Silicic acid and silica gel are normally activated by drying at 130°C for 36 to 48 h prior to use. Deactivation with water as for Florisil is accomplished by adding appropriate amounts of water and shaking. Sulfuric acid treated silicic acid or silica gel (usually ≈40 percent H_2SO_4 by weight used for lipid cleanup) is prepared by adding successive small aliquots of ultrapure H_2SO_4 to the heat treated silica and shaking vigorously until a uniform coating is achieved.

4. Standards

Standard solutions of commercially produced Aroclors® 1232, 1242, 1016, 1254, and 1260 are available free of charge from the U.S. Environmental Protection Agency through Radian Corporation, Houston, TX as solutions 5000 μg/ml in methanol. Standard Aroclors solutions are also available from Ultra Scientific, Hope, RI; Applied Science, State College, PA; and Analabs, North Haven, CT. Steichen et al.[133] showed that the composition of these various Aroclors® marketed for GC calibration are similar to reference Aroclors® reported in the literature.[134,135] But characterized materials are preferable for PCB calibration. About 80 PCB congeners are available commercially from Ultra Scientific. While all of the 209 isomers have been synthesized[136] most congeners are not readily available.

C. Extraction

1. Water

Procedures for isolation of PCBs from water fall into three categories: (1) liquid-liquid extraction (LLE) with organic solvents, (2) adsorption using columns packed with various adsorbents, and (3) steam distillation. The preferable solvents for LLE are hexane and methylene chloride. Extraction can be carried out in a separatory funnel with a minimum volume of 1 l of water extracted 3 times with 50 ml portions of CH_2Cl_2.[137] Alternatively, the water sample (one gallon in a solvent bottle) can be extracted with hexane ($\approx$75 ml) by stirring overnight with a Teflon® stir bar.[138] Recoveries by both procedures are in the 80 to 90% range from spiked distilled water. The latter procedure is preferable for whole water samples containing particulates, since a better extraction from the particles will occur with longer solvent contact times.[138] Recently the design of a large volume, 200 l, water extractor has been reported for reducing the detection limits or organics in ambient waters.[139] The extractor consists of a 50-g stainless steel drum equipped with a solvent recycling pump which continuous recycles CH_2Cl_2 through the water sample as a fine spray. The successful use of this extractor for recovering PCBs from open lake samples has recently been accomplished.[140]

If dissolved PCB concentration is required the sample must be prefiltered or centrifuged to remove particulates prior to analysis. This can be accomplished using a pressure filter with 0.45 μ glass fiber filter paper (use of suction filtration may lead to losses of the more volatile PCBs). Another approach for removing particulates by centrifugation (small samples) or continuous-flow centrifugation (large samples).[141] From separate analysis of the ''particle free'' water and the particulate, fraction the PCB in the sample can be classified into ''dissolved'' and ''bound'' fractions.

Activated carbon and graphitized carbon[142] are effective for extracting PCBs from water samples. Unfortunately, it is difficult to recover the PCBs from the carbon because the adsorption is partially irreversible.[142] Polyurethane foams have been used to adsorb PCBs from water but recoveries are generally low and variable.[143,144] The most promising adsorbent for PCBs appears to be the XAD resins. XAD-2 efficiently adsorbs PCBs from water and the PCBs are easily recovered from the resin with organic solvents.[145,146]

Steam distillation has recently been employed with excellent success for recovery of PCBs from water. Veith and Kiwus[147] reported recoveries in the 98 to 100% range for Aroclor® 1016, 1242, 1248, and 1254. Godefroot et al.[148] reported recoveries in the 81 to 100% range for 30 peaks in Aroclor® 1260. The major advantage of steam distillation is that the PCBs are extracted into a small volume of solvent (isooctane, toluene, or pentane) and, because of the low volatility of potential interfering compounds such as lipids and humic materials, the extract is reportedly clean enough to be analyzed without further cleanup.

2. Sediments

Soxhlet extraction is usually considered the most exhaustive procedure for extraction of chlorinated contaminants from sediments. The azeotropic mixture 41:59, V/V, hexane/acetone is an excellent solvent system and its low boiling point, 50°C, minimizes volatilization losses. Recoveries for PCBs using this method are usually >90%.[138] Several other solvent systems have also been employed.[149] In trace organics work, a glass thimble with a sintered glass filter is usually used. Approximately 3 cm of celite is added to the thimble to prevent the filter from becoming plugged and contaminated with sediment. Usually the whole apparatus plus celite is extracted for 2 to 3 h with the solvent. This solvent is replaced by new solvent prior to sample addition. The samples are usually extracted wet ($\approx$10 to 15 g dry weight, 20 to 30 g wet weight) but excess water should be decanted from the sediment prior to adding it to the thimble, since excess water can impede the extraction and cause low recoveries.

Ultrasonic extraction of sediments with various solvent combinations has been successfully employed.[150] Extraction is effected by immersing the ultrasonic probe in the sediment/solvent (acetone/hexane, 50:50) slurry and sonicating for 30 s. After settling the solvent is poured off and replaced with fresh solvent and the procedure repeated. The combined extracts are filtered prior to cleanup. Recoveries by this method are reportedly almost as good as the soxhlet procedure (80 to 100%). A solvent mixture of methlene chloride-methanol (2:1) has also been shown to provide excellent recoveries.[151]

Various blending, tumbling, and liquid-liquid extractions[152,153] have also been employed for sediments but recoveries by these procedures are usually not as high as by the soxhlet or ultrasonic methods. Steam distillation is another potential method for sediment extraction. Only a limited amount of data on sediments (which are extracted as aqueous slurries) is available, but the PCB recoveries reported by Veith and Kiwus[147] show this method has potential.

3. Biota

Biota and fish samples are first homogenized in a blender, the mixture (10 to 20 g) is then dried by either grinding or blending with anhydrous Na_2SO_4 (3 g Na_2SO_4/g tissue) until free-flowing before soxhlet, column, or batch extraction. Soxhlet extraction for a period between 4 and 16 h using hexane/acetone (41:59),[138,154] methylene chloride/hexane (1:1),[155] hexane,[156,157] and light petroleum[149] have been reported. Although few papers report recoveries, they should be in the 90 to 100% ranges particularly for the mixed solvent systems.[138]

Column extractions are carried out in glass columns ≈50 cm long × 2 cm I.D. equipped with Teflon® stopcocks and sintered glass filters at the bottom. The column is loosely packed with the dried free-flowing biota sample and ≈200 ml of the appropriate solvent is run through the column at a flow rate of ≈3 to 6 ml/min. Various combinations of solvents have been used: 5% diethyl ether in petroleum ether,[158] hexane,[159,160] diethyl ether,[161] cyclohexane,[162] hexane/acetone (2:1),[163] petroleum ether.[164] None of the authors have done detailed studies on the recoveries of PCBs or other pesticides using this technique. Hattula[165] has compared the PCB, DDT, and DDE analysis of several fish samples by the cold column extraction method with the hot soxhlet extraction technique using four solvent combinations: diethyl ether, diethyl ether/pentane (1:1), hexane/acetone/diethyl ether/petroleum ether (2.5:5.5:1:9), and methanol/chloroform (1:1). She found that considerably lower results were obtained from the cold column method for all solvent systems. It is doubtful that PCB concentrations determined by cold column extraction are quantitative.

Direct solvent extraction of Na_2SO_4-dried biota have been carried out by several investigators. Mills[166] used acetone, then hexane for fish; Ford et al.,[167] hexane for invertebrates and hexane/isopropyl alcohol (3:1) for fish, birds, and mammals; Clayton et al.,[168] hexane/acetone (2:1) for algae and zooplankton; Musial et al.,[169] hexane/acetone (1:1) for fish; Bjerk and Brevik,[170] diethyl ether for fish and invertebrates; Shaw and Connell,[171] hexane for fish; and Tuinstra et al.,[172] chloroform for fish. In general, the procedures use about a 10 to 20 to 1 ratio (V:W) of solvent to sample, and samples are blended from 1 to 4 times, usually twice, with the appropriate solvent. No recovery data are reported by any of these investigators. Porter et al.[173] performed recovery studies on several pesticides (not PCBs) from fish using solvent extraction with petroleum ether and found excellent recoveries, >90%. They also found the method was in good agreement with soxhlet extraction with that solvent. The method requires validation for PCBs and for the solvent combinations used above before it can be recommended.

Saponification or alkaline hydrolysis has been used to extract PCBs from biota. Samples were refluxed for 1 h (at ≈90°C) with various alcoholic KOH or NaOH solutions: (1) 10 ml, 10N KOH + 20 ml ethanol;[174] (2) 200 ml, 2% NaOH in ethanol;[175] (3) 34 ml ethanol + 6 ml H_2O/10 g KOH.[176] The saponified solution is then extracted twice with 10 ml hexane (method 1), 3 times with 50 ml hexane (method 2), or once with 30 ml hexane (method 3). The hexane extracts are washed with distilled water or a 2% aqueous NaCl solution, and dried with Na_2SO_4. PCB recoveries from spiked fish were over 95% (method 2) and about 90% (method 3) so the method does seem to work well. Because saponification makes lipids more water soluble and less soluble in organic solvents such as hexane, an added advantage

(a) (b) (c)

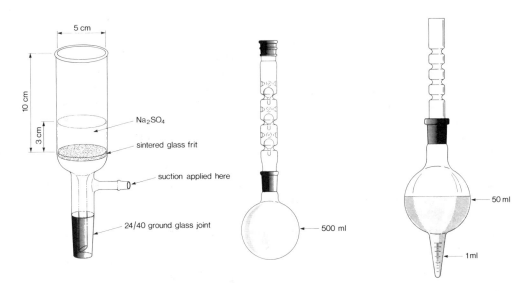

FIGURE 6.   (a) Filter funnel for drying extracts; (b) round bottom flask with Snyder condenser; and (c) evaporation flask with Kuderna Danish condenser.

of the procedure is that the extract contains smaller quantities of lipids compared to other methods such as soxhlet extraction. This makes subsequent cleanup procedures somewhat easier.

Acid digestion prior to solvent extraction has also been successfully employed for analysis of fish and other biota. Rote and Murphy[177] used a perchloric-acetic acid mixture followed by extraction with hexane for PCB quantitation. But the most common method is HCl digestion followed by hexane/methylene chloride (3:1) extraction.[178]

Steam distillation, as mentioned for water and sediment analysis, appears to have considerable potential for PCB determinations in fish and other biota,[147,148] but insufficient data are available for a proper evaluation of this technique.

Extraction procedures for water birds and bird eggs[179] and for human tissue samples such as adipose,[180] human milk,[181,182] and blood[183] are similar to those described above.

D. Concentration and Cleanup

1. Extract Drying and Concentration

For most extraction or cleanup procedures considerable volumes of solvent are used and so the extract must be reconstituted to an appropriate small volume prior to analysis. The extract must be dried to remove the water prior to concentration and this is usually accomplished by filtering through 3 cm of anhydrous Na_2SO_4 (see Figure 6a). For evaporation of large volumes (>100 ml) of dried solvent, the solvent should be evaporated to 20 to 25 ml in a round bottom flask equipped with a 3-ball Snyder condenser using a heating mantle of appropriate dimensions (Figure 6b). The solvent is then transferred to a graduated evaporation flask equipped with a Kuderna-Danish condenser (Figure 6c) for evaporation to 0.5 to 1 ml using a water bath. Recoveries for PCBs using these evaporation procedures are greater than 90%.[138]

Evaporation of solvents can also be accomplished using a rotary evaporator at 35 to 40°C, but a keeper such as isooctane or toluene (≈3 ml) should be utilized. A stream of purified nitrogen can be used for solvent evaporation when a keeper is employed. For all procedures care must be taken *not* to take the sample to dryness otherwise severe losses will be encountered.

2. Removal of Bulk Lipids

Water and sediment extracts usually have sufficiently low concentrations of interfering coextractants that they can be cleaned up in a single step. Fish extracts, on the other hand, can contain large concentrations of lipids, which may exceed the capacity of most cleanup columns, so some type of precleanup step to remove bulk lipids is sometimes required. Commercially available solid supports that have high adsorptive capacity for oils and fats have been used for this purpose.[154,184] The method involves mixing the extracted fat ≈30 g with 15 g of Micro Cel-E (a synthethic calcium silicate) until free flowing. The fat coated powder is then extracted with 250 ml 5% acetone in acetonitrile in a blender. A solvent transfer to petroleum ether is performed prior to further cleanup. This method was shown to remove ≈90% of the fat and give recoveries of ≈85% for several pesticides. Recovery of PCBs should be similar.

Liquid-liquid partitioning has also been employed for bulk lipid removal. Acetonitrile and dimethylformamide (DMFA) are the two most common solvents used for this purpose. A sample of 5 to 10 g of oil is dissolved in 50 ml of hexane and the hexane is extracted 4 times with 15 ml of acetonitrile or DMFA. The acetonitrile or DMFA extracts are combined and, after the addition of 100 ml of water (+ 2% NaCl), are extracted three times with 20 ml portions of hexane. Both solvent systems lead to a 90% reduction in lipid content, but PCB recoveries were only 45 to 60% for acetonitrile but were >95% for DMFA.[185]

Direct treatment of fat containing extracts with concentrated H_2SO_4 has been employed for removal of lipids.[138,165,170,186,187] Approximately 1 g of fish fat in 30 ml of hexane is shaken in 5 ml of high purity H_2SO_4 for 30 to 60 s. The hexane layer is then dried by passage through ≈3 cm of Na_2SO_4 supported in a sintered glass filter. Approximately 90% of the lipids are destroyed by this procedure and PCB recoveries are over 90%.[138,186] The only disadvantage of the method is that acid-labile compounds such as dieldrin are destroyed. If such compounds are also of interest, this method would be inappropriate.

Saponification can also be used as a pretreatment method. The detailed procedures for extract treatment by this technique are described in Section II.C.3 of this chapter.[174,175,176]

Low temperature precipitation, in which the extract is cooled to − 78°C in a dry ice-methanol bath for 30 min, has been used for bulk lipid removal.[188] The method recovery for several insecticides, fungicides and herbicides is good (80 to 116%) and fat removals are about 95%. No recovery studies for PCBs have been performed for this method.

The simplest and most reliable of the above methods for PCB determinations is direct H_2SO_4 treatment.

3. Gel Permeation Chromatography

Stalling et al.[189] developed a gel permeation chromatographic technique for separation of pesticides from lipids. Lipids, with their comparatively large molecular size (MW 600-1500) are excluded from the resin pores while pesticides and PCBs with smaller molecular size (MW 200-400) can enter the resin pores. As a consequence, the pesticides travel longer distances than the lipids, so take longer to reach the end of the column. A typical separation between lipids (the first fraction) and pesticides and PCBs (the second fraction) is shown in Figure 7. The method removes >98% of the lipids and gives PCB recoveries of virtually 100%.[189,190] Columns (2.5 cm I.D. × 60 cm long) are packed with Bio-Beads SX-2 or SX-3 resin (copolystyrene, 2 or 3% divinylbenzene) which have been preswelled overnight in the solvent. The solvent of choice is usually cyclohexane,[154,189,190] although a 50:50 mixture of cyclohexane and dichloromethane has recently been used.[191] The sample (5 ml containing

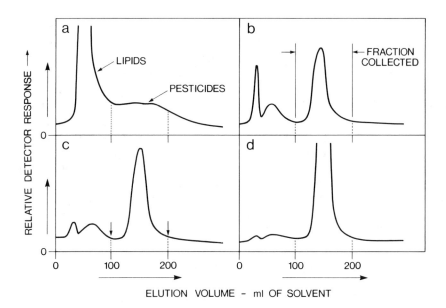

FIGURE 7.   Gel permeation chromatograms of pesticides in fish extracts after (a) extraction; (b) Micro Cel-E cleanup; (c) gel permeation cleanup; and (d) final gel permeation cleanup. (Reproduced from Webb, R. G. and McCall, A. C., *J. Chromatogr. Sci.*, 11, 371, 1973. By permission of Preston Publications, Inc.)

≤0.5 g lipids) is injected using a sample loop and the pesticide fraction collected. A major advantage of this system is that lipids are excluded from the column and so the column can be continually reused. Also the system can be automated[192] — a commercially produced automated GPC unit, Autoprep 1001, is available (Analytical Biochemistry Laboratories, Inc., Columbia, MO).

4. Liquid-Solid Chromatography
a. Florisil
Florisil has been widely used for the cleanup of environmental extracts.[157,164,169,193,194,195] Most procedures use either activated Florisil (dried at 130°C) or Florisil deactivated with 1 or 2% water. Depending on the capacity required various column sizes can be employed. The petroleum ether-diethyl ether solvent system has been largely replaced by the hexane and hexane-dichloromethane system because of improved separations with this solvent system.[178] Columns are normally eluted first with hexane to recover the PCBs and several of the pesticides.[178] Further elution with hexane-dichloromethane mixtures elutes the rest of the pesticides. In general, Florisil provides a good separation between interfering compounds such as lipids and PCBs, but does not effectively separate PCBs from many common classes of pesticides.[196]

b. Alumina
Holden and Marsden[132] were one of the first to employ alumina for cleanup in organochlorine residue analysis. Telling et al.[197] extended the earlier study to include PCBs. Columns of various sizes have been used and hexane is the preferred elution solvent.[149,160,197] Similar separations to those on Florisil can be achieved on alumina, and the capacity of alumina (5% deactivated) for lipids is reported to be higher than for Florisil. If a small volume of hexane is collected the PCBs, HCB, mirex, and p,p'-DDE are included in this fraction[197] by increasing the collection volume slightly all the DDT family are included in the first fraction.[197] Continuing the elution with hexane eventually removes the rest of the OCs — the BHCs, dieldrin, endrin, methoxychlor, etc. — from the column.[197]

c. Silicic Acid and Silica Gel

Silicic acid (silica gel) has been used for cleanup of extracts in environmental analysis. The capacity of silicic acid for lipids is somewhat lower than either alumina or Florisil,[130] but it does have somewhat better selectivity than the other adsorbents. For this reason silicic acid is mostly used in combination with other adsorbents or other lipid removal methods to provide a separation between organochlorine pesticides and PCBs.[198,199] In at least one investigation p,p'-DDE was separated from Aroclor® 1254 using silica.[200] But, because of difficulties obtaining reproducible deactivation and chromatography on silica, most chemists use silica to separate the other DDT isomers from PCBs with p,p' DDE remaining in the PCB fraction. More recently PCBs and toxaphene have been separated using silicic acid.[201,202] Usually silicic acid is deactivated with 3% water prior to use.[198,199] Earlier investigators used silicic acid/celite mixtures.[198] This practice was abandoned because of difficulties obtaining a homogeneous mixture of these two solids.[199] Petroleum ether was the solvent of choice in earlier studies but currently hexane followed by benzene is the most widely used solvent system.

Silicic acid has also been impregnated with various reagents for residue analysis.[203] Ten percent $AgNO_3$ on silica has been used to separate p,p'-DDE from PCBs.[204] Also 40% H_2SO_4 on silica has been used as a final polishing step to oxidize residue lipids prior to capillary gas chromatography.[138]

d. Carbon

Charcoal and charcoal suspended on foam have been used to separate PCBs from chlorinated pesticides[149,205] and from mirex and photomirex,[206] respectively. Foam-charcoal mixtures have also been used to separate the more toxic nonortho substituted PCBs from ortho substituted congeners,[207] and for separation of PCBs from polychlorodibenzodioxins and polychlorodibenzofurans.[208] Carbon dispersed on glass fibers has been demonstrated to be useful for the same purposes.[209]

5. Miscellaneous

a. Sulfur Removal

Sediment extracts, in particular, usually contain large amounts of sulfur containing coextractants which interfere with PCB analysis. The two best methods for removing sulfur are shaking the extract with activated copper powder (see reagents section)[128] passing the extract through a column packed with 50 g of activated copper,[210] or stirring it vigorously on a vortex stirrer with metallic mercury.[127] The treatment with mercury should be repeated until the mercury remains shiny.

b. Chemical Conversion of Interferences

DDT and its metabolites can interfere with quantitation of PCBs (particularly when packed GC columns are used) because they elute at similar retention times to some PCB isomers. One method to eliminate DDT interferences is to convert DDT and its analogs to dichlorobenzophenone (DCBP) which can easily be separated from PCBs using a Florisil column.[166,211-214] DDT and DDD (o,p' and p,p') are first dehydrochlorinated to DDE using DBU (1,5-diazobicyclo [5,4,0] undec-5-ene) in benzene. The DDE is then oxidized to DCBP by reaction with chromic acid. Recent studies have shown that considerable losses of PCB congeners occur during chromic acid digestion[215] so this procedure would seem to be of little use in PCB analysis.

c. PCB Conversions

Because of the complexity of PCB quantitation using GC procedures, investigators have attempted to convert PCBs to more easily measurable compounds. Usually the PCBs are either dechlorinated to biphenyl[216,217] or perchlorinated to decachlorobiphenyl, DCB,[205,218,219] prior to analysis by either liquid or gas chromatography. Dechlorination can be accomplished directly on a 3% palladium or 5% platinum catalyst packed in the gas chromatographic injector[216] or by reacting the PCBs with LiA1H$_4$ in a sealed tube at 180°C with subsequent LC or GC determination of the biphenyl in the extract.[217] The detection limit of the GC detection of biphenyl using a flame ionization detector is considerably higher than conventional PCB determinations using the electron capture detector. But a considerable improvement in sensitivity has been reported using an LC with an ultraviolet detector.[217] Perchlorination of PCBs is usually accomplished by reaction with antimony pentachloride in CHCl$_3$ in a sealed tube at ≈ 170°C.[218] The gas chromatographic analysis of the resulting decachlorobiphenyl is simple and the detection limits are excellent.

A critical evaluation of both procedures for a variety of environmental samples showed that dechlorination gave results in good agreement with the normal GC/ECD procedure for all sample types.[175] On the other hand, perchlorination was found to give PCB values 2 to 30 times higher than conventional methods for certain sample types.[175] The major interfering compounds were hydrogenated terphenyls, although many other classes of compounds yielded some DCB on perchlorination.[175] Thus, of these two methods, dechlorination is the only one which has some potential for estimating total PCBs. But so much information is lost applying either method that we hesitate to recommend either one. All information regarding specific PCB congeners present, the degree of chlorination of the PCBs, and type of Aroclor® present is lost.

E. Quantification

The quantification of PCBs is normally carried out using either packed column or capillary column gas chromatography with a halogen sensitive detector, such as the electron capture detector, or a mass spectrometer in the electron impact or chemical ionization mode. A limited amount of work has been carried out with high-speed liquid chromatography[220] but this method appears to be useful only for industrial Aroclor® mixtures and is not sensitive enough for environmental samples.

1. Packed Column Gas Chromatography

Several different columns have been tested for PCB analysis;[13] in general, columns with nonpolar liquid phases (e.g., SE30, OV101) give better separations than columns with polar phases because of higher solubility of PCBs in the former. For general purposes, analysis of PCBs and pesticides mixed phase columns are often used.[149,170]

The most common detector used for PCB analysis is the electron capture detector. This detector provides the highest sensitivity, but it presents some problems because it exhibits different responses to various PCB isomers.[14] For example, the response factor for mono or dichloro-substituted PCBs is 1/100 to 1/1000 that of the higher chlorinated congeners such as hepta- or octachlorobiphenyls. Also, response factors are significantly different for PCBs with the same number of chlorines but having different substitution patterns.[14]

In order to minimize these effects most PCB analyses are carried out using industrial PCB mixtures (such as the Aroclors®) as standards. Most environmental samples of fish and sediments contain the more highly chlorinated mixtures such as Aroclors® 1254 or 1260. To perform a PCB analysis one would prepare weighted standards of, for example, Aroclor 1254. The GC pattern for the standard at various concentrations would be determined and the peaks integrated. Then a calibration graph of Aroclor® concentration versus total peak area of the PCB envelope would be constructed. This simple procedure has been used by some investigators.[161,176]

Sawyer[221] demonstrated from an interlab comparison study that the individual peak method developed by Webb and McCall[134] was a better technique than total area. Webb and McCall's

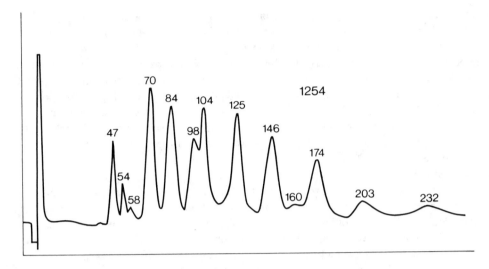

FIGURE 8.   EC chromatogram of Aroclor® 1254 on SE30. Peak identification numbers correspond to the retention time relative to p,p'-DDE = 100. (From Veith, G. D., Kuehl, D. W., and Rosenthal, J., *J. Assoc. Off. Anal. Chem.*, 58, 1, 1975. With permission.)

chromatogram for Aroclor® 1254 on a 3% SE-30 on 80/100 Gas Chrom Q column (carrier gas 95% argon, 5% methane, 80 to 100 ml/min) operated isothermally at 200°C is shown in Figure 8. The numbers on the chromatogram peaks are retention times relative to p,p'-DDE, which is arbitrarily assigned a value of 100.[134] Webb and McCall measured the weight percent of PCB present in each GC peak by finding the empirical formula of the compounds in each peak by GC/MS and by measuring the absolute amount of chlorine in the peak using a Coulson conductivity detector. With this knowledge it is possible to find a detector response factor for each individual peak in the Aroclor® mixtures. A limited amount of standard solution used for calibration in the above method is available from Webb and McCall.[222] More recently, Sawyer[135,223,224] has also standardized a series of Aroclor® mixtures that can be used for calibration and Steichen et al.[133] have calibrated several commercially available Aroclor® standards.

The procedure essentially involves running a series of standardized Aroclors® on the GC using the column and conditions in the previous paragraph. From the weight percent in each peak from the above papers, a calibration plot for each peak in the Aroclor® is established. Next the samples are run using the flow chart[225] (Figure 9) which identifies the Aroclor® mixture most closely matching the sample. This procedure is more accurate than the total area method but is still not exact because Aroclor mixtures are differentially weathered and metabolized in the environment. Therefore, a perfect match to the original Aroclor® is not possible for environmental samples.

An alternative approach using packed GC columns is the GC/MS technique.[226,227,228] GC/MS data for the molecular ions of each PCB group (PCBs with the same molecular weight) are simultaneously recorded. Areas of PCB peaks in the mass chromatograms are used to determine concentrations. An internal standard, such as 2,4,6 tribromobiphenyl, TCB, is used to improve the precision.[226] The ions used for quantification are shown in Table 3. The coefficient of variation of the GC/MS procedure is claimed to be ±20%. In addition to total PCBs the method also gives the concentration of PCBs in each isomeric group, e.g., tetras, hexas, etc. The high cost of the instrumentation and difficulties automating the analysis make the adoption of this method for routine PCB analysis unlikely. But it is an excellent method for confirmatory analysis.

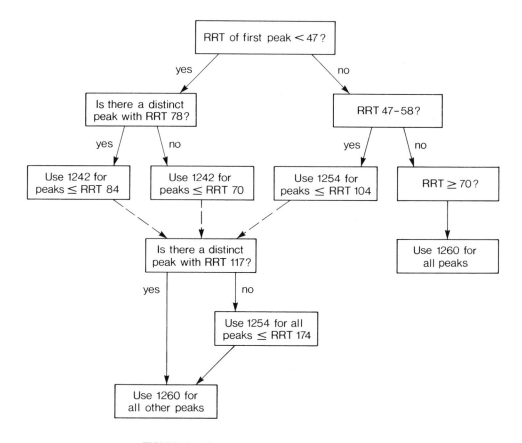

FIGURE 9. Chromatogram partitioning flow diagram.

Table 3
IONS USED FOR GC/MS
QUANTITATION

PCB group	Ions	PCB group	Ions
C_1	188(190)[a]	C_6	360(364)[a]
C_2	222(224)	C_7	394(398)
C_3	258	C_8	430(432)
C_4	292(294)	C_9	464(468)
C_5	326(328)	C_{10}	498
TBB	390		

[a] Ion for quantification to minimize $M^+ - HCl$ interferences.

2. Capillary Column Gas Chromatography

Capillary gas chromatography provides the potential for separation and quantification of all 209 PCB congeners. In practice it has not been possible to separate all PCB components on a single capillary column, but at least one investigator[9] has claimed to have separated all isomers using four capillary columns of differing polarity. Quantification of all the individual PCBs and total PCBs by this technique has been hampered because not all 209 congeners are available. In spite of these difficulties the technique has considerable potential and is a significant improvement over other methodologies. Figure 10 shows a comparison

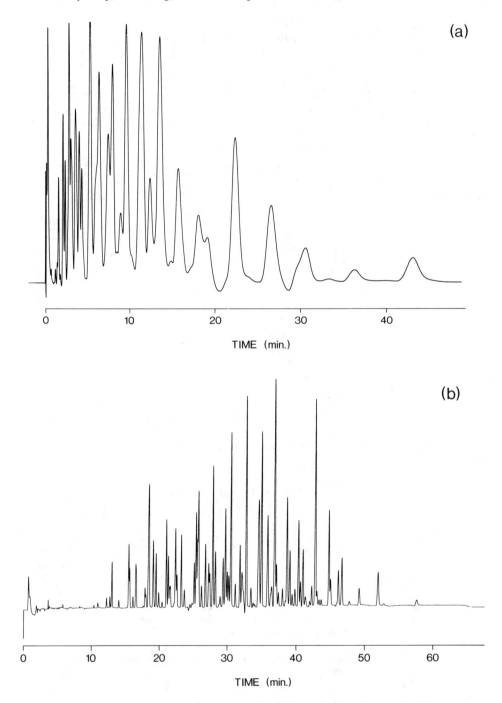

FIGURE 10.   (a) Packed column chromatogram of 1:1:1 mixture of Aroclors® 1242:1254:1260; and
(b) Capillary column chromatogram of 1:1:1 mixture of Aroclors® 1242:1254:1260.

of a 1:1:1 mixture of Aroclors® 1242, 1254, 1260 run on a packed column and on a capillary
column. The greatly improved resolution and larger number of peaks in the capillary chromatogram are apparent.

Earlier PCB separations were accomplished using glass wall coated capillaries,[9,229-232] but
recent advances in column technology have provided the much more stable and durable

fused silica capillaries with bonded liquid phases.[233,234] The bonded phases are produced by crosslinking the liquid phase polymers using free radicals produced chemically or photo-chemically and provide a longer life column with a higher temperature limit and with lower liquid phase bleeding. Column dimensions vary but columns are usually 30 or 60 m long and have internal diameters between 0.25 and 0.33 mm. As with packed columns lower polarity liquid phases usually provide the best PCB separations. Apiezon L or M have been demonstrated to provide the best PCB resolution,[233] but other common liquid phases such as OV1, SE54 or equivalent substitutes have also been shown to provide adequate separations.[136,234] Thinner stationary phase films have been shown to be preferable to thicker coatings for PCB resolution.[233] However, the use of very thin films, <0.1 μm, is not recommended, since this leads to reduced column stability and capacity.

The preferred carrier gases used for capillary chromatography are hydrogen and helium.[235] Linear gas velocities in the range 35 to 40 cm/s for hydrogen and 18 to 20 cm/s for helium, measured by injecting freon at the highest column temperature, are used. However, because the use of hydrogen with an electron capture detector can lead to lower sensitivity and component losses due to adsorption,[236] we recommend helium. To achieve better separations the columns are temperature programmed. Rates between 1 and 10°C/min have been utilized with lower temperature programming rates providing improved separations but also resulting in longer analysis times. The initial column temperature is set at ≈15 to 20°C below the boiling point of the solvent when splitless injection (1 to 2 μl) is used. Split injection can be employed for samples containing sufficiently high PCB concentrations. For either splitless or split injection the injector temperature is normally set at about 250°C. On-column injection has been suggested as a way of improving the reproducibility of PCB analysis (percent RSD ± 10% splitless, ±2% on-column).[237] As with packed columns the most common detector used is the ECD, although a mass spectrometer has been used.[238-240] Mass spectrometry using selected ion monitoring (SIM) has recently been proposed to improve the sensitivity of the MS approach and to give quantitative information on the homologue composition of the PCBs.[241-243] The flame ionization detector[9] has been utilized for studying commercial PCB mixtures but is not sensitive enough for environmental applications.

Because all 209 PCB congeners are not available commercially, calibration of the capillary GC system can be difficult. The 1/2 RI method[7,13] has been used to tentatively identify the other PCBs in technical formulations. This method assumes that the RI for a PCB is an additive function of two parts consisting of two chlorosubstituted phenyl groups. For example, the approximate RI for 2,5-dichlorobiphenyl can be found by adding 1/2 (RI for biphenyl) + 1/2 (RI for 2,5,2′,5′-tetrachlorobiphenyl). More recent research using a larger number of PCB pure congeners has shown that peak assignments made using this method are incorrect for considerable number of PCBs.[234] For this reason peak assignments made using 1/2 RI indices must be considered questionable.

Several methods have been advanced for estimating total PCBs from capillary ECD chromatograms. The ECD response factors for unavailable PCBs are usually estimated by averaging the RFs for available isomers with the same number of chlorines[172,232,234] (if GC/MS information is available) or by averaging the RFs of adjacent known peaks in the chromatograms[153] or by using homologous series (di, tri, etc.,) grouping.[244] Now that all 209 congeners have been synthesized,[136] more sophisticated approaches to assignment and quantification of the individual congeners are possible. Mullin et al.[136] have compiled the retention indices for all 209 congeners on a SE54 capillary column, and Mullin et al.[136] and Pellizzari et al.[245] have compiled lists of the relative response factors for the EC detector for all congeners. Thus it is possible to identify the congeners by running Aroclor® mixtures (available from EPA) with an SE54 capillary column using the same conditions as employed by Mullin et al.[136] In order to make the proper assignments and to develop response factors it is necessary to know the concentrations of the individual congeners in the Aroclors®.

Capel et al.[246] have recently reported the weight composition of congeners in Aroclors® 1242, 1254, and 1260. Mullin[247] has compiled composition data for Aroclors® 1221, 1232, 1016, 1242, 1248, 1254, 1260, 1262, and 1268. Thus the Aroclors® can be used as secondary standards to both identify and quantify the major PCB congeners which are likely to be present in environmental samples. As recommended by Mullin[247] we are currently using a mixture of four Aroclors® 1221, 1016, 1254, and 1262 in a ratio of 10:5:3.5:3 for instrument calibration, since this mixture contains measurable concentrations of all congeners which are likely to be present in the environment.

Although all 209 PCBs are not commercially available, about 80 congeners can be obtained from Ultra Scientific, Hope, RI. Recently, four PCB solutions containing 14 or 15 congeners (51 in total), which are well separated by capillary GC, have become available at nominal cost ($100 per set) from the National Research Council of Canada, Marine Analytical Chemistry Standard Program, Atlantic Research Laboratory, Halifax, Nova Scotia. Some of these pure congeners (particularly the major congeners present in the Aroclors) should be obtained and run on your GC to improve confidence in the assignments and response factors.

A completely different calibration approach that has been proposed recently is the use of commercially-available surrogate PCB standards. Cooper et al.[248] has developed a set of 31 available congeners each representing a group of congeners with similar ECD response factors for instrument calibration. The validity of this approach for diverse commercially-available instruments has not yet been substantiated. Similarly Gebhart et al.[242] have suggested a set of nine commercially-available congeners that could be used as surrogates for GC/MS (SIM) analysis of PCBs.

In all of the above methods the concentrations of the available PCB congeners and the estimated concentrations of the unavailable congeners are summed to obtain total PCBs. Total PCB concentrations by capillary GC are probably as good if not better than those from packed column GC because the former is less subject to interferences from pesticides.[244] The primary advantage of capillary GC over the other techniques is that it provides the potential for studying the behavior of individual PCB compounds in various compartments of the aquatic environment.[146,153,163,172,195]

III. RECOMMENDED ANALYTICAL METHODS

A. Extraction

1. Water

Using a magnetic stirrer, extract water sample by stirring vigorously with 50 ml of CH_2Cl_2 for 30 min in the original sample bottle. Transfer mixture to a separatory funnel. After layers separate, drain CH_2Cl_2 (lower) layer to a 500-ml round bottom flask (RBF) and aqueous layer back to original sample bottle. Repeat extraction twice with two fresh 50 ml portions of CH_2Cl_2. After the last extraction, determine the volume of water sample using a measuring cylinder. Filter the combined CH_2Cl_2 extract through 50 g of anhydrous Na_2SO_4 in a filter column. Rinse column with 50 ml of CH_2Cl_2 and apply vacuum to remove the solvent.

Collect filtrate in a 500 ml RBF, attach a 3-ball Snyder condenser (Figure 6b), add a few boiling chips, and (using either a heating mantle or water bath at $\approx 80°C$) evaporate solvent down to 20 to 25 ml. Transfer concentrated extract to a small volume evaporation flask equipped with Kuderna-Danish (K-D) condenser (Figure 6c), add a few boiling chips, and evaporate extract to 1 to 2 ml with a water bath. Add 40 ml of hexane to concentrated extract and repeat evaporation to 1 ml. Alternately, add 3 ml of isoctane keeper to dried filtrate and evaporate solvent down to 5 ml using a rotary evaporator with a water bath temperature of $\leq 40°C$. Add 50 ml of hexane to concentrate and reevaporate to 5 ml. Repeat the addition of 50 ml hexane and evaporate solvent down to 3 ml.

Note: Although CH_2Cl_2 is a more efficient extractant for general organics, hexane may be used if only PCBs, chlorobenzenes and similar compounds are of interest. In this case the reevaporation steps used to remove CH_2Cl_2 in the above procedures can be omitted.

2. Sediments

Place 3 cm of celite in a glass extraction thimble and accurately weigh thimble plus celite on a balance. Place the thimble into a soxhlet extractor and extract with 300 ml 41:59 hexane/acetone for 2 to 3 h. After extraction, drain solvent inside extractor back into receiving flask and discard. Add wet sediment (10 to 15 g dry weight) to thimble, place in soxhlet extractor with 300 ml of fresh solvent, and extract overnight ($\approx$16 h). Freeze-dried sediments can also be used since our studies have shown no losses of PCBs on freeze-drying. Cool, then drain solvent from extractor back into receiving flask. Transfer extract to 2 l separatory funnel containing 1 l of organics-free water and shake for 1 min. After layers separate, discard aqueous (lower) layer and filter hexane extract through 3 cm of anhydrous Na_2SO_4. Rinse separatory funnel with $\approx$50 ml hexane and pass this hexane rinse through the Na_2SO_4. Evaporate the combined extract as described for water samples above. Remove thimbles containing the sediment from soxhlet apparatus, dry them overnight at 130°C, transfer them to a dessicator to cool to room temperature, and then weigh them on a balance to obtain the sediment dry weight.

3. Biota

Weighed, homogenized fish and other biota samples (10 to 15 g) are ground with anhydrous Na_2SO_4 (30 to 45 g) in a mortar and pestle until free-flowing. The samples are then soxhlet-extracted and the extracts concentrated the same as the sediments (above).

B. Cleanup

1. Bulk Lipid Removal

(Omit this step for water, sediment and other low lipid content samples). Place extract (20 to 25 ml) in a 50 ml vial and add 5 ml of concentrated H_2SO_4. Seal vial with a Teflon®-lined septum using a screw or crimp top and shake vigorously for 1 min. Allow layers to separate (several hours) or centrifuge at 600 rpm for 10 min. Filter hexane layer through 3 cm of anhydrous Na_2SO_4. Evaporate this extract to a small volume ($\approx$1 ml) using the procedure described for the water samples. Alternatively bulk lipids may be removed using the automated gel premeation chromatography apparatus if available. These procedures remove only 90 to 98% of the lipids from the extract and a further polishing cleanup step is required before they are ready for gas chromatographic determination. Proceed to Florisil cleanup step.

2. Florisil Column Cleanup

Activate Florisil by heating the adsorbent in a 130°C oven for 16 h. Prepare a cleanup column by plugging a 600 × 20 mm I.D. (or similar size) chromatographic tube with a piece of silanized glass wool. Fill the column with 20.0 g of activated Florisil and top the adsorbent layer with 2 cm of anhydrous Na_2SO_4. Rinse column with 100 ml of hexane (or petroleum ether) and discard. When the hexane level just drops into the Na_2SO_4 layer, transfer the concentrated extract quantiatively to the column with a Pasteur pipette and several hexane rinsings. Elute column with ~150 ml of hexane and collect the eluate. Determine the exact volume of hexane required by running spike recovery experiments using PCB standards. Evaporate to the appropriate volume using a K-D evaporator or add 3 ml of isooctane to the eluate and evaporate solvent down to a volume of 3 ml using a rotary evaporator.

3. Sulfur Removal

Activate copper powder by washing with 2N HCl with subsequent rinsings of distilled water (until pH is neutral), acetone and hexane. Use activated copper immediately. Treat the concentrated extract after Florisil cleanup with 100 mg of activated copper. Repeat if copper completely turns black. Alternatively, shake the sample extract with a drop of triple-distilled mercury in a test tube repeat if necessary until the metal is shiny.

C. Detection and Quantification

1. Packed Column Method

Use a gas chromatograph equipped with a Ni[63] electron-capture detector operating in the pulsed, constant current mode, a 1.8 m × 2 mm I.D. 3% OV-101 glass column and a heated, on-column injection port. The column oven is kept at ∼190°C and flow rate is ∼30 ml/min. Obtain Aroclor® standards of which the weight-percent compositions have been previously determined by the Webb-McCall approach (Section II.E.1). For quantification, use an Aroclor standard which closely resembles the sample (see Figure 9). For many environmental samples, a 1:1:1 mixture of Aroclor® 1242, 1254, and 1260 has been found to be satisfactory. To quantitate PCBs, chromatograph known amounts of the standard. Measure peak area for each peak and determine the response factor by using the weight percent factor. Chromatograph the sample and measure the area of each peak. Multiply the area of each peak by the response factor for that peak. Add the amount of PCBs found in each peak to obtain the total PCB content.

The precision of this procedure for PCB analysis of sediments has been shown to be good with a coefficient of variation of ±3.6% for 97 replicate determinations on a sediment reference material.[249] The recovery efficiency of the method from spiked sediment samples is also excellent, >99%.[249] However, interlaboratory studies have shown that there are considerable difficulties applying this procedure.[250] Also, it is not possible to establish the absolute accuracy of the procedure for environmental samples since weathering processes can change the relative composition of the PCBs comprising each GC peak in the packed column chromatogram and thus change the peak response factors.

2. Capillary Column Method

Attach a 30-m or 60-m fused silica capillary column (I.D. 0.25 mm) with bonded SE54 or OV1 liquid phase, 0.1 μm or 0.25 μm film thickness, to a split/splitless capillary injector and an electron capture detector. Set the helium flow rate through the column to 18 to 20 cm/s at the upper column temperature (250°C) using a freon, or similar low molecular weight halogenated compound, injection. Set injector temperature to 250°C and Ni[63] electron capture detector to 350°C. If hexane is the solvent, use an initial temperature of 50°C (80°C for isooctane), program the temperature of 250°C at a rate of 1°C /min, and hold at the final temperature of 250°C for 30 min. Inject a volume of 1 μl (splitless) of Aroclor® mixture 1221(10): 1016(5): 1254(3.5): 1262(3) in hexane or isooctane with either an autosampler (preferable) or manually with a syringe. Assign peaks by comparison to literature information (as described earlier) and calculate response factors from peak areas. If one wishes to use relative retention times, octachloronaphthalene or PCB 204 (which is not present in the Aroclors®) can be added as an internal standard to the final mixed Aroclor® standard. Run sample extracts using the same conditions to obtain concentration of individual PCB congeners.

The precision of this method for individual PCB congeners is better than ±10%. The higher error compared to the packed column method is due to the fact that splitless capillary injections are not as reproducible as packed column injections. The absolute accuracy of the method for individual PCB congeners will also be in the neighborhood of ±10%. Interlab studies have shown coefficients of variation for reproducibility of 11 to 24% at the 40 to 300 ng/g level per congener for cleaned eel-fat extracts.[251]

The precision and accuracy of the capillary method for total PCBs are more difficult to define. The more congeners available the better will be the quantification, because there will be fewer components to be approximated. Using the secondary standard approach described here, analyses to within ± 10% of the target values for the various Aroclors® run at different total concentrations are obtained.

D. Confirmation

Confirmation of the presence of PCBs is usually carried out using the GC/MS technique.[226,227] Some characteristic PCB ions are listed in Table 3. If it is necessary to obtain only total PCB's or homolog PCB concentrations, and a mass spectrometer or mass selective detector is available, the method described by Gebhart et al.[242] can be used. This method can also be used for PCB confirmation. Alternatively, dechlorination and perchlorination can also be used to confirm the presence of PCBs. These procedures have been discussed in detail in Section II.D.5. Dechlorination appears to be less subject to interferences than perchlorination.[175]

E. Future Research Needs

The analytical procedures for PCBs have been vastly improved in recent years with the increased use of capillary gas chromatography. Even so, no single column has been shown to be capable of separating all 209 congeners. Perhaps with the use of finer bore columns (such as the 0.1 mm I.D. columns now commercially available) separation of all PCB components on a single column will be possible. Improved separations may also be possible with the use of hydrogen as the carrier gas, but this will require redesign of conventional electron capture detectors to minimize adsorptive losses.[236] Quantification of PCBs would also be advanced if all congeners were available. More complete information on the composition and chromatography of the industrial PCB mixtures is required to help refine their usage as secondary standards for PCB analyses.

There is a dearth of information on the behavior of individual PCB congeners. With the development of simple procedures for quantifying individual PCBs, research on environmental pathways, toxicological and carcinogenic properties, and hazard assessments of specific PCBs can be tackled, and some firm conclusions reached. New computer techniques will be required to deal with the vast quantities of data that will be generated from specific congener PCB analyses and to aid in data interpretation.[252]

REFERENCES

1. **Fishbein, L.,** Chromatographic and biological aspects of polychlorinated biphenyls, *J. Chromatogr.,* 68, 345, 1972.
2. **Neely, W. B.,** Reactivity and environmental peristence of PCB isomers, in *Physical Behaviour of PCB's in the Great Lakes,* MacKay, D., Paterson, S., Eisenreich, S. J., and Simmons, M. S., Eds., Ann Arbor Science, Ann Arbor, MI, 1983, 71.
3. **Peakall, D. B.,** PCB's and their environmental effects, *Crit. Rev. Environ. Control,* 5, 469, 1975.
4. **Kaiser, K. L. E.,** On the optical activity of polychlorinated biphenyls, *Environ. Pollut.,* 7, 93, 1974.
5. **Tucker, E. S., Saeger, V. W., and Hicks, O.,** Activated sludge primary biodegradation of polychlorinated biphenyls, *Bull. Environ. Contam. Toxicol.,* 14, 705, 1975.
6. **Willis, D. E. and Addison, R. F.,** Identification and estimation of the major components of a commercial polychlorinated biphenyl, Aroclor 1221, *J. Fish. Res. Board Can.,* 29, 592, 1972.
7. **Sissons, D. and Welti, D.,** Structural identification of polychlorinated biphenyls in commercial mixtures by gas-liquid chromatography, nuclear magnetic resonance and mass spectrometry, *J. Chromatogr.,* 60, 15, 1971.

8. **Albro, P. W. and Parker, C. E.,** Comparison of the composition of Aroclor 1242 and Aroclor 1016, *J. Chromatogr.,* 169, 161, 1979.

9. **Albro, P. W., Corbett, J. T., and Schroeder, J. L.,** Quantitative characterization of polychlorinated biphenyl mixtures (Aroclors 1248, 1254, and 1260) by gas chromatography using capillary columns, *J. Chromatogr.,* 205, 103, 1981.

10. **Brinkman, U. A. Th. and de Kok, A.,** Production, properties and usage, in *Halogenated Biphenyls, Terphenyls, Naphthalenes, Dibenzodioxins and Related Products,* Kimbrough, R. D., Ed., Elsevier/North Holland Biomedical Press, Amsterdam, 1980, 1.

11. **Hutzinger, O., Safe, S., and Zitko, V.,** *The Chemistry of PCB's,* CRC Press, Boca Raton, FL, 1974.

12. **Webb, R. G. and McCall, A. C.,** Identities of polychlorinated biphenyl isomers in Aroclors, *J. Assoc. Off. Anal. Chem.,* 55, 746, 1975.

13. **Albro, P. W., Haseman, J. K., Clemmer, T. A., and Corbett, B. J.,** Identification of the individual polychlorinated biphenyls in a mixture by gas-liquid chromatography, *J. Chromatogr.,* 136, 147, 1977.

14. **Cairns, T. and Siegmund, E. C.,** PCB's regulatory history and analytical problems, *Anal. Chem.,* 53, 1183A, 1981.

15. **Addison, R. F.,** PCB replacements in dielectric fluids, *Environ. Sci. Technol.,* 17, 486A, 1983.

16. **Ackerman, D. G., Scinto, L. L., Bakshi, P. S., Delumyea, R. G., Johnson, R. J., Richard, G., Takata, A. M., and Sworzyn, E. M.,** *Destruction and Disposal of PCBs by Thermal and Non-Thermal Means.* Noyes Data Corporation, Park Ridge, N.J., U.S.A., 1983, 417.

17. **Kokoszka, L., and Kuntz, G.,** Methods of PCB disposal, *Water Air Soil Pollut.,* 25, 41, 1985.

18. **Piver, W. T. and Lindstrom, F. T.,** Waste disposal technologies for polychlorinated biphenyls, *Environ. Health Persp.,* 59, 163, 1985.

19. **Johnston, L. E.,** Decontamination and disposal of PCB wastes, *Environ. Health Persp.,* 60, 339, 1985.

20. **Ahling, B.,** Destruction of chlorinated hydrocarbons in a cement kiln, *Environ. Sci. Technol.,* 13, 1377, 1979.

21. **Hunt, G. T., Wolf, P., and Fennelty, P. R.,** Incineration of polychlorinated biphenyls in high-efficiency boilers: a viable disposal option, *Environ. Sci. Technol.,* 18, 171, 1984.

22. **Buser, H.-R.,** Formation, occurrence, and analysis of polychlorinated dibenzofurans, dioxins, and related compounds, *Environ. Health Persp.,* 60, 259, 1985.

23. **Erickson, M. D., Gorman, P. G., and Heggem, D. T.,** Relationship of destruction parameters to the destruction/removal efficiency of PCBs, *J. Air Pollut. Control Assoc.,* 35, 603, 1985.

24. **Hutzinger, O., Choudry, G. G., Chittim, B. G., and Johnson, L. E.,** Formation of polychlorinated dibenzofurans and dioxins during combustion, electrical equipment fires and PCB incineration, *Environ. Health Persp.,* 60, 3, 1985.

25. **Kornel, A. and Rogers, C.,** PCB destruction: a novel dehalogenation reagent, *J. Hazardous Materials,* 12, 161, 1985.

26. **Jensen, S.,** The PCB story, *Ambio,* 1, 123, 1972.

27. **Subramanian, B. R., Tanabe, S., Hidaka, H., and Tatsukawa, R.,** DDTs and PCB isomers and congeners in Antarctic fish, *Arch. Environ. Contam. Toxicol.,* 12, 621, 1983.

28. **Kuratsune, M., Yoshimura, T., Matsuzaka, J., and Yamaguchi, A.,** Epidemiological study on Yusho, a poisoning caused by ingestion of rice oil contaminated with a commercial brand of polychlorinated biphenyls, *Environ. Health Perspect.,* 1, 119, 1972.

29. **Miller, S.,** The PCB imbroglio, *Environ. Sci. Technol.,* 17, 11A, 1983.

30. **Miller, S.,** The persistent PCB problem, *Environ. Sci. Technol.,* 16, 98A, 1982.

31. **Nisbet, I. C. T. and Sarofim, A. F.,** Rates and routes of transport of PCB's in the environment, *Environ. Health Persp.,* 1, 21, 1972.

32. **Giam, C. S., Atlas, E., Chan, H. S., and Neff, G. S.,** Phthalate esters, PCB and DDT residues in the Gulf of Mexico atmosphere, *Atm. Environ.,* 14, 65, 1980.

33. **Atlas, E. and Giam, C. S.,** Global transport of organic pollutants: ambient concentrations in the remote marine atmosphere, *Science,* 211, 163, 1981.

34. **Eisenreich, S. J., Looney, B. B., and Hollod, G. J.,** PCBs in the Lake Superior atmosphere 1978-1980, In *Physical Behaviour of PCB's in the Great Lakes,* MacKay, D., Paterson, S., Eisenreich, S. J., and Simmons, M. S., Eds., Ann Arbor Science, Ann Arbor, MI, 1983, 115.

35. **Singer, E., Jarv, T., and Sage, M.,** Survey of polychlorinated biphenyls in ambient air across the Province of Ontario, in *Physical Behaviour of PCB's in the Great Lakes,* MacKay, D., Paterson, S., Eisenreich, S. J., and Simmons, M. S., Eds., Ann Arbor Science, Ann Arbor, MI, 1983, 115.

36. **Harvey, G. R. and Steinhauer, W. G.,** Atmospheric transport of polychlorobiphenyls to the North Atlantic, *Atmos. Environ.,* 8, 777, 1974.

37. **Eisenreich, S. J., Capel, P. D., and Looney, B. B.,** PCB dynamics in Lake Superior waters, in *Physical Behaviour of PCB's in the Great Lakes,* MacKay, D., Paterson, S., Eisenreich, S. J., and Simmons, M. S., Eds., Ann Arbor Science, Ann Arbor, MI, 1983, 181.

38. **Eisenreich, S. J., Hollod, G. J., Johnson, T. C., and Evans, J. E.,** PCB and other microcontaminant-sediment interactions in Lake Superior, in *Contaminants and Sediments,* Vol. 1, Baker, R. A., Ed., Ann Arbor Science, Ann Arbor, MI, 1980, 67.

39. **Burgomeister, G., Aswold, K., Machado, L., Mowrer, J., and Tarradellas, J.,** Concentrations en PCBs et DDT des sédiments superficiels de la rive Suisse du lac Leman, *Schweiz. Z. Hydrol.,* 45, 223, 1983.

40. **Ballschmiter, K., Zell, M., and Neu, H. J.,** Persistence of PCBs in the ecosystem; will some PCB-components "never" degrade?, *Chemosphere,* 7, 173, 1978.

41. **Schulte, E. and Acker, L.,** Identifizierung und Metabolisierbarkeit von polychlorierten Biphenylen, *Naturwissenschaften,* 61, 79, 1974.

42. **Jensen, S. and Sundström, G.,** Structures and levels of most chlorobiphenyls in two technical PCB products and in human adipose tissue, *Ambio,* 3, 71, 1974.

43. **Zell, M., Neu, H. J., and Ballschmiter, K.,** Single component analysis of polychlorinated biphenyl (PCB) and chlorinated pesticide residues in marine fish samples, *Fresenius Z. Anal. Chem.,* 292, 97, 1978.

44. **Mowrer, J., Åswold, K., Burgomeister, G., Michado, L., and Tarradellas, J.,** PCB in a Lake Geneva ecosystem, *Ambio,* 11, 355, 1982.

45. **Falkner, R. and Simonis, W.,** Polychlorierte Biphenyle (PCB) in Lebensraum Wasser (Aufnahme und Anreicherung durch Organismen — Probleme der Weitergabe in der Nahrungspyramide), *Arch. Hydrobiol. Beih. Ergebn. Limnol.,* 17, 1, 1982.

46. **Bunce, N. J.,** Photodechlorination of PCB's: current status, *Chemosphere,* 11, 701, 1982.

47. **Zabik, M. J.,** The photochemistry of PCBs, in *PCBs — Health and Environmental Hazards,* D'Itri, F. M. and Kamrin, M. A., Eds., Butterworth Publishers, Boston, 1983, 141.

48. **Bunce, N. J., Kumar, Y., and Brownlee, B. G.,** An assessment of the impact of solar degradation of polychlorinated biphenyls in the aquatic environment, *Chemosphere,* 7, 155, 1978.

49. **Carey, J. H., Lawrence, J., and Tosine, H. M.,** Photodechlorination of PCB's in the presence of titanium dioxide in aqueous suspensions, *Bull. Environ. Contam. Toxicol.,* 16, 697, 1976.

50. **Furukawa, K.,** Microbial degradation of polychlorinated biphenyls (PCBs), in *Biodegradation and Detoxification of Environmental Pollutants,* Chakrabarty, A. M., Ed., CRC Press, Boca Raton, FL, 1982, 33.

51. **Parsons, J., Veerkamp, W., and Hutzinger, O.,** Microbial metabolism of chlorobiphenyls, *Toxicol. Environ. Chem.,* 6, 327, 1983.

52. **Furukawa, K. and Chakrabarty, A. M.,** Involvement of plasmids in total degradation of chlorinated biphenyls, *App. Environ. Microbiol.,* 44, 619, 1982.

53. **Kong, H. L. and Sayler, G. S.,** Degradation and total mineralization of monohalogenated biphenyls in natural sediment and mixed bacterial culture, *Appl. Environ. Microbiol.,* 46, 666, 1983.

54. **Shields, M. S., Hooper, S. W., and Sayler, G. S.,** Plasmid-mediated mineralization of 4-chlorobiphenyl, *J. Bacteriol.,* 163, 882, 1985.

55. **Furukawa, K., Tomizuka, N., and Kamibayashi, A.,** Effect of chlorine substitution on the bacterial metabolism of various polychlorinated biphenyls, *App. Environ. Microbiol.,* 38, 301, 1979.

56. **Bedard, D. L., Unterman, R., Bopp, L. H., Brennan, M. J., Haberl, M. L., and Johnson, C.,** Rapid assay for screening and characterizing microorganisms for the ability to degrade polychlorinated biphenyls, *Appl. Environ. Microbiol.,* 51, 761, 1986.

57. **Furukawa, K., Tonomura, K., and Kamibayashi, A.,** Metabolism of 2,4,4'-trichlorobiphenyl by *Acinetobacter* sp. P6, *Agric. Biol. Chem.,* 43, 1577, 1979.

58. **Sylvestre, M., Massé, R., Messier, F., Fauteux, J., Bisaillon, J.-G., and Beaudet, R.,** Bacterial nitration of 4-chlorobiphenyl, *Appl. Environ. Microbiol.,* 44, 871, 1982.

59. **Hankin, L. and Sawhney, B. J.,** Microbial degradation of polychlorinated biphenyls in soils, *Soil Science,* 137, 401, 1984.

60. **Brunner, W., Sutherland, F. H., and Focht, D. D.,** Enhanced biodegradation of polychlorinated biphenyls in soil by analogue enrichment and bacterial inoculation, *J. Environ. Quality,* 14, 324, 1985.

61. **Focht, D. D. and Brunner, W.,** Kinetics of biphenyl and polychlorinated biphenyl metabolism in soil, *Appl. Environ. Microbiol.,* 50, 1058, 1985.

62. **Eaton, D. C.,** Mineralization of polychlorinated biphenyls by *Phanerochaete chrysosporium:* a lignolytic fungus, *Enzyme Microb. Technol.,* 7, 194, 1985.

63. **Bumpus, J. A., Tien, M., Wright, D., and Aust, S. D.,** Oxidation of persistent environmental pollutants by a white rot fungus, *Science,* 228, 1434, 1958.

64. **Kelley, R. L. and Reddy, C. A.,** Identification of glucose oxidase activity as the primary source of hydrogen peroxide production in lignolytic cultures of *Phanerochaete chrysosporium, Arch. Microbiol.,* 144, 248, 1986.

65. **Baxter, R. M. and Sutherland, D. A.,** Biochemical and photochemical processes in the degradation of chlorinated biphenyls, *Environ. Sci. Technol.,* 18, 608, 1984.

66. **Pal, D., Weber, J. B., and Overcash, M. R.,** Fate of polychlorinated biphenyls (PCBs) in soil-plant systems, *Residue Rev.,* 74, 45, 1980.

67. **Baxter, R. M.,** Bacterial formation of humus-like materials from polychlorinated biphenyls (PCBs), *Water Poll. Res. J. Canada,* 21, 1, 1986.

68. **Kaiser, K. L. E. and Wong, P. T. S.,** Bacterial degradation of polychlorinated biphenyls. I. Identification of some metabolic products from Aroclor 1242, *Bull. Environ. Contam. Toxicol.,* 11, 291, 1974.

69. **Brown, J. F., Wagner, R. E., Bedard, D. L., Brennan, M. J., Carnahan, J. C., May, R. J., and Rofflemire, T. J.,** PCB transformations in upper Hudson sediments, *Northeastern Environ. Sci.,* 3, 166, 1984.

70. **Bruggeman, W. A., Matron, L. B. J. M., Kooiman, D., and Hutzinger, O.,** Accumulation and elimination kinetics of di-, tri-, and tetra-chlorobiphenyls by goldfish after dietary and aqueous exposure, *Chemosphere,* 10, 811, 1981.

71. **Niimi, A. J. and Oliver, B. G.,** Biological half-lives of polychlorinated biphenyl (PCB) congeners in whole fish and muscle of rainbow trout *(Salmo gairdneri), Can. J. Fish. Aquat. Sci.,* 40, 1388, 1983.

72. **Safe, S.,** Halogenated hydrocarbons and aryl hydrocarbons identified in human tissues, *Toxicol. Environ. Chem.,* 5, 153, 1982.

73. **Kuroki, H. and Masuda, Y.,** Structures and concentrations of the main components of polychlorinated biphenyls retained in patients with Yusho, *Chemosphere,* 6, 469, 1977.

74. **Yakushiji, T., Watanabe, I., Kuwabara, K., Yoshida, S., Koyama, K., Hara, I., and Kunita, N.,** Long-term studies of the excretion of polychlorinated biphenyls (PCBs) through the mother's milk of an occupationally exposed worker, *Arch. Environ. Contam. Toxicol.,* 7, 495, 1978.

75. **Sundström, G., Hutzinger, O., and Safe, S.,** The metabolism of chlorobiphenyls — a review, *Chemosphere,* 5, 267, 1976.

76. **Jansen, B., Jensen, S., Olssen, M., Renberg, L., Sundstöm, G., and Vaz, R.,** Identification by GC-MS of phenolic metabolites of PCB and p,p'-DDE isolated from Baltic Guillemot and seal, *Ambio,* 4, 93, 1975.

77. **Jensen, S. and Sundström, G.,** Metabolic hydroxylation of a chlorobiphenyl containing only isolated unsubstituted positions — 2,2',4,4',5,5'-hexachlorobiphenyl, *Nature,* 251, 219, 1974.

78. **Mizutani, T., Yamamoto, K., and Tajima, K.,** Sulfur-containing metabolites of chlorobiphenyl isomers, a comparative study, *J. Agric. Food Chem.,* 26, 862, 1978.

79. **Mio, T. and Sumino, K.,** Mechanism of biosynthesis of methylsulfones from PCBs and related compounds, *Environ. Health Persp.,* 59, 129, 1985.

80. **Jensen, S. and Jansson, B.,** Methyl sulfone metabolites of PCB and DDE, *Ambio,* 5, 257, 1976.

81. **Buser, H.-R. and Müller, M. D.,** Methylthio metabolites of polychlorobiphenyls identified in sediment samples from two lakes in Switzerland, *Environ. Sci. Technol.,* 20, 730, 1986.

82. **Safe, S., Wyndham, C., Parkinson, A., Purdy, R., and Crawford, A.,** Halogenated biphenyl metabolism, in *Hydrocarbons and Halogenated Hydrocarbons in the Aquatic Environment,* Afghan, B. K. and Mackay, D., Eds., Plenum Press, New York, 1980, 537.

83. **Haraguichi, K., Kuroki, H., Masuda, Y., Koga, N., Kuroki, J., Hokama, Y., and Yoshimura, H.,** Toxicological evaluation of sulfur-containing metabolites of 2,5,2',5'-tetrachlorobiphenyl, *Chemosphere,* 14, 1755, 1985.

84. **Jerina, D. M. and Daly, J. W.,** Arene oxides: a new aspect of drug metabolism, *Science,* 185, 573, 1974.

85. **Ullrich, V.,** Cytochrome P450 and biological hydroxylation reactions, in *Topics in Current Chemistry, 83: Biochemistry,* Springer-Verlag, Berlin, 1979, 67.

86. **Paine, A. J.,** Hepatic cytochrome P-450, in *Essays in Biochemistry,* Vol. 17, Campbell, P. N. and Marshall, R. D., Eds., Academic Press, London, 1981, 85.

87. **Goldstein, J. A.,** The structure-activity relationships of halogenated biphenyls as enzyme inducers, *Ann. N.Y. Acad. Sci.,* 320, 164, 1979.

88. **Alvares, A. P., Bickers, D. R., and Kappas, A.,** Polychlorinated biphenyls: a new type of inducer of cytochrome P-448 in the liver, *Proc. Natl. Acad. Sci. U.S.A.,* 70, 1321, 1973.

89. **Alvares, A. P. and Kappas, A.,** Heterogeneity of cytochrome P-450s induced by polychlorinated biphenyls, *J. Biol. Chem.,* 252, 6373, 1977.

90. **Poland, A. and Glover, E.,** Chlorinated biphenyl induction of aryl hydrocarbon hydroxylase activity: a study of the structure-activity relationship, *Mol. Pharmacol.,* 13, 924, 1977.

91. **Parkinson, A., Cockerline, R., and Safe, S.,** Polychlorinated biphenyl isomers and congeners as inducers of both 3-methylcholanthrene- and phenobarbitone-type microsomal enzyme activity, *Chem. Biol. Interact.,* 29, 277, 1980.

92. **Parkinson, A. and Safe, S.,** Aryl hydrocarbon hydroxylase induction and its relationship to the toxicity of halogenated aryl hydrocarbons, *Toxicol. Environ. Chem. Rev.,* 4, 1, 1981.

93. **Safe, S., Bandiera, S., Sawyer, T., Robertson, L., Safe, L., Parkinson, A., Thomas, P. E., Ryan, D. E., Reik, L. M., Levin, W., Denomme, M. A., and Fujita, T.,** PCBs: structure function relationships and mechanism of action, *Environ. Health Persp.,* 60, 47, 1985.

94. **Poland, A. and Knutson, J. C.**, 2,3,7,8-tetrachloro-dibenzo-p-dioxin and related halogenated aromatic hydrocarbons: examination of the mechanism of toxicity, *Ann. Rev. Pharmacol. Toxicol.*, 22, 517, 1982.

95. **Rifkind, A. B. and Muschick, H.**, Benoxaprofen suppression of polychlorinated biphenyl toxicity without alteration of mixed function oxidase function, *Nature*, 303, 524, 1983.

96. **Jensen, R. K., Sleight, S. D., Aust, S. D., Goodman, J. I., and Trosko, J. E.**, Hepatic tumor-promoting ability of 3,3',4,4',5,5'-hexabromobiphenyl: the interrelationship between toxicity, induction of hepatic microsomal drug metabolizing enzymes, and tumor-promoting ability, *Toxicol. Appl. Pharmacol.*, 71, 163, 1983.

97. **McKinney, J. D., Chae, K., McConnell, E. E., and Birnbaum, L. S.**, Structure-induction versus structure-toxicity relationships for polychlorinated biphenyls and related aromatic compounds, *Environ. Health Persp.*, 60, 57, 1985.

98. **Connell, D. W. and Miller, G. J.**, *Chemistry and Ecotoxicology of Pollution*, John Wiley & Sons, New York, 1984.

99. **Passino, D. R. M.**, Biochemical indicators of stress in fishes: an overview, in *Contaminant Effects on Fisheries*, Cairns, V. W., Hodson, P. V., and Nriagu, J. O., Eds., John Wiley & Sons, New York, 1984, 37.

100. **Hornshaw, T. C., Avlerich, R. J., and Johnson, H. E.**, Feeding Great Lakes fish to mink: effects on mink and accumulation of and elimination of PCBs by mink, *J. Toxicol. Environ. Health*, 11, 937, 1983.

101. **Massé, R., Martineau, D., Tremblay, L., and Béland, P.**, Concentrations and chromatographic profiles of DDT metabolites and polychlorobiphenyl (PCB) residues in stranded Beluga Whales *(Delphinapterus leucas)*, from the St. Lawrence Estuary, Canada, *Arch. Environ. Contam. Toxicol.*, 15, 567, 1986.

102. **Bergman, A. and Ollson, M.**, Pathology of Baltic Grey Seal and Ringed Seal females with special reference to adrenocortical hyperplasia: is environmental pollution the cause of a widely distributed disease syndrome, *Finish Game Res.*, 44, 47, 1985.

103. **Reijnders, P. J. H.**, Reproductive failure in common seals feeding on fish from polluted coastal waters, *Nature*, 324, 456, 1986.

104. **Bleavins, M. R. and Aulerich, R. J.**, Immunotoxicological effects of polychlorinated biphenyls on the cell-mediated and humoral immune systems, *Residue Rev.*, 90, 57, 1983.

105. **Martini, O.**, Polychlorinated biphenyls: is their bad rap deserved?, *Can. Consult. Eng.*, September 1985, 28.

106. **Masuda, Y., Yamaryo, T., Haraguchi, K., Kuratsune, M., and Hsu, S. T.**, Comparison of causal agents in Taiwan and Fukuoka PCB poisonings, *Chemosphere*, 11, 199, 1982.

107. **Gaffrey, W. R.**, The epidemiology of PCBs, in *PCBs: Human and Environmental Hazards*, D'Itri, F. M., and Kamrin, M. A., Eds., Butterworth Publishers, Boston, 1983, 279.

108. **Sawney, B. L. and Hankin, L.**, Polychlorinated biphenyls in food: a review, *J. Food Prot.*, 48, 442, 1985.

109. **Bennett, B. G.**, Exposure of man to environmental PCBs — an exposure commitment assessment, *Sci. Total. Environ.*, 29, 191, 1983.

110. **Steele, G., Stehr-Green, P., and Welty, E.**, Estimates of the biological half-life of polychlorinated biphenyls in human serum, *N. Engl. J. Med.*, 314, 926, 1986.

111. **Parkinson, A., Robertson, L. W., and Safe, S.**, Reconstituted human breast milk PCBs as potent inducers of aryl hydrocarbon hydroxylase, *Biochem. Biophys. Res. Commun.*, 96, 882, 1980.

112. **Jensen, A. A.**, Chemical contaminants in human milk, *Residue Rev.*, 89, 1, 1983.

113. **Fein, G. G., Jacobson, J. L., Jacobson, S. W., Schwartz, P. M., and Dowler, J. K.**, Prenatal exposure to polychlorinated biphenyls: effects on birth size and gestational age, *J. Pediatr.*, 105, 315, 1984.

114. **Jacobson, J. L., Jacobson, S. W., Schwartz, P. M., Fein, G. G., and Dowler, J. K.**, Prenatal exposure to an environmental toxin: a test of the multiple effects model, *Dev. Psychol.*, 20, 523, 1984.

115. **Sherma, J.**, *Manual of Analytical Quality for Pesticides and Related Compounds in Human and Environmental Samples*, EPA-600/1-79-008, U.S. Environmental Protection Agency report, Research Triangle Park, NC, 1979.

116. **Hutchinson, G. E.**, *A Treatise on Limnology*, Vol. 1, John Wiley & Sons, New York, 1975, chap. 7.

117. **Strachan, W. M. J. and Huneault, H.**, Evaluation of an organic automated rain sampler, *Environ. Can. Tech. Bull.*, No. 128, 1982.

118. **Harvey, G. W.**, Microlayer collection from the sea surface: a new method and initial results, *Limnol. Oceanogr.*, 11, 608, 1966.

119. **Platford, R. F., Maguire, R. J., Tkacz, R. J., and Madsen, N.**, A note on the NWRI microlayer sampler, *Environ. Can. Tech. Bull.*, in press,

120. **Baxter, M. S., Farmer, J. G., McKinley, I. G., Swan, D. S., and Jack, W.**, Evidence of the unsuitability of gravity coring for collecting sediment in pollution and sedimentation rate studies, *Environ. Sci. Technol.*, 15, 843, 1981.

121. **Bouma, A. H.**, *Methods for the Study of Sedimentary Structures*, Wiley-Interscience, New York, 1969.

122. **Lagler, K. F.,** Capture, sampling and examination of fishes, in *International Biological Programme Handbook, No. 3, Methods for Assessment of Fish Production in Fresh Waters,* Ricker, W. E., Ed., Blackwell, Oxford, 1971, chap. 2.
123. **Tonolli, V.,** Methods of collection, in *International Biological Programme Handbook No. 17, A Manual on Methods for the Assessment of Secondary Productivity in Fresh Waters,* Edmondsen, W. T. and Winberg, G. G., Eds., Blackwell, Oxford, 1971, chap. 1.
124. **Schwoerbel, J.,** Ed., *Methods of Hydrobiology,* Pergamon Press, London, 1970, chaps, 2—4.
125. **Gossett, R. W., Brown, D. A., and Young, D. R.,** Predicting the bioaccumulation of organic compounds in marine organisms using octanol/water partition coefficients, *Mar. Pollut. Bull.,* 14, 387, 1983.
126. **Fowler, S. W., Polikarpov, G. G., Elder, D. L., Parsi, P., and Villeneuve, J. P.,** Polychlorinated biphenyls: accumulation from contaminated sediments and water by the polychaete *Nereis diversicolor, Marine Biology,* 48, 303, 1978.
127. **Goerlitz, D. F. and Law, L. M.,** Note on removal of sulfur interferences from sediment extracts for pesticide analysis, *Bull. Environ. Contam. Toxicol.,* 6, 9, 1971.
128. **Blumer, M.,** Removal of elemental sulfur from hydrocarbon fractions, *Anal. Chem.,* 29, 1039, 1957.
129. **Junk, J. A., Richard, J. J., Grieser, M. D., Witiak, D., Witiak, J. L., Arguello, M. D., Vick, R., Svec, H. J., Fritz, J. S. and Calder, G. V.,** Uses of macroreticular resin in the analysis of water for trace organic contaminants, *J. Chromatogr.,* 99, 745, 1974.
130. **Claeys, R. R. and Inman, R. D.,** Adsorption chromatographic separation of chlorinated hydrocarbons from lipids, *J. Assoc. Off. Anal. Chem.,* 57, 399, 1974.
131. **Mills, P. A.,** Variation of Florisil activity: simple method for measuring adsorbent capacity and its use in standardizing Florisil columns, *J. Assoc. Off. Anal. Chem.,* 51, 29, 1968.
132. **Holden, A. V. and Marsden, K.,** Single-stage cleanup of animal tissue extracts for organochlorine residue analysis, *J. Chromatogr.,* 44, 481, 1969.
133. **Steichen, J. J., Tucker, R. G., and Mechon, E.,** Standardization of Aroclor lots for individual-peak gas chromatographic calibration, *J. Chromatogr.,* 236, 113, 1982.
134. **Webb, R. G. and McCall, A. C.,** Quantitative PCB standards for electron capture gas chromatography, *J. Chromatogr. Sci.,* 11, 366, 1973.
135. **Sawyer, L. D.,** Quantitation of polychlorinated biphenyl residues by electron capture gas-liquid chromatography: reference material characterization and preliminary study, *J. Assoc. Off. Anal. Chem.,* 61, 272, 1978.
136. **Mullin, M. D., Pochini, C. M., McCrindle, S., Romkes, M., Safe, S. H., and Safe, L. M.,** High-resolution PCB analysis: synthesis and chromatographic properties of all 209 PCB congeners, *Environ. Sci. Technol.,* 18, 468, 1984.
137. **Environment Canada,** *Analytical Methods Manual,* Water Quality Branch, Inland Waters Directorate, Ottawa, Canada, 1983.
138. **Oliver, B. G. and Nicol, K. D.,** Gas chromatographic determination of chlorobenzenes and other chlorinated hydrocarbons in environmental samples using fused silica capillary columns, *Chromatographic,* 16, 336, 1982.
139. **McCrea, R. C.,** *Development of an Aqueous Phase Liquid-Liquid Extractor,* Interim Report, Inland Waters Directorate, Ontario Region, Water Quality Branch, Burlington, Ontario, 1982.
140. **Oliver, B. G. and Nicol, K. D.,** Field testing of a large volume liquid-liquid extraction device for halogenated organics in natural waters, *Intern. J. Environ. Anal. Chem.,* 25, 275, 1986.
141. **Kuntz, K., Chan, C. H., Clignett, A. H., and Boucher, R.,** *Water Quality Sampling Methods at Niagara-on-the-Lake,* Interim Report, Inland Waters Directorate, Ontario Region, Water Quality Branch, Burlington, Ontario, 1982.
142. **Bacaloni, A., Goretti, G., Lagana, A., Petronio, B. M., and Rotatori, M.,** Sorption capacities of graphitized carbon black in determination of chlorinated pesticide traces in water, *Anal. Chem.,* 52, 2033, 1980.
143. **Uthe, J. F., Reinke, J., and Gesser, H.,** Extraction of organochlorine insecticides from water by porous polyurethane coated with selective absorbent, *Environ. Lett.,* 3, 117, 1972.
144. **Musty, P. R. and Nickless, G.,** The extraction and recovery of chlorinated insecticides and polychlorinated biphenyls from water using porous polyurethane foams, *J. Chromatogr.,* 100, 83, 1974.
145. **Coburn, J. A., Valdmanis, I. A., and Chau, A. S. Y.,** Evaluation of XAD-2 for multiresidue extraction of organochlorine pesticides and polychlorinated biphenyls from natural waters, *J. Assoc. Off. Anal. Chem.,* 60, 224, 1977.
146. **LeBel, G. L. and Williams, D. T.,** Determination of low μg/l levels of polychlorinated biphenyls in drinking water by extraction with macroreticular resin and analysis using a capillary column, *Bull. Environ. Contam. Toxicol.,* 24, 397, 1980.
147. **Veith, G. D. and Kiwus, L. M.,** An exhaustive steam-distillation and solvent-extraction unit for pesticides and industrial chemicals, *Bull. Environ. Contam. Toxicol.,* 17, 631, 1977.

148. **Godefroot, M., Stechele, M., Sandra, P., and Verzele, M.,** A new method for the quantitative analysis of organochlorine pesticides and polychlorinated biphenyls, *J. High Resolut. Chromatogr. Chromatogr. Commun.,* 5, 75, 1982.

149. **Teichman, J., Bevenue, A., and Hylin, J. W.,** Separation of polychlorobiphenyls from chlorinated pesticides in sediment and oyster samples for analysis by gas chromatography, *J. Chromatogr.,* 151, 155, 1978.

150. **Johnsen, R. E. and Starr, R. I.,** Ultrarapid extraction of insecticides from soil using a new ultrasonic technique, *J. Agr. Food. Chem.,* 20, 48, 1972.

151. **Grimalt, J., Marfil, C., and Albaiges, J.,** Analysis of hydrocarbons in aquatic sediments, *Intern. J. Environ. Anal. Chem.,* 18, 183, 1984.

152. **Goerlitz, D. F. and Law, L. M.,** Determination of chlorinated insecticides in suspended sediments and bottom material, *J. Assoc. Off. Anal. Chem.,* 57, 176, 1974.

153. **Kerkhoff, M. A. T., de Vries, A., Wegmann, R. C. C., and Hofstee, A. W. M.,** Analysis of PCB's in sediments by glass capillary gas chromatography, *Chemosphere,* 11, 165, 1982.

154. **Veith, G. D., Kuehl, D. W., and Rosenthal, J.,** Preparative method for gas chromatographic/mass spectral analysis of trace quantities of pesticides in fish tissue, *J. Assoc. Off. Anal. Chem.,* 58, 1, 1975.

155. **Veith, G. D., De Foe, D. L., and Bergstedt, B. V.,** Measuring and estimating the bioconcentration factor of chemicals in fish, *J. Fish. Res. Board Can.,* 36, 1040, 1979.

156. **Wharfe, J. R. and Van Den Broek, W. L. F.,** Chlorinated hydrocarbons in macroinvertebrates and fish from the Lower Medway Estuary, Kent, *Mar. Pollut. Bull.,* 9, 76, 1978.

157. **Pastel, M., Bush, B., and Kim, J. S.,** Accumulation of polychlorinated biphenyls in American shad during their migration in the Hudson River, spring 1977. *Pestic. Monit. J.,* 14, 11, 1980.

158. **Hesselberg, R. J. and Johnson, J. L.,** Column extraction of pesticides from fish, fish food and mud, *Bull. Environ. Contam. Toxicol.,* 7, 115, 1972.

159. **Schwartz, T. R. and Lehmann, R. G.,** Determination of polychlorinated biphenyls in plant tissue, *Bull. Environ. Contam. Toxicol.,* 28, 723, 1982.

160. **Donkin, P., Mann, S. V., and Hamilton, E. I.,** Microcoulometric determination of total organochlorine pesticide and polychlorinated biphenyl residues in grey seal *(Halichoerus grypus)* blubber, *Anal. Chim. Acta,* 88, 289, 1977.

161. **Linko, R. R., Kaitaranta, J., Rantamäki, P., and Eronen, L.,** Occurrence of DDT and PCB compounds in Baltic herring and pike from the Turko Archipelago, *Environ. Pollut.,* 7, 193, 1974.

162. **Stalling, D. L., Tindle, R. C., and Johnson, J. L.,** Cleanup of pesticide and polychlorinated biphenyl residues in fish extracts by gel permeation chromatography, *J. Assoc. Off. Anal. Chem.,* 55, 32, 1972.

163. **Ballschmiter, K. and Zell, M.,** Baseline studies of the global pollution. I. Occurrence of organohalogens in pristine European and Antarctic environments, *Intern. J. Environ. Anal. Chem.,* 8, 15, 1980.

164. **Erney, D. R.,** Rapid screening method for analysis of chlorinated pesticide and polychlorinated biphenyl residues in fish, *J. Assoc. Off. Anal. Chem.,* 57, 576, 1974.

165. **Hattula, M. L.,** Some aspects of the recovery of chlorinated residues (DDT-type compounds and PCB) from fish tissue using different extraction methods, *Bull. Environ. Contam. Toxicol.,* 12, 301, 1974.

166. **Miles, J. R. W.,** Conversion of DDT and its metabolites to dichlorobenzophenones for analysis in the presence of polychlorinated biphenyls, *J. Assoc. Off. Anal. Chem.,* 55, 1039, 1972.

167. **Ford, J. H., McDaniel, C. A., White, F. C., Vest, R. E., and Roberts, R. E.,** Sampling and analysis of pesticides in the environment, *J. Chromatogr. Sci.,* 13, 291, 1975.

168. **Clayton, J. R., Pavlov, S. P., and Breitner, N. F.,** Polychlorinated biphenyls in coastal marine zooplankton: bioaccumulation by equilibrium partitioning, *Environ. Sci. Technol.,* 11, 676, 1977.

169. **Musial, C. J., Uthe, J. F., Wiseman, R. J., and Matheson, R. A.,** Occurrence of PCB residues in burbot *(Lota lota)* and lake trout *(Salvelinus namaycush)* from the Churchill Falls power development area, *Bull. Environ. Contam. Toxicol.,* 23, 256, 1979.

170. **Bjerk, J. E. and Brevik, E. M.,** Organochlorine compounds in aquatic environments, *Arch. Environ. Contam. Toxicol.,* 9, 743, 1980.

171. **Shaw, G. R. and Connell, D. W.,** Relationships between steric factors and bioconcentration of polychlorinated biphenyls (PCB's) by the sea mullet *(Mugil cephalus linnaeus)*, *Chemosphere,* 9, 731, 1980.

172. **Tuinstra, L. G. M. Th., Driessen, J. J. M., Keukens, H. J., Van Munsteren, T. J., Roos, A. H., and Traag, W. A.,** Quantitative determination of specified chlorobiphenyls in fish with capillary gas chromatography and its use for monitoring and tolerance purposes, *Intern. J. Environ. Anal. Chem.,* 14, 147, 1983.

173. **Porter, M. L., Young, S. J. V., and Burke, J. A.,** A method for the analysis of fish, animal and poultry tissue for chlorinated pesticide residues, *J. Assoc. Off. Anal. Chem.,* 53, 1300, 1970.

174. **Castelli, M. G., Martelli, G. P., Spagone, C., Cappellini, L., and Fanelli, R.,** Quantitative determination of polychlorinated biphenyls (PCB's) in marine organisms analyzed by high resolution gas chromatography selected ion monitoring, *Chemosphere,* 12, 291, 1983.

175. **De Kok, A., Geerdink, R. B., Frei, R. W., and Brinkmann, U. A. Th.**, Limitations in the use of perchlorination as a technique for the quantitative analysis of polychlorinated biphenyls, *Intern. J. Environ. Anal. Chem.*, 11, 17, 1982.

176. **Simmons, M. S., Sweetman, J. A., Miller, T. J., and Jude, D. J.**, A saponification procedure for the determination of some chlorinated hydrocarbons in fish, *J. Great Lakes Res.*, 8, 587, 1982.

177. **Rote, J. W. and Murphy, P. G.**, A method for the quantitation of polychlorinated biphenyl (PCB) isomers, *Bull. Environ. Contam. Toxicol.*, 6, 377, 1971.

178. Ontario Ministry of the Environment, *Handbook of Analytical Methods,* Toronto, Ontario, 1984.

179. **Hallett, D. J., Norstrom, R. J., Onuska, F. I., Comba, M. E., and Sampson, R.**, Mass spectral confirmation and analysis by the Hall detector of mirex and photomirex in herring gulls from Lake Ontario, *J. Agric. Food Chem.*, 24, 1189, 1976.

180. **Smrek, A. L. and Needham, L. L.**, Simplified cleanup procedures for adipose tissue containing polychlorinated biphenyls, DDT, and DDT metabolites, *Bull. Environ. Contam. Toxicol.*, 28, 718, 1982.

181. **Mes, J., Davies, D. J., and Lau, P. Y.**, The effect of extraction technique on the fat content, polychlorinated biphenyl level and tri- to octabiphenyl distribution in human milk, *Chemosphere*, 9, 763, 1980.

182. **Mes, J.**, Experiences in human milk analysis for halogenated hydrocarbon residues, *Intern. J. Environ. Anal. Chem.*, 9, 283, 1981.

183. **Needham, L. L., Burse, V. W., and Price, H. A.**, Temperature-programmed gas chromatographic determination of polychlorinated and polybrominated biphenyls in serum, *J. Assoc. Off. Anal. Chem.*, 64, 1131, 1981.

184. **Rogers, W. M.**, The use of a solid support for the extraction of chlorinated pesticides from large quantities of fats and oils, *J. Assoc. Off. Anal. Chem.*, 55, 1053, 1972.

185. **Seidl, G. and Ballschmiter, K.**, Isolation of PCB's from vegetable oils: recovery and efficiency of "cleanup" methods, *Chemosphere*, 5, 363, 1976.

186. **Murphy, P. G.**, Sulfuric acid for the cleanup of animal tissues for analysis of acid-stable chlorinated hydrocarbon residues, *J. Assoc. Off. Anal. Chem.*, 55, 1360, 1972.

187. **Lunde, G. and Ofstad, E. B.**, Determination of fat-soluble chlorinated compounds in fish, *Z. Anal. Chem.*, 282, 395, 1976.

188. **McLeod, H. A. and Wales, P. J.**, A low-temperature cleanup procedure for pesticides and their metabolites in biological samples, *J. Agric. Food Chem.*, 20, 624, 1972.

189. **Stalling, D. L., Tindle, R. C., and Johnson, J. L.**, Cleanup of pesticide and polychlorinated biphenyl residues in fish extracts by gel permeation chromatography, *J. Assoc. Off. Anal. Chem.*, 55, 32, 1972.

190. **Griffitt, K. R. and Craun, J. C.**, Gel permeation chromatographic system: an evaluation, *J. Assoc. Off. Anal. Chem.*, 57, 168, 1974.

191. **Kuehl, D. W. and Leonard, E. N.**, Isolation of xenobiotic chemicals from tissue samples by gel permeation chromatography, *Anal. Chem.*, 50, 182, 1978.

192. **Tindle, R. C. and Stalling, D. L.**, Apparatus for automated gel permeation cleanup for pesticide residue analysis, application to fish lipids, *Anal. Chem.*, 44, 1768, 1972.

193. **Mills, P. A., Bong, B. A., Kamps, L. R., and Burke, J. A.**, Elution solvent system for Florisil column cleanup in organochlorine pesticide residue analyses, *J. Assoc. Off. Anal. Chem.*, 55, 39, 1972.

194. **Norwicki, H. G.**, Application of azulene as a visual aid to monitor column chromatographic fractionation of samples for pesticide and polychlorinated biphenyl determination, *J. Assoc. Off. Anal. Chem.*, 64, 16, 1981.

195. **Zell, M. and Ballschmiter, K.**, Baseline studies of the global pollution. III. Trace analysis of polychlorinated biphenyls (PCB) by ECD glass capillary gas chromatography in environmental samples at different trophic levels, *Fresenius' Z. Anal. Chem.*, 304, 337, 1980.

196. **Bevenue, A. and Ogata, J. N.**, A note on the use of Florisil adsorbent for the separation of polychlorobiphenyls from chlorinated pesticides, *J. Chromatogr.*, 50, 142, 1972.

197. **Telling, G. M., Sissons, D. J., and Brinkham, H. W.**, Determination of organochlorine insecticide residues in fatty foodstuffs using a clean-up technique based on a single column of activated alumina, *J. Chromatogr.*, 137, 405, 1977.

198. **Armour, J. A. and Burke, J. A.**, Method for separating polychlorinated biphenyls from DDT and its analogs, *J. Assoc. Off. Anal. Chem.*, 53, 761, 1970.

199. **Matsumoto, H. T.**, Study of the silicic acid procedure of Armour and Burke for the separation of polychlorinated biphenyls from DDT and its analogs, *J. Assoc. Off. Anal. Chem.*, 55, 1092, 1972.

200. **Griffin, D. A., Marin, B., and Deinger, M. L.**, Optimization of chromatographic conditions for the separation of p,p'-DDE from Aroclor 1254 on silica, using azulene as indicator, *J. Assoc. Off. Anal. Chem.*, 63, 959, 1980.

201. **Bidleman, T. F., Matthews, J. R., Olney, C. E., and Rice, C. P.**, Separation of polychlorinated biphenyls, chlordane, and p,p'-DDT from toxaphene by silicic acid column chromatography, *J. Assoc. Off. Anal. Chem.*, 61, 820, 1978.

202. **Tai, H., Williams, M. T., and McMurtrey, K. D.,** Separation of polychlorinated biphenyls from toxaphene by silicic acid column chromatography, *Bull. Environ. Contam. Toxicol.,* 29, 64, 1982.

203. **Lamparski, L. L., Nestrick, T. J., and Stehl, R. H.,** Determination of part-per-trillion concentrations of 2,3,7,8-tetrachlorodibenzo-p-dioxin in fish, *Anal. Chem.,* 51, 1453, 1979.

204. **Needham, L. L., Smrek, A. L., Head, S. L., Burse, V. W., and Liddle, J. A.,** Column chromatography separation of polychlorinated biphenyls from dichlorodiphenyl-trichloroethane and metabolites, *Anal. Chem.,* 52, 2227, 1980.

205. **Berg, O. W., Diosady, P. L., and Rees, G. A. V.,** Column chromatographic separation of polychlorinated biphenyls from chlorinated pesticides, and their subsequent gas chromatographic quantitation in terms of derivatives, *Bull. Environ. Contam. Toxicol.,* 7, 338, 1972.

206. **Chau, A. S. Y. and Babjak, L. J.,** Column chromatographic determination of mirex, photomirex and polychlorinated biphenyls in lake sediments, *J. Assoc. Off. Anal. Chem.,* 62, 107, 1979.

207. **Huckins, J. N., Stalling, D. L., and Petty, J. D.,** Carbon-foam chromatographic separation of non-o,o'-chlorine substituted PCB's from Aroclor mixtures, *J. Assoc. Off. Anal. Chem.,* 63, 750, 1980.

208. **Huckins, J. N., Stalling, D. L., and Smith, W. A.,** Foam-charcoal chromatography for analysis of polychlorinated dibenzodioxins in herbicide orange, *J. Assoc. Off. Anal. Chem.,* 61, 32, 1978.

209. **Smith, L. M.,** Carbon dispersed in glass fibers as an adsorbent for contaminant enrichment and fractionation, *Anal. Chem.;* 53, 2152, 1981.

210. **Czuczwa, J. M. and Hites, R. A.,** Environmental fate of combustion-generated polychlorinated dioxins and furans, *Environ. Sci. Technol.,* 18, 444, 1984.

211. **Mulhern, B. M., Cromartie, E., Reichel, W. L., and Belisle, A. A.,** Semiquantitative determination of polychlorinated biphenyls in tissue samples by thin layer chromatography, *J. Assoc. Off. Anal. Chem.,* 54, 548, 1971.

212. **Haller, H. L., Bartlett, P. D., Drake, N. L., Newman, M. S., Cristol, S. J., Baker, C. M., Hayes, R. A., Kilmer, G. W., Magerlein, B., Mueller, G. P., Schneider, A., and Wheatley, W.,** Chemical composition of technical DDT, *J. Am. Chem. Soc.* 67, 1591, 1945.

213. **Trotter, W. J.,** Removing the interference of DDT and its analogs in the analysis for residues of polychlorinated biphenyls, *J. Assoc. Off. Anal. Chem.,* 58, 461, 1975.

214. **Underwood, J. C.,** Separation of polychlorinated biphenyls from DDT and its analogs using chromic acid and silica gel, *Bull. Environ. Contam. Toxicol.,* 21, 787, 1979.

215. **Szelewski, M. J., Hill, D. R., Spiegel, S. J., and Tifft, E. C.,** Loss of polychlorinated biphenyl homologues during chromium trioxide extraction of fish tissue, *Anal. Chem.,* 51, 2405, 1979.

216. **Cooke, M., Nickless, G., Prescott, A. M., and Roberts, D. J.,** Analysis of polychlorinated naphthalenes, polychlorinated biphenyls and polychlorinated terphenyls via carbon skeleton gas-liquid chromatography, *J. Chromatogr.,* 156, 293, 1978.

217. **De Kok, A., Geerdink, R. B., Frei, R. W., and Brinkman, U. A. Th.,** The use of dechlorination in the analysis of polychlorinated biphenyls and related classes of compounds, *Intern. J. Environ. Anal. Chem.,* 9, 301, 1981.

218. **Armour, J. A.,** Quantitative perchlorination of polychlorinated biphenyls as a method for confirmatory residue measurement and identification, *J. Assoc. Off. Anal. Chem.,* 56, 987, 1973.

219. **Trotter, W. J. and Young, S. J. V.,** Limitation on the use of antimony pentachloride for perchlorination of polychlorinated biphenyls, *J. Assoc. Off. Anal. Chem.,* 58, 466, 1975.

220. **Brinkman, U. A. Th., Seetz, J. W. F. L., and Reymer, H. G. M.,** High-speed liquid chromatography of polychlorinated biphenyls and related compounds, *J. Chromatogr.,* 116, 353, 1976.

221. **Sawyer, L. D.,** Collaborative study of the recovery and gas chromatographic quantitation of polychlorinated biphenyls in chicken fat and polychlorinated biphenyl-DDT combinations in fish, *J. Assoc. Off. Anal. Chem.,* 56, 1015, 1973.

222. **Webb, R. G. and McCall, A. C.,** Southeast Environmental Research Laboratory, National Environmental Research Center — Corvallis, Environmental Protection Agency, Athens, GA, U.S., 30601.

223. **Sawyer, L. D.,** Food and Drug Administration, 240 Hennepin Ave., Minneapolis, MN, U.S., 55401.

224. **Sawyer, L. D.,** Quantitation of polychlorinated biphenyl residues by electron capture gas-liquid chromatography: collaborative study, *J. Assoc. Off. Anal. Chem.,* 61, 282, 1978.

225. **Chau, A. S. Y. and Sampson, R. C. J.,** Electron capture gas chromatographic methodology for the quantitation of polychlorinated biphenyls: survey and compromise, *Environ. Lett.,* 8, 89, 1975.

226. **Tindall, G. W. and Wininger, P. E.,** Gas chromatography-mass spectrometry method for identifying and determining polychlorinated biphenyls, *J. Chromatogr.,* 196, 109, 1980.

227. **Cairns, T. and Siegmund, E. G.,** Determination of polychlorinated biphenyls by chemical ionization mass spectrometry, *Anal. Chem.,* 53, 1599, 1981.

228. **Collard, R. S. and Irwin, M. M.,** GC/MS determination of incidental PCB's in complex chlorinated-hydrocarbon process and waste streams, *Talanta,* 30, 811, 1983.

229. **Onuska, F. I. and Comba, M.,** Identification and quantitative analysis of polychlorinated biphenyls on WCOT glass capillary columns, in *Hydrocarbon and Halogenated Hydrocarbons in the Aquatic Environment,* Afghan, B. K. and MacKay, D., Eds., Plenum Press, New York, 1980, 285.

230. **Ballschmiter, K. and Zell, M.,** Analysis of polychlorinated biphenyls (PCB) by glass capillary gas chromatography, *Fresenius' Z. Anal. Chem.,* 302, 20, 1980.

231. **Mullin, M. D. and Filkins, J. C.,** Analysis of polychlorinated biphenyls by glass capillary and packed-column chromatography, in *Advances in the Identification and Analysis of Organic Pollutants in Water,* Keith, L. W., Ed., Ann Arbor Press, MI, 1981, 187.

232. **Bush, B., Connor, S., and Snow, J.,** Glass capillary gas chromatography for sensitive, accurate poly-chlorinated biphenyl analysis, *J. Assoc. Off. Anal. Chem.,* 65, 555, 1982.

233. **Moseley, M. A. and Pellizzari, E. D.,** Development and evaluation of wall coated open-tubular columns for GE analysis of individual polychlorinated biphenyl isomers, *J. High Resolut. Chromatogr. Chromatogr. Commun.,* 5, 404, 1982.

234. **Duinker, J. C. and Hildebrand, M. T. J.,** Characterization of PCB components in Clophen formulations by capillary GC-MS and GC-ECD techniques. *Environ. Sci. Technol.,* 17, 449, 1983.

235. **Jennings, W.** *Gas Chromatography with Glass Capillary Columns,* 2nd ed., Academic Press, New York, 1980.

236. **Wells, G.,** A micro-volume electron capture detector for use in high resolution capillary gas chromatography, *J. High Resolut. Chromatogr. Chromatogr. Commun.,* 6, 651, 1983.

237. **Onuska, F. I., Kominar, R. J., and Terry, K.,** An evaluation of splitless and on-column injection techniques for the determination of priority micropollutants, *J. Chromatogr. Sci.,* 21, 512, 1983.

238. **Krupcik, J., Leclercq, P. A., Simova, A., Suchanek, P., Collak, M., and Hrivnak, J.,** Possibilities and limitations of capillary gas chromatography and mass spectrometry in the analysis of polychlorinated biphenyls, *J. Chromatogr.,* 119, 271, 1976.

239. **Krupcik, J., Leclercq, P. A., Garaj, J., and Simova, A.,** Analysis of alkylated mixtures of polychlorinated biphenyls by capillary gas chromatography-mass spectrometry, *J. Chromatogr.,* 191, 207, 1980.

240. **Tausch, H., Stehlik, G., and Wihlidal, H.,** Analyse von Organochloropestizid- und PCB-Rückständen in Fischen mittels Kapillargaschromatographie/Massenspektrometrie, *Chromatographia,* 14, 403, 1981.

241. **Liu, R. H., Ramesh, S., Liu, J. Y., and Kim, S.,** Qualitative and quantitative analyses of commercial polychlorinated biphenyl formulation mixtures by single ion monitoring gas-liquid chromatography/mass spectrometry and multiple regression, *Anal. Chem.,* 56, 1808, 1984.

242. **Gebhart, J. E., Hayes, T. L., Alford-Stevens, A. L., and Budde, W. L.,** Mass Spectrometric determination of polychlorinated biphenyls as isomer groups, *Anal. Chem.,* 57, 2458, 1985.

243. **Slivon, L. E., Gebhart, J. E., Hayes, T. L., Alford-Stevens, A. L. and Budde, W. L.,** Automated procedures for mass spectrometric determination of polychlorinated biphenyls as isomer groups, *Anal. Chem.,* 57, 2464, 1985.

244. **Onuska, F. I., Kominar, R. J., and Terry, K. A.,** Identification and determination of polychlorinated biphenyls using high resolution gas chromatography, *J. Chromatogr.,* 279, 111, 1983.

245. **Pellizzari, E. D., Moseley, M. A., and Cooper, S. D.,** Recent advances in the analysis of polychlorinated biphenyls in environmental and biological media, *J. Chromatogr.,* 334, 277, 1985.

246. **Capel, P. D., Rapaport, R. A., Eisenreich, S. J., and Looney, B. B.,** PCBQ: computerized quantification of total PCB and congeners in environmental samples, *Chemosphere,* 14, 439, 1985.

247. **Mullin, M. D.,** U.S. Environmental Protection Agency, Congener Specific PCB Workshop, Large Lakes Research Station, Grosse Ile, Michigan, June 1985.

248. **Cooper, S. D., Moseley, M. A., and Pellizzari, E. D.,** Surrogate standards for the determination of individual polychlorinated biphenyls using high-resolution gas chromatography with electron capture detection, *Anal. Chem.,* 57, 2469, 1985.

249. **Chau, A. S. Y. and Lee, H. B.,** Analytical reference materials. III. Preparation and homogeneity test of large quantities of wet and dry sediment reference materials for long term polychlorinated biphenyl quality control studies, *J. Assoc. Off. Anal. Chem.,* 64, 947, 1980.

250. **Alford-Stevens, A. L., Budde, W. L., and Bellar, T. A.,** Interlaboratory study on determination of polychlorinated biphenyls in environmentally contaminated sediments, *Anal. Chem.,* 57, 2452, 1985.

251. **Tuinstra, L. G. M. Th., Roos, A. H., Griepink, B., and Wells, D. E.,** Interlaboratory studies of the determination of selected chlorobiphenyl congeners with capillary gas chromatography using splitless- and on-column injection techniques, *J. High Resolut. Chromatogr. Chromatogr. Commun.,* 8, 475, 1985.

252. **Dunn, W. J., III, Stalling, D. L., Schwartz, T. R., Hogan, J. W., Petty, J. D., Johansson, E., and Wold, S.,** Pattern recognition for classification and determination of polychlorinated biphenyls in environmental samples, *Anal. Chem.,* 56, 1308, 1984.

Chapter 3

ANALYSIS OF TOXAPHENE IN ENVIRONMENTAL SAMPLES

D. B. Sergeant and F. I. Onuska

TABLE OF CONTENTS

I. INTRODUCTION

Toxaphene is another example of history repeating itself. When DDT use was banned in December 1972 because of its adverse environmental effects[1] toxaphene use increased dramatically to fill the void in pest control programs. In turn, toxaphene itself was discovered to be an environmental problem.[2] Today, most uses of toxaphene are banned in Canada, the U.S., and in many other countries. However, the legacy of its pesticide use remains as it continues to recycle between various compartments in the environment. It even appears in samples from remote regions where it was never used. Where PCBs and DDT went, toxaphene appears to be following.

It is within this context that the authors undertook this review of toxaphene — its manufacture, use, environmental fate, and mobility — and in particular its accurate analysis in various environmental matrices. Being located also on the largest system of freshwater lakes in the world, the Great Lakes Basin, we are actively interested in toxaphene in these lakes and ultimately on its effects on fish and their habitat.

II. NOMENCLATURE, MANUFACTURE, AND PROPERTIES

Millions of pounds of toxaphene have been manufactured,[3-5] sold, and used under a variety of trade names or synonyms (Table 1). The most common synonym is camphechlor, which is widely used in Europe.[6] Toxaphene was an attractive insecticide because of its broad

Table 1
COMMON AND TRADE NAMES OF TOXAPHENE

Trade Names — ™		Common Names
		Toxaphene
Allotox	Phenacide	Camphechlor
Estonox	Toxakil	Chlorinated Camphene
Chem-Phene	Toxyphen	Polychlorocamphene
Geniphene	Melipax	Octachlorocamphene
Gy-Phene	Toxaphen	Polychlorocamphene
Phenacide	Estonox	
Phenatox	Attac	
Toxadust		
Toxadust 10		
Toxaspra		
Hercules 3956		
Synthetic 3956		
Fasco Terpene		
Penphene		

Table 2
PROPERTIES OF TOXAPHENE[6,7,12,14,15]

Molecular formula:	$C_{10}H_{10}Cl_8$
Molecular weight:	414
Melting point:	amber (yellow)-colored waxy solid melts at 65 to 90°C
Density:	1.65 (25°C)
Vapor pressure:	6.7×10^{-6} mbar at 20°C
Solubility:	3 mg/l in water at room temperature; soluble in most organic solvents
Corrosivity:	Corrosive to metals under moist conditions and at elevated temperatures
Incompatibilities:	Incompatible with alkaline formulations; iron, aluminum, and heat catalyze decomposition
Toxicity:	Toxic to fish, mammals, and birds; not toxic to bees
CAS Registry number:	8001-35-2
Uses:	Wide variety of crops and as veterinary insecticides
Partition coefficient (octanol / water)	825
Safety:	Avoid dermal and inhalation contact and minimize exposure

spectrum of activity. This is evident in the fact that over a billion pounds of toxaphene were used in the U.S. alone.[7] Toxaphene was first registered there in 1947[8] and was introduced into the marketplace in 1948 as a contact insecticide,[6] where it eventually became the most heavily used pesticide. Its use increased when DDT was banned. However, toxaphene too became a problem resulting in the U.S. Environmental Protection Agency (E.P.A.) issuing a rebuttable presumption against registration (RPAR)[9] on May 25, 1977.

This RPAR was based upon oncogenicity[9,10] and reductions in nontarget species. In 1982,[8] most uses of toxaphene were banned in the U.S., however, stocks could be used up until 1986.

Toxaphene, a mixture of at least 250 polychlorinated compounds, polychloroboranes, and polydihydrocamphenes, is produced from reaction of camphene and chlorine.[12] The reaction takes place at atmospheric pressure under catalysis by UV irradiation. As the reaction is exothermic some form of cooling is required to reach completion. The final product contains 67 to 69% chlorine forming a yellow waxy product, which is subsequently formulated into emulsifiable concentrates, wetable powders, and dusts.

Despite the uncontrolled (nonuniform) feedstock the product is fairly uniform in composition over the years.[4] Key properties of toxaphene are recorded in Table 2. Since toxaphene is not a discrete chemical compound no structural diagram is presented. General structures are cited in the literature.[6,9,16.]

III. ENVIRONMENTAL DISTRIBUTION

For ease of discussion we have divided the environment into distinct compartments. Each compartment contains all aspects of levels of toxaphene in that compartment as well as toxicity. Toxaphene metabolism, degradation, and persistence has been treated as a separate issue and is discussed in Section IV.

A. Atmospheric Transport

Toxaphene continues to be a pesticide of concern in air and water pollution due to its atmospheric transport. Although toxaphene use has been almost eliminated in North America and most other industrialized countries, it is continually identified everywhere — rainwater, sediment, biota, fish, etc. Toxaphene may have been long-range transported to the remote regions of our globe[17,203] but remained undetected in samples because of analysis problems and high detection limits.

1. Precipitation

Toxaphene has been found in precipitation samples both close to major use points (southern U.S. cotton belt) and far away where use has been low.[17-19] Harder et al.[18] found seasonal levels in South Carolina rose from not detectable (N.D.) to 150 ng/kg in rainwater. The higher level correlated with heavy agricultural use in June to August. Mean toxaphene concentration in rainwater was found to be 45 ng/kg. The authors also analyzed rain from off the coast of Delaware and found levels of 50 ng/kg.

At sampling sites well removed from the southern U.S., there have been measurements of toxaphene in rainwater.[19] However, Stachan[20] found no toxaphene at either Caribou Island or Isle Royale in Lake Superior in 1983. He later documents[21] that, up to 1984, values for atmospheric toxaphene had not been reported, although toxaphene was being reported for brown and lake trout in the Great Lakes. Rice and Evans[17] in the same time frame published some of their data on toxaphene in the Great Lakes Basin for a variety of matrices. They found 9.2 ng/l in a composite sample taken along the eastern shore of Lake Michigan.

Snow also has been analyzed for toxaphene at several very remote locations. Measurements at the North Pole and North, Central, and South Finland[22] indicated no polychlorinated hydrocarbons, phenols, guiacols, or catechols at different sampling times in the 1983 to 1985 time period. Snow as an indicator of pollution has not been studied intensively to date.[22] In addition the 1 l sample size for melted snow may have been insufficient to permit detection of the compounds of interest. Five small unknown peaks appeared in the North Pole electron capture detector (ECD) chromatogram while flame ionization detector (FID) chromatogram showed none. Central Finland samples showed more ECD peaks and 2 FID peaks. ECD limit of detection was reported as 0.5 ng/l for each chlorohydrocarbon. Chlorophenols were found in some of these samples however.

Munson[23] found wide variation in toxaphene concentrations of eight samples taken in Maryland. Values ranged from 14 to 280 ppm.

2. Air

A number of studies have been conducted to determine toxaphene and other organics in air.[3,24-30] The collection techniques and trapping efficiency of toxaphene is similar to that ot DDT, PCBs, phthalates, etc. A variety of trapping techniques and collection efficiencies were described by Bidleman et al.[19]

Toxaphene levels in air samples varied greatly in concentration ranges at a South Carolina estuary.[18] Toxaphene ranged from 0.33 to 7.2 ng/kg in air. At 2 of 3 southern locations in U.S. the levels ranged from 16 to 2,520 ng/m³. Similarly, Arthur et al.[30] found toxaphene at 1,746.5 ng/m³ in weekly air samples in the Mississippi Delta area of the U.S. In the continental U.S. toxaphene ranked second to malathion in maximum concentration found in the air samples, however the frequency at which toxaphene was found was the lowest. Mean toxaphene values were 1,890 ng/m³ with a maximum concentration of 8,700 ng/m³ in 1970 to 1972 and 1,756.5 ng/m³ for 1972 to 1974. The authors concluded that toxaphene should be placed in the same category as DDT, PCBs, and other compounds capable of long range transport and atmospheric cycling.

Bidleman et al.[19] summarized data for chlorinated hydrocarbons (p,p'-DDT, o,p'-DDT, p,p'-DDE, dieldrin, chlordane, toxaphene, and PCB) in continental and marine air.

Multi-component chlorinated compounds such as PCBs, chlordane and toxaphene were measured less frequently in the studies cited. No doubt this was due to the greater difficulty in detection of toxaphene concentrations in continental air. Toxaphene in continental air appeared to exceed marine air concentrations by roughly 100 times. More recently, Bidleman et al.[203] determined toxaphene, other organochlorine pesticides, and polychlorinated biphenyls in air samples at Stockholm and at Aspvreten (a rural site) in Southern Sweden. Toxaphene was quantitated by GCMS and ranged from 4 to 15 and 5 to 225 pg/m³ for Stockholm and Aspvreten, respectively. Rice and Evans[17] found nondetectable to 5.5 ng/m³ in the Great Lakes along the eastern shore of Lake Michigan.

Analysis of peat cores has yielded valuable information on historical toxaphene deposition.[31,32] A direct correlation between input and production of toxaphene over time was found by analysis of peat cores. A comparison with DDT, which was banned before toxaphene was, shows a similar increase in input to the environment with DDT starting to fall off before toxaphene does. Whereas DDT input has fallen to almost zero, toxaphene is still falling. Toxaphene inputs were reported to be two to four times those of PCBs (industrial) and DDT (pesticide).

B. Water

Values for concentration of toxaphene in water are scarce in the literature. In a national (U.S.) surface water monitoring program toxaphene occurred in 0.01% of the 1482 samples analyzed. The highest value observed was 1.65 ppb (µg/l). Nicholson et al.[34] studied municipal and treated water in an Alabama river basin. Carbon adsorption and gas chromatography were employed in part for analyses. Toxaphene was found in both treated and untreated water. Treatment did not remove toxaphene from municipal water. Goldberg[35] similarly monitored New Orleans water supply and did not detect toxaphene. Intense use in nearby states above New Orleans was noted. The author also reported that surveys of stream and surface water in the U.S. found no toxaphene at a detection limit of 1 ppb in water. Similarly, Schafer et al.[36] found no toxaphene in 63 samples taken in the mid 1960s, despite the fact that chlordanes were found. Pollock and Kilgore[3] referenced a study by Hughes and co-workers where 2.1 to 13.7 ppb toxaphene was found in lakes where rough fish control had occurred.

A safe value for toxaphene in water has been recommended by the International Joint Commission (IJC).[37] They recommended "The concentration of toxaphene in water should not exceed 0.008 micrograms per liter for the protection of aquatic life" and provided a rationale for this level based upon toxicity tests.[37]

Lorber and Mulkey[38] compared three pesticide runoff models for their ability to predict the movement of toxaphene and atrazine. The models were tested for predictive ability by using toxaphene and atrazine runoff and erosion data. All models were able to predict the field data.

C. Biota

Coinciding with concerns about toxicity, bioaccumulation, and other adverse affects of toxaphene on fish and birds in particular, several researchers have investigated toxaphene effects on foodchain organisms. An early reference by Schoettger and Olive[39] investigated technical and the wettable powder formulations of toxaphene. Their research was spawned by the possibility of toxaphene use as a piscicide (fish erradicant). Residual toxicity in these waters had been observed leading them to question whether fish-food organisms could have accumulated pesticide. No analyses were performed and fish toxicity was used to determine residual effects. Sublethal doses of toxaphene in fish-food organisms were insufficient to produce mortality. Sublethal concentrations were reported for several species: 0.0036 ppm for green sunfish and kakanee salmon (96 h); 0.01 ppm for shiners (96 h); 0.03 ppm for *Daphnia pulex* and *Daphnia magna* (168 and 120 h, respectively); and 0.004 ppm for damselfly nymphs.

At the same time Hoffman and Olive[40] investigated the effect of rotenone and toxaphene on plankton in two Colorado reservoirs. Plankton were divided into three groups for study purposes: *Entomostraca, Rotatoria, and Protozoa.* All three groups were adversely affected by the pesticides as evidenced by a decline in numbers and, in the case of the protozoa, a decline in growth.

Needham[41] and Henegar[42] subsequently investigated effects of toxaphene on plankton and aquatic invertebrates and minimum lethal levels on fish in North Dakota Lakes. Under natural and controlled conditions, Needham[41] did not observe marked reductions of a dominant zooplankton at treatment levels of 5 to 34 ppb toxaphene for fish control. Phytoplankton populations also were unaffected. Tolerance levels for various species were determined experimentally. Henegar' objective [42] was to determine what level of toxaphene was necessary to erradicate all fish from small lakes prior to restocking them. Concentrations of 0.025 to 0.035 ppm were found to meet this need.

Toxaphene was also assessed for use to eradicate sea lampreys.[43] Treatment level of 100 ppb was chosen. Lack of analytical equipment precluded toxaphene determinations over time. Although the treatment effectively eliminated the lamprey ammocetes, the effect on local fish was devastating — 50 to 80 large rainbow trout, 25 northern pike and 3,000 to 4,000 yellow perch were also killed.

Toxaphene was toxic at 18, 7 and 2.3 ppb in 24, 48, and 96 h. LC_{50} toxicity tests for *Pteronarcys california* naiads.[44] Similarly, testing for *Pteronarcella badia* and *Claassenia sabalosa* produced LC_{50}s of 9.2, 5.6, and 3.0 ppb and 6.0, 3.2, and 1.3 ppb, respectively.

A tubificid worm[45] *(Branchiura sowerbyi)* exhibited complete mortality when exposed to 1 ppm of toxaphene for 72 h at 3 different temperatures (4.4, 21, and 32.2°C). Toxaphene was also toxic to the Southern Leopard Frog *(Rana sphenocephala).*[46] LC_{50} was determined to be 0.378 to 0.790 ppm toxaphene for subadults and 0.032 to 0.054 ppm toxaphene for larval. A bioconcentration factor of 100 was determined and adverse behaviorial and growth effects were observed.

In a recent study,[47] the toxicity of toxaphene was assessed for two cladocerans, *Bosima longirostris* and *Daphnia spp.* For the former a 48-h LC_{50} of 1.4 µg/l was determined and for the later 8.2 and 11.3 µg/l was determined for *Daphnia pulex* and *Daphnia magna,* respectively.

Uptake has been assessed for both microrganisms[48] and in several estuarine organisms.[48,49] In the case of microrganisms a bacteria and a fungus were tested. Toxaphene was extracted from whole cultures and its octanol/water partition coefficient was determined to be (3.3 ± 2.5) × 10^3. Field and laboratory distribution coefficient (K_d) data agreed well with field sample value being (6.6 ± 0.2) × 10^3. The authors indicated that octanol/water partition coefficient comparison with K_d should forecast bioaccumulation potential and their data shows this. Schimmel et al.[49] not only reported uptake but determined 96-h LC_{50}s based on

Table 3
BIOACCUMULATION AND TOXICITY OF TOXAPHENE
IN ESTUARINE ORGANISMS[49]

Species	96-h LC$_{50}$ mg/l	96-h BCFs (tissue/water)
Pink shrimp	1.4	3,100—20,600
Penaeus duorarum		
Grass shrimp	4.4	3,100—20,600
Palaemantes pugio		
Sheepshead minnow	1.1	400—1,200
Cypriondon variegatus		
Pinfish	0.5	400—1,200
Langodan rhomboides		
Longnose killifish	0.9—1.4[a]	4,200—60,000
Fundulus similis		(whole body)

[a] 28 days.

analytically determined concentrations. Table 3 summarizes some of their findings. The authors also reported a 50% reduction in shell growth for American oysters exposed to 16 ppb toxaphene.

Toxaphene toxicity to daphnids was assessed and compared with that of organophosphate insecticides.[50] Toxaphene and several other organochlorinated pesticides were less toxic to the daphnids than were the organophosphate pesticides. An opposite trend was found when toxaphene and rotenone (a fish eradicant pesticide) were compared as to their effects on the bottom fauna of two Colorado reservoirs. Toxaphene had greater adverse effects.[51]

D. Fish

Toxaphene has long been of interest to fisheries biologists,[52,53] not only for its acute toxicity to fish, but also because it was found to accumulate in their tissues.[54,55] Toxaphene sediment degradation products remain toxic to fish after 10 months in the environment.[56] Toxaphene toxicity posed severe problems when used in fish-eradication programs because of its persistence and toxicity. In several cases when restocking with desirable fish occurred, toxicity and bioaccumulation took their toll. In deep lakes, fish could not be restocked for about six years.[55] Accidental eradication of fish has occurred in South Africa.[57] Careless flushing of a cattle dip tank was determined to be the cause. The populations of top carnivores in the foodchain of the Hluhluwe River, finfoots, *(Podica senegalensius)* and crocodiles, *(Crocodylas niloticus)* appeared to decline, as did most fish species populations. Levels of 1.68 to 10.87 ppm in finfoot livers were explained on the basis of the birds eating either contaminated fish or aquatic fauna. A limited number of fish and birds were analyzed after the spill. Average whole body toxaphene levels in 3 fish were 0.49 ppm. Dissolved oxygen levels were normal and the cause of the fish kill was attributed to toxaphene. Water concentration was 0.0032 ppm 5 days after the kill. No sediment analyses were reported but three species of mussels were analyzed. Mussel levels ranged from 0.07 to 0.31 ppm average toxaphene concentration.

Toxaphene has been studied in a model ecosystem[58] using nine toxaphene fractions. These fractions were isolated by dividing a dry silica column chromatographic separation into nine fractions of roughly the same amount of ^{14}C-toxaphene. All nine fractions and unfractionated toxaphene were then tested in a recirculating static model ecosystem. Environmentally relevant quantities (0.1 to 1.0 ppm) of toxaphene were adsorbed into soil which were then placed in model ecosystems with test organisms. The two central fractions were the most toxic. All organisms (algae, snails, daphnids, and fish) accumulated about the same concentration of toxaphene.

Table 4
TOXICITY COMPARISON FOR FRESHWATER SPECIES[63]

Species	Static 24-L LC_{50} toxaphene (ng/l)
Daphnia magna	23
Daphnia pulex	76
Coho salmon	17
Rainbow trout	5.0—17
Brown trout	>14
Carp	7.1
Fathead minnow	20—24
Green sunfish	24
Bluegill	6.5—25
Largemouth bass	5.6—13
Yellow perch	18
Fowlers toad	560
Western chorus frog	5600

Note: 100% Technical Material. Fish size for most part was about 1.0 g.

Earlier Nelson and Matsumura[59] used a combination of separation techniques involving Sephadex® LH-20/methanol column chromatography, normal TLC, and reversed phase TLC to isolate and characterize two previously identified toxic components of toxaphene. Toxicant 'A' was found to be a mixture of two octachloroboranes[60] 2,2,5-endo, 6-exo, 8,8,9,10- and 2,2,5-endo, 6-exo, 8,9,9,10-octachloroborane, while toxicant 'B' was identified as 2,2,5-endo-6-exo, 8,9,10,heptachloroborane.

Subsequent to Schaper and Crowder[61] reporting uptake of ^{37}Cl-toxaphene by mosquitofish *(Gambusin affinis)*, Moffet and Yarbrough[62] studied disruption of succinic dehydrogenese activity in the mitochondria of liver and brain tissues. Toxaphene, along with DDT and dieldrin, inhibited this enzyme for intact susceptible mitochondria. However, only toxaphene inhibited the enzyme in intact mitochondria from resistant fish. When mitochondria with disrupted membranes were studied all three pesticides inhibited the enzyme. The authors concluded that a membrane barrier exists in insecticide-resistant mosquitofish. Toxaphene resistance of fish, shrimp, and other aquatic animals has been reported in Pollock and Killgore's review.[3]

With respect to toxicity of pesticides to fish and other aquatic organisms, a very helpful manual has been published recently.[63] The authors stress the species sensitivity plays an important part in scientists' assessment of the potential danger of a pesticide to an ecosystem. Ideally, in hazard assessment, one should seek to protect the most sensitive species in the ecosystem. If this were done then there should be no foodchain or habitat destruction. In their study, 63 species were tested against 174 chemicals, including toxaphene. Toxaphene was one of several chemicals tested against various life stages of fishes in a 96-h LC_{50} study. At the various life stages of channel catfish toxaphene toxicity was 0.008, 0.001, and 0.004 mg/l at the yolk-sac fry; swim-up fry; and advanced fry stages. An examination of the species toxicity data for toxaphene shows that rainbow trout is the most sensitive species the majority of the time. Table 4 shows the toxicity of toxaphene to a variety of species. Comparison to other organochlorine (OC) pesticides shows toxaphene is about as toxic as DDT, but more toxic than compounds like heptachlor, lindane, chlordane, mirex, and Arochlor 1254.

The effects of toxaphene on growth and bone composition of fish has been extensively investigated at the U.S. Fish and Wildlife Service by Mayer, Mehrle, and others.[5,64-69] In the 1977 summary[5] of their work on toxaphene, three species (brook trout, fathead minnow, and channel catfish) were continuously exposed to toxaphene in water. Adult, egg, and fry life stages were exposed. For some fish radiographs were examined for alterations in vertebral structure and bone development assessed by determining collagen, calcium, and phosphorus concentrations. Adult brook trout were the only species to exhibit mortality during the study. At spawning all adult trout died at 502 ng/l toxaphene concentration and 50% died at an exposure of 288 ng/l. Mayer and Mehrle believed that this was due to the dual stresses of pesticide exposure and spawning. Growth of fathead minnows was significantly lowered but at lower concentrations (97 and 173 ng/l toxaphene) than for trout. Catfish growth was not affected.

Brook trout reproductive success was inversely proportional to toxaphene concentration as evidenced by number of eggs spawned and egg viability. Effects on reproduction were not observed for fathead minnows. Hatchability of channel catfish eggs were reduced at 670 ng/l toxaphene.

Fry growth and survival was affected by toxaphene concentrations for all species. Brook trout fry started to respond adversely at 39 ng/l, fathead minnow fry at 97 ng/l, and channel catfish at about 299 ng/l (no effect level not determined) toxaphene concentrations.

Reduced collagen synthesis and increased mineralization (calcium and phosphorus) were observed for fry of all three species. Again, the trend was towards catfish being least affected.

Residues of toxaphene in eggs, fry and adults were measured for the species exposed at the 6 concentrations (0, 39, 68, 139, 288, and 502 ng/l) used in the study. Residues increased with concentration in all cases. Bioconcentration factors were 76,000, 55,000 to 69,000 and 27,000 to 50,000, respectively for trout, minnows, and catfish. Excretion of toxaphene in adult fish of all three species was slow.

In a later paper,[66] Mayer, Mehrle, and Crutcher researched the effects of dietary vitamin C on channel catfish exposed to varying vitamin C and toxaphene concentrations. Mucous cell number and epidermal thickness were reduced in fish exposed to toxaphene and fed 63 and 670 ng/kg dietary vitamin C. Spinal deformities decreased with increasing dietary vitamin C as did whole-body toxaphene residues. Tolerance to chronic effects of toxaphene on growth, bone development, and skin lesions increased with increasing dietary vitamin C. In 1981, Hamilton et al.[68] reported that the mechanical properties (vertebral elasticity) of bone in channel catfish were reduced for a vitamin C-deficient diet.

Toxaphene is not the only contaminant capable of affecting survival of fish stocks due to spinal deformities. PCBs, cadmium, and lead might also weaken vertebral structure.[69]

With the intensive use of organochlorine pesticides in agriculture researchers became interested in determining the mechanism of fish toxicity for these compounds. Several researchers[70-72] investigated inhibition of various adenosine triphosphatases (ATPases) by toxaphene and other chlorinated pesticides, herbicides, and polychlorinated biphenyls. Davis et al.[70] found ATPase activity varied slightly for Na^+, K^+, and Mg^{++} ATPases and with concentration of toxaphene. Mg ATPase was 66% inhibited at 40 ppm while NaKMg and NaK ATPases were only 49 and 36% inhibited, respectively. That is, Mg ATPase was the most sensitive for rainbow trout gill microsome fraction. At 4 ppm toxaphene the percent inhibition was lower as expected. Desaiah and Koch[71] followed up this work by determining the inhibition of these ATPases in brain, kidney, and gill tissues of catfish. Mg ATPase was inhibited in all three cases. The relative order of inhibition was that brain and gill tissue were about equally inhibited with kidney tissue being the least affected. Inhibition once again increased with increasing toxaphene concentration. In another study reported the same year Yap et al.[72] studied blue gill fish brain ATPases for several cyclodiene insecticides as well as several other insecticides including toxaphene. Mg ATPase of sensitive and insensitive fish once again exhibited greater inhibition than the NaK-ATPase for most compounds tested. The need for further studies on individual toxaphene isomers[71] and metabolites [72] were identified.

Toxaphene levels are reported throughout literature studies on the pesticide but for the

Table 5
TEMPORAL TOXAPHENE
CONCENTRATIONS IN
FRESHWATER FISH — 1970—1981[a]

Year	Geometric mean (μg/g wet wt) of toxaphene in fish
1970	Not available
1971	0.01
1972	0.13
1973	0.17
1974	0.17
1976—1977	0.36[a] (0.35)[b]
1978—1979	0.32 (0.29)[b]
1980—1981	0.27

[a]   Data for period 1976—1977 and 1978—1979 are in References 18 and 20.
[b]   Reference 18.
[c]   Reference 20.

most part, with the exception of the U.S.'s Pesticide Monitoring Program, the values are for reference only (that is to say, to demonstrate that the compound is present at a quantifiable level). Even the National Pesticide Monitoring Study does not have a complete data set for toxaphene. Analysis for this pesticide in fish was not available prior to 1971[73] and even then the values that were reported were qualified with a footnote indicating they were not quantitative.[74] This no doubt relates to the difficulties encountered in attempts to analyze this very complex mixture. In the last available results we examined,[75] for the period 1980 to 1981, toxaphene concentration in freshwater fish had maximized and begun to decrease slightly. Continual increase had been observed throughout the 1970s. Table 5 shows this process. During this period (1970 to 1974) Schmitt et al.[76] reported increased occurrence of toxaphene in freshwater fish. Positive results for 74 National Pesticide Monitoring Program stations ranged from 9.5 to 14.9% of stations from 1971 to 1974 where a jump to 60.9% for 1976 to 1977 occurred.

Although the mean values in Table 5 are low, they are significant. Ludke and Schmitt[73] reported that Columbia National Fisheries Research Laboratory found toxaphene residues of 1.0 μg/g which may be associated with impaired growth and developmental abnormalities in young fish and that residues exceeding this value were not uncommon. In an earlier report, Schmitt and co-authors[77] reported wet weight levels of 5 to 10 μg/g in Lake Michigan lake trout. These values are very similar to the 2.9 to 10.7 μg/g wet weight values reported by Gouch and Matsumura.[78] These levels were as high as those in fish from the most heavily contaminated stations in the Cotton Belt. Lake Superior values ranged from 2 to 7 μg/g wet weight and Lake Huron had a single data point at 9.0 μg/g wet weight.[77] This was and remains a point of great concern and other than explanation by atmospheric transport from southern U.S. cannot be accounted for at present.

In a suspected case of fish poisoning, levels of toxaphene were measured at 1.71 to 1.76 μg/g in the serum[79] of dead catfish. Control (different site) and measurements at the same site a year later showed levels of <0.10 μg/g in serum. In agricultural areas such as the San Joaquin Valley of California and the Lower Rio Grande Valley in Texas, toxaphene has been measured in fish at levels of 3.123 mg/kg[80] and N.D. to 31.5 mg/kg[81] wet weight in carp and a variety of fish species, respectively. The single positive toxaphene[80] value in only 1 of 8 samples of carp is surprising as Bowes[82] indicated that toxaphene was 15th of the top 15 pesticides used in 5 counties of this area of California, hence it should be found frequently. Toxaphene has also been found at 0.1 ppm in fish in Alaska,[75] and in Swedish fish[83] at 9 to 13 ppm on a lipid basis (toxaphene was never used in Lake Vättern or in Sweden) and in Canadian East Coast Marine Fish[87] at levels of 0.4 to 1.1 ppm on a wet weight basis. This can be compared to commercially raised catfish in the U.S. where toxaphene levels ranged from N.D. to 16.74 ppm.[85]

Both toxaphene and its degradation products have been shown to be taken up by spot and mullet in 96-h bioassays. Whole body residues varied in proportion to the aqueous toxaphene concentrations in the low parts-per-billion range. Bioconcentration factors ranged from 2×10^3 to 6×10^3 for most fish tested for both forms of toxaphene. Harder and his co-authors[86] concluded that research is desirable: to determine the rate of reductive dechlorination of toxaphene in sediments; on the sublethal effects of degradation products; to determine the isomeric distribution in the mixture; and analyses should determine degraded toxaphene fingerprints in environmental samples.

If fish are given a choice they will avoid toxaphene as well as endrin and parathion.[87] If avoidance does not occur then the LC_{50s} listed by Johnson and Finley[88] would apply to the various exposed fish species.

In light of all the fish data and other data on environmental levels and distribution of toxaphene, we were somewhat amused to find a 1983 publicity article entitled "Getting the Jump on Toxaphene".[89] Based on what is known, the shoe would appear to be on the other foot and we will have to run to catch up.

E. Birds, Mammals, and Humans

These species represent the top of the foodchain above fish, game, domestic animals, fruits, and vegetables. An exception exists where birds of prey or mammals prey upon one another. As such, all these species are likely to show evidence of accummulation of residues of toxaphene or ill-effects resulting from exposure to the pesticide or its metabolites.

Keith[90] in 1960 reported a case of unusually high mortality in fish-eating birds of Tule Lake and Lower Klanath Refuges, part of an agricultural irrigation system in north-eastern California and Oregon. Major species affected were white pelicans, American egrets, gulls, black-crowned night herons, and western grebes. The authors indicated that analysts had difficulty quantitating DDT and DDT-metabolites when toxaphene was present in equal or greater amounts. Paper chromatography was used in 1960 and 1961, while paper chromatography and gas chromatography were analyzed in 1962. Unfortunately, carcass sample (whole body) data were for skinned birds so some toxaphene burden was possibly lost. Liver, kidney, and a composite sample (heart, liver, kidney, breast muscle, and brain) were analyzed to a "trace" limit of 0.1 ppm. Values ranged from nondetectable (zero) through 31.5 ppm in 2 adipose samples of the western grebe. No toxaphene was detected in the 1962 samples. DDT residues were present in all tissue samples and were about ten times the toxaphene residue data. Whole fish sampled in 1960 to 1961 had toxaphene residues ranging from 0.1 ppm to 8.0 ppm, while invertebrates collected at that time contained 0.0 to 0.2 ppm. The author did not unequivocally relate the death of the birds to pesticide poisoning. He did perform a toxicity feeding study and established that 50 ppm of toxaphene in diet was an effect level and that toxaphene effectively eliminated parasites from birds.

Brown pelican eggs were analyzed[91] for a range of organohalogen pesticides in a study on eggshell thickness 1969 to 1976. Toxaphene data did not cover the entire time frame and were mainly for 1974 and 1975 at two collection sites. The geometric mean of substancial data sets for the two years showed levels in 1975 twice those of 1974. Levels in dead birds 0.48 to 0.56 ppm wet weight in 1975 were observed whereas in 1974 no toxaphene residues were found. These levels are about one-tenth of the highest levels found by Keith.[90] It is impossible to determine whether there was any method modification between the two years that may have affected results.

Causey et al.[92] studied quail, rabbits, and deer in or near selected Alabama soybean fields and control areas. Levels of toxaphene in quail paralleled closely to the total DDT and ranged from 0.001 ppm to 88.9 ppm on lipid basis. Rabbit and deer residues taken at the same site appear not to correlate with the quail levels, that is, they were less than 0.001 ppm. Control areas and species were virtually free of toxaphene.

Toxaphene has been fed to white leghorns.[4] The effects were different than those reported by Blus[93] for brown pelicans in that no deaths were reported. However, renal lesions were observed for 50 and 100 ppm feeding levels. Shell strength and hatchability were not adversely affected at even the 100 ppm level. Adipose tissue residue levels increased with increasing dietary toxaphene with 50% accounted for in this tissue. Toxaphene levels in adipose tissue reached three to four times that in the feed. Tissue half-lives of toxaphene, when birds were placed on a zero toxaphene diet, were 20.0, 25.2, and 41.5 days after initial feeding at levels of 100, 50, and 5 ppm, respectively. Egg half-lives were 15.7, 15.6, and 20 days, respectively.

Similarly, toxaphene has been fed to black ducks[94] to study the effects upon reproduction. Reproduction was not adversely affected at 50 ppm maximum feeding level. At this 50 ppm feeding level males accumulated 28 ppm toxaphene in their carcasses and females 11 ppm. The low body residues were felt to be an indication that toxaphene was metabolized and excreted without accumulation in the internal organs. This complimented and confirmed the feeding studies of Mehrle et al.[95] in which black ducks were assessed for dietary toxaphene effects on bone development. Backbone development was adversely affected and growth reduced. However, reproduction and survival were not affected.

In two following studies[96,97] on brown pelicans, Blus and co-workers studied reproductive success of these birds. Residue toxaphene levels were not consistently identified by the method used. Highest mean tissue level found was 0.32 ppm in 1973.

Winemeyer and co-workers[98] screened 1974 osprey egg samples for toxaphene. They were investigating the population crash of this bird and its lack of reproductive success. Although the study covered the period 1970 to 1974 only the 1974 samples were analyzed for toxaphene. Only 3 eggs were collected in 1974 and mean toxaphene concentration of 0.03 ppm was found.

Toxaphene has been reported as toxic to wildlife by U.S. Department of the Interior.[99] LD_{50s} at 95% confidence limit were variable. Adult mallards were half as susceptible as mallard ducklings at average LD_{50} of 70.7 mg/kg; pheasants were comparable to ducklings at 40.0 ppm; and bobwhite quail were about as susceptible as the Fulvous tree duck at 85 to 99 mg/kg. Domestic goats and mule deer required higher doses in the range of 139 to 240 mg/kg to exhibit this toxic effect.

Two recent studies[100,101] identified traces of toxaphene in Scandinavian individuals. Vas and Blomkvist[100] reported polychlorinated terpenes in a 10 kg composite human milk sample using GC/MS methodology. Both electron impact (EI) and negative ion chemical ionization (NICI) mass spectrometric detection were used with NICI being very helpful for quantitation of toxaphene in the presence of other OCs. It was noted that the 0.1 mg/kg milk fat result could have been affected by differences between the chromatograms of the toxaphene analytical standard and that of the human milk sample, which was altered due to metabolism. The authors could find no previously reported data for human milk or tissues. However, Massalo-Rauhamao et al.[102] had performed earlier analyses on human adipose tissues. Metabolized toxaphene appeared to be present in all the samples at levels comparable to chlordane. Precise determination of the clorinated perpenes was not possible because sufficient resolution could not be achieved.

Also in 1985, Pyysalo and Antervo[101] reported their GLC-SIM comparison of Finnish environmental samples to both toxaphene and Eastern European "dark", which contains many common congeners. Human milk was estimated to contain 1 to 10 μg/kg of toxaphene on a whole milk basis. This is about the same as the level Vaz and Blomkvist[100] determined on a fat basis. Lower chlorinated terpenes were not observed indicating partial metabolism had occurred. Finnish adipose tissue toxaphene levels were reported to be in the order of 0.01 to 0.1 mg/kg.

F. Soils, Sediments, and Crops

Given the major use of toxaphene as an insecticide on crops, it is not unexpected to find toxaphene in the crops, incorporated into the surrounding soil and in the sediments of adjacent rivers and lakes because of runoff. These later two compartments are probably the ultimate sinks for this compound and its degradation products. Pollock and Kilgore,[3] as part of their review of toxaphene also examined the toxaphene residue issue and found much confusion. This was to attributed to the residue methodology problems of researchers trying to accurately and unequivocally identify toxaphene. This problem of unequivocally identifying toxaphene has continued up to now and further discussions will follow in the analysis section of our review.

Claborn et al.[103] studied the residues of toxaphene and other insecticides in cows milk after toxaphene application as a spray. They did not indicate whether analysis of the feed was made as a control. Both toxaphene and strobane persisted for 14 d at the low ppm level. Zweig et al.[104] in a later study attempted to establish feed levels of toxaphene that would not result in detectable levels in milk and to confirm the work of Claborn et al.[103] Their toxaphene detection limit was 0.02 ppm for milk, however their method was a total chloride method which is nonspecific and would have responded to any chlorinated insecticides present. In fact, the feed contained 0.05 ppm chlorinated insecticide. The safe level in feed was calculated to be about 1 ppm toxaphene.

Minyard and Jackson[105] surveyed toxaphene in some commercial animal feeds. They reported difficulties in determining both chlordane and toxaphene in the presence of other OCs. From the published sample chromatograms presented we can certainly appreciate the scope of their problem. Approximately 15% of feeds analyzed contained 60 to 530 ppb of toxaphene, compared to 98% contamination with DDT at 10 to 448 ppb.

Klein and Link[106] studied the weathering of toxaphene on kale using gas chromatography. Residue composition remained stable up to day 7, but by day 14 the composition was changing and about 1% of the pesticide remained. By day 28 levels had fallen to nondetectable levels.

Subsequently, toxaphene was determined in clover as part of a study to improve toxaphene analysis.[107] Method of quantitation was colorimetric, after extensive cleanup, and might also have been interferred with by other OCs as well as by aromatic amines. Minimum residue level reported was 0.8 ppm at recoveries of 85 ± 2.4%

Along the same lines Yang and his co-workers[108] analyzed sugarbeet pulp and molasses for several OCs and organophosphates (OPs). OC residues found in sugar beet pulp were DDT, dieldrin, and toxaphene. As these pest control products had been routinely applied to soil or seed prior to planting they were able to relate them as a cause and effect of a pesticide application. Toxaphene ranged from N.D. to 0.34 ppm. The high value was approximately seven times the highest DDT value and twice the DDE value. Molasses, soybean oil, and tallow contained no detectable levels of toxaphene.

As part of the National Soil Monitoring Program of the U.S., Carey et al.[109] collected composite samples of soils and mature crops from 37 States. Toxaphene was not found in 13 of the 29 crop groups surveyed. Arithmetic mean of positive detections ranged from 0.01 to 10.2 ppm on a dry weight basis over these crops with the highest value being found on cotton stalks for 70.5% of the samples.

A comprehensive two-part study[110,111] of pesticides, heavy metals, and related compounds in infant and toddler diet samples was performed in the late 1970s. Sampling spanned the time from August 1976 to September 1978. Analyzed matrices were divided into 11 food classes ranging from water and beverages through fruits, vegetables, and meats. Toxaphene was found in both studies at a low frequency in composite samples and occurred in the fats and oils group. Ten of 117 samples were found to be positive, 2 were recorded as trace; and others ranged from 0.068 to 0.280 ppm. Of significant concern however, was the indication that toxaphene was increasing in both toddler and infant diets during the period studied.[111] Strangely enough, toxaphene was not found in samples of the meat, fish, and poultry group.

Extension of this study to adult food samples[112] showed a totally different distribution of dietary toxaphene. Four of 240 food composites tested positive with 2 of these reported as trace. Two of the four positives were in the meat, fish and poultry grouping; one in the garden fruits, and the remaining one in the oils, fats, and shortening group. Again a similar increase in daily dietary intake was observed, but keep in mind the number of positive samples was very low. In the earlier part of this study[113] 10 of 300 food composites tested positive for toxaphene. Two were in the meat, fish, and poultry group and 4 each were in the garden fruits and oils, fats, and shortenings group.

Weirsma et al.[114] in 1969 surveyed toxaphene levels in the soils of eight cities across the U.S. Composites of 50 random samples were analyzed by gas chromatography based on 4 toxaphene peaks. Of a variety of chlorinated pesticides measured toxaphene and endrin occurred with the least frequency and were less than the DDT group in concentrations on a ppm basis. Highest value found was 1.34 ppm for Miami, FL. An expanded follow up study was carried out for 14 cities in 1970[115] but the list of cities are mutually exclusive and toxaphene was not detected in most samples (3 of 28 sites at Greenville, MS were positive). The next study[116] in 1971 used 5 more cities and only Macon, GA tested positive for 25% of the samples with an arithmetic mean of 0.24 ppm. These positives were attributed to nearby on-site or agricultural application.

A National Soils Monitoring Program[117] in 1969 analyzed cropland soils from 43 States. Noncropland soil samples were collected for 11 other States. Toxaphene was found at a mean residue level of 0.07 ppm in 4.2 percent of cropland samples. Values did range as high as 11.72 ppm but the site was not identified. Noncropland samples had 0.01 mean residue level at 0.5% of sites. As no range was given the 0.52 ppm reported is for one site only. A cropland study in 1971[109] used 37 states and 93.8% of the samples were negative (N.D.) for toxaphene. However, 6.7% of the remaining samples exceeded 10 ppm toxaphene on a dry weight basis. Lowest value reported was 0.18 and highest was 36.33 ppm. For relativity, the total DDT group was found in 24% of samples with highest value being 388.16 ppm or ten times greater.

Compared to soils relatively little sediment toxaphene data is reported in the literature. Durant et al.[118] found 32.8 ppm toxaphene in a single composited sediment core sample near a Georgia pesticide plant. Near the outfall from the plant levels approached 1900 ppm in the top sediment layer. Dredging disturbances did not raise the levels of toxaphene in local oysters or sediments in the estuary.

Veith and Lee[119] examined the part played by lake sediments in toxaphene. Toxaphene was transported downward in sediment to about the 15 cm mark at a rate of 0.4 to 1.1 cm/d in three lakes studied. This toxaphene was strongly sorbed to the sediment and could not be leached off by lake water in the laboratory (flask shaking method with filtration and analysis of water and sediment separately). Hence, once sorbed toxaphene should not re-enter the water column. It is unclear, however, how long the samples were stored frozen prior to analysis. In one lake, toxaphene 0 to 5 cm sediment concentration increased up to 90 ppb at the 50 d mark and decreased to 7.1 ppb 2 yr post treatment of lake for rough fish control. The 50 d level was attributed to direct sorption onto the sediment and co-deposition with toxaphene-ladened algal blooms and other particulate matter. This did not agree with other reported toxaphene studies.

In a southern Florida study, Mattraw[120] found only 3.2% of soil and sediment samples taken between 1969 and 1972 contained detectable toxaphene. Water samples were virtually N.D. for 146 samples.

IV. METABOLISM, DEGRADATION, AND PERSISTENCE

One identified toxic component of toxaphene is 2,2,5-endo, 6-exo, 8,9,10-heptachloroborane.[121-124] Its preparation[122] and toxicity are important, however its degradation products need to be determined. Casida et al.[121] realized the importance of determining persistance and environmental fate of toxaphene in 1974 and devised a procedure to isolate individual toxaphene components using silica gel-hexane column chromatography coupled with preparative GLC and sublimation and crystallization. The identity of the toxic component was determined by X-ray crystallography and NMR. Rat feeding studies with ^{36}Cl and ^{14}C-labeled toxaphene indicated that dechlorination was an important metabolic route. Subsequent studies[125,126] examined metabolism of toxaphene in mammals. New compounds were identified by NMR and CI-MS after TLC and GLC isolation.[126] Chromatograms of toxaphene and toxaphene plus metabolites were given for fat, liver, and feces of rats fed toxaphene. Metabolites were most evident in the feces fraction which had a very different chromatographic profile than that of standard toxaphene.

Persistence and half-life of toxaphene have been evaluated in other studies.[127,128] At 14 ppm organic chloride level toxaphene half-life was 4 years in a California soil.[128]

Toxaphene appears to degrade quickest under flooded anaerobic conditions and least under moist aerobic conditions.[129] The following percent decompositions were observed:

Flooded stirred anaerobic soil	98%
Moist anaerobic soil	90%
Flooded unstirred aerobic soil	50%
Moist aerobic soil	0%

Williams and Bidleman[130] examined toxaphene degradation in estuarine sediments and found breakdown did not occur if the iron redox couple system was not present. Anoxic salt marsh sediments provided rapid breakdown of toxaphene. The ultimate breakdown products were not determined nor was their toxicity.

Seiber et al.[131] found that the major loss of toxaphene from leaves in a cotton field was intact vaporization of the pesticide. A similar trend was observed for aerated top soil with the exception that one component was significantly degraded. Anaerobic reduction, resulting in extensive toxaphene degradation was found in soil core and irrigation ditch sediment samples.

Lee et al.[56] found evidence that toxaphene is partially degraded in the aquatic environment. A sediment sample taken 10 mo after a lake was treated with toxaphene for rough fish control was extracted and bioassayed with bluegills. In a comparison with a toxicity study using the original formulation the sediment extract was found to be less toxic. No attempt to identify metabolites or degradation products was made.

Cairns et al.[132] examined metabolized toxaphene in dairy milk fat samples. Cows were fed toxaphene at 5 ppm level. No pesticides were evident in their milk prior to feeding and feed was carefully monitored. Therefore, no other organochlorines would be present to complicate the analysis. Electron capture detection gas chromatograms of metabolized toxaphene did not resemble those of the reference toxaphene fed to the cows. In chemical ionization (CI) mass spectral study of gas chromatographic column effluent using methane as the reagent gas C_{10}-polychloro cations at m/z 199, 235, 271, 307, 343, 377, and 411 dominated the elution profile. Lower chlorinated polychlorinated boranes and borenes became more pronounced. Hydroxylation and/or dehydrochlorination were suggested as source of the metabolism. For low level detection of toxaphene and its metabolites in milk careful selection of ions was recommended. Most abundant species should respond on m/z 159, 199, 235, 271, 307, 243, and 377. The study was solely qualitative and did not attempt to quantify or identify metabolic residues. Some residue levels were determined by ECD-GC and CI-GCMS-SIM and compared. Those of ECD-GC were lower, as expected, due to decreased ECD response of metabolized compounds. CI-GCMS-SIM was thought to be closer to the true values in the samples as response factors of reference and altered residues should have been about equal. If this approach were to be coupled to studies like those of Nash[127] and Oswald et al.[133] a great deal more information might be derived about metabolites and degradation products. However, the multicomponent nature of toxaphene will make this complex and time consuming.

The effects of thermal pocessing and storage on residue levels of selected OCs, OPs and carbamate pesticides applied to spinach and apricots was investigated by Elkins et al.[134] Substrates were canned prior to processing. Analysis was performed by gas chromatography confirmed by thin layer chromatography. Toxaphene was initially present at 6.5 ppm in spinach and 6.8 ppm in apricots. Losses on thermal processing were initially 27 and 7%, respectively. Storage at ambient for 1 yr showed losses of 60 and 35%, respectively. Storage at 100°F for 1 yr resulted in almost complete pesticide loss — 95 and 92% respectively. Losses of other chlorinated pesticides were similar but varied. However, losses of OPs and carbamates were very high under these conditions. Apricots tended to retain more unaltered pesticide than spinach.

V. SAMPLING AND SAMPLE STORAGE

A. Air

A common method of sampling for airborne toxaphene is to draw a known volume of air at a predetermined rate through a glass-fiber filter and then through a polyurethane foam plug.[18,24,25] Less than 5% of the toxaphene remains on the glass-fiber filters.[25] Tenax-GC resin was compared to polyurethane foam by Billings and Bidleman[135] for its efficiency to capture chlorinated hydrocarbons including toxaphene in high-volume sampling experiments. For toxaphene the difference between the 2 trapping media was less than 5% for samplers run side-by-side in the field. Seiber et al.[131] utilized XAD-4 macroreticular resin as the trapping medium in their high volume air sampling to determine toxaphene residues in a California cotton field. Toxaphene was leached from the resin with acetone and the extract was further cleaned up on Florisil prior to capillary column gas chromatographic analysis

Stanley et al.[28] used a modified Greenburg-Smith impinger with hexylene glycol (2-methyl 2,4-pentanediol) as the trapping liquid, an alumina trapping tube and glass-cloth filter. All three media were extracted and the extracts pooled to form a single sample which was analyzed, after a Florisil cleanup step, by packed-column gas chromatography with ECD detection.

B. Water

Samples can be taken by a wide variety of methods as described in Chapter 2. All materials contacting the sample must be scrupulously cleaned and solvent rinsed. Because of the volatility of many of the toxaphene isomers[17,19,20,24,25,30] extraction should take place as soon as possible after sample collection. Use of XAD resin extraction techniques[136,137] in the field should be employed when delays in transporting samples back to the laboratory are anticipated. Elution of the XAD column and storage of the extracts in tightly-sealed upright glass/Teflon® containers would permit columns to be reused after regeneration. This would facilitate processing greater numbers of samples in the field.

C. Sediments

Sediment samples may be taken by a variety of dredges and manual sampling methods. The entire sample should be returned for analysis or great care taken to ensure homogeneity of the sample prior to taking and preserving a representative subsample. Sediment samples are best preserved by freezing in glass foil-lined bottle (precleaned). Care must be exercised to leave expansion room for freezing. Failure to do this will result in cracked bottles and selective volatilization of sample components. When a sample is thawed for analysis it will have to be well mixed to reincorporate any water that separated out on freezing. In samples where a large volume of water overlies the sediment sample the water can be decanted off and analyzed separately from the sediment.

Ideally, to achieve the greatest possible accuracy, one would want to weigh out a representative (10 to 25g) subsamples in the field and analyze the entire sample back at the laboratory. Unfortunately, power supplies for balances are not always available and most balances, with exception of very modern electronic ones, do not travel well.

D. Fish and Biota

Fish are usually caught by electrofishing or netting. The fish should be sacrificed, wrapped whole in solvent rinsed aluminum foil, and placed on ice or frozen until returned to laboratory for analysis. To achieve rapid freezing dry ice is recommended. A polethylene plastic bag may be necessary to preclude contact with the water of the melting ice. In laboratory experiments fish may not need to be stored but can be macerated, subsampled and immediately extracted by pretested acceptable methodology.[164] If fish are small, 3 to 5 gm, grinding with sodium sulfate and coarse sand in a mortar and pestel followed by immediate extraction is easiest. Fish larger than 5-g size require the use of food grinder, homogenizers, blenders, etc.

Plankton and algae are removed from water by filtration through different mesh-sized nets. Sediment-dwelling organisms will probably have to be removed by hand-sorting to preclude damaging of the organism. Preservation is achieved by freezing as described above.

VI. GENERAL ANALYTICAL PROCEDURES

A. Overview

The older analytical techniques of toxaphene analysis will not be discussed in this review. These methods tended to be nonspecific[4] and suffer from elevated detection limits making them unsuited for environmental analyses. These techniques were thin layer chromatography,[166] combustion methods,[167] infrared methods,[168] and the spectrophotometric diphenylamine methods.[169] While examined globally, the modern methods for analysis for organochlorine pesticides, PCBs, PAHs, dioxins, and dibenzofurans, and many other environmental contaminants are very similar. Slight changes in extraction, cleanup, determination, and confirmation are made to fine-tune a method for a contaminant or a group of related contaminants (the so-called multiresidue methods). Polarity of the analyte and the sample matrix itself dictates the approach to be taken. Required reading for the novice analyst should be the methods manuals of various government organizations[170-172] as well as the Official Methods of Analysis of the Association of Official Analytical Chemists (A.O.A.C.).[173] These manuals and Moye's book on pesticide residue analysis,[174] several of which are becoming out-dated, provide excellent background information on the rationale of how and why we as organic analysts perform various procedures. In the area of water analysis, for an more up-to-date review of these procedures, we refer the reader to Minear and Keith's 1984 book on organic analysis.[175]

Given that the experienced analyst gains from their own past experiences and those of others as documented in the scientific literature we chose to pay particular attention to some of these published recommendations. In 1971 Guyer et al.,[4] in their report to the EPA, drew several conclusions regarding analytical techniques and residue data evaluation for toxaphene. In particular they identified a need to specify detection limits; to improve the specificity and senstivity of methods; to retain and re-examine, where possible, environmental samples previously analyzed by other techniques (in particular the nonspecific total chlorine and spectrophotometric methods); and to use dehydrochlorination as an identification aid. Even now these needs remain relevent. Compared to today's high-powered analytical instrumentation and advances in clean-up techniques, how do the early gas chromatographic analyses of toxaphene compare? How much confidence can be placed on this data? Especially when only 30 to 40 principal constituents were thought to be present. This leads us to the problems discussed by Pollock and Kilgore[3] in their 1978 review:

1. Sensitivity
2. Sample size limitations
3. Interferences
4. Lack of resolution of chromatographic peaks
5. Need for knowledge of sample history before GC chromatogram of toxaphene could be believed
6. Analyst experience factor
7. Necessity of a nitration step to eliminate interferences

In the years since their review much has happened to permit more accurate and, therefore, more reliable toxaphene analyses:

1. Column technology has improved greatly for packed columns.
2. Capillary columns are in use in most environmental laboratories with ability now to resolve greater than 200 toxaphene peaks.
3. Detector technology has improved — linear ECD detectors with wide linear ranges and low pirogram sensitivity exist.
4. GCMS has become available as a routine analytical tool and confirmation device.
5. Improvements have been made to column chromatographic cleanup techniques.
6. Automated lipid removal is possible using commercial or self-designed equipment.
7. The "experience factor" is in effect in that knowledge has increased over time and the scientific literature on toxaphene has increased somewhat, but not to the same extent for toxaphene that it has for dioxin in recent years.

When examined in this context analysts should be able to perform selective, accurate, and precise analyses of toxaphene in water, sediment, fish, and biota. This is not to say that there is not room for improvement. If we examine the three key areas outlined by Bowman in 1974:[176] extraction of residues; separation of residues from coextractives; and measurement of extracted residues we find a rather weak procedural area. Although we can recover surrogate spikes of analyte from most matrices precisely, can we do this accurately? Unfortunately for the moment we really cannot determine this with 100% certainty. We would also, in the case of most determinations, like to add to Bowman's list one further item — confirmation of results. This is essential to generation of data used in research and monitoring studies and for decision-making.

B. Reagents
1. Water, Solvents, Chemicals, and Adsorbents
The purity of these materials must be checked before and during any series of analyses.

This is discussed in Chapter 2. As the same precautions apply in toxaphene analysis their contribution on this subject should be consulted for details.

2. Standards

Crystalline technical toxaphene "standard" mixtures are available from the U.S. EPA Pesticides and Industrial Chemicals Repository (MD-8), Research Triangle Park, NC, U.S. or similar agency such as Agriculture Canada. Identity of individual components is uncertain as this standard is a mixture of chlorinated camphenes of 67 to 69% chlorine content. Standards are also available from most major suppliers of chromatographic columns and associated supplies.

Stock standard solution should be carefully prepared by precise weighing out of a known amount of toxaphene on a precalibrated analytical balance. Successive, accurate, volumetric dilutions of this stock can then be made to prepare working-level chromatographic standards or recovery spiking solutions of known concentration.

Chromatographic responses of working standard must be maintained over time. It is, therefore, advisable to remake the working-level standard monthly or bi-monthly and compare to the standard in current use. By frequently remaking the gas chromatographic standard there should be little chance of the standard going "off" and inaccurate analyses being performed.

C. Extraction, Concentration, and Cleanup

Extraction techniques and solvents vary depending upon the matrix being analysed. A number of extraction techniques for various matrices and solvent combinations are described in the literature.[25,34,49,57,79,83,84,90,92,93,98,100,101,105,118, 119,131,134,149,153,156-159,173,196-199] These are routine techniques that have been used and range in complexity from using no equipment to using moderately costly capital items. For water the main extraction technique is liquid-liquid partitioning using an appropriate solvent and pH. Resin extraction methods have potential as well. We have routinely achieved toxaphene recoveries in excess of 80% in the Ultratrace Laboratory. Another possible water extraction method would be the use of polyurethane foam. Sediment and fish can be extracted by soxhlet extraction, polytron extraction, mixing with anhydrous sodium sulfate, packing mixture into a glass chromatographic column, and eluting with an extraction solvent, or simply leaching the sample in organic solvent for an extended period of time. Nash[138] used steam distillation to extract toxaphene from environmental samples with limited success. Toxaphene recoveries were low and were also lower than the comparison soxhlet extraction method.

Sample extracts are concentrated usually by rotoevaporation to 2-3 ml final volume and lower for application to silica and alumina cleanup columns. A keeper is usually added to prevent loss of analyates. A 3-ball Snyder column in conjunction with a Kuderna-Danish can also be used for volume reduction as described in Chapter 2.

1. Gel Permeation Chromatography (GPC)

This important approach to lipid removal is described in detail in the literature.[139-146] Excellent recoveries for a wide range of pollutants were obtained while simultaneously greater than 98% of the lipids were removed from sample extracts. Lipid removal is essential to the success of column chromatographic cleanups using adsorbents such as florisil, silica gel, alumina, and celite. These adsorbents have a low capacity for lipids and when overloaded with them coelution of interferences with analytes occurs.[139]

As indicated in Chapter 2, an automated GPC unit is commercially available. The capacity of the Autoprep 1002 A is 23 five-ml loops. Unattended lipid removal from 23 samples is therefore possible. The analyst is left free to attend to other cleanup and determination steps. In the authors' laboratories this equipment has been routinely used for several years for lipid removal steps in OC, PCB, PAH, dioxin, and dibenzofuran methods for sediment, fish, and biota samples. However, the 23 samples must be rotary-evaporated down prior to further cleanup steps. This is laborious and slow. Analytical Biochemistry Laboratories has recently addressed this problem by introducing an add-on automated evaporator.

To make sample loading easier we have added a flow-through valve to the front of the loading loop of the Autoprep. This permits loading of several loops from a high lipid sample without constantly reloading the syringe. The normal mess of sample solution dripping back out of the tubing when the syringe is removed is eliminated. In the Department of Fisheries and Oceans (DFO) Ultratrace Laboratory, we use a Hamilton HV inert valve for this purpose.

Centrifugation or filtration through a porosity 0.45 mm membrane filters is required prior to loading the loops. Particulate matter in sample extracts will easily plug the very fine holes in the samples loading "waffers" of the Autoprep. Careful dissassembly and cleaning is time-consuming and sample loss can occur. Filtration of the 1:1 methylene chloride-cyclohexane mobile phase is also recommended as added insurance against stray particulate matter.

It is not essential that an ABC Autoprep be used. Using laboratory HPLC equipment, automated loop injector, and a low-pressure glass HPLC column of appropriate size packed with Bio Beads SX-3 (Biorad Laboratories) an automated system can be constructed and optimized. It is also possible to manually perform lipid removals.[147]

2. Chemical Conversion for Interference Removal

The multicomponent chemical composition of toxaphene leads to a relatively complex capillary column gas chromatogram. Interferences, therefore, could pose problems during quantitation of extracts containing coextracted pollutants not removed by cleanup columns. Packed column GC analysis of toxaphene also suffered from this problem. Despite the lower resolving power of packed columns, interferences showed up as contributions to the continuum under the toxaphene peaks. To solve this potential interference analysts resort to various chemical reactions to destroy the interferences with minimal destruction of toxaphene components.

Nitration was one of the earliest methods used to remove interferences in toxaphene analysis.[148] In that paper fuming sulfuric acid: sulfuric acid (fat and oil removal) was combined with nitration (1:1 V/V solution fuming sulfuric and nitric acids) to clean up sample extracts. Chromatograms were run before and after treatment so the DDT group pesticides could be determined before they were destroyed as interferences. This step remains part of modern methods of toxaphene analysis.[84] It was also reported[149] that chlordane remains as an interference after this nitration treatment. Reynolds[150] examined nitration within the context of PCB analysis in which DDT was an interference. His approach was to separate the PCBs and pesticides on a Florisil column because some PCBs nitrated and the nitration approach yielded more complex chromatograms.

Several other researchers preferred to use a different chemical interference removal step, basic alcoholic hydrolysis.[151-153] Partial dehalogenation of interfering pesticides occurred yielding an earlier eluting interference envelope.[152] This envelope still overlapped the early part of the toxaphene chromatogram but it left the later two-thirds clean permitting quantitation of toxaphene based on later-eluting peaks by packed column GC.

Miller and Wells[154] and Smith et al.[155] chose to perform their chemical conversions in the gas chromatograph itself during analysis. Miller et al.[154] used an alkaline pre-column to remove crop interferences. For toxaphene the chromatogram became easier to interpret with conversion of coextracted materials and pesticides to earlier eluting compounds. Lower toxaphene values were obtained when samples were analysed with and without the precolumn. Smith et al.,[155] used platinum and palladium as catalyst materials to catalytically hydrodechlorinate toxaphene. The catalyst was supported on the same solid support as the analytical packed column. Temperatures in excess of 175°C were required for the dechlorination to take place. Results, however, were not quantiative as further degradation occurred before complete reduction of toxaphene occurred. Milder reduction conditions were suggested.

3. Cleanup Columns

Standard column chromatographic cleanup adsorbents have been used for removal of interfering compounds in lipid-free sample extracts for toxaphene analysis. Florisil,[84,130,150,152,153,156,157] alumina,[11,157] and silica gel[83,100,158-164] are the adsorbents more frequently used in cleanup columns. Hughes et al.[157] used two different two-adsorbent columns to clean up specific sample matrix extracts. For water or extracts of solid materials a magnesium oxide-Florisil column was employed and for plankton extracts a 1% water deactivated alumina-Florisil-celite (4:1) was used. Fish extracts were cleaned up by first passing through a 1% water deactivated alumina column followed by an acid celite column and sediments by passing through acid celite followed by Florisil columns.[165]

D. Quantification

1. Packed Column Gas Chromatography

Packed column gas chromatography usually with either a microcoulometric or electron capture detectors and sometimes a flame ionization detector has been used extensively for toxaphene analysis.[11,17,86,130,156,157,160,177-185]

The ability to separate the various toxaphene components improved progressively from 1959 when Coulson et al.[182] published a gas chromatogram of toxaphene in which no peak resolution was evident, although the chromatogram indicated a multicomponent mixture was eluting. Bevenue[183,184] concluded that toxaphene, chlordane, and strobane could not be characterized by gas chromatography because they were a mixture of closely related compounds. In 1966 nanogram level sensitivity was possible[178,180,186] and chromatographic resolution had increased. Gaul[186] described a quantitation routine to determine toxaphene in the presence of DDTs and DDE. This method involved establishing baselines for multiple residues in the presence of toxaphene.

Early packed column toxaphene separations were carried out on metal columns[182] with single phases at high percent coatings.[179,182,183] The introduction of glass columns; new and mixed phase packing materials; and improvements to chromatographic instrumentation (better column oven stability; temperature programming); led to improved resolution. Detector linearity for ECD is improved as did the sensitivity of the detectors. However, despite these advances only a few of the components known to be present in toxaphene could be separated and all the peaks in the chromatograms could not be used for sample analyses due to matrix and other contaminants present as interferences. By 1975 the capillary column with its higher resolving capability was becoming an attractive alternative to the packed column. Seiber et al.[11] analyzed toxaphene on an OV-101 nonbonded capillary column and concluded that it would be a valuable tool for isolating and comparing the constituents of the toxaphene mixture.

2. Capillary Column Gas Chromatography

Capillary column chromatography or wall-coated open tubular column chromatography (WCOT) with its increased resolving power and better mass spectrometer flow compatibility is rapidly replacing packed column GC for most environmental analyses. In an evaluation of capillary GC for analysis of environmental samples Bellar et al.[187] briefly discussed toxaphene analysis in commercial products but not in environmental samples. They felt packed column chromatography was more efficient and presented less data handling workload, that is, capillary column GC produced too many peaks, which had to be processed.

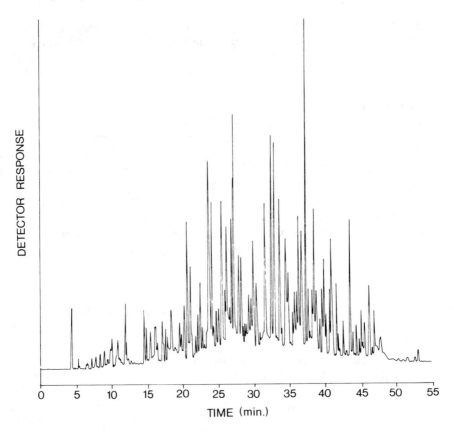

FIGURE 1.   Chromatogram of 1.0 ng/μl toxaphene standard run under the following conditions: on-column cryogenic injector programmed from 80 to 170°C at 20°C/min with 10 min hold; column SE-30 WCOT; chemically bonded; 25m × 0.1 mm I.D., 2-m deactivated fused silica retention gap; temperature programmed from 115 to 180°C at 20°C/min, 5 min hold, after at 2°C/min to 250°C, 5 min hold; hydrogen carrier gas at 60 psi; with electron capture detection.

Capillary chromatography continues to progress as evidenced by the periodic reviews, such as that of Karasek et al.[188] A large number of the papers reviewed for the period of the mid-1970s through 1987 utilized capillary column gas chromatography to separate the components of toxaphene prior to detection by ECD or GCMS.[83,84,100,101,139,158,159,161,189,190,191] In about half these citations peak selection or retention window selection was employed to reduce the interference by other organochlorine insecticides or added internal standards.[83,84,100,158,159] In another paper[161] capillary chromatography was used and careful selection of masses monitored in EI or NICI capillary GCMS was made to overcome the interferences problem.

To separate toxaphene components temperature programming is necessary. Selection of type of injector (split/splittless, on-column, etc.) and variations to temperature programs will provide the analyst with different resolutions of peaks and varied analysis times. Figure 1 is an example of a very good separation using a slightly longer than normal analysis time.

The determination of toxaphene in environmental samples by HRGC-ECD is a formidable analytical problem, which is due to possible alteration of its congeners by environmental biodegradations. Interferences of other chlorinated pesticides and PCB can further complicate the determination of toxaphene.

Detection limit of 100 pg/μl toxaphene can be routinely obtained. Analysis of toxaphene samples in fish by fused silica WCOT column gas chromatography in tandem with electron impact ionization mass spectrometry and single ion monitoring at m/z 159 provides a highly selective and sensitive technique for analysis of toxaphene and toxaphene- like contaminants in the presence of other chlorinated hydrocarbon insecticides (see Section 3.b.i). The results indicated that selection of a particular ion (m/z 159) can be of great value in determining whether or not an observed residue by electron capture detector should be attributed to an altered residue of toxaphene. A more in-depth discussion of the various mass spectrometric quantitation techniques is given in the following section.

3. Mass Spectrometry Techniques

The shifting emphasis in environmental studies in the use of mass spectrometry from the strictly qualitative to quantitative applications has been taking place. Unlike qualitative studies which often have as their aim, the identification of unknown compounds and thus benefit from the information provided by extensive fragmentation, quantitative determination presupposes this knowledge, so that the presence of a compound may often be deduced by the existence of a particular characteristic ion, molecular ion or fragment, particularly when this is coupled with appropriate retention data from a chromatographic method. The amount of solute present is deduced from the intensity of that ion.

Much emphasis at present is placed rather on detection limits, and in this respect mass spectrometry provides an uniquely sensitive method. While radioimmunoassay techniques may reach similar detection limits, they often fail to discriminate between congeners and metabolites. This combination of sensitivity and selectivity is the unique capability of mass spectrometry.

Two methods of quantitation, one based on selected ion monitoring which is particularly adaptable to GCMS, the other on a total integrated ion current which may enable the elimination of one or several prepurification steps, are in general use. Because the instrumental parameters are so varied and as it is virtually impossible to maintain them all at a constant level of performance over a reasonable period of time, it is in practice always necessary to use comparative methods with an added standard and calibration curves.

a. Direct Inlet Techniques

A large number of organic compounds are relatively thermally stable solids or liquids with molecular weights in the approximate range of 100 to 600. The majority of such samples may be introduced into the ion source by means of a direct insertion device. This device allows the introduction of samples which have insufficient thermal stability to be heated quickly to 200 to 300°C or have very low vapor pressures even at these temperatures. Advantage is taken of the low pressures which can be achieved in the ion source. The sample is introduced directly into the ion source on the cooled end of a probe. The end of the probe can be heated because of its proximity to the electron beam or, in addition, by means of a heater wire in the probe as it is in our case. An increased vapor pressure of about 10 torr permits a mass spectrum to be obtained using this direct probe technique on a sample of toxaphene of as little as a 500 pg, and quantitation can be performed in both EI and CI modes of ionization.

b. General Mass Spectrometric Behavior of Toxaphene

i. Electron Impact Fragmentation of Toxaphene

The electron impact (EI) mass spectra of toxaphene is dominated by characteristic toxaphene fragments. It has been noted that a primary fragmentation characterizing the mass spectra of toxaphene congeners involves elimination of the chlorine of HCL of the toxaphene moiety. This important reaction of the toxaphene gives rise to the cleavage of the carbon-chlorine linkage and the subsequent breakdown produces abundant ions derived from the toxaphene moiety. Thus the fragments due to the decomposition of toxaphene moiety appear in every congener. The structure and origin of the fragments of chlorinated bornanes and bornenes have been discussed in detail by Turner.[60]

The principal ions resulting from the fragmentation of the toxaphene congeners are shown in Figure 2. Examination of the mass spectra of the toxaphene revealed the presence of most abundant fragment ions of m/z 125, 159, 161, 197, 293, 335, 341, 377, and 413, for all of the congeners in the mixture. The relative intensity of the fragment at m/z 159, which is generally present at approximately 70 percent of the most abundant ion (m/z 335) has been used for the selected ion monitoring (SIM) toxaphene residue analysis. It is of the most importance that polychlorinated biphenyls (PCB) and all of the organochlorine hydrocarbon insecticides (OCs) did not show any interferences at m/z 159.[192]

ii. Chemical Ionization Fragmentation of Toxaphene

In chemical ionization (CI) mass spectrometry, ionization of the molecule under investigation is effected by reactions between the molecules of the analyte and a set of ions which serve as ionizing reactant ions.[193]

The essential reactions in positive ion chemical ionization (PICI) are:

$$R + e^- \rightarrow R_I^+ \tag{1}$$

$$R + A^- \rightarrow A_I^+ + N_1 \tag{2}$$

$$A_I^+ \rightarrow A_2^+ + N_2 \tag{3}$$

$$A_I^+ \rightarrow A_3^+ + N_3$$

$$R_I^+ + M \rightarrow (M-H)^+ + R_1H \tag{4}$$

$$RH^+ + M \rightarrow MH^+ + R \tag{5}$$

Where reaction 5 is resulting in initial formation of the protonated molecule MH^+ and represents by far the most common CI reaction. R is the reactant ion and A is the additive ion.

An alternative mode of ionization is that of charge-exchange:

$$R^+ + M \rightarrow M^+ + R$$

In this mode the odd-electron molecular ion, characteristic of EI-ionization is the initial product ion.

Most studies to date have used methane as the source of reactant ions (m/z 131,164,182). Electron impact on the methane molecules produces largely CH_4^+ and CH_3^+ ions which undergo ion molecule reactions with neutral methane molecules to produce CH_5^+ or $C_2H_5^+$. Chemical ionization spectra have also been determined using isobutane and ammonia as the ionizing reagents. At a source pressure of 1 torr the predominant ion is the tert.-butyl ion $(C_4H_9)^+$ formed by hydrogen abstraction from the isobutane molecule. The isobutane induced mass spectrum is virtually monoionic and the $C_4H_9^+$ ion behaves as a mild Broensted acid, protonating the sample molecules. The tendency to protonate is not as high for $C_4H_9^+$ as for the CH_5^+ ion.

Examination of toxaphene by PICI-MS with direct probe resulted in mass spectra characterized by multiple ion clusters. These cluster ions represent isomeric series of $C_{10}H_5Cl_9$ with variable numbers of hydrogen atoms for each chlorine substitution as shown in Figure 3. A brief review of the molecular formulas of several of the possible empirical formulas for increasing chlorination of major constituents of chlorinated camphene having general

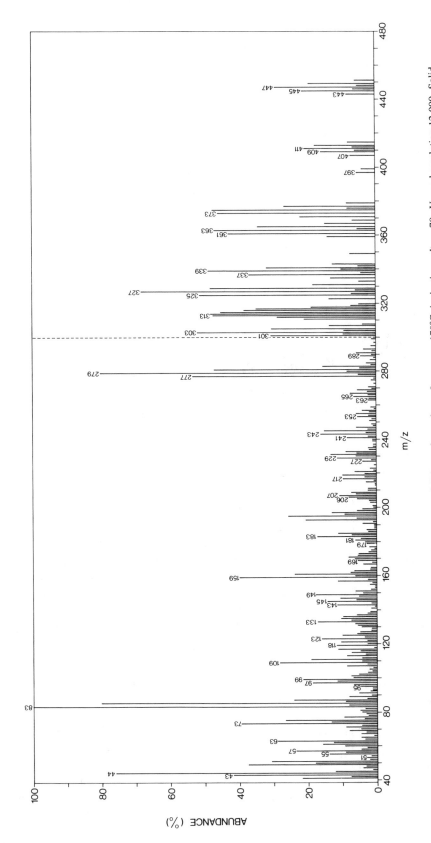

FIGURE 2. Direct probe EI full scan (m/z 40 to 400) spectrum of 500 pg of toxaphene. Ion source 170°C; ionization voltage 70 eVs; and resolution 12,000. Solid probe was heated up to 200°C in 5 s.

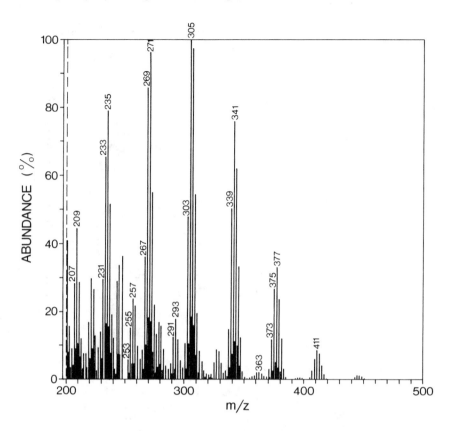

FIGURE 3.   Direct probe positive ion CI of 10 ng toxaphene full scan (m/z 200-600). Methane was used as CI reagent gas and electron multiplier voltage 2.4 kV; 0.2 mA filament current.

formula $C_{10}H_{14}$-nCln and bornenes-$C_{10}H_{16}$nCln, and bornadienes, $C_{10}H_{14}$nCln are summarized in Table 6.

An estimation of the total composition of toxaphene mixture is calculated from intensities, as measured from the probe CI-MS (Table 7), data reveals that in every case the chemical ionization mass spectra are considerably less complex then their electron inpact counterparts. The most significant observation in this study is the relative unimportance of the protonated molecular ion formation, or, perhaps more precisely, the relative instability of $(MH)^+$ which plays a major role in the CI mass spectra of other classes of compounds. In the mass spectra of toxaphene and its congeners this role it taken by the $(M - Cl)^+$ base peak fragments. Although not directly providing molecular weight information, the fact may nevertheless be of analytical diagnostic value.

iii. Negative Ion Chemical Ionization Mass Spectrometry

The technique of negative ion chemical ionization (NICI) mass spectrometry is an extension of the nonreactive gas enhancement of negative ion mass spectra which has advanced rapidly in the recent years.[194]

In contrast to positive ion chemical ionization (PICI), the major route for negative ion chemical ionization, ion formation involves electron capture, either dissociative or nondissociative. It is known that the use of methane as the reactant gas results in large yields of thermal energy electrons, while the addition of small amounts of water will lead respectively to OH^- as the major reactant ion. Using methane as the reactant ion, the major feature in mass spectra of chlorinated pesticides is the group of peaks due to chlorine attachment $(M + Cl)^-$.

Table 6
CALCULATED VALUES OF
BORNANES AND BORNENES

Name	Formula	Formula weight
Bornanes		
Camphene	$C_{10}H_{18}$	138.140
Dichlorobornanes	$C_{10}H_{16}Cl_2$	206.062
Trichlorobornanes	$C_{10}H_{15}Cl_3$	240.023
Tetrachlorobornanes	$C_{10}H_{14}Cl_4$	273.984
Pentachlorobornanes	$C_{10}G_{13}Cl_5$	307.945
Hexachlorobornanes	$C_{10}H_{12}Cl_6$	341.906
Heptachlorobornanes	$C_{10}H_{11}Cl_7$	375.867
Octachlorobornanes	$C_{10}H_{10}Cl_8$	409.826
Nonachlorobornanes	$C_{10}H_9Cl_9$	443.789
Decachlorobornanes	$C_{10}H_8Cl_{10}$	477.750
Bornenes		
Dichlorobornenes	$C_{10}H_{14}Cl_2$	204.046
Trichlorobornenes	$C_{10}H_{13}Cl_3$	271.968
Pentachlorobornenes	$C_{10}H_{12}Cl_4$	305.930
Hexachlorobornenes	$C_{10}H_{11}Cl_5$	339.891
Heptachlorobornenes	$C_{10}H_{10}Cl_6$	373.851
Octachlorobornenes	$C_{10}H_9Cl_7$	407.812
Nonachlorobornenes	$C_{10}H_8Cl_8$	441.773
Decachlorobornenes	$C_{10}H_7Cl_9$	475.734

Table 7
COMPOSITION OF TOXAPHENE
BY DIRECT PROBE CI-MS

Group	Percentage
Bornanes	52.8
Bornenes	30.5
Bornandienes	16.7

The negative CI mass spectrum of toxaphene is presented in Figure 4. In this spectrum methane at approximately 100 Pa (0.75 torr) was used as reactant gas.

Principal positive and negative ions present in the toxaphene and the most common interfering organochlorine hydrocarbon pesticides are summarized in Tables 8 and 9.

From the data reported, the NICI-MS under chemical ionization conditions has proved to be an important new technique to the analytical confirmation of toxaphene under the experimental conditions used. The technique is complimentary to PICI mass spectrometry and may technically be used on the same sample by appropriate quadrupole system.

The negative ion mass spectra reported here, were recorded on a VG quadrupole mass spectrometer, Model MM 12000, and Finnigan 4500 Model quadrupole mass spectrometer operating in the negative ionization mode. The samples were examined by direct probe inlet techniques by the authors. Methane at approximately 100 Pa (0.75 torr) ion source pressure was used as reactant gas.

c. Selected Ion Monitoring (SIM)

The defined objectives for quantitative analysis have brought forth an abundance of papers dealing with quantitation and some of these are used to introduce the methodology being

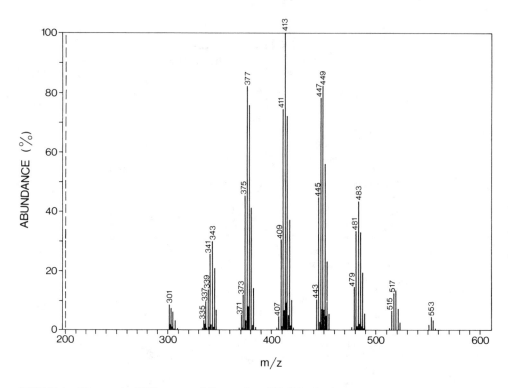

FIGURE 4.   Direct probe NICI spectum, full scan, (m/z 200-600) of 1.0 ng toxaphene. Methane reagent gas.

Table 8

PRINCIPAL TOXAPHENE IONS IN POSITIVE ION CHEMICAL IONIZATION MASS SPECTRA

5Cl		6 Cl		7 Cl		8 Cl		9 Cl	
m/z	%	m/z	%	m/z	%	m/z	%	m/z	%
303	70	341	86	377	65	409	21	445	6
305	67	343	100	379	76	411	29	447	10

Table 9

IMPORTANT TOXAPHENE IONS IN NEGATIVE ION CHEMICAL IONIZATION MASS SPECTRUM

	m/z										
Toxaphene[a]	483	481	449	447	415	413	381	379	345	343	311	309
Chlordanes	336	334	302	300	266	264	239	237			

[a]   Toxaphene ions at m/z 449, 447, 415, and 413 ought to be checked carefully because they are prone to interference from chlordane components.

applied to environmental research.[195] The process of switching the mass spectrometer to one or more preselected ions for the purpose of recording the changing intensity is called selected ion monitoring (SIM) and the output, usually represented by the rate of change of abundance of ions is called the mass fragmentogram. Low resolution is taken as R≤1500 resolving power, and high resolution is taken to be at R≥7000.

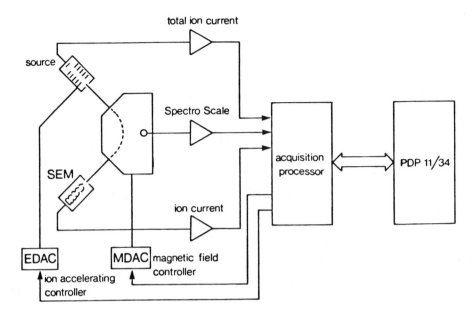

total ion current

source

Spectro Scale

SEM

ion current

EDAC

MDAC magnetic field
controller

ion accelerating
controller

acquisition
processor

PDP 11/34

FIGURE 5.   Schematic of a magnetic sector instrument.

Knowing the characteristic ions of the mixture, it is possible to perform single or multiple
ion recording at low or high resolution so that the mass spectrometer serves as a very
selective and sensitive detector (Figure 5). The mass spectrometer is equipped with a peak
selector, which instructs the mass spectrometer to measure and monitor only preselected
masses. This method is much more sensitive than mass fragmentography. It is performed
with a computer and mass spectrometer in the cyclic ''jump mode'' employing a suitable
dwell time on the particular ion. However, this needs careful handling so as to prevent
misreading of the mass which is close to some interfering ions in the mass spectrum. The
situation is simple when the masses to be used are characteristic of those of the matrix in
which the trace components are hidden. If not, interference of the characteristic masses with
the same nominal masses from other compounds occurs. To decrease the chances of interference, increasing the resolving power of the mass spectrometer is a great help because
then one can tune the instrument not only to the nominal mass of interest but even to the
elemental composition of the m/z value that one is searching for.

i. Single Ion Monitoring

The selected ion monitoring technique can monitor one ion ion-single ion recording (SIR)
or more than one ion-multiple ion recording (MIR). Single ion recording can be used for
the confirmation (using high resolution MS) and quantification of a single compound or
isomers by focusing the mass spectrometer on an ion characteristic of that compound. An
internal standard possessing the same ion but with different GC retention time must be used.
Such standards are generally readily available synthetically. The chemical similarity of the
standard and the substance under investigation ensures similar extraction properties and
renders them excellent as standards. Alternatively, a structurally distinct molecule which
possesses the same fragment ion as the investigated compound has been used. Because the
mass spectrometer remains permanently focused on the one ion, instabilities associated with
voltage switching are removed, and also the output from the electron multiplier can be
integrated over a longer period of time. The mass spectrometer is thus at its most stable
condition. This technique represents the most sensitive of all mass spectrometric means of
detection.

ii. Multiple Ion Recording (MIR)

When more than one ion is recorded by MIR, the sampling rate is determined by the dwell times that is, the times spent integrating the ion current at a selected mass. The optimum combination of dwell time, electron multiplier gain and frequency filter must be found empirically to obtain the best signal-noise ratio.

d. Electron Impact Ionization — Direct Probe Analysis

Examination of toxaphene by EI-MS with direct probe (DP) access resulted in very complex mass spectrum which cannot be easily interpreted. However, it is possible to observe that dominating ion in this mixture is $C_7H_5Cl_2^+$ having m/z value 158.97683 as discussed by Saleh.[192] Especially when medium resolution (R = 5000) is employed many interferences can be eliminated and quantitative determination would improve. Evaluation of this technique is given below. Other significant masses in EI-mode are observed at m/z 83, 125, 159, and 197.

The temperature effect on the MS-SIM response by solid probe analysis is significant. It is important to keep the sample at low (ambient) temperature even after the sample is introduced into the ion source.

The limit of detection in 10 g fish samples varied depending both on the toxaphene degradation and the specific sample. Using criteria of signal to noise of 3 to 1, *it was necessary to have at least 250 pg/g* toxaphene in a sample[181] for detection.

This technique would allow to analyze up to 30 samples per day, a significant improvement from 5 to 6 samples when HRGC/MS is used. Significant advantage of this technique is that a larger amount of sample can be introduced into the ion source after solvent evaporation and so increased response is observed.

Regarding interferences, if there is a nominal mass m/z 159 close to the characteristic fragment of toxaphene with the elemental composition $C_7H_5Cl_2^+$, there may be interference observed. Fortunately, there are not too many compounds exhibiting this ion.

Perhaps, the closest compound with the molecular ion m/z 159 is 1-methyl, 2-oxide quinoline having m/z 159.0634 corresponds to the molecular formula $C_{10}H_9NO$. In this case, even low resolution mass spectrometry would be able to separate the interfering ion from the toxaphene fragment, because R = 500 is quite sufficient.

> Sensitivity: SIR (m/z 159) 350 pg
> MIR (m/z 125, 159, 231) 500 pg
> Full scan 10 ng of toxaphene at R
> = 1000

e. Chemical Ionization —Direct Probe Analysis

Examination of toxaphene by CI-MS with direct probe provides mass spectrum characterized by multiple ion clusters, Figures 6 and 7. These clusters represent isomeric series of $C_{10}H_{10}Cl_{6-10}$ with variable numbers of hydrogen atoms for each degree of chlorine substitution as shown in Figure 6.

The volatilization curve based on all scanned masses in the 200 to 500 amu range is similar to the total ion current (TIC) trace with any slight fractionation of toxaphene observed. Quantitative analysis is performed by means of: (a) computer-controlled MIR, (b) hardware-controlled MIR at R = 1000 to 3000.

By comparing the sample with a standard toxaphene by means of recording ions at m/z 307.0, 342.9, and 376.9 using methane PICI-MS-MIR, one can quantify and confirm toxaphene rapidly. By comparing with three ions, quantitation is very simple. However, cleaned up extracts are required.

Detection limit for solid probe methane PICI is about 5 ng. A 5 replicate analysis gave a RSD of 16 percent for m/z 307, 29% for m/z 376.9 and 25% for m/z 342.9. Samples

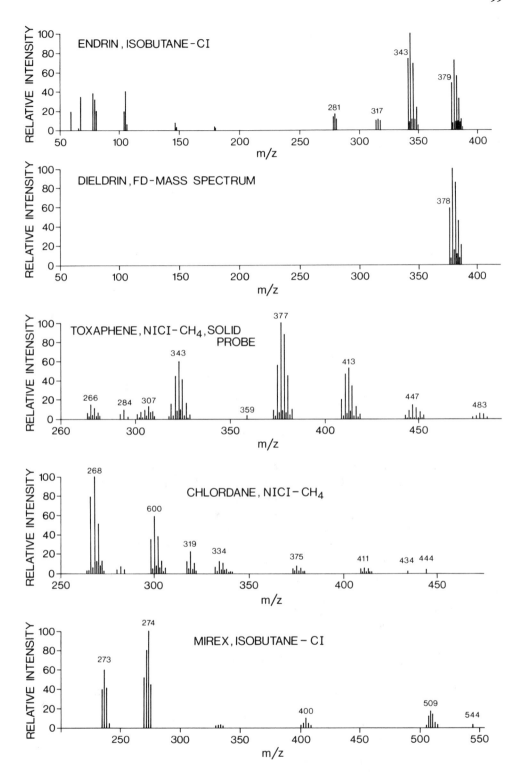

FIGURE 6.   NICI-CH$_4$ mass spectrum of toxaphene run by direct probe on a quadrupole mass spectrometer.

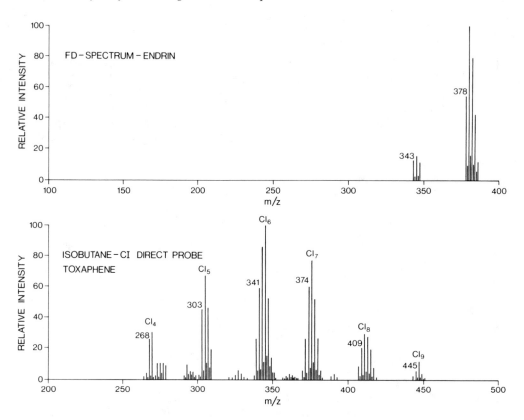

FIGURE 7.   Isobutane-CI mass spectrum of toxaphene run by direct probe on a quadrupole mass spectrometer.

were run at R = 1000 @ 10% valley, electron multiplier at 2.4 kV and 0.2 mA filament current; using methane as a reactant gas by computer-controlled MIR.

It is evident that the inclusion of as many toxaphene peaks as possible in the calculations is necessary to avoid a significant error due to differences in individual peak areas and to allow reliable comparison among differences caused by biodegradation or alteration of samples. It is obvious that the chlordane, DDT and toxaphene peaks overlap each other partially and obstruct quantitative determinations. The peaks monitored are given in Table 10. The interferences from chlordane are removed by the omission of the RT-window where interferences can be present.

For the quantitative analysis of toxaphene in fish tissue, all three techniques were evaluated on the same samples and replicate analyses were performed. The results are given in Table 11.

As is evident from Table 11, the estimated levels differ not only between different gas chromatographic analyses but also, and more understandable, between different techniques.

Toxicants A and B, previously discussed, also produce distinct and useful EI and PICI mass spectra, Figures 8 to 11. Although we have not used them for quantitation selection and use of characteristic ions of these toxicants should permit their accurate quantitation in DP, SIR, and MIR modes.

VII. RECOMMENDED ANALYTICAL METHODS

A. E.P.A. Toxaphene Methods

Two relatively accessible toxaphene methods are available in the U.S. Federal Register.[196,197] These are Method 608 for organochlorines and PCBs[196] and Method 625 for

Table 10
CHARACTERISTIC PEAKS FOR MONITORING
TOXAPHENE RELATIVE RETENTION TIMES WERE
MEASURED AGAINST *TRANS*-NONACHLOR

Toxaphene peak No.	Ret. Time (min)	RRT	Peak no.	Ret. time (min)	RRT
1	32.72	1.040	16	40.87	1.299
2	32.96	1.048	17	41.48	1.318
3	34.51	1.097	18	41.63	1.323
4	34.79	1.106	19	42.50	1.351
5	35.67	1.134	20	42.71	1.358
6	36.27	1.153	21	43.28	1.376
7	36.75	1.168	22	43.96	1.397
8	36.89	1.173	23	44.66	1.420
9	37.28	1.185	24	44.99	1.430
10	37.52	1.193	25	45.51	1.447
11	38.43	1.222	26	47.26	1.502
12	38.85	1.235	27	48.28	1.535
13	39.54	1.257	28	48.53	1.543
14	40.08	1.274	29	48.96	1.556
15	40.42	1.285	*trans*-nonachlor	31.52	1.000

Table 11
RESULTS FROM THE ANALYSES OF TOXAPHENE IN
FISH TISSUE SAMPLES USING THREE DIFFERENT
TECHNIQUES

Sample no.	GC/ECD (ng/μl)	Solid probe (ng/μl)	GC/MS-SIM @ m/z 159 (ng/μl)
1001 A	0.165	0.150 (15.1)	below DL
1004 A	8.54	11.4 (8.7)	13.5
1005 B	1.36	1.74 (12.3)	1.14
1006 A	0.521	0.670 (11.2)	0.725
2000	1.15	1.12 (12.0)	0.687
80 WHS	3.01	4.13 (9.0)	3.36
80 SEV-1	16.8	19.0 (7.4)	17.1

base/neutrals and acids.[197] Both methods provide extensive detail on scope of method; quality control; sample collection, preservation and handling; sample extraction and cleanup; and gas chromatographic analysis by packed columns. In Method 625 the chromatographic detector is a mass spectrometer and primary and secondary electron impact masses are given for use in screening for a wide variety of acid and base/neutral compounds. The analyst for exercise of these methods is given some flexibility to each "professional judgement" but should modify only with extensive validation in legal proceedings.

B. A.O.A.C. Toxaphene Methods
A second set of reliable and universally accepted methods for toxaphene are the general organochlorine and organophosphate methods described in the A.O.A.C. Official Methods of Analysis.[173] Since the A.O.A.C. has given official sanction to these methods they are known to be reliable as established through interlaboratory collaborative studies. Analysis of extracts containing fats, oils, and plant materials from foods and soils is discussed.

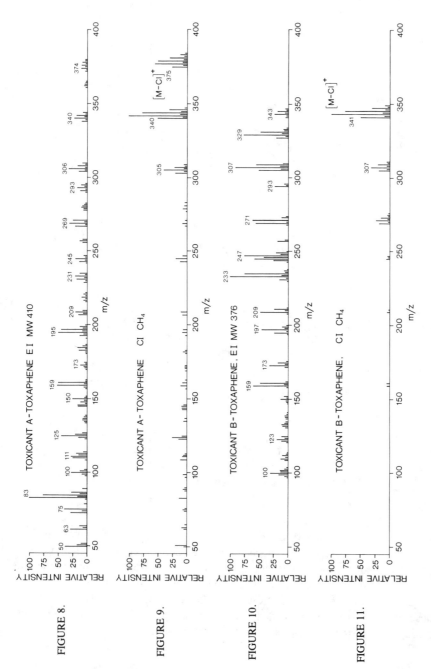

FIGURE 8.

FIGURE 9.

FIGURE 10.

FIGURE 11.

FIGURE 8. EI mass spectrum of toxicant "A" run by direct probe on a quadrupole mass spectrometer. FIGURE 9. CI-CH₄ mass spectrum of toxicant "A" run on a quadrupole mass spectrometer. FIGURE 10. EI mass spectrum of toxicant "B" run by direct probe on a quadrupole instrument. FIGURE 11. CI-CH₄ mass spectrum of toxicant "B" run by direct probe on a quadrupole system.

Techniques for gel permeation and adsorption column chromatographic clean up are outlined. Quantitation by thin layer and gas chromatography are detailed.

C. U.S. Department of the Interior Fish Toxaphene Method

The method as described permits the analyst to quantitate toxaphene residues in fish down to 0.1 μg/g.[198] Although the method is specific for fish with careful modifications and validation it could be applied to water and most sediments. Sediments heavily contaminated with industrial organics and those with high organic content may however pose severe difficulties and alternate methods may be necessary.

The following brief summary of the method is presented to enable the reader to decide whether they possess the necessary instrumentation and laboratory equipment or supplies necessary for successful application of the method. Homogenized fish tissues are extracted with dichloromethane (DCM) after mixing with anhydrous sodium sulfate and packing into a chromatographic column. Extraction is by elution of the column to collect fish oil (lipid) and contaminants. Lipid is removed on a automated GPC and the extract concentration and cleaned up by Florisil and silica gel column chromatography. A portion of the toxaphene fraction from the silica gel column is treated with concentrated sulfuric and fuming nitric acid mixture to remove interferences. However, this nitration step only eliminated interferences from DDT group pesticides. Chlordane components remained as interferences and toxaphene concentration was reduced by about 5%. Samples are analysed by both standard packed column gas chromatography and capillary GC with ECD detection. A total of 29 toxaphene peaks were selected to perform quantitation during capillary column analysis. Packed column analysis presented greater difficulty as the separation was less and coelution of toxaphene and interferences occured. Only 15 to 20 peaks were obtained. In addition an arbitrary baseline had to be used and only the later part of the chromatogram could be used for quantitation. Packed column analyses yielded lower spike values than did capillary analyses. Both NIMS and EIMS were evaluated as confirmation tools and NIMS was found to be about 100 times more sensitive. NIMS also eliminated chlordanes as a source of interference in that chlordane produced different ions than toxaphene. In EIMS this is not the case as there was overlap of the $(M-Cl)^+$ ion cluster for toxaphene and chlordane components with identical numbers of chlorines. Selection of alternate mass fragments not containing major interferences was done at the expense of not be able to use some peaks for quantitations.

D. D.F.O. Toxaphene Method

This method, from the Department of Fisheries and Oceans Ultratrace Laboratory, is suitable for the analysis of water, sediments, biota, and fish. Variations on extraction and cleanup are required for different matrices. That is, water does not require GPC and sediment GPC may be omitted.

Water samples are sequentially extracted with 3 × 100 ml of methylene chloride (dichloromethane, DCM). The pooled methylene chloride extracts are dried through anhydrous sodium sulfate in an Allihn filter funnel and collected in an appropriately sized round bottom flask. Volume is reduced to 3 to 4 ml with isooctane added as a "keeper". The sample is then fractionated in the same manner as described below for fish.

Fish and sediment samples are extracted by mixing a 5-g homogeneous subsample with approximately 5 g of anhydrous sodium sulfate and 30 ml in DCM a 50 ml stainless steel extraction tube. Stainless steel is preferred to glass for safety reasons. Polytron homogenation is performed for 1.5 to 2.0 min to ensure that all portions of the sample are extracted. The sample is allowed to settle and the supernatant DCM is decanted through anhydrous sodium sulfate in an Allihn filter funnel into a round bottom flask. The extraction is repeated twice more. For the final extraction no settling is required as the entire contents of the tube are transferred with washings to the Allihn funnel. Suction is applied to remove all the DCM from the filter-cake and the walls of the funnel and the filter-cake are washed with sufficient DCM to ensure quantitative transfer of the extract to the receiving flask. The combined extracts are rotary evaporated to 2 to 3 ml (do not let contents of round bottom go to dryness nor use water bath temperatures in excess of 40°C during evaporation). The extract is then transferred with DCM rinses to a 15 ml centrifuge tube and the volume adjusted to 5 ml by either addition of DCM or by reducing the volume by blowing the sample down under a gentle stream of purified nitrogen. The volume is next brought to 10 ml with cyclohexane and the sample is centrifuged at 2000 rpm for 5 min to settle out particulates. The extract is now ready for gel permeation chromatography (GPC) for lipid removal.

A 10 ml syringe with luer-lock adaptor and wide-bore hypodermic needle is used withdraw greater than 7 ml of sample extract. Care must be taken to ensure not to take up any particulate material. Particulate material will block the very small orfices of the waffers of the Autoprep 1002 necessitating dissassembly and cleaning and resulting in loss of all samples loaded into loops to that point, ie., reanalysis of loaded samples. The 5 ml sample loop of the Autoprep is loaded with the extract and the counter indexed to the next loop ready for the next sample. A wash between loop-loading steps is advisable to minimize possible cross-contamination of samples. Once the desired number of loops are loaded the system is started and the lipid removed by collection of predetermined toxaphene fraction. Eluting solvent is the standard 50:50 mixture of DCM:cyclohexane. Samples are collected in individual 250 ml round bottom flasks. Isooctane is added as a keeper and the extract reduced to approximately 3 ml volume prior to silica gel cleanup.

The final cleanup and fractionation step is performed on 3% water deactivated silica gel. This is prepared by activating silica gel (Woelm, 100 to 200, meshaktiv, neutral) at 130°C overnight and deactivating with 3% weight/weight organic-free distilled water. Fifteen grams of 3% deactivated silica gel are sandwiched between two 2-cm layers of anhydrous solium sulfate in a 10 cm × 1.5 cm I.D. chromatographic column (no frit) equipped with a stopcock. Silanized glass wool supports the column materials. The column is prewet with 50 ml of hexane (discarded) and the lipid free extract is applied with 3 × 2 ml hexane rinses. The column is eluted with 60 ml hexane, including the rinses, and a Fraction A collected. A second fraction, Fraction B, is collected by eluting the column with 50 ml of benzene. As benzene is a carcinogenic solvent all operations using it must be carried out in a fumehood. If good separation of Fraction A and B compounds into discrete fractions, as described below, cannot be achieved modification of the polarity of Fraction A eluting solvent and both elution volumes should be made as required.

Fraction A contains PCDPE's (polychlorinated diphenylethers), DDE, PCB, HCB, and mirex. Fraction B contains the organochlorine pesticides dieldrin, endrin, chlordanes, heptachlor epoxide, DDTs, and BHCs as well as toxaphene. Fraction B is analyzed by gas chromatography with electron capture detection on a Varian 6000 gas chromatograph under the following conditions:

Column:	J&W DB-5 or DB-1
	30 m × 0.25 mm I.D. × 0.25 μm
	film thickness capillary column
Injector Temperature:	230°C
Column Temperature:	90°C for 2 min
Temperature programmed:	90 to 160°C at 10°C/min
	160 to 280°C at 4°C/min
	10 min hold at 280°C
Detector temperature:	330°C
Total run time:	49 min

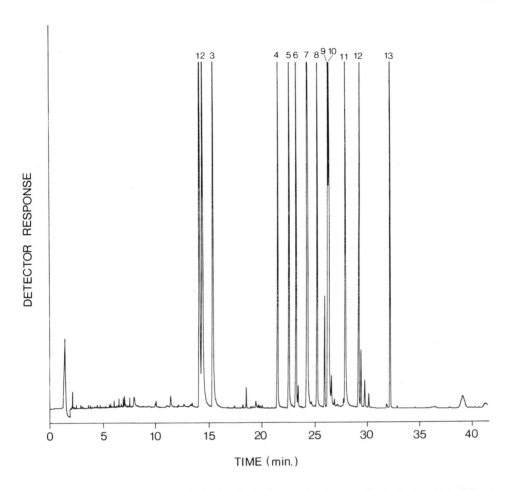

FIGURE 12.   Chromatogram of a 2-μl injection of mixed organochlorine pesticide standard containing 50 pg/ μl of each of the following compounds: (1) α-BHC; (2) hexachlorobenzene; (3) γ-BHC; (4) heptachlor epoxide; (5) γ-chlordane; (6) α-chlordane; (7) pp′DDE and dieldrin; (8) endrin; (9) pp′TDE; (10) o,p′-DDT; (11) p,p′-DDT; (12) photomirex; and (13) mirex. Numbers correspond to numbering of peaks on the chromatogram.

In the D.F.O. Ultratrace Laboratory samples are usually processed in batches of 10 to 20 samples including method blanks and recovery spikes. An autosampler is employed to inject samples and quantitation standards. The order of loading is a calibration standard followed by four samples and a repeat of this order until the tray is full or all samples are loaded. A wash vial is inserted between every injection vial to ensure that the possibility of carryover is minimized. Data is fed to a chromatographic data system where calibration factors are stored, updated on re-injection of the calibration standard, and applied to peak areas or peak heights of each sample. Prior to commencing any analyses, and periodically thereafter the setup procedure outlined below must be performed.

A well-conditioned DB-5 capillary column is installed in the instrument, leak tested, and its performance verified using an organochlorinated pesticide standard, Figure 12. If responses are acceptable, when compared to previous chromatograms on other columns, verification of the quantitation procedure takes place.

To quantitate toxaphene in environmental samples interference by routinely analyzed halogenated contaminants must not occur. Thus analytical standards of toxaphene, PCBs (mixture of 1242, 1254, and 1260 in a single injection standard), organochlorinated pesticides, and polychlorinated diphenyl ethers are analyzed. Detailed retention time comparisons are carried out for the chromatograms and overlap of peaks with toxaphene determined. This

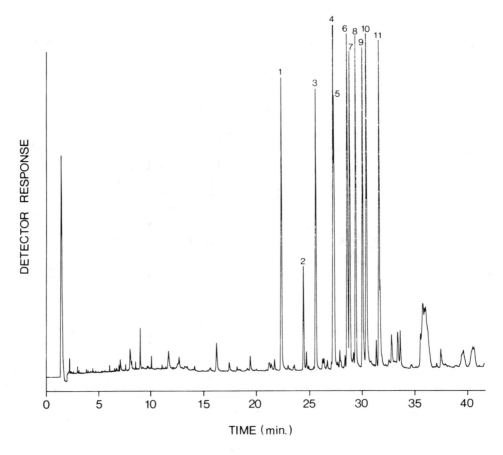

FIGURE 13.   Chromatogram of a 2-µl injection of mixed chlorinated diphenyl ether standard containing 50 pg/µl of each of the following CDPEs: (1) 2,2′,4,4′-tetrachlorodiphenyl ether; (2) 2,2′,4,4′,6-pentachlorodiphenyl ether; (3) 2,2′,2,4′,5-pentachlorodiphenyl ether; (4) 2,2′,3,4,4′,6-hexachlorodiphenyl ether; (5) 2,2′,3,4,4′-pentachlorodiphenyl ether; (6) 2,2′,4,4′,5,5′-hexachlorodiphenyl ether; (7) 2,2′,4,4′,5,6′-hexachlorodiphenyl ether; (8) 2,2′,3,4,4′,6′-hexachlorodiphenyl ether; (9) 2,2′,3,4,4′,5-hexachlorodiphenyl ether; (10) 2,2′,3,4,4′,5′-hexachlorodiphenyl ether; and (11) 2,2,3,4,4,5,6′-heptachlorodiphenyl ether. Numbers correspond to numbering of peaks on the chromatogram.

is portrayed in Figures 12-15. Figure 16 is a real cleaned up fish extract shown for comparison purposes. In our experience we have found 13 major toxaphene peaks that would not coelute with any of the potential interferences. These 13 peaks were then selected as the basis for toxaphene quantitation in our data systems. The total area response of these peaks is equated to the total toxaphene concentration in the standard. This then yields a concentration factor. Samples are then analysed and the total area of the selected peaks is determined. The total area is then divided by the concentration factor to obtain the concentration of toxaphene in the sample extract. Final toxaphene concentration in the original sample is calculated by taking into consideration dilutions, individual initial sample weights and aliquots taken at various cleanup steps (i.e., GPC). This is usually expressed as an equation:

$$\text{Total toxaphene} = \frac{\text{Asam} + \text{Vsam} + \text{Cstd}}{\text{Astd} \times \text{Wsam} \times \text{D.F.} \times 1000} = \mu g/g \quad \text{or} \quad \mu g/l$$

where Asam = summed area of the 13 toxaphene peaks in the sample; Vsam = sample extract volume in milliliters; Wsam = sample weight in grams; Astd = summed area of

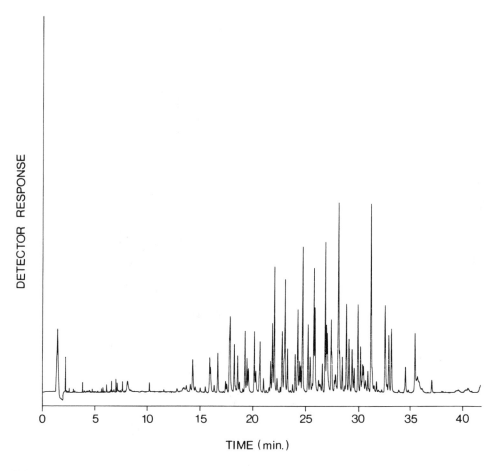

FIGURE 14.   Chromatogram of a 2-μl injection of mixed polychlorinated biphenyl standard which is 1:1:1
Aroclor® 1248, 1254, 1260, at 200 pg/μl.

the 13 toxaphene peaks in the standard; 1000 = conversion factor from microliters to milli
liters; Cstd = concentration of standard in pg/μl; and D.F. = dilution factor.

Very narrow retention time windows are employed to ensure accurate peak selection by
the data system for quantitation. A Spectra Physics SP 4000 and a Waters 840 Chromatographic Data System have been employed using windows of 0.2% of retention time and
0.02 min, respectively. The later is a much tighter window. Interference peaks are, therefore, eliminated and because the chromatographic system exhibits long-term retention
time stability, rejection of real toxaphene peaks will not occur.

Detection limits of 50 μg/kg for fish and <15 μg/l for water are achieved for extract
final volumes of 10 ml. Depending upon matrix contributions and cleanup efficiency it may
be possible to lower these detection limits by further evaporation of the extracts.

Confirmations can then be performed by either dehydrochlorination or GCMS as needed.

E. Other Recently Published Methods

As toxahene has taken on increased importance as a Great Lakes contaminant, its determination in foodchain organisms, fish in particular, has been subject of intensive investigation. In addition to the work performed in our laboratories, Ryan and Scott[199] modified
and validated National Water Quality Laboratory methodology for OCs to incorporate toxaphene analysis. Amounts in excess of 2×10^{-10} g/g samples could be quantitated, but
10^{-11} g of toxahene alone could be detected. Range, precision, and reproducibility of the

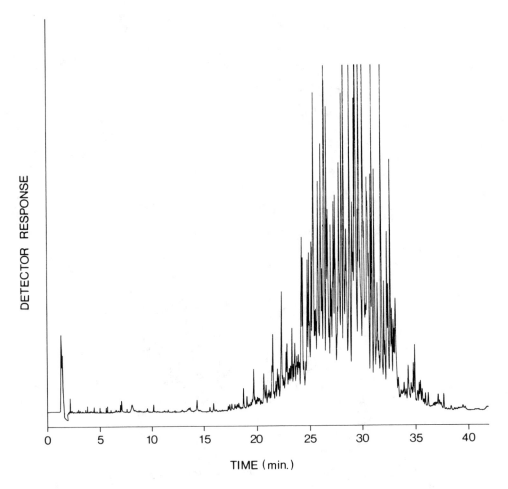

FIGURE 15.   Chromatogram of a 2-μl injection of toxaphene analytical standard at concentration of 1.47 ng/μl.

analysis were assessed through the cleanup procedures used. Quanitation of sample extracts by ECD capillary gas chromatography after GPC and silica gel cleanup was based upon 11 toxaphene peaks.

Another good fish method is that of Musial and Uthe[84] on the East Coast of Canada. They analyzed toxaphene in herring *(Clupea harengus harengus)* by ECD capillary chromatography and forming nitric acid-sulfuric acid cleanup coupled with the same ECD quantitation. Negative interferences in chromatograms were found to disappear after treatment.

In laboratories with GCMS capabilities the recently published NICI quantitation of Swackhammer et al.[159] can be used for cleaned up environmental samples. Detection limit of 75 pg was achieved and linearity of NICI response was 4 orders of magnitude. The major interferent, chlordane, did not adversely affect accuracy or precision in their method.

VIII. QUALITY ASSURANCE

To our knowledge Certified Standard Reference Materials (SRMs) exist for toxaphene in water, fish, sediment, or biota and are available from the U.S. E.P.A. depository. These should be used by analysts to ensure the validity of their data. Reagent blanks, method blanks, method spike recoveries, surrogate sample matrix spikes, and duplicate samples are all mandatory. Addition of an internal standard as described by Widequist et al.[158] and Wells

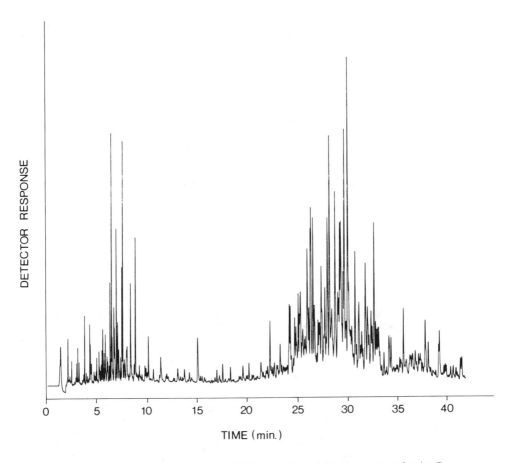

FIGURE 16.   Chromatogram of a 2-μl injection of a real fish tissue extract fraction B.

and Cowan[190] would aid quantitation. They employed Dechlorane 603, a high-boiling chloroalicyclic compound and reported toxaphene down to 10 ng/g extracted fat.

Other quality assurance protocols are mainly good analytical practice and common sense:

1.   Use of high quality analytical standards.
2.   Accurate weighing and diluting of standards to desired "known" concentrations.
3.   Continual assessment of gas chromatographic analytical standards for: (a) Response factors; (b) Retention times; (c) Peak shapes; and (d) Linearity to identify composition changes over time and column degradation.
4.   Periodic replacement of standards with comparison of working levels (and other standards if required) prior to discarding the old standard.
5.   Development and periodic analysis of an in-house pseudo-SRM. Long-term method stability or variation can then be assessed.
6.   Frequent spot-check confirmation tests should be performed for both positive and negative (N.D.) samples. Reanalysis of randomly chosen high and low samples ensures that no chromatographic anomalies are generated or reported.
7.   If toxaphene analysis is a new analytical program for the laboratory all positive values should be confirmed by reanalysis on a second column of different polarity or by GCMS in either full scan, levels permitting, or by SIM in either EI or NICI modes.
8.   All chromatograms and data system outputs must be scrutinized. Anomalous peak responses and chromatographic pattern changes could indicate the presence of inter-

ferences, which are artificially raising the final toxaphene value. If this is observed the samples should be reanalysed to confirm initial finding and then sent for GCMS analysis.

Recently a collaborative study on analysis of toxaphene in compounds and tissue homogenates[200] was performed. Results indicate that data from the laboratories of highly skilled and experienced analysts can be compared.

IX. DATA COMPARISON

From our observations during our literature search on toxaphene we concluded that varying levels of confidence must be placed on the published data. These levels of confidence arise from the constant improvements in analytical methods since the 1950s. The scientist interpreting the large data set over this time frame must be aware of problems (which may have occurred despite the great care of the analysts) of: identification; detection limits; interferences (PCB's, chlordane); and to use or not to use certain data accordingly.

Toxaphene analysis pre-1960 was nonspecific. Total chlorine analysis (TCA) and spectrophotometric methods predominated. In TCA any chlorinated compound would have contributed to take total toxaphene value due to this nonspecificity. Similarly interferences are to be expected in the spectrophotometric method by other chromophores that is coextractives, not completely removed during cleanup and the pesticides aldrin, dieldrin and chlordane.[169]

Interferences of course would have been minimal in experimental set-ups where only toxaphene was used and "before" and "after" samples were analyzed. In the period 1960 to 1963 the analysis of organo-halogen pesticides started to be performed by packed-column gas chromatography using the relatively new technology of the electron capture detector. Toxaphene chromatographed as a series of baseline humps of low amplitude.[169,171] In fact, the analyst was doing well if seven peaks were attained. Although toxaphene could be quantitated detection limits were high relative to today and interferences could have been present. As early column were nonreproducible depending on the skill of the individual who coated and packed them, and ECDs were nonlinear a significant variability of data is to be expected in these data sets.

Improved phases and column packing techniques resolved more and more toxaphene peaks and improved resolution and peak sharpness. Today with the advent of WCOT columns and frequency pulsed linear ECDs the quantitation of toxaphene is much more reliable. Interferences still present quantitation problems, but the experienced analyst should spot and eliminate this source of error.

Therefore, data must be placed in the context of suitability of the method for toxaphene analysis and the changing technology employed. Precision, accuracy, and detection limits, if determined, as well as confirmatory techniques must be key elements of data assessment.

X. RECOMMENDATIONS

In the course of this review we took particular note of areas where further research is necessary. The recommendations for future research listed below are intended to improve data quality for environmental surveys of toxaphene and fill data gaps that exist at present.

1.   Collaborative studies need to be performed on the analytical methods for determining toxaphene in environmental matrices — water, sediments, fish, birds.
2.   Methods for confirmation of toxaphene residues need to be assessed/developed to ensure quality of data so that true extent of the problem can be determined.
3.   Toxaphene analysis should be performed by capillary column GC and confirmed via

GCMS and nitration cleanup is recommended. Both total residues and Toxicant "A" and "B" must be determined.

4. The application of HPLC for analysis of toxaphene and its metabolites needs to be researched. Use of photodiode array detectors or mass spectrometric detector in tandem with HPLC or supercritical fluid chromatography could simplify analyses.

5. The two key toxic components of toxaphene, toxicants A and B need to be synthesized, certified, and made available for use in quantitating residues. Large-scale TLC of preparative HPLC could be an alternate route to synthesis.

6. Storage and loss rate studies are necessary for all environmental matrices. These are essential for data interpretation.

7. Environmental samples should be examined for metabolic toxaphene profiles.

8. Toxicological and environmental significance of other degradation products of toxaphene should be assessed.[202]

9. An intensive monitoring study should be mounted to determine the distribution of toxaphene and its metabolites present, ie., from the bottom to the top of food chain, including man, and all environmental compartments and sinks. This study should take place as soon as practicable because of the ban on toxaphene use in North America. T = 0 of disappearance has already occurred.

10. There is a need to consider and investigate whether other sources of toxaphene or toxaphene-like compounds exist through inadvertent chlorination of waste streams (industrial or municipal).

11. Disposal by land spreading[201] should be discouraged.

REFERENCES

1. **Reimold, R. J. and Shealy, M. H., Jr.,** Chlorinated hydrocarbon pesticides and mercury in coastal young-of-the-year finfish, South Carolina and Georgia, 1974-74, *Pestic. Monit. J.,* 9, 170, 1976.
2. **Eisler, R. and Jackson, J.,** Toxaphene hazards to fish, wildlife, and invertebrates: a synoptic review. *U.S. Fish Wildl. Serv. Biol. Rep.,* 85(1.4), 26, 1985.
3. **Pollock, G. A. and Kilgore, W. W.,** Toxaphene, *Residue Rev.,* 69, 87, 1978.
4. **Guyer, G. E., Adkisson, P. L., DuBois, K., Menzie, C., Nicholson, H. P., Zweig, G., and Dunn, C. L.,** *Toxaphene Status Report. Special Report to the Hazardous Materials Advisory Committee Environmental Protection Agency,* November, 1981.
5. **Mayer, F. L. and Mehrle, P. M.,** Toxicological aspects of toxaphene in fish: a summary, in *Environmental Contaminants: Wildlife, Fish and Public Health,* Transactions of the 42nd North American Wildlife and Natural Resources Conference, Wildlife Management Institute, Washington, D.C., 1977, 365.
6. Toxaphene, in *Handbook of Toxic and Hazardous Chemicals and Carcinogens,* 2nd ed., Sittig, M., Ed., Noyes Publications, Parkridge, NJ, 1985, 875.
7. **Mashburn, S. A.,** *Health and Environmental Effects of Toxaphene — A Literature Compilation, 1962-1978,* ORNL/TIRC — 78/5, October 1978.
8. **Anon.,** EPA bans most uses of pesticide toxaphene, *C&EN,* 6, October 25, 1982.
9. Toxaphene, in *Pesticide Manufacturing and Toxic Materials Control Encyclopedia,* Sittig, M., Ed., Noyes Data Corp., Parkridge, NJ, 1980, 734.
10. **Sax, N. I.,** Toxaphene, in *Cancer Causing Chemicals,* Van Norstrand Reinhold, New York, 1981, 466.
11. **Seiber, J. N., Landrum, P. F., Madden, S. C., Nugent, K.D., and Winterlin, W. L.,** Isolation and gas chromatographic characterization of some toxaphene components, *J. Chromatogr.,* 114, 361, 1975.
12. **Hartley, D. and Kidd, H., Eds.,** *The Agrochemicals Handbook,* The Royal Society of Chemistry, Nottingham, England, 1983.
13. **Martin, H. and Worthing, C. R., Eds.,** *Pesticide Manual Basic Information on the Chemicals Used as Active Components of Pesticides,* 4th ed., British Crop Protection Council, 1974, 75.
14. **Brooks, E. T.,** *Chlorinated Insecticides, Vol. I, Technology and Application,* CRC Press, Boca Raton, FL, 1974, 207.
15. **Brooks, G. T.,** *Chlorinated Insecticides, Vol. II, Biological and Environmental Aspects,* CRC Press, Boca Raton, FL, 1974, 130.

16. **Packer, K., Ed.,** *Nanogen Index. A Dictionary of Pesticides and Chemical Pollutants,* Nanogens International, Freedom, CA, 1975, 101.
17. **Rice, C. P. and Evans, M. S.,** Toxaphene in the Great Lakes, in *Toxic Contaminants in The Great Lakes,* Vol. 14, Nriagu, J. O. and Simmons, M. S., Eds., John Wiley & Sons/Interscience, New York, 1984.
18. **Harder, H. W., Christensen, E. C., Matthews, J. R., and Bidleman, T. F.,** Rainfall input of toxaphene to a South Carolina estuary, *Estuaries,* 3, 142, 1980.
19. **Bidleman, T. F., Rice, C. P., and Onley, C. E.,** High molecular weight chlorinated hydrocarbons in the air and sea: rates and mechanisms of air/sea transfer, in *Marine Pollutant Transfer,* Windom, H. L. and Duce, R. A., Eds., Lexington Books, Lexington, MA, 1981.
20. **Strachan, W. M. J.,** Environmental distribution of toxic chemicals, in *Current Research,* Environmental Contaminants Division, NWRI, Environment Canada, 1984,42.
21. **Strachan, W. M. J. and Edwards, C. J.,** Organic pollutants in Lake Ontario, in *Toxic Contaminants in the Great Lakes,* Nriagu, J. O. and Simmons, M. S., Eds., John Wiley & Sons/Interscience, New York, 1984, 239.
22. **Paasivirta, J., Knuutila, M., and Paukku, R.,** Study of organochlorine pollutants in snow at North Pole and comparison to the snow at North, Central and South Finland, *Chemosphere,* 14, 1741, 1985.
23. **Munson, T. O.,** A note on toxaphene in environmental samples from the Chesapeake Bay region, *Bull. Environ. Contam. Toxicol.,* 16, 491, 1976.
24. **Bidleman, T. F. and Onley, C. E.,** Chlorinated hydrocarbons in the Argasse Sea atmosphere and surface water, *Science,* 183, 516, 1974.
25. **Bidleman, T. F. and Onley, C. E.,** Long range transport of toxaphene insecticide in the atmosphere of the western North Atlantic, *Nature,* 257, 475, 1975.
26. **Eisenreich, S. J., Looney, B. B., and Thournton, J. D.,** Airborne organic contaminants in the Great Lakes ecosystem, *Environ. Sci. Technol.,* 15(1), 30, 1981.
27. **Giam, C. S., Chan, H. S., and Neff, G. S.,** Concentrations and fluxes of phthalates, DOT's, and PCB's to the Gulf of Mexico, in *Marine Pollutant Transfer,* Windom, H. L. and Duce, R. A., Eds., Lexington Books, MA, 375, 1981.
28. **Stanley, C. W., Barney, J. E., II, Helton, M. R., and Yobs, A. R.,** Measurement of atmospheric levels of pesticides, *Environ. Sci. Technol.,* 5, 430, 1971.
29. **Seiber, J. N., Kim, Y. H., Wehner, T., and Woodrow, J. E.,** Analysis of xenobiotics in air, in *Pesticide Chemistry: Human Welfare and the Environment,* Proc. 5th Int. Congr. Pesticide Chemistry, Kyoto, Japan, 29 August to 4 September, 1982, 1983, 3.
30. **Arthur, R. D. et al.,** Atmospheric levels of pesticides in the Mississippi Delta, *Bull. Environ. Contam. Toxicol.,* 15, 129, 1976.
31. **Eisenreich, S. J. and Rapaport, R. A.,** Historical atmospheric concentrations and fluxes of chlorinated hydrocarbons in eastern North America, in *Abstracts 192nd National Meeting, American Chemical Society,* Division of Environmental Chemistry, American Chemical Society, Anaheim, CA, 1986, 180.
32. **Rapaport, R. A. and Eisenreich, S. J.,** Atmospheric deposition of toxaphene to eastern North America derived from pentaccumulation, *Atmos. Environ.,* 20, 2367, 1986.
33. **Kutz, F. W.,** Chemical exposure monitoring, *Residue Rev.,* 85, 277, 1983.
34. **Nicholson, H. P. et al.,** Water pollution by insecticides in an agricultural river basin. I. Occurrence of insecticides in river and treated municipal water, *Limnol. and Oceanogr.,* 9, 310, 1964.
35. **Goldberg, L.** Water quality implications of HMAC (EPA) reports on agricultural chemicals, in *Water Quality Proceedings of an International Forum.* Coulston, F. and Mrak, E., Eds., Academic Press, New York, 257, 1977.
36. **Schafer, M. L., Peeler, J. T., Gardner, W. S., and Campbell, J. E.,** Pesticides in drinking water. waters from the Mississippi and Missouri Rivers, *Environ. Sci. Technol.,* 3, 1261, 1969.
37. *New and Revised Great Lakes Water Quality Objectives, Vol. III,* International Joint Commission, Canada and U.S., 1977, 30.
38. **Lorber, M. N. and Mulkey, L. A.,** An evaluation of three pesticide runoff loading models, *J. Environ. Qual.,* 11, 519, 1982.
39. **Schoettger, R. A. and Olive, J. R.,** Accumulation of toxaphene by fish-food organisms, *Limnol. Oceanogr.,* 6, 216, 1961.
40. **Hoffman, D. A. and Olive, J. R.,** Accumulation of toxaphene by fish-food organisms, *Limnol and Oceanogr.,* 6, 216, 1961.
41. **Needham, R. G.,** Effects of toxaphene on plankton and aquatic invertebrates in North Dakota Lakes, in *Resource Publication 8,* U.S. Department of the Interior, Fish and Wildlife Service, Washington, D.C., January 1966.
42. **Henegar, D. L.,** Minimum lethal levels of toxaphene as a pesticide in North Dakota Lakes, in *Resource Publication 7,* U.S. Department of the Interior, Fish and Wildlife Service, Washington, D.C., January, 1966.

113

43. **Gaylord, W. E.,** Treatment of East Bay, Alger County, Michigan with toxaphene for control of sea lampreys, in *Resource Publication 11,* U.S. Department of the Interior, Fish and Wildlife Service, Washington, D. C., January, 1966.
44. **Sanders, H. O. and Cope, O. B.,** The relative toxicities of several pesticides to naoids of three species of stoneflies, *Limnol. Oceanogr.,* 13, 112, 1968.
45. **Naqvi, S. M. Z.,** Toxicity of twenty-three insecticides to a tubificial worm *Branchiura sowerbyi* from the Mississippi Delta, *J. Econ. Entomol.,* 66, 70, 1973.
46. **Hall, R. J. and Swinefore, D.,** Toxic effects of endrin and toxaphene on the southern leopard frog *Nana sphenocepalia, Environ. Pollut. Ser. A,* 23, 53, 1980.
47. **Novak, A. J. and Passino, D. R. M.,** Toxicity of toxaphene to *Bosima longinostris* and *Daphnia* spp. (Crustacea), *Bull. Environ. Contamin. Toxicol.,* 37, 719, 1986.
48. **Paris, D. F. et al.,** Bioconcentration of toxaphene by microorganisms, *Bull. Environ. Contam. Toxicol.,* 17, 564, 1977.
49. **Schimmel, S. C. et al.,** Uptake and toxicity of toxaphene in several estuarine organisms, *Arch. Environ. Contam. Toxicol.,* 5, 353, 1977.
50. **Sanders, H. O. and Cope, O. B.,** Toxicities of several pesticides to two species of cladocerans, *Trans. Am. Fish Soc.,* 95, 165, 1966.
51. **Cushing, C. E., Jr. and Olive, J. R.,** Effects of toxaphene and rotenance upon the macroscopic button fauna of two northern Colorado reservoirs, *Trans. Am. Fish. Soc.,* 86, 294, 1957.
52. **Lawrence, J. M.,** Toxicity of some new insecticides to several species of pondfish, *Prog. Fish Cult.,* V12, 141, 1950.
53. **Stringer, G. E. and McCann, R. G.,** Three years' use of toxaphene as a fish toxicant in British Columbia, *Can. Fish Cult.,* 28, 37, 1960.
54. **Hughes, R. A. and Lee, G. F.,** Toxaphene accumulation in fish in lakes treated for rough fish control, *Environ. Sci. Technol.,* 7, 934, 1973.
55. **Terriere, L. C. et al.,** The persistence of toxaphene in lake water and its uptake by aquatic plants and animals, *J. Agric. Food. Chem.,* 14, 66, 1966.
56. **Lee, G. F., Hughes, R. A., and Veith, G. D.,** Evidence for partial degradation of toxaphene in the aquatic environment, *Water Air Soil Pollut.,* 8, 479, 1977.
57. **Brooks, P. M. and Gardner, B. D.,** Effect of cattle dip containing toxaphene on the fauna of a South African river, *J. Limnol. Soc. South Africa,* 6, 113, 1980.
58. **Isensee, A. R., Jones, G. E., McCann, J. A., and Pitcher, F. G.,** Toxicity and fate of nine toxaphene fractions in an aquatic model ecosystem, *J. Agric. Food. Chem.,* 27, 1041, 1979.
59. **Nelson, J. O. and Matsumura, F.,** Separation and toxicity of toxaphene components, *J. Agric. Food. Chem.,* 23, 984, 1975.
60. **Turner, W. V., Khalifa, S., and Casida, J. E.,** Toxaphene toxicant a mixture of 2,2,5-endo, 6-exo, 8,8,9,10-octochloroborne and 2,2,5-endo, 6-exo, 8,9,9,10-octachoroborane, *J. Agric. Food. Chem.,* 23, 991, 1975.
61. **Schaper, R. A. and Crowder, L. A.,** Uptake of ^{36}CI-toxaphene in mosquito fish, *Gambusia affinia, Bull. Environ. Contam. Toxicol.,* 15, 581, 1976.
62. **Moffett, G. B. and Zarbrough, J. D.,** The effects of DDT, toxaphene and dieldrin on succinic dehydrogenose activity in insecticide-resistant and susceptible *Gambusia affinia, J. Agric. Food. Chem.,* 20, 558, 1972.
63. **Mayer, F. L. and Ellersieck, M. R.,** *Manual of Acute Toxicity: Interpretation and Data Base for 410 Chemicals and 66 Species of Freshwater Animals,* in Resource Publication 160, U.S. Department of the Interior, Fish and Wildlife Service, Washington, D.C., 1986.
64. **Mehrle, P. M. and Mayer, F. L., Jr.,** Toxaphene effects on growth and bone composition of fathead minnows, *Pimephales promelesm, J. Fish. Res. Board Can.,* 32, 593, 1975.
65. **Mehrle, P. M. and Mayer, F. L., Jr.,** Toxaphene effects on growth and development of brook trout *(Salvelinus fontinolis), J. Fish. Res. Board Can.,* 32, 609, 1975.
66. **Mayer, F. L., Mehrle, P. M., and Crutcher, P. L.,** Interactions of toxaphene and vitamin C in channel catfish, *Trans. Am. Fish. Soc.,* 107, 326, 1978.
67. **Mayer, F. L., Mehrle, P. M., and Schoettger, R. A.,** Collagen metabolism in fish exposed to organic chemicals, recent advances, in *Fish Toxicology a Symposium,* EPA-600/3-77-085, U.S. Environmental Protection Agency, Washington, D.C., 1977, 31.
68. **Hamilton, S. J., Mehrle, P. M., and Mayer, F. L.,** Mechanical properties of bone in Channel catfish as affected by vitamin C and toxaphene, *Trans. Am. Fish. Soc.,* 110, 718, 1981.
69. **Mehrle, P. M., Haines, T. A., Hamilton, S., Ludke, J. L., Mayer, F. L., and Ribick, M. A.,** Relationship between body contaminants and bone development in East Coast striped bass, *Trans. Am. Fish. Soc.,* 111, 231, 1982.
70. **Davis, P. W. et al.,** Organochlorine insecticide, herbicide and polychlorinated biphenyl (PCB) inhibition of NaK-ATP use in rainbow trout, *Bull. Environ. Contam. Toxicol.,* 8, 69, 1972.

71. **Desaiah, D. and Koch, R. B.**, Toxaphene inhibition of ATP use activity in catfish, *Ictaburus punctatus,* tissues, *Bull. Environ. Contam. Toxicol.,* 13, 238, 1975.

72. **Yap, H. H., Desaiah, D., Cutkomp, L. K., and Koch, R. B.**, *In vitro* inhibition of fish brain ATP use activity by cyclodiene insecticides and related compounds, *Bull. Environ. Contam. Toxicol.,* 14, 163, 1975.

73. **Ludke, J. L. and Schmitt, C. J.**, Monitoring contaminant residues in freshwater fishes in the United States: the national pesticide monitoring program, in Proceedings of the 3rd USA-USSR Symposium, in *The Effects of Pollutants Upon Aquatic Ecosystems: Theoretical Aspects of Aquatic Toxicology,* EPA-60019-80-34, Swain, W. R. and Shannon, V. R., Eds., U.S. Environmental Protection Agency, Duluth, MN

74. **Schmitt, C. J., Ludke, J. L., and Walsh, D. F.**, Organochlorine residues in fish: national pesticide monitoring program, 1970-74, *Pestic. Monit. J.,* 14, 136, 1981.

75. **Schmitt, C. J., Zajicek, J. L., and Ribick, M. A.**, National pesticide monitoring program: residues of organochlorine chemicals in freshwater fish, 1980-81, *Arch. Environ. Contam. Toxicol.,* 14, 225, 1985.

76. **Schmitt, C. J., Ludke, J. L., and Walsh, D. L.**, Organochlorine residues in fish: national pesticide monitoring program, 1970-74. *Pestic. Monit. J.,* 14, 136, 1981.

77. **Schmitt, C. J. et al.**, National pesticide monitoring program: organochlorine residues in freshwater fish, 1976-79, in *Resource Publication 152,* U.S. Department of the Interior, Fish, and Wildlife Service, Washington, D.C., 1983.

78. **Gooch, J. W. and Matsumura, F.**, Evaluation of the toxic components of toxaphene in Lake Michigan lake trout, *J. Agric. Food. Chem.,* 33, 844, 1985.

79. **Plumb, J. A. and Richburg, R. W.**, Pesticide levels in sera of moribund channel catfish from a continuous winter mortality, *Trans. Am. Fish. Soc.,* 106, 185, 1977.

80. **Saiki, M. K. and Schmitt, C. J.**, Organochlorine chemical residues in bluegills and common carp from the irrigated San Joaquin Valley Floor, California, *Arch. Environ. Contam. Toxicol.,* 15, 357, 1986.

81. **White, D. H., Mitchell, C. A., Kennedy, H. D., Krynitsky, A. J., and Ribick, M. A.**, Elevated DDE and toxaphene residues in fishes and birds reflect local contamination in the Lower Rio Grande Valley, Texas, *Southwest Nat.,* 28, 325, 1983.

82. **Bowes, G. W.**, The incorporation of toxic organic chemicals into food chains, *Biomed. Mass Spectrom.,* 8, 419, 1981.

83. **Jansson, B., Vaz, R., Blomkvist, G., Jensen, S., and Olssoar, M.**, Chlorinated toxaphene and chlordane components found in fish, guillemot and seal from Swedish waters, *Chemosphere,* 4, 181, 1979.

84. **Musial, C. J. and Uthe, J. F.**, Widespread occurrence of the pesticide toxaphene in Canadian east coast makine fish, *Intern. J. Environ. Anal. Chem.,* 14, 117, 1983.

85. **Hawthorne, J. C. and Ford, J. H.**, Residues of mirex and other chlorinated pesticides in commercially raised catfish, *Bull. Environ. Contam. Toxicol.,* 11, 258, 1974.

86. **Harder, H. W., Carter, T. V., and Bidleman, T. F.**, Acute effects of toxaphene and its sediment-degraded products on estuarine fish, *Can. J. Fish. Aquat. Sci.,* 40, 2119, 1983.

87. **Kynard, B.**, Avoidance behavior of insecticide susceptible and resistant populations of mosquito fish to four insecticides, *Trans. Am. Fish. Soc.,* No. 3, 557, 1974.

88. **Johnson, W. W. and Finley, M. T.**, Handbook of acute toxicity of chemicals to fish and aquatic invertebrates, in *Resource Publication 137,* U.S. Dept. of Interior Fish and Wildlife Service, Washington, D.C., 1980.

89. Getting the jump on toxaphene, *Upwellings,* 5(No.2), 1, 1983.

90. **Keith, J. O.**, Insecticide contaminations in wetland habitats and their effects on fish-eating birds, *J. Appl. Ecol.,* Suppl., 3, 71, 1966.

91. **Blus, L. J., Lamont, T. G., and Neely, B. S., Jr.**, Effects of organochlorine residues on eggshell thickness, reproduction and population status of brown pelicans in South Carolina and Florida, 1969-76, *Pestic. Monit. J.,* 12, 172, 1979.

92. **Causey, K., McIntyre, S. C., Jr., and Richburg, R. W.**, Organochlorine insecticide residues in quail, rabbits and deer from selected Alabama soybean fields, *J. Agric. Food Chem.,* 20, 1205, 1972.

93. **Bush, P. B., Kiker, J. T., Page, R. K., Booth, N. H., and Fletcher, O. J.**, Effects of graded levels of toxaphene on poultry residue accumulation, egg production, shell quality and leghorns hatchability in white leghorns, *J. Agric. Food Chem.,* 25, 928, 1977.

94. **Haseltine, S. D., Finley, M. T., and Cromartie, E.**, Reproduction and residue accumulation in black ducks fed toxaphene, *Arch. Environ. Contamin. Toxicol.,* 9, 461, 1980.

95. **Mehrle, P. M. et al.**, Bone development in black ducks as affected by dietary toxaphene, *Pestic. Biochem. Physiol.,* 10, 168, 1979.

96. **Blus, L., Cromartie, E., McNease, L., and Joanen, T.**, Brown pelican: population status, reproductive success and organochlorine residues in Louisiana, 1971-1976, *Bull. Environ. Contam. Toxicol.,* 22, 128, 1979.

97. **Blus, L. J. et al.**, The brown pelican and certain environmental pollutants in Louisiana, *Bull. Environ. Contam. Toxicol.,* 13, 646, 1975.

98. **Wiemeyer, S. N., Swineford, D. M., Spitzer, P. R., and McLain, P.D.,** Organochlorine residues in New Jersey osprey eggs, *Bull. Environ. Contam. Toxicol.,* 56, 1978.
99. **Tucker, R. K.,** *Handbook of Toxicity of Pesticides to Wildlife,* U.S. Department of the Interior, Denver, CO, 1970.
100. **Vaz, R. and Blomkvist, G.,** Traces of toxaphene components in Swedish breast milk analysed by capillary GC using ECD, electron impact and negative ion chemical ionization MS, *Chemosphere,* 14, 223, 1985.
101. **Pyysalo, H. and Antervo, K.,** GC-Profiles of chlorinated terpenes (toxaphenes) in some Swedish environmental samples, *Chemosphere,* 14, 1723, 1985.
102. **Mussalo-Rauhamaa, H., Pyysalo, H., and Moilanen, R.,** Influence of diet and other factors on the levels of organochlorine compounds in human adipose tissue in Finland, *J. Toxicol. Environ. Health,* 13, 689, 1984.
103. **Claborn, H. V., Bushland, R. C., Mann, H. D., Ivey, M. C., and Radeleff, R. D.,** Meat and milk residues from livestock sprays, *J. Agric. Food Chem.,* 8, 439, 1960.
104. **Zweig, G., Pye, E. L., Sitlani, R., and Peoples, S. A.,** Residues in milk from dairy cows fed low levels of toxaphene in their daily ration, *J. Agric. Food. Chem.,* 11, 71, 1963.
105. **Minyard, J. P. and Jackson, E. R.,** Pesticide residues in commercial animal feeds, *J. Assoc. Off. Anal. Chem.,* 46, 843, 1963.
106. **Klein, A. K. and Link, J. D.,** Field weathering of toxaphene and chlordane, *J. Assoc. Off. Anal. Chem.,* 50, 586, 1967.
107. **Adamovic, V. M. and Hus, M.,** Determination of toxaphene residues in clover, *Mikrochim. Acta,* 12, 1969.
108. **Yang, H. S. C., Wiersma, G. B., and Mitchell, W. G.,** Organochlorine pesticide residues in sugarbeet pulps and molasses from 16 states, 1971, *Pestic. Monit. J.,* 10, 41, 1976.
109. **Carey, A. E., Gowen, J. A., Tai, H., Mitchell, W. G., and Wiersma, G. B.,** Pesticide residue levels in soils and crops, 1971, national soils monitoring program. III, *Pestic. Monit. J.,* 12, 117, 1978.
110. **Johnson, R. D. et al.,** Pesticide, heavy metal and other chemical residues in infant and toddler total diet samples, III, Aug. 76 - Sept. 77, *J. Assoc. Off. Anal. Chem.,* 67, 145, 1984.
111. **Podrebarac, D. S.,** Pesticide, heavy metal and other chemical residues in infant and toddler total diet samples, IV, Oct. 77 - Sept. 78, *J. Assoc. Off. Anal. Chem.,* 67, 166, 1984.
112. **Podrebarac, D. S.,** Pesticide, heavy metal and other chemical residues in adult total diet samples, XIV, Oct. 77 - Sept. 78, *J. Assoc. Off. Anal. Chem.,* 67, 176, 1984.
113. **Johnson, R. D., Manske, D. D., New, D. H., and Podresbarac, D. S.,** Pesticide, heavy metal and other chemical residues in adult total diet samples, XIII, Aug. 76 - Sept. 77., *J. Assoc. Off. Anal. Chem.,* 67, 154, 1984.
114. **Wiersma, G. B., Tai, H., and Sand, P. F.,** Pesticide residues in soil from eight cities — 1969, *Pestic. Monit. J.,* 6, 126, 1972.
115. **Carey, A. E., Wiersma, G. B., and Tai, H.,** Pesticide residues in urban soils from 14 United States cities, 1970, *Pestic. Monit. J.,* 10, 54, 1976.
116. **Carey, A. E., Douglas, P., Tai, H., Mitchell, and Wiersma, G. B.,** Pesticide residue concentrations in soils of five United States cities, 1971 — urban soils monitoring program, *Pestic. Monit. J.,* 13, 17, 1979.
117. **Wiersma, G. B., Tai, H., and Sand, P. F.,** Pesticide residue levels in soils, FY 1969 — national soils monitoring program, *Pestic. Monit. J.,* 6, 194, 1972.
118. **Durant, C. J. et al.,** Effects of estuarine dredging of toxaphene contaminated sediments in Terry Creek, Brunswich, GA — 1971, *Pestic. Monit. J.,* 6, 94, 1972.
119. **Veith, G. D. and Lee, G. F.,** Water chemistry of toxaphene — role of lake sediments, *Environ. Sci. Technol.,* 5, 230, 1971.
120. **Mattraw, H. C., Jr.,** Occurrence of chlorinated hydrocarbon insecticides, Southern Florida, 1968-72, *Pestic. Monit. J.,* 9, 106, 1975.
121. **Casida, J. E., Holmstead, R. L., Khalifa, S., Knox, J. R., Ohsawa, T., Palmer, K. J., and Wong, R. Y.,** Toxaphene insecticide: a complex biodegradable mixture, *Science,* 183, 520, 1974.
122. **Turner, W. V., Engel, J. L., and Casida, J. E.,** Toxaphene components and related compounds: preparation and toxicity of some hepta-, octa- and nonachloroboranes, hexa- and heptachloroboranes, and a hexachloroboradiene, *J. Agric. Food. Chem.,* 25, 1394, 1977.
123. **Chandurkar, P. S., Matsumura, F., and Ikeda, T.,** Identification and toxicity of toxicant Ac, a toxic component of toxaphene, *Chemosphere 2,* 123, 1978.
124. **Seiber, J. N., Landrum, P. F., Madden, S. C., Nugent, K. D., and Wintercin, W. L.,** Isolation and gas chromatographic characterization of some toxaphene components, *J. Chromatogr.,* 114, 361, 1975.
125. **Oshawa, T., Knox, J. R., Khalifa, S., and Casida, J. E.,** Metabolic dechlorination of toxaphene in rats, *J. Agric. Food. Chem.,* 23, 98, 1975.
126. **Saleh, M. A. and Casida, J. E.,** Reductive dechlorination of the toxaphene component 2,2,5-endo-6-exo, 8,9,10- heptachloroborane in various chemical, phtochemical and metabolic systems, *J. Agric. Food. Chem.,* 26, 583, 1976.

127. **Nash, R. G.,** Determining environmental fate of pesticides with microagroecosystems, *Residue Rev.,* 85, 199, 1983.

128. **Hermanson, H. P., Gunther, F. A., Anderson, L. D., and Garber, M. J.,** Installment application effects upon insecticide residue content of a California soil, *J. Agric. Food. Chem.,* 19, 722, 1971.

129. **Parr, J. F. and Smith, S.,** Degradation of toxaphene in selected anaerobic soil environments, *Soil Science,* 121, 52, 1976.

130. **Williams, R. R. and Bidleman, T. F.,** Toxaphene degradation in estuarine sediments, *J. Agric. Food. Chem.,* 26, 280, 1978.

131. **Seiber, J. N., Madden, S. C., McChesney, M. M., and Winterlin, W. L.,** Toxaphene dissipation from treated cotton field environments: component residual behaviour on leaves and in air, soil and sediments determined by capillary gas chromatography, *J. Agric. Food. Chem.,* 27, 284, 1979.

132. **Cairns, T., Siegmund, E. G., and Froberg, J. E.,** Chemical ionization mass spectrometric examination of detabolized toxaphene from milk fat, *Biomed. Mass. Spectrom.,* 8, 569, 1981.

133. **Oswald, E. O., Albro, P. W., and McKinney, J. D.,** Utilization of gas-liquid chromatography coupled with chemical ionization mass spectrometry for the investigation of potentially hazardous environmental agents and their metabolites, *J. Chromatogr.,* 98, 363, 1974.

134. **Elkins, E. K. et al.,** The effect of heat processing and storage or pesticide residues in spinach and apricots, *J. Agric. Food. Chem.,* 20, 286, 1972.

135. **Billings, W. N. and Bidleman, T. F.,** Field comparison of polyurethane foam and tenax -GC resin for high-volume air sampling of chlorinated hydrocarbons, *Environ. Sci. Technol.,* 14, 679, 1980.

136. **Junk, G. A., Richard, J. J., Grieser, M. D., Witiak, E., Witiak, J. L. Arguello, M. D., Vitch, R., Svec, H. J., Fritz, J. S., and Calder, G. V.,** Use of macroreticular resins in the analysis of water for trace organic contaminants, *J. Chromatogr.,* 99, 745, 1974.

137. **Stepan, S. F., Smith, J. F., Flego, U., and Reukers, J.,** Apparatus for on-site extraction of organic compounds from water, *Water Res.,* 12, 447, 1978.

138. **Nash, R. G.,** Extraction of pesticides from environmental samples by steam distillation, *J. Assoc. Off. Anal. Chem.,* 67, 199, 1984.

139. **Ricick, M. A., Smith, L. M., Dubay, G. R., and Stalling, D. L.,** Applications and results of analytical methods used, in *Monitoring Environmental Contaminants in Aquatic Toxicology and Hazard Assessment: Fourth Conference,* ASTM STP 737, Branson, D. R. and Dickson, K. L., Eds., American Society for Testing Materials, 249, 1981.

140. **Hopper, M. L.,** Gel permeation system for removal of fats during analysis of foods for residues of pesticides and herbicides, *J. Assoc. Off. Anal. Chem.,* 64, 720, 1981.

141. **Veith, G. D. and Kuehl, D. W.,** Automated gel permeation system for removal of lipids in gas chromatography/mass spectrometric analysis of fatty tissues for xenobiotic chemicals, *Anal. Chem.,* 53, 1132, 1981.

142. *Elution Profiles of Compounds Using the ABC Autoprep 1002A GPC/SC-3 BioBeads Eluted with Methylene Chloride: Cyclohexane (15:85 or 1:1),* Analytical Bio Chemistry Laboratories, Colombia, MI.

143. **Tessari, J. D., Griffin, L., and Aaronseaon, M. J.,** Comparison of two cleanup procedures (Mills, Onley, Gaither vs. Automated Gel Permeation) for residues of organochlorine pesticides and polychlorinated biphenyls in human adipose tissue, *Bull. Environ. Contam. Toxicol.,* 25, 59, 1980.

144. **Stalling, D. L., Tindle, R. C., and Johnson, J. L.,** Cleanup of pesticide and polychlorinated biphenyl residues in fish extracts by gel permeation chromatography, *J. Assoc. Off. Anal. Chem.,* 55, 32, 1972.

145. **Tindle, R. C. and Stalling, D. L.,** Apparatus for automated gel permeation cleanup for pesticide residue analysis — application to fish lipids, *Anal. Chem.,* 44, 1768, 1972.

146. **Griffitt, K. R. and Craun, J. C.,** Gel permeation chromatographic system: an evaluation, *J. Assoc. Off. Anal. Chem.,* 57, 168, 1974.

147. **Sergeant, D. B., Zitko, V., and Burridge, L. E.,** *The determination of fenitrothion in bivalves.* Technical Report No. 906, Can. Tech. Report of Fisheries and Aquatic Sciences, Fisheries and Marine Service

148. **Erro, F., Bevenue, A., and Beckman, H.,** A method for the determination of toxaphene in the presence of DDT, *Bull. Environ. Contam. Toxicol.,* 2, 372, 1967.

149. **Klein, A. K. and Link J. D.,** Elimination of interferences in the determination of toxaphene residues, *J. Assoc. Off. Anal. Chem.,* 53, 524, 1970.

150. **Reynolds, L. M.,** Polychlorinated biphenyls (PCB's) and their interference with pesticide residue analysis, *Bull. Environ. Contam. Toxicol.,* 4, 128, 1969.

151. **Crosby, D. G. and Archer, T. E.,** A rapid analytical method for persistent pesticides in proteinaceons samples, *Bull. Environ. Contam. Toxicol.,* 1, 16, 1966.

152. **Gomes, E. D.,** Determination of toxaphene by basic alcoholic hydrolysis and florisil separation, *Bull. Environ. Contam. Toxicol.,* 17, 456, 1977.

153. **Crist, H. L., Harless, R. L., Moseman, R. F., and Callis, M. H.,** Application of dehydrochlorination to the determination of toxaphene in soil and identification of the major gas chromatographic peak, *Bull. Environ. Contam. Toxicol.,* 24, 231, 1980.

154. **Miller, G. A. and Wells, C. E.,** Alkaline pre-column for use in gas chromatographic pesticide residue analysis, *J. Assoc. Off. Anal. Chem.,* 52, 548, 1969.

155. **Smith, K. F. et al.,** Heterogeneous hydrodechlorination of toxaphene, *Bull. Environ. Contam. Toxicol.,* 24, 866, 1980.

156. **Boshoff, P. R. and Pretorius, V.,** Determination of toxaphene in milk, butter and meat, *Bull. Environ. Contam. Toxicol.,* 22, 405, 1979.

157. **Hughes, R. A., Veith, G. D., and Lee, G. F.,** Gas chromatographic analysis of toxaphene in natural waters, fish and lake sediments, *Water Res.,* 4, 547, 1970.

158. **Widequvist, U., Jansson, B., Reutergardh, L., and Sundstrom, G.,** The evaluation of an analytical method for polychlorinated terpenes (PCC) in biological samples using an internal standard, *Chemosphere,* 13, 367, 1984.

159. **Swackhammer, D. L., Charles, M. J., and Hites, R. A.,** Quantitation of toxaphene in environmental samples using negative ion chemical ionization mass spectrometry, *Anal. Chem.,* 59, 913, 1987.

160. **Holmstead, R. L. et al.,** Toxaphene composition analysed by combined gas chromatography — chemical ionization mass spectrometry, *J. Agric. Food. Chem.,* 22, 939, 1974.

161. **Jansson, B. and Wideqvist, U.,** Analysis of toxaphene (pcc) and chlordane in biological samples by NCI mass spectrometry, *Intern. J. Environ. Anal. Chem.,* 13, 309, 1983.

162. **Tai, H., Williams, M. T., and McMurtrey, K. D.,** Separation of polychlorinated biphenyls from toxaphene by silicic acid column chromatography, *Bull. Environ. Contam. Toxicol.,* 29, 64, 1982.

163. **Bidleman, T. F. et al.,** Separation of polychlorinated biphenyls, chlordane, and pp'DDT from toxaphene by silicic acid column chromatography, *J. Assoc. Off. Anal. Chem.,* 61, 820, 1980.

164. **Stalling, D. L. and Huckins, J. N.,** *Analysis and GC-MS Characterization of Toxaphene in Fish and Water,* EPA-602/3-76-076, U.S. Environmental Protection Agency, Washington, D.C..

165. **Underwood, J. C.,** Charcoal column separation of chlordane and toxaphene, *Bull. Environ. Contamin. Toxicol.,* 20, 445, 1978.

166. **Kovacs, M. F.,** Thin-layer chromatography for chlorinated pesticide residue analysis, *J. Assoc. Off. Anal. Chem.,* 46, 884, 1963.

167. **Hudy, J. A. and Dunn, C. L.,** Determination of organic chlorides and residues from chlorinated pesticides by combastian analysis, *J. Agric. Food Chem.,* 5, 351, 1957.

168. **Dunn, C. L., Toxaphene,** in *Analytical Methods for Pesticides and Plant Growth Regulators and Food Additives, Vol. II, Insecticides,* Sweig, G., Ed., Academic Press, New York, 523, 1964.

169. **Graupner, A. J. and Dunn, C. L.,** Determination of Toxaphene by a spectrophotometric diphenylamine procedure, *J. Agric. Food Chem.,* 8, 287, 1960.

170. **Watts, R. R., et al. Eds.,** *Manual of Analytical Methods for the Analysis of Pesticides in Humans and Environmental Samples,* EPA-600/8-80-038, U.S. Environmental Protection Agency, Health Effects Research Laboratory, Environmental Toxicology Division, Research Triangle Park, NC, 1980.

171. *Analytical Methods Manual 1979,* Environment Canada, Inland Waters Directorate, Water Quality Branch, Ottawa, Canada, August, 1979, Revised, 1981.

172. **McMahan, B. M. and Sawyer, L. D.,** *Pesticide Analytical Manual, Volume I, Methods Which Detect Multiple Residues,* U.S. Department of Health, Education, and Welfare, Food and Drug Administration, Revised, June 1979. (available NTIS, U.S. Dept. of Commerce, 5285 Part Royal Rd., Springfield, VA, 22161).

173. **Corneliussen, P.E., McCully, K. A., McMahon, B., and Newsome, W. H.,** Pesticide and industrial chemical residues, in *Official Methods of Analysis of the Association of Official Analytical Chemists,* 14th ed., 1984, 533.

174. **Moye, H. A.,** Ed., *Analysis of Pesticide Residues,* John Wiley & Sons, New York, 467, 1981.

175. **Minear, R. A. and Keith, L. H.,** *Water Analysis, Volume III, Organic Species,* Academic Press, Orlando, FL, 1984, 456.

176. **Bowman, M. C.,** Pesticides, their analysis and the AOAC: an overview of a century of progress (1884-1984), *J. Assoc. Off. Anal. Chem.,* 67, 204, 1984.

177. **Kongovi, R.,** Quantitation of toxaphene in water, *Am. Lab.,* 16, 128, 1984.

178. **Archer, T. E. and Crosby, D. G.,** Gas chromatographic measurement of toxaphene in milk, fat, blood and alfalfa harp, *Bull. Environ. Contamin. Toxicol.,* 1, 70, 1966.

179. **Archer, T. E.,** Quantitative measurement of combinations of aramite, DDT, toxaphene and endrin in crop residues, *Bull. Environ. Contam. Toxicol.,* 3, 71, 1968.

180. **Bevenue, A. and Beckman, H.,** The examination of toxaphene by gas chromatography, *Bull. Environ. Contam. Toxicol.,* 1, 1, 1966.

181. **Onuska, F. I. and Terry, K.,** *Toxaphene — Analytical Techniques for Determining Low Concentrations of Toxaphene Residues,* Environment Canada, National Water Research Institute, Inland Waters Directorate Report, Burlington, Ontario.

182. **Coulson, D. M. et al.,** Gas chromatography of pesticides, *J. Agric. Food. Chem.,* 7, 250, 1959.

183. **Bevenue, A.,** Gas chromatography, in *Analytical Methods for Pesticides, Plant and Growth Regulators and Food Additives, Vol. I, Principles, Methods and General Applications,* Academic Press, New York, New York, 189, 1963.

184. **Bevenue, A.,** Gas chromatography, in *Analytical Methods for Pesticides, Plant Growth Regulators and Food Additives,* Zweig, G., Ed., Academic Press, New York, 1967, 3.

185. **Ismail, R. J. and Bonner, F. L.,** New, improved thin layer chromatography for polychlorinated biphenyls, toxaphene and chlordane components, *J. Assoc. Off. Anal. Chem.,* 57, 1026, 1974.

186. **Gaul, J. A.,** Quantitative calculation of gas chromatographic peaks in pesticide residue analysis, *J. Assoc. Off. Anal. Chem.,* 49, 389, 1966.

187. **Bellar, T. A., Stemmer, P., and Lichtenberg, J. J.,** *Evaluation of Capillary Systems For The Analysis of Environmental Extracts,* EPA-600/4-84-004, U.S. Environmental Protection Agency, Washington, D.C., 1984.

188. **Karasek, F. W., Onuska, F. I., Yang, F. J., and Clement, R. E.,** Gas chromatography, *Anal. Chem. Reviews,* 56, 174R, 1984.

189. **Saleh, M. A. and Casida, J. E.,** Consistency of toxaphene composition analysed by open tubular column gas-liquid chromatography, *J. Agric. Food. Chem.,* 25, 63, 1977.

190. **Wells, D. E. and Cowan, A. A.,** Quantitation of environmental contaminants by fused-silica capillary column gas chromatography-mass spectrometry with multiple internal standards and on-column injection, *J. Chromatogr.,* 279, 209, 1983.

191. **Ribick, M. A. and Zajicek, J.,** Gas chromatographic and mass spectrometric identification of chlordane components in fish from Manda Stream, Hawaii, *Chemosphere,* 12, 1229, 1983.

192. **Saleh, M. A.,** Capillary gas chromatography — electron import and chemical ionization mass spectrometry of toxaphene, *J. Agric. Food. Chem.,* 31, 748, 1983.

193. **Harrison, A. G.,** *Chemical Ionization Mass Spectrometry,* CRC Press, Boca Raton, FL, 1983.

194. **Benson, W. R. and Damico, J. N.,** Mass spectra of some carbonmates and related ureas. II, *J. Assoc. Offic. Anal. Chem.,* 51, 347, 1968.

195. **Onuska, F. I. and Thomas, R. D.,** *A Review of Toxaphene Methodologies,* Environment Canada, NWRI — Internal Report, 1980, Anal. Methods Div., Burlington, Ontario.

196. **Anon.,** Method 608, organochlorine pesticides and PCBs, in Federal Register Vol. 49, No. 209, October 26, 1984, 43321.

197. **Anon.,** Method 625, base/neutrals and acids, in Federal Register Vol. 49, No. 209, October 26, 1984, 43385.

198. **Ribick, M. A., Dubay, G. R., Petty, J. D., Stalling, D. L., and Schmitt, C. J.,** Toxaphene residues in fish: identification, quantification and confirmation at parts per billion levels, *Environ. Sci. Technol.,* 16, 310, 1982.

199. **Ryan, J. F. and Scott, B. F.,** Analytical Method for Toxaphene in Fish Tissue, Environment Canada, Analytical Methods Division, National Water Research Institute, Canada Centre for Inland Waters, Burlington, Ontario, Canada, NWRI, Contribution 86—96.

200. **Scott, B. F. and Ryan, J. F.,** Toxaphene Methodology Validation, Environment Canada, Analytical Methods Division, National Water Research Institute, Canada Centre For Inland Waters, Burlington, Ontario, Canada, NWRI Contribution 86—97.

201. **Anon.,** NCA, frustrated with EPA over toxaphene disposal takes matter to Bush, in *Pestic. Toxic Chem. News,* 11, 12, 1983.

202. **Apsimon, J. W., Atkinson, A. J., and Byrne, D. S.,** Toxaphene Report, Project 221348, (Contract 1057, Health and Welfare Canada) Department of Chemistry, Carlton University, 1983.

203. **Bidleman, T. F., Wideqvist, U., Jasson, B., and Soderlund, R.,** Organochlorine pesticides and polychlorinated biphenyls in the atmosphere of southern Sweden, *Atmos. Environ.,* 21, 641, 1987.

Chapter 4

ENVIRONMENTAL ASPECTS AND ANALYSIS OF PHENOLS IN THE AQUATIC ENVIRONMENT

J. M. Carron and B. K. Afghan

TABLE OF CONTENTS

I. INTRODUCTION

A. Occurrence

Phenols occur in the aquatic environment due to their wide spread usage. Degering[1] has estimated a yearly consumption of phenol to be in several hundred million kilograms. In 1975 the amount of chlorinated phenols used in the U.S. alone was estimated to be approximately 80×10^6 kg.[2] Environment Canada estimated the production of 3 chlorophenols (2,4-dichlorophenol, 2,3,4,6-tetrachlorophenol, and pentachlorophenol) for 1981 to be approximately 4×10^6 kg in 1980 and a further 2.6×10^6 kg were also imported. However, the use of chlorophenols has been declining since the late 1970s.[3]

Some phenols are products of natural origin such as thymol in thyme oil and methyl salicylate, an insect attractant, in oil of wintergreen. Others are generated as by-products in manufacturing processes such as coke production in the steel industry, paper and pulp processing, and coal gas liquification. Phenols are also manufactured for use as fungicides, antimicrobials, wood preservatives, and as reactants or intermediates for many synthetic products such as pharmaceuticals, dyes, and pesticides.

Misuse of these compounds, improper disposal practices, and the possibility that chlorophenols may contain polychlorinated dibenzo-*p*-dioxins (PCDDs), polychlorinated dibenzofurans (PCDFs), and hexachlorobenzenes (HXB) as impurities has led to an increased interest in the measurement of phenols and related compounds in the aquatic environment.

The monitoring of industrial effluents and natural waters for trace concentration of phenols is also carried out because these compounds can adversely affect the potability of water at low concentration. Chlorinated phenols are also determined in air, water, sediment, and biota to assess their environmental impact on the aquatic ecosystem.

B. Structure, Nomenclature, and General Properties

The compounds containing hydroxyls attached to carbon the atom of the benzene ring are called phenols. The simplest of the uninuclear monohydroxy derivative of benzene is called phenol or carbolic acid as is shown below:

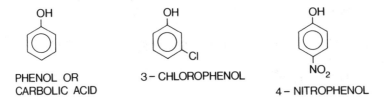

PHENOL OR   3 – CHLOROPHENOL
CARBOLIC ACID

4 – NITROPHENOL

Phenols are named by the IUPAC system in which the suffix -ol is added to the name of the parent hydrocarbon. Thus the hydroxy derivatives of benzene which is an aromatic alcohol is given the name phenol. Similarly, naphthol, xylenol, and phenanthrol are named by replacing the suffix of the respective hydrocarbons by -ol. Three theoretically possible hydroxybenzenes are known. They contain one, two, or three hydroxy functional groups and are known as mono, dihydric, and trihydric phenols, respectively. IUPAC rules as they apply to the system of nomenclature of organic compounds also apply for phenols. The phenyl radical imparts acidic properties to the hydroxy group to phenols. The acidity of the phenol hydroxyl becomes more pronounced under the influence of electronegative groups such as nitro. The nitro-phenols are stronger acids than phenol or its homologs. Phenols can be halogenated, nitrated, or sulfonated by replacing hydrogen in the benzene ring by halogen, nitro, or SO_3H group, respectively.[4]

Phenols and their substituted derivatives are the most important industrial semiproducts. For examples, phenols are used in the manufacture of plastics, picric acid, salicylic drugs, dyes, etc.[4] Chlorinated phenols have very wide applications. These are used as insecticides, fungicides, mold inhibitors, antiseptics, disinfectants, as starting material for pesticides, and also as dye pigment.[3,5] Derivatives of nitrophenols are used as crop herbicides. For example, dinoterb (2-tert-butyl-4,6-dinitrophenol) is used as crop herbicide and dinoseb (2-sec-butyl-4,6-dinitrophenol) when made in acetate form becomes a potent insecticide and weed killer. Similarly, the acetate derivative (2-tert-butyl-4,6-dinitro phenyl acetate) is used as a pre-emergent herbicide and nematocide and is also used as a dormant spray on fruit trees and for the control of red spider mites.[6]

Phenol is a colorless crystalline substance, has a melting point of 42.3°C, and a boiling point of 182°C. Phenol has a characteristic highly intense odor and can be oxidized with time, when standing exposed to air. Phenols are slightly soluble in water, however, they dissolve easily when treated with alkali. All phenols are strong bactericidal agents even in very weak solutions. Lysol®, a very frequently used disinfectant, is a soap solution mixture of o-, m-, and p-cresols prepared from coal-tar and from the products of the dry distillation of wood by heat.[4]

C. Synthesis and Production of Phenols and Substituted Phenols

There are several methods by which phenols may be synthesized. These include:

1. Fusion of sulfonic acid salts with alkalis.
2. Decomposing diazonium salts of various anilines under certain conditions.
3. Treating halogen derivatives of benzene hydrocarbons with alkali solutions at a high temperature.

The latter method, known as the DOW Process, is used to obtain phenol as shown below:

Chlorophenols are manufactured by the direct liquid-phase chlorination of phenol. The process known as the Boehringer Process involves the use of iron salt as catalyst with low heat, 157°C and high pressure, 19.5 atmospheres:

A second major industrial process for the manufacture of pentachlorophenol is the DOW Process. The starting material for this process is hexachlorobenzene which undergoes alkaline hydrolysis at 180°C as follows:

II. ENVIRONMENTAL ASPECTS

A. Routes of Entry of Phenols into the Aquatic Environment

Phenols occur throughout the plant and animal kingdom. Phenols may also enter in aquatic environment through misuse, accidental release, or intentional disposal at the manufacturing or use sites. They may also enter the aquatic environment because of domestic or agricultural use of many commercial products containing these compounds.

The demand for lower cost energy has necessitated that the petroleum industry get the maximum yields from the natural petroleum/fossil fuel (coal) that are available. Vogt et al.[7] reference coal gasification as a process which has the potential of increasing the usable coal energy return by as much as four times. The coal gasification involves low temperature and partial combustion and gasification (pyrolysis) of the coal to produce a combustible gas which can be withdrawn from the coal mine (or gas well). Phenols are among the many by-products that are produced by this process. Those compounds, including phenols, that remain in the gas well and cannot be removed are a threat to the underground water system and may affect the quality of the ground water. Orazio et al.[8] state that there is a potential threat to the purity of ground water supplies from the by-products which stay under ground following *in situ* coal gasification.

B. Environmental Exposure and Fate

In one form or another we are, or will be, exposed to phenol-related products. We are constantly exposed to several phenol derivatives. The food we eat may have been treated with herbicides and insecticides while it was growing. Preservation and storage of food requires some form of preservation against spoilage and may require the addition of phenol derivatives as a suitable preservating agent. Chlorination of drinking water and sewage treatment may lead to the production of chlorinated phenols and other substituted phenolic compounds. We use wood products that may have been treated with preservatives such as tetra and pentachlorophenols. The burning of treated wood can produce dioxins, furans, and diphenylethers thus affecting air and water environments and human health.[9]

Phenols enter and are transported in the environment via adsorption, diffusion, volatilization, leaching, surface movement, and atmospheric movement.[4,5] The worldwide use of phenols has resulted in extensive and wide distribution of these substances throughout the environment.

The environmental distribution and cycling of chlorophenols can affect air, water, soil, aquatic organisms, wildlife, and man as shown in Figure 1.[10] Chlorinated phenols including selected degradation products and neutral impurities such as PCDDs and PCDFs are known to bioaccumulate. Human exposure to chlorophenols is shown in Figure 2.[11] A complete discussion and documentation of the manufacture, importation, use, and environmental residues of chlorinated phenols in Canada is reviewed by Jones[12] and Gilman et al.[3]

C. Toxicity

Phenols have a broad spectrum of toxicity due to their varied uses and the nature of the substituents in environmentally important phenols. Some phenolic compounds are known to impart objectionable odor and taste to water at the ppb level. Others such as 2,4-dichlorophenol, 2,4,6-trichlorophenol, and pentachlorophenol are classified as suspected carcinogenic compounds in drinking water.[13] Gilman et al.[3] has documented the impact of chlorophenols on man and his environment. Environment Canada[14] has tabulated the toxicity of chlorinated phenols to organisms in the aquatic environment as shown in Table 1. This table indicates that the acute toxicity of chlorophenols is high to aquatic organisms and less so to terrestrial biota.

Renberg[15] has classified and summarized the toxicity of environmentally important phen-

123

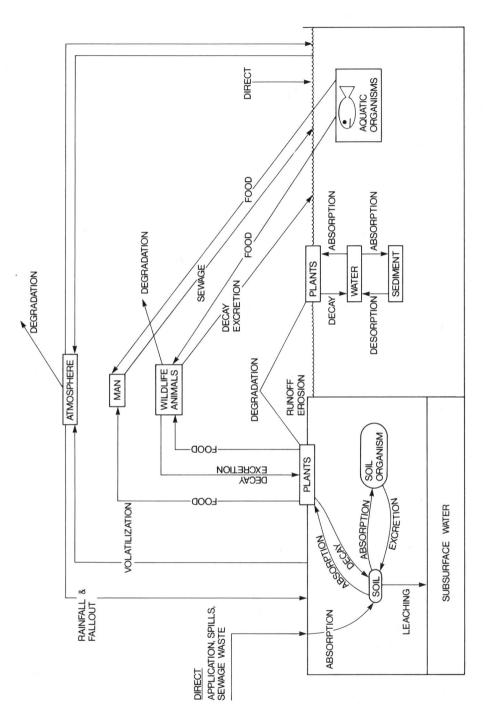

FIGURE 1. The environmental distribution and cycling of chlorophenols.

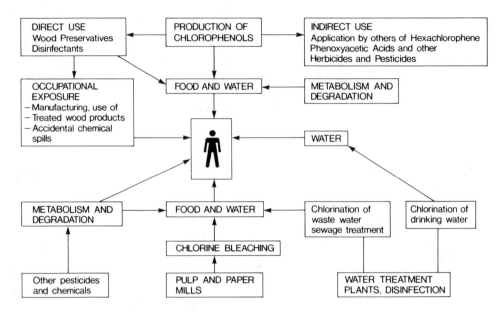

FIGURE 2.   Humane exposure to chlorophenols.

Table 1
ACUTE TOXICITY OF CHLOROPHENOLS (CPs) IN AQUATIC ORGANISMS

CP	Organism	Test[a]	Concentration (ppm)
NaPCP	Juvenile chinook salmon	96 h-LC$_{50}$	0.078
NaPCP	Developing embryos - eastern oyster	48 h-EC$_{50}$	0.040
PCP	Rainbow trout	96 h-LC$_{50}$	0.05—0.10
2,3,4,6-TTCP	Soft-shelled clam	96 h-LT	11.8
2,3,4-TCP	Shrimp	96 h-LT	2.0
CP	Shrimp	96 h-LT	4.6

[a]   LC, lethal concentration; EC, effective concentration; LT, lethal threshold.

ols, including their persistence, bioaccumulation, and other properties. The following general statements on the toxicity, persistence, biodegradation, and bioaccumulation were made[15]:

- The hydrocarbon based phenols are usually less toxic than chloro or nitrophenols.
- Chlorinated phenols are more toxic and persistent.
- Nitrophenols are the most toxic among the phenols, however, these seem to readily biodegrade through reduction to corresponding amines.
- Phenol toxicity increases with increasing acid strength.
- Alkylphenols are relatively biodegradable with the exception of those phenols that have sterically hindered hydroxy groups which makes them more persistent to biodegradation.
- Persistence and biodegradation of phenols increase with increasing lipophilicity as shown in Figure 3.[15]
- Highly chlorinated phenols are readily accumulated in aquatic organisms. Biomagnification factors ranging between 200 to 300 have been reported for highly substituted phenols. This points out the need for sensitive analytical methods for various compartments of the aquatic environment. Figure 3 provides a general comparison for three types of phenols.

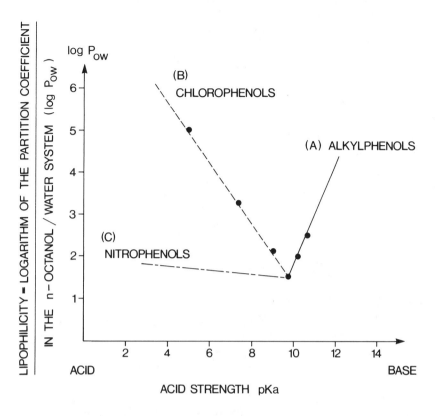

FIGURE 3. A general comparison for three types of phenols.

A clear understanding of the toxicology of chlorinated phenols is complicated by the presence of several impurities in technical grade formulation of these products.[3] The literature data on toxicology of phenols is further complicated by the fact the toxicology data was obtained using both purified and commercial grade chlorophenols with varying degrees of purification. The commercial grade chemicals contain several impurities such as PCDDs and PCDFs which are known to be very toxic. The DOW Process may generate dioxins if the reaction temperature exceeds 230° C and goes to 410°C. The Boehringer Process for the production of pentachlorophenol has the following predioxin and isopredioxin impurities:

PREDIOXIN   ISOPREDIOXIN

Of the two above procedures for the synthesis of chlorophenol, the DOW Process for the formation of 2,4,5-trichlorophenol may favor the formation of 2,3,7,8-TCDD tetrachloro dibenzo-*p*-dioxin (2,3,7,8-TCDD) as shown below:

Therefore, the reported data on toxicity of chlorinated phenols must be reviewed to establish whether purified or commercial formulation were used before making a judgement on the toxicity of individual substances.

III. PHENOL METHODOLOGY

A review of methodology for phenols prior to 1971 was done by Regnier and Watson.[16] A bibliography of 100 references is given for the determination of phenols in a variety of sample matrices. A review of the methodology for chlorinated phenols in environmental and other matrices, for the period 1968 to 1981, is provided by the National Research Council of Canada Publication No. 18578.[17] Information on other methods of analysis for chlorophenols prior to 1968 can be found in the reviews by Bevenue and Beckman,[18] Rao,[19] and Jones.[5]

This section will cover selected references for sampling, preconcentration, and cleanup techniques for quantitative analysis and confirmation of phenols in various compartments of the aquatic environment.

A. Sampling and Storage

There are very few references covering the sampling and storage of phenols in detail. Sampling for phenols is carried out in a similar manner as described in this book in chapters for PCBs and PNAs except that water and sediment samples after collection are either acidified or made basic to prevent microbial degradation. West and West[20] and Fichnolz et al.[21] have reported high losses of trace contaminants by absorption, and earlier workers[22] have also reported that some loss of phenol occurred during storage. Afghan et al.[23] investigated the loss of phenol on plastic and glass container. During this study, ^{14}C-labeled phenol was spiked at 1, 5, and 10 mg/l levels in deionized water, filtered, and unfiltered lake water. Spiked samples were stored in plastic and glass bottles and the radioactivities of these samples were measured at 1, 2, 4, 8, and 16-d intervals. Unfiltered samples were filtered through 0.45-μm filter paper before the measurement of radioactivity. The results indicated that the samples stored in glass bottles suffered no apparent loss of phenol during a 16-d period. However, there was marked decrease in radioactivity of the samples stored in plastic bottles during the first 2 d; a plateau was reached after 4 d. Therefore, the authors recommended the samples be collected and stored in glass bottles. Phenols are also known to photodecompose, therefore, brown glass bottles for sampling and storage are to be used.

B. Preservation

Samples containing low levels of phenols are generally preserved to inhibit any biode-gradation or chemical oxidation during storage. The photodegradation of phenols can easily be controlled by using brown bottles/jars or wrapping the solid biota samples with aluminum foil.

The various preservation techniques to inhibit biodegradation include the addition of copper sulfate plus phosphoric acid,[24] sodium hydroxide,[25] hydrochloric acid.[23] The effectiveness of the above preservation techniques was investigated by Afghan et al.[23] The authors concluded that all preservatives were able to kill bacteria known to degrade phenols, however, the use of either hydrochloric acid or sodium hydroxide was found to be more effective even when the contact time was only 1 min.

C. Extraction

Organic solvents are often used to extract phenols from environmental samples. The preferred extraction solvent should combine a high solubility of phenol, a low miscibility with water, and a low boiling point to facilitate removal of the solvent from extracted compounds. Extraction techniques used include shaking in separate funnel, vigorous stirring, the use of mixer/homogenizer, soxhlet apparatus, and continuous flow extractors.[26]

The National Research Council Publication No. 18578[17] has summarized various extraction procedures for different matrices. The primary extraction and partition system included blending, heat, and use of solvents such as benzene, hexane, pentane, acetone, ether, and mixture of solvents. The matrices cover water, sediment, soil, fish, air, plant tissues, etc.

Afghan et al.[23] evaluated the efficiency of various solvent extraction techniques for trace levels of phenols in natural waters. The study concluded that petroleum ether, benzene, and chloroform were found to extract quantitatively highly substituted phenols such as pentach-lorophenol and trimethylphenol. These solvents did not quantitatively extract phenol, o-, m-, and p-cresols. n-Butyl acetate or iso-amyl acetate was found to quantitatively extract phenol, cresols, and chlorinated phenols. Various pertinent extraction variables such as time, number of extractions, pH, and the volume of organic solvent necessary to extract phenols were also studied. Up to 90% of phenol could be extracted with a 50 ml volume of organic solvent over a period of 2 to 3 min. Similar recoveries were also obtained for various ratios of organic:aqueous phase up to 1:25.

D. Isolation and Cleanup Procedures

A liquid/liquid partition step using bases is used to isolate and separate phenol compounds from other co-extractants.[22,23,27] Sulfur and mercaptans are effectively removed from the organic extract containing phenols by shaking with acidic Bismuth.[23] This technique was found to be more effective than the use of metallic mercury.

Column chromatography has been used sparingly; alumina and gel permeation systems have been used to a lesser extent than Florisil, silica gel, ion exchange resin and HPLC. These techniques are sometimes employed after derivatization of phenols prior to analysis by gas chromatography.[28,29] National Research Council Publication No. 18578 lists some cleanup steps for varous matrices.[17]

E. Colorimetric and Fluorescence Spectrophotometry

Earlier methodologies for the determination of simple phenols were carried out by col-orimetric techniques with visual color matching. Smith[30] determined over 100 individual phenols using 6 different reagents (see Tables 2 to 4). The author also used derivatization technique, using tretracyanoethylene, prior to formation of color complexes for 131 phen-ols.[31] These simple colorimetric tests using these techniques could prove quite useful in checking out spills or other accidental contamination of the environment. Simple field kits

Table 2
COLORIMETRIC TECHNIQUES FOR SIMPLE PHENOLS

Phenol structure: positions 5, 6 / 4, 3, 2 with —OH

No.		Ferric chloride in ether-pyridine	Ferric chloride in ether	Reagent — Quinone chloroimide	Sodium periodate	Potassium permanganate	Ammoniacal silver nitrate
1	No substit.	V ring	GB ring	dBU(3 —15m)	dBR(30 s)	IY(i)	IBR(20 m)
2	$CH_3(2)$	V ring	B ring	dBU(15 s — 2m)	IBR(15 s)	Y(i)	IBR(20 m)
3	$CH_3(3)$	V ring	BG ring	BuG(1 — 20 m)	IBY(15 s)	W(i)	Neg.
4	$CH_3(4)$	V ring	B ring	Neg.	IBR(10 s)	W(5 s)	Neg.
5	$C_3H_{17}(4)$	GrV ring	GrB ring	Neg.	IBR(15)	W(1 m)	Neg.
6	$tert.-C_4H_9(4)$	GrV ring	Neg.	Neg.	GY(1 m)	W(30 s)	Neg.
7	$tert.-C_5H_{11}(4)$	GrV ring	GrV ring	Neg.	BGY(30 s)	W(30 s)	IGr(30 m)
8	$CH_3(2,3)$	Gr ring	BYGr ring	dBU(5 — 30s)	IBR(30s)	W(10 s)	IGr(10 m)
9	$CH_3(2,4)$	GrV ring	BY ring	IBU(2 m)¹	dBR(2 s)	W(5 s)	BuGr(2 m)
10	$CH_3(2,5)$	GrV ring	Y ring	dBu(5 — 30 s)	IBR(10 s)	W(10 s)	IGr(6 m)
11	$CH_3(2,6)$	Gr ring	BY ring	dBuV(10 s)	GY(5 s)	W(10 s)	GrG(2 m)
12	$CH_3(3,4)$	GrV ring	BY ring	Neg.	IBR(5 s)	W(10 s)	GrBIV(5 m)
13	$CH_3(3,5)$	Gr ring	BYGr ring	IBuG(30 s)	IY(1 m)	W(10 s)	Neg.
14	$iso-C_3H_7(2)CH_3(5)$	GrV ring	Neg.	dBu(5 — 30 s)	IBR(30 s)	W(15 s)	Gr ring(10 m)
15	$iso-C_3H_7(5)CH_3(2)$	GrV ring	Neg.	dBu(5 — 30 s)	IBY(5 s)	IY(10 s)	GrG(2 m)
16	$CH_3(2,3,5)$	GrV ring(t)	BY ring	BuG(10 s)	IBR(5 s)	W(10 s)	GrBl(10 m)
17	$CH_3(2,4,6)$	Gr ring	Gr ring	IBu(3 m)¹	dBR(5 s)	W(10 s)	BuGr(20 s)
18	$tert.-C_4H_9(2,6)CH_3(4)$	IGr ring	Neg.	Neg.	Neg.	IB(1 m)	IV(15 m)
19	$tert.-C_4H_9(2,4)CH_3(5)$	Gr ring	BY ring	IBu(1 m)	IBR(20 s)	IB(30 s)	Neg.
20	$p-HOC_6H_4C(CH_3)_2(4)$	GrV ring	BG ring	Neg.	BY(20 s)	W(20 s)	IGr ring(30 m)
21	$HOCH_2(2)$	RV→P spot	RV→BuG→Gr spot	BuG(30 s — 3m)	BY(10 s)	Y(i)	Neg.
22	$HOCH_2(3)$	V→P ring	B ring→BY spot	BuG(30 s — 2 m)	IB(10 m)	W(5 s)	IGr(17 m)
23	$HOCH_2(4)$	GrV→P ring	B ring→BY spot	BuG(15 s — 2 m)	IBY(30 s)	W(5 s)	IB ring(7 m)
24	$HOCH_2(4)CH_3(2,5)$	GrV ring	G→BY ring→B spot	dBu(3 s — 1 m)	IBY(1.5 m)	W(2 s)	BBl ring(1 m)
25	$HOCH_2(4)CH_3(2,6)$	Gr ring	C ring→spot	dBuV(1 s)	BR(5 s)	W(5 s)	V(1 m)
26	$CH_2 = CHCH_2(2)$	V ring	BY ring	dBu(10 s — 2 m)	RB(10 s)	Y(10 s)	Neg.
27	$CH=CHCOOH(4)CH_3O(2)$	IBR ring→GrB spot	O→B spot	dBuV(5 s — 1 m)	RB(5 s)	B(5 s)	GrGP(2 m)
28	$C_6H_5(2)$	V ring	V→GrB ring	GBu(30 s — 20 m)	BR(30 s)	dBY→P(15 s)	Neg.
29	$C_6H_5(4)$	GrV ring	B ring	Neg.	Precipitate	W(40 s)	IGr ring(40 m)

No.	Compound						
30	o-HOC$_6$H$_4$(2)	V ring→GrV spot→BP spot	BGr ring	BuV(2 — 5 h)	BBl(i)	GrBlY(10 s)	Neg.
31	Cl(2)	V ring→spot	Neg.	BuG(10 m — 2 h)	IY(3 m)	Y(5 s)	Neg.
32	Cl(3)	V ring→spot	BY ring	BuG(10 m — 2 h)	IY(3 m)	W(2 s)	Neg.
33	Cl(4)	V ring	BY ring	Neg.	IY(2 m)	W(25 s)	Neg.
34	Cl(2,4)	V ring→spot	Neg.	Neg.	IY(5 m)	W(25 s)	Neg.
35	Cl(2,4,5)	V spot	V ring(t)	Neg.	IY(8 m)	W(1 m)	Neg.
36	Br(2,4,6)	P ring→PV spot	RB ring(t)	Neg.	Neg.	W(50 s)	O ring(i)
37	Cl(2,3,4,5,6)	P ring→RV spot	RV ring(t)	Neg.	Neg.	B(50 s)	Y→O(i)
38	Cl(2)CH$_3$(5)	V ring→spot	Neg.	BuG(2 m — 1 h)	IY(3 m)	W(10 s)	Neg.
39	Cl(4)CH$_3$(3)	V ring→spot	YG ring	Neg.	IY(3 m)	W(20 s)	Bl ring(6 m)

1) These compounds contained impurities which could not be removed completely.

From Smith, B., Persmark, U., and Edman, E., *Acta Chem. Scand.*, 17, 709, 1963. With permission.

Table 3
COLORIMETRIC TECHNIQUES FOR SIMPLE PHENOLS

No.		Ferric chloride in ether-pyridine	Ferric chloride in ether	Reagent Quinone chloroimide	Sodium periodate	Potassium permanganate	Ammoniacal silver nitrate
40	CHO(2)	P spot	RV→dV spot	GY(2 h)	Neg.	IY(5 s)	IGrY(i)
41	CHO(3)	P ring	B ring	1G(1 h)	Neg.	W(5 s)	Neg.
42	CHO(4)	RV→P ring	dB ring→spot	Neg.	Neg.	W(5 s)	Neg.
43	COCH$_3$(2)	P spot	V→BuV spot	BuG(45 m — 2 h)	Neg.	IBY(10 s)	Neg.
44	COCH$_3$(4)	RV→P ring	BY spot	Neg.	Neg.	W(10 s)	Neg.
45	COC$_2$H^5(4)	RV→P ring	dB ring	Neg.	Neg.	W(10 s)	Neg.
46	COC$_2$H$_5$(4)CH$_3$(2,5)	V ring	IBY ring	Neg.	Neg.	W(30 s)	IBY ring(20 m)
47	COOH(2)	V→P spot	V→B spot	GY(10 s — 2 m)	Neg.	W(10 s)	Neg.
48	COOH(3)	IB spot	Y spot	BuG(30 s — 2 m)	Neg.	W(3 s)	Neg.
49	COOH(4)	IOY spot	dOY spot	IGBu(3 — 10 m)	Neg.	W(10 s)	Neg.
50	COOCH$_3$(2)	P ring→spot	V→BIY→BuV spot	IBu(15 — 45m)	Neg.	W(20 s)	Neg.
51	COOC$_4$H$_9$-iso(2)	P ring→spot	RV→BIY→BuV spot	IBu(10 — 45 m)	Neg.	IBY(20 s)	IBR(7 m)
52	COOCH$_2$C$_6$H$_5$(2)	P ring→spot	V→Y→BuV spot	IBu(10 — 40 m)	Neg.	YB(1.5 m)	Neg.
53	COOC$_6$H$_5$(2)	P ring→spot	V→Y→BuV spot	BuG(7 — 30 m)	Neg.	IYB(1 m)	IGr ring(7 m)
54	COOC$_6$H$_5$(4)	RV ring	BY ring	Neg.	Neg.	W(1 m)	Neg.
55	OCOCH$_3$(3)	PB ring→spot	RV→BV spot	P(1 — 2 m)	Y(1 m)	W(5 s)	GrBI(45 s)
56	CH$_3$(2)	V ring	BuG→BuBI spot	dBu(15 s — 3m)	dBR(2 s)	dBY→P(2 s)	BI(10 s)
57	CH$_3$(3)	RV ring→spot	RV ring→spot→BIB spot	dBu(15 s — 2 m)	BY(20 s)	W(2 s)	Gr(5 m)
58	CH$_3$(4)	BuG→GrV spot	IB ring→spot	dBuG(30s — 2 m)	BR(2 s)	W(2 s)	BuBI(2 s)
59	CH$_3$O(2,6)	BuGr ring→spot	BuG→BR spot	dBu(i — 15 s)	dBR(i)	dBY→BR(2 s)	BuBIO(i)
60	C$_6$H$_5$CH$_2$O(4)	IGrV spot	BY ring	BuG(20 s — 3 m)	BR(2 s)	W(10 s)	BuBI(7 m)
61	C$_6$H$_5$O(2)	V ring→spot	Neg.	BuG(1 — 5 m)	BR(10 s)	IBY→1BR(15 s)	Gr(23 m)
62	C$_6$H$_5$O(4)	GrV ring	B ring	IBuG(2 — 40 m)	BR(2 s)	W(15 s)	BI(2 m)
63	NO$_2$(2)	BR ring→spot	BR ring(t)	Neg.	Neg.	Y(10 s)¹)	BY(i)
64	No$_2$(3)	BR ring→spot	IBY spot	Neg.	Neg.	W(10 s)	OY(i)
65	NO$_2$(4)	BR ring→spot	IB ring(t)	Neg.	Neg.	W(20 s)	GY(i)
66	NO$_2$(2,4)	O spot	O ring(t)	Neg.	Neg.	RY(1 m)¹)	IY(i)
67	NO$_2$(2,4,6)	GY spot	BYR ring→spot	Neg.	Neg.	dBY(20 s)¹)	GY(i)

¹) The ether solutions of these phenols produced yellow spots on filter paper.

From Smith, B., Persmark, U., and Edman, E., *Acta Chem. Scand.*, *17*, 709, 1963. With permission.

could be put together for on-site testing of contaminated areas, dump sites, or phenol contamination of open waters during spills or other accidents involving phenols.

Houghton and Pelly[32] described a colorimetric method suitable for determination of phenols in water at concentrations of 10 ppb and above. The procedure involves a 200 ml aliquot of water sample and the use of diamine reagent. The resultant solution is titrated with sodium hypochlorite until the pale pink tinge disappears and the solution is starting to turn blue, indicating the formation of indophenol. The resultant solution is then extracted with carbon tetrachloride. The quantity of phenol in the sample is determined by visual color matching of the anhydrous carbon tetrachloride extract with the known quantities of phenol standards passed through the same procedure.

A method involving extraction procedures for concentration of phenols and subsequently determination of trace quantities of phenols by 4-aminoantipyrine (4-AAP), UV spectrophotometry, and/or spectrofluorometry was described by Afghan et al.[23] The sensitivities of these procedures for phenol, cresols, and selected chlorophenols are also provided. A new method for preservation of phenols in natural water is described which could preserve the samples for a period of 2 to 3 wk without any noticeable loss from biodegradation, chemical transformation and/or absorption. These methods have been evaluated in terms of sensitivity, selectivity, precision, and ability to determine different phenols. The 4-AAP method was suitable for determination of phenol, however, direct UV spectrophotometry was applicable to phenol, cresols, and chlorinated phenols. The solvent extraction fluorimetric technique was found to be the most sensitive and accurate for the determination of phenol and cresols; the detection limit is 0.1 μg/l per individual phenols. The use of acid as preservative was recommended and the samples were collected and stored in brown glass bottles.

F. Chromatographic Techniques

A survey of chromatographic separations of phenols using gel, absorption, and ion exchange chromatography is reported by Churacek and Coupek.[33] This publication provides a detailed discussion on the different procedures applicable to the separation of phenols including the type of chromatographic technique, compounds analyzed, sorbent, eluent, and tabulated references for individual separations.

The most widely used current methods of analysis involve the use of gas chromatography and liquid chromatography. The gas chromatographic methods utilize both packed column-flame ionization detection and capillary column-electron capture detection for separation and quantitative analysis. The majority of gas chromatographic methods rely on a derivatization step to convert phenols to more volatile species enabling the separation by gas chromatography.

The majority of liquid chromatographic methods employ reverse phase column with UV and/or electrochemical detection for the analysis of phenols. Ion-pair high performance liquid chromatography is also used to analyze phenolic compounds. Tomlinson et al.[34] has provided a detailed discussion on ion-pair high performance liquid chromatography. Factors which can affect solute retention in ion-pairing systems are discussed. Paired ion chromatography allows ionic compounds to be separated by reverse phase chromatography while eliminating problems of precise pH, temperature control, reproducibility, and short column life associated with ion exchange.[35]

High pressure liquid chromatography (HPLC) though not as popular as gas chromatography, offers a great potential for the analysis of phenols, other semivolatile and nonvolatile compounds for several reasons. Because the mobile phase is a liquid and thus often similar in nature to the sample matrix or final aliquot for analysis, a larger variety of samples and size can be handled compared to gas chromatography. The separation power of HPLC is potentially greater than that of GC in that two parameters affecting separation, the stationary and mobile phases, can be varied over a wide range of polarities and chemical functionalities.

Table 4

COLORIMETRIC TECHNIQUES FOR SIMPLE PHENOLS

No.	Phenol	Ferric chloride in ether-pyridine	Ferric chloride in ether	Quinone chloroimide	Sodium periodate	Potassium permanganate	Ammoniacal silver nitrate
				Reagent			
68	OH(2)	BuR→V spot	BuG spot	P(5 — 30 s)	BBI(i)	IB(i)	BuBI(i)
69	OH(3)	V→BR ring	B ring→spot	P(5 — 30 s)	IBY(3 m)	W(i)	BI(1 m)
70	OH(3)CH₃(5)	V→BR ring	BY ring→spot	P(5 — 10 s)	BY(2 m)	W(3 s)	RB(20 s)
71	OH(4)	GrV ring	BR ring→Bu spot	BR (30 s — 2 m)	BR(i)	W(i)	BuBI(i)
72	OH(2,3)	Bu→BuV spot	V→GrV spot	BR(5 — 30 s)	dBR(i)	Y(i)	BBI(i)
73	OH(3,5)	V→BR ring	IBY spot	P(5 — 30 s)	IBY(3 m)	W(i)	GrBI(2 m)
74	OH(2)CHO(4)	GrG→IBu→P→PGr spot	Gr→GrG→BIG spot	—	dBR(i)	IY(1 s)	GBI(i)
75	OH(2)COOH(4)	GGr→BuV spot	GrG→Gr spot	Y or GY(15 s)	dBR(2 s)	W(i)	BBI(i)
76	OH(2)CH₃O(3)	Bu→BuV spot	BuV→BuBI→GrBI spot	B(10 s)	dBR(i)	dY→GrG(i)	BI(i)
77	OH(3)CHO(4)	P spot	BIV spot	P(5 — 30 m)	Neg.	W(10 s)	BR(15 m)
78	OH(3)COCH₃(4)	P ring→spot	B1V spot	P(45 m)	Neg.	W(10 s)	Neg.
79	OH(3)COOH(5)	IB spot	IBY spot	RV(3 — 30 s)	Neg.	W(i)	IB ring(7 m)
80	OH(4)COOH(2)	Bu→V spot	Bu→BuG→GrG spot	IBR(2 s — 2 m)	RB(3 s)	W(2 s)	BI(i)
81	OH(4)CH₃O(2)	BuGr spot	PBV spot	RV(30 s — 3 m)	BY(1 s)	IY(2 s)	GBI→BBI(i)
82	OH(2,3)COOH(5)	dBu→dV spot	dBu→BuG→BuBI spot	See below¹)	dBR(i)	W(i)	BI(i)
83	NH²(2)	P→BY→BP spot	RV ring→dV spot	YGB(5 s)	BBI(i)	BY(i)	BIG(i)
84	NH₂(3)	Gr→B ring	RB spot→ring	dV(5 — 15 s)	BBI(5 s)	IBY(i)	GrB(3 m)
85	NH₂(4)	IB spot	G ring→GrBIG spot	RV(i — 1 m)	dRV or BV(i)	W(i)	BIBuGrB(i)
86	CH₃O(2)HOCH₃(5)	V→PGr ring→spot	G→BuG→Bu→BI ring	dBu(10 s — 2 m)	dR(3 s)	W(2 s)	BuBI(50 s)
87	CH₃O(2)HOCH₃(6)	RV ring→spot	Bu→BuG→Gr spot	dBu(10 s — 2 m)	BR(3 s)	BY→BR(2s)	BBI(3 s)
88	CH₃O(2)C₂H₅OCH₃(4)	Bu→V spot	BuG→GrG spot	IRB(1h)	RB(3 s)	W (3 m)	BuBI(25 s)
89	CH₃O(2)CH₃CH=CH(4)	BuG→BuGr spot	BuG→GY spot	Neg.	RB(i)	W(5 s)	BuBIGrt(6 s)
90	CH₃O(2)CH₂=CHCH₂(4)	Bu ring→spot→BuGr spot	BuG→BuBI spot	Neg.	RB(2 s)	IY→IBR(5 s)²)	BuBI(30 s)
91	CH₃O(2)Br(5)	V ring→spot	Bu→BuG ring→GrG spot	BuG(45 s — 10 m)	BR(5 s)	W(10 s)	GrBI(2 m)
92	CH₃O(2)Br(6)	RV ring→spot	Bu→BuG→GrY spot	BuG(30 s — 10 m)	dBR(5 s)	BY→RB(5 s)	GrB(40 s)
93	CH₃O(2)CHO(4)	V ring→P spot→V spot	Gr→B ring→GrBI spot	Neg.	IBR(6 m)	W(5 s)	GrV(9 m)

94	$CH_3O(2)CHO(5)$	PV spot	B ring→spot	IGY(2 m)	BY(3 m)	W(5 s)	GrV(5 m)
95	$CH_3O(2)CHO(6)$	V spot	Bu spot→BuG ring→BuBl ring	Neg.	BY(3 m)	W(30 s)	Gr ring(9 m)
96	$CH_3O(2)COCH_3(5)$	V→RV ring	GrG→dGrBu ring	GBu(2 — 10 m)	BR(1 m)	W(3 s)	BlV(5 m)
97	$CH_3O(2)COC_2H_5(4)$	PV ring→spot	Bu→BuGr ring→BuBl spot	Neg.	IBR(6 m)	W(5 s)	Gr ring(20 m)
98	$CH_3O(2)COOH(4)$	RV spot	O→B spot	IBu(3 — 10 m)	BY(4 m)	W(5 s)	Neg.
99	$CH_3O(2)COOCH_3(4)$	PV ring→spot	Bu ring→BuBl spot	Neg.	BY(3 m)	W(5 s)	IGr ring(20 m)
100	α-Naphthol	GrG→GrBl ring	B ring→BGr spot	V(10 — 30 s)	BBl(5 s)	IBY→IBR(3 s)	Bu(2 m)
101	β-Naphthol	PGr ring	lB ring	GY(30 s — 2 m)	YB(10 s)	lY(3 s)	GrBl(4 m)

¹) The alkaline solution of this phenol was grey-green. No colour change was observed on addition of the quinone chloroimide reagent.

²) The ether solution of this phenol produced a yellow spot on filter paper.

From Smith, B., Persmark, U., and Edman, E., *Acta Chem. Scand.*, 17, 709, 1963. With permission.

Explanations to Tables 2,3, and 4

Ferric chloride tests. The color obtained is, whenever suitable, only repeated once. Thus, V ring = spot means V spot. Further O = B spot means O spot = B spot, and V = P ring means V ring = P ring.

Quinone chloroimide tests. The color given is the final stable one and it corresponds to the last time notation within parentheses. The first time notation denotes the time for the first change in color. That a test is negative means that no color change was observed within 2 h. The notation 3-15 means 3 to 15 min and 5-30 s in the same way means 5 to 30 s. When only one time notation is given the solution attained its final color already at the first color change.

Sodium periodate. The color given is the color after 15 min. The time notation within parentheses denotes the time for the first color change. If no color change had occurred within 15 min the test was considered to be negative.

Potassium permanganate. The time notation within parentheses denotes the time when the red permanganate color had vanished.

Ammoniacal silver nitrate. The time notation within parentheses denotes the time when the first change in color was discernible. The color given corresponds to this time. The test was considered to be negative if no color change had occurred within 30 min.

Abbreviations used. B = brown, Bl = black, Bu = blue, C = cerise, G = green, Gr = gray, O = orange, P = pink, R = red, V = violet, W = white, Y = yellow, d = dark, h = hour(s), i = immediately, l = light, m = minute(s), s = second(s), t = transient.

Since mobile phase is a liquid, other techniques can be interfaced, both before and after the separation has been performed, allowing a wide variety of materials and matrices to be analyzed and confirmed.

Numerous methods, employing chromatographic techniques, are reported in the current literature for the analysis of phenols in environmental samples. Some of them will be discussed below.

1. High Performance Liquid Chromatographic Methods

High performance/pressure liquid chromatography is used to analyze phenols and chlorophenols. The main advantage of the technique is that it can be used without time consuming derivatization steps which are often necessary when utilizing gas chromatography. The use of small particle sizes for the solid phase and use of multidetectors has enabled analysts to achieve the necessary resolution, sensitivity and specificity.

Ogan and Katz[36] used reverse phase HPLC columns to compare the response of alkylphenols using UV and fluorescence detection. They found that the elution order of 19 alkylphenols was in the order of increasing total carbon number of the alkyl substituent. The fluorescence intensity was not strongly influenced by the chain length of the alkyl substituents, however, the fluorescence intensity decreased by alkyl substitution at several sites on the phenol ring. Fluorescence detection was generally more sensitive and selective than UV detection at 254 nm. Realini[37] was able to extract phenols, alkyl phenols, chlorophenols, and nitrophenols by adding tetrabutyl ammonium chloride to the water sample and extracting the phenols at pH 14. The separation was achieved using a 30 cm × 4 mm Micro Pak MCH 5 μ C_{18} column with UV detection at 254 and 280 nm. Recoveries of phenols ranged from 75 to 99%. The standard response ratios of absorbance at 280/254 nm and the correct retention time, were used to confirm the presence of individual phenols.

Ion pair high performance liquid chromatography was used by Hoffsommer et al.[38] to analyze trace levels of alkyl-, hydroxy-, amino- and nitrophenols in aqueous samples. Detection limits of 0.03 mg/l was reported using UV detection at 254 nm. Wegman and Wammes[39] determined nitrophenols in water samples using a reverse phase HPLC with ion-pairing technique. The best results were obtained when 0.48% w/v cetrimide were added to a 1-l water sample, adjusted to pH 8 with phosphate buffer and extracted with methylene chloride. The concentrated extracts were analyzed using ODS 5 μm column and mobile phase containing 3:1 v/v methanol/water. The detection limit, using 1-l water samples, was found to be 0.1 μg/l at 365 nm.

HPLC with electrochemical detection offers an excellent method for analysis of phenols and related compounds. An excellent review on electrochemical detection and application notes are provided by Bioanalytical System Inc.[40] Phenol compounds in environmental samples have been determined by number of authors using electrochemical detection.[41-43]

2. Gas Chromatographic Methods

Corcia et al.[44] resolved the priority pollutant phenols using a column packed with 0.4% trimesic acid and 1% PEG 20M. The GC-FID was used to resolve and carry out quantitative analysis. Eleven free phenols were resolved within 20 min. However, the majority of the methods for analysis of phenols are based on derivatization of parent phenols followed by appropriate cleanup and analysis by GC-FID or GC-ECD. Acetylation and ether formation using a wide variety of reagents are also used for dervitization.

Direct acetylation of chlorophenols and nitrophenols was the basis for a chromatographic method proposed by Mathew and Elzerman.[45] The derivatives were separated using a 6-foot packed column of 1% SP-1240 DA and analyzed using a flame ionization detector. The recovery for the entire method was over 90% for each phenol at the ppb level. Chau and Coburn[22] analyzed several substituted phenols by acetylation in carbonate solution. This

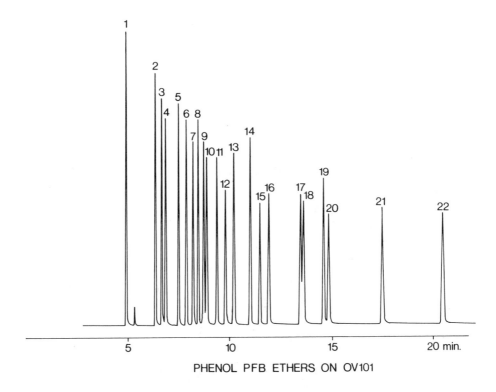

FIGURE 4.   Chromatogram of 22 resolved chlorophenol PFB ether derivatives.

method is currently being used as a standard in many routine laboratories for the analysis of environmental samples. Lamparski and Nestrick[46] reported a method for analysis of trace levels of phenols in industrial effluents and natural waters. This method is based on the formation of heptafluorobutyl derivatives of phenols and analysis by GC-EC using a packed nitro-DEGS column with temperature programming. Sixteen phenols were resolved within 12 min. Lehtonen[47] converted phenols to their 2,4-dinitrophenyl ethers prior to analysis by capillary column gas chromatography with electron-capture detection. Twenty-five phenols were separated using SP2100 using temperature programming. A detection limit, using splitless mode, of 10 to 90 pg was achieved using the above technique. Recently Lee et al.[29,49] have developed methods for analysis of the priority phenols and substituted chloro-phenols using pentafluorobenzyl (BFB) ethers as a derivatization agent. They have also provided the separation of 32 phenols using six different columns — excellent resolution is obtained using a capillary OV-101 column. A method for the analysis of phenol and 21 chlorinated phenols was also developed by Lee et al.[29] using pentafluorobenzyl ether deriv-atives. Detection limit down to 0.1 ppb using a 1-l water sample is achieved by the authors with recoveries of 80%. This method is stated as being suitable for simultaneous screening of nonchlorinated and chorinated phenols. Figure 4 and Table 5 shows the elution order and response of 22 chlorophenol PFB ether derivatives.

G. Mass Spectrometry and High Resolution Gas Chromatography/Mass Spectrometry

Mass spectrometry (MS) and high resolution gas chromatography/mass spectrometry (HRGC/MS) are used extensively for identification and quantitative analysis of phenols in environmental samples. The use of the technique is escalating with time. The detection limits as well as degree of confidence for identification is increasing with time as new instrumentation and techniques are developed each year.

The choice of a suitable column, to separate parent phenols or their derivatives, is most

Table 5

RETENTION TIMES AND RELATIVE RETENTION TIMES FOR PHENOL PFB ETHERS ON SIX COLUMNS

No.	Parent phenol	3% OV-1	3% OV-17	Ultrabond 20M	OV-101 FSCC	SE-54 FSCC	Carbowax 20M FSCC
1[a]	Phenol	1.30 (0.54)	1.17 (0.43)	1.14 (0.57)	5.07 (2.45)	8.96 (3.04)	4.74 (2.66)
2[a]	2-Chloro-	2.30 (0.95)	2.25 (0.83)	2.26 (1.14)	6.54 (3.16)	11.28 (3.83)	6.45 (3.63)
3	3-Chloro-	2.59 (1.07)	2.42 (0.90)	2.55 (1.28)	6.86 (3.32)	11.66 (3.96)	6.73 (3.78)
4	4-Chloro-	2.74 (1.13)	2.60 (0.96)	2.84 (1.43)	7.05 (3.41)	11.96 (4.05)	7.10 (3.99)
5	2,6-Dichloro-	3.56 (1.47)	3.42 (1.27)	2.95 (1.48)	8.03 (3.89)	13.49 (4.58)	7.27 (4.09)
6	2,5-Dichloro-	4.08 (1.68)	4.28 (1.59)	4.59 (2.31)	8.68 (4.20)	14.30 (4.85)	8.86 (4.98)
7[a]	2,4-Dichloro-	4.34 (1.79)	4.36 (1.62)	4.84 (2.44)	8.96 (4.34)	14.68 (4.98)	9.12 (5.13)
8	3,5-Dichloro-	4.49 (1.85)	4.12 (1.53)	4.64 (2.33)	9.09 (4.40)	14.68 (4.98)	8.80 (4.95)
9	2,3-Dichloro-	4.96 (2.05)	5.49 (2.04)	6.26 (3.15)	9.62 (4.65)	15.64 (5.31)	10.42 (5.86)
10	3,4-Dichloro-	5.42 (2.23)	5.61 (2.08)	6.70 (3.37)	10.05 (4.86)	16.06 (5.45)	10.76 (6.05)
11[a]	2,4,6-Trichloro-	5.84 (2.41)	5.44 (2.02)	4.30 (2.16)	10.37 (5.02)	16.57 (5.62)	8.51 (4.78)
12	2,3,6-Trichloro-	6.74 (2.78)	6.97 (2.58)	6.17 (3.10)	11.20 (5.42)	17.76 (6.03)	10.25 (5.76)
13	2,4,5-Trichloro-	7.89 (3.25)	8.50 (3.15)	9.46 (4.76)	12.17 (5.89)	18.81 (6.38)	12.46 (7.00)
14	2,3,5-Trichloro-	8.05 (3.32)	8.42 (3.12)	9.47 (4.76)	12.20 (5.90)	18.86 (6.40)	12.48 (7.02)
15	2,3,4-Trichloro-	10.03 (4.13)	11.79 (4.37)	14.74 (7.41)	13.79 (6.67)	20.94 (7.11)	15.61 (8.78)
16	3,4,5-Trichloro-	10.31 (4.25)	10.68 (3.96)	13.42 (6.75)	13.90 (6.73)	20.82 (7.07)	14.84 (8.34)
17	2,3,5,6-Tetrachloro-	11.69 (4.82)	11.97 (4.44)	9.40 (4.73)	14.77 (7.15)	22.12 (7.51)	12.46 (7.00)
18	2,3,4,6-Tetrachloro-	12.29 (5.07)	12.54 (4.65)	9.69 (4.87)	15.03 (7.27)	22.45 (7.62)	12.77 (7.18)
19	2,3,4,5-Tetrachloro-	17.24 (7.11)	20.59 (7.63)	24.23 (12.19)	17.80 (8.61)	25.59 (8.68)	19.03 (10.70)
20[a]	PCP	24.26 (10.00)	26.97 (10.00)	19.88 (10.00)	20.67 (10.00)	29.47 (10.00)	17.79 (10.00)
21[a]	2,4-Dimethyl-	2.30 (0.95)	2.00 (0.74)	1.74 (0.88)	6.60 (3.19)	11.20 (3.80)	5.39 (3.03)
22	2,3-Dimethyl-	2.64 (1.09)	2.30 (0.85)	1.97 (0.99)	6.95 (3.36)	11.74 (3.98)	5.74 (3.23)
23	2-Chloro-5-methyl-	3.22 (1.33)	3.04 (1.13)	2.98 (1.50)	7.66 (3.71)	12.81 (4.35)	7.11 (4.00)
24[a]	4-Chloro-3-methyl-	3.86 (1.59)	3.59 (1.33)	3.66 (1.84)	8.38 (4.05)	13.76 (4.67)	7.84 (4.41)
25	2-Chloro-4-tert-butyl-	7.30 (3.01)	6.84 (2.54)	6.05 (3.04)	11.63 (5.63)	17.95 (6.09)	9.07 (5.10)
26	Thiophenol	1.89 (0.78)	1.82 (0.67)	1.54 (0.78)	5.91 (2.86)	10.44 (3.54)	5.36 (3.01)
27	4-Cyano-	5.22 (2.15)	7.57 (2.81)	15.24 (7.67)	9.85 (4.77)	16.10 (5.46)	16.38 (9.21)
28[a]	2-Nitro-	5.08 (2.09)	7.54 (2.80)	14.10 (7.09)	9.92 (4.80)	16.16 (5.48)	15.82 (8.89)

29[a]	4-Nitro-	8.29 (3.42)	12.12 (4.49)	24.83 (12.49)	12.62 (6.11)	19.56 (6.64)	20.31 (11.42)
30	3-Methyl-4-nitro-	10.14 (4.18)	14.35 (5.32)	26.44 (13.30)	13.95 (6.75)	20.92 (7.10)	20.15 (11.33)
31[a]	2,4-Dinitro-	22.45 (9.25)	51.80 (19.21)	—[b]	21.05 (10.18)	29.32 (9.95)	—[c]
32[a]	4,6-Dinitro-2-methyl-	21.29 (8.78)	37.50 (13.90)	—[b]	19.83 (9.59)	28.38 (9.63)	—[c]

[a] U.S. EPA priority pollutants.

[b] Did not chromatograph at 155°C in 60 min.

[c] Did not chromatograph under the conditions used.

Table 6
GC/MS CONFIDENCE LEVELS FOR COMPOUND IDENTIFICATION

Level I	Computer interpretation. The raw data generated from the analysis of samples are subjected to computerized deconvolution/library search. Compound identification made using this computerized approach has the lowest level of confidence. In general Level I is reserved for only those cases where verification of tentative compound identification is the primary intent of the qualitative analysis.
Level II	Manual interpretation. The plotted mass spectra are manually interpreted by a skilled interpreter and compared to those spectra in a data compendium. In general a minimum of 5 to 8 masses and intensities ($\pm 20\%$) should match between the unknown and library spectrum. This level does not utilize any further information such as retention time since, for many compounds, the authentic compound may not be available for establishing retention times.
Level III	Manual interpretation plus retention time/boiling point of compound. In addition to the criteria attained under Level II, the retention time of the compound is compared to the retention time of a standard which has been derived from previous chromatographic analysis. Also the boiling point of the identified compound is compared to the boiling points of other compounds in the near vicinity of the one in question when a capillary coated with a nonpolar phase has been used.
Level IV	Manual interpretation plus retention time and matching spectrum of authentic compounds. Under this level, the authentic compound has been chromatographed on the same capillary column using identical operating conditions and the mass spectrum and retention times should correspond $\pm 10\%$. The ultimate confirmation of identification consists of coinection of authentic standard and observation of no change in mass spectrum or retention time.
Level IVa	Independent confirmation techniques. This level utilizes other physical methods of analysis such as GC/Fourier transform IR, GC/high resolution mass spectrometry, or NMR analysis in the absence of authentic standard. This level constitutes the highest degree of confidence in the identification of organic compounds when no standard is available.

important when analyzing phenols by mass spectrometry. Electron impact and chemical ionization have been employed to analyze and confirm phenols. The amount of material required for GC/MS should be in the order of 25 ng for each component in order to obtain a good response for full scan spectra, however, it can be considerably less if single ion or multiple ion monitoring is used.

HRGC/MS can be utilized for confirmation after the samples are first analyzed using GC-ECD or GC-FID. The confirmation is carried out using similar GC conditions as those used for GC-ECD or GC-FID runs. The retention times and spectra of each peak are observed and a library is made using this data. The sample peaks within a given window ($\pm 1\%$ of the GC retention time) are then searched against the user created library. A match of sample and standard at the same retention time constitutes a confirmation. Table 6 from the U.S. E.P.A. master analytical scheme outlines five levels of confidence that can be used for GC/ MS compound identification.[49]

A few examples of the use of mass spectrometry and related techniques are given below.

Kueh and Dougherty[50] reported the analysis of pentachlorophenol in the environment by negative chemical ionization mass spectrometry. The method employed steam distillation and a concentrated sample extract was analyzed by direct probe mass spectrometry using isobutane doped with 2% oxygen as the reagent gas. A detection limit of 10 ng was stated.

In the authors' laboratory[51] both electron impact (EI) and chemical ionization (CI) GC/ MS techniques are used for quantitative and confirmatory analysis of phenols. Figures 5 to 7 show three reconstructed ion chromatograms (RICs) of the priority pollutant phenols using a Finnigan 4000 GC/MS and Super Incos software. Table 7 shows the results of two runs, the first EI and the second the simultaneous acquisition of $-$ve and $+$ve CI data (PPINICI). The above figures show redrawn chromatograms for three different runs. The $+$ve ion EI and $-$ve ion CI were made using similar conditions and a standard mixture supplied from Radian Corp.[52] The third chromatogram, $+$ve CI was run under different GC/MS conditions using standard mixture supplied by Supelco.[53]

139

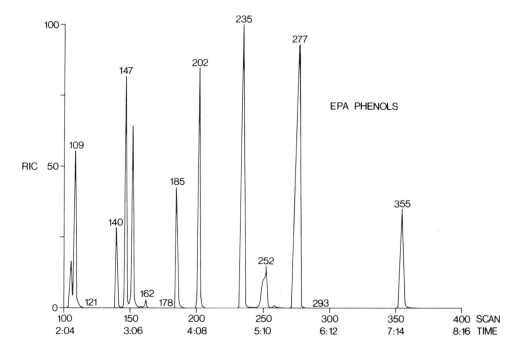

FIGURE 5.   GC/MS EI run of phenols.

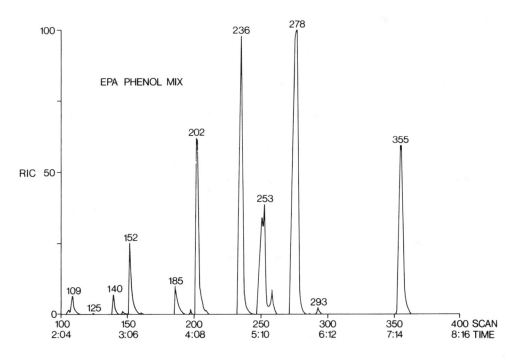

FIGURE 6.   GC/MS −ve CI run of phenols.

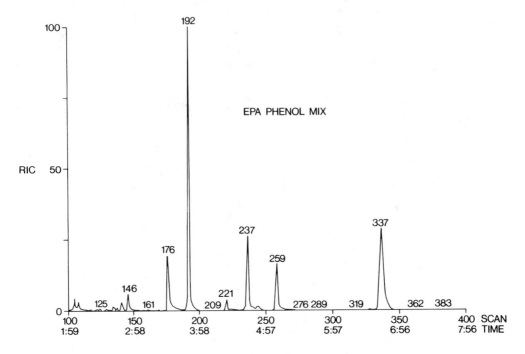

FIGURE 7.   GC/MS +ve CI run of phenols.

The above data indicates that both EI and CI (+ ve and − ve) modes of mass spectrometry are the desirable techniques for the analysis of free phenols and their derivatives provided that suitable GC conditions are used to obtain the necessary resolution.

IV. STANDARD METHODS FOR ANALYSIS OF PHENOLS

The choice of method for analysis will depend upon the ultimate use of the data, the required sensitivity, the availability of the techniques being used in the laboratory carrying out analysis, and finally, the actual cost of the analysis to the user.

For monitoring programs requiring data of submicrogram per liter levels of phenol in water, automated methodology, Goulden et al.[54] is recommended as a low cost method of analysis. For specific phenols, a very simple colorimetric method by Smith[30,31] could be used that would have low equipment costs and would be very applicable for field use. Technical surveys or research studies may require specific or target phenolic compounds at trace levels requiring specific methods that are more comprehensive. To meet specific or specialized requirements, the methods should employ appropriate preservation, concentration, clean up, and analytical instrumentation that will provide chromatographic analysis at trace levels. The methodology should be extensively validated and documented in order to provide the user with precise and accurate results that are within defined and demonstrated confidence intervals at specified levels.

The following methods are extensively validated and documented to provide precise and accurate results for specified phenols.

The most comprehensive method to date for phenols in water is the EPA document[49] entitled "Master Analytical Scheme for the Analysis of Organic Compounds in Water". For specific chlorophenol analysis in water, the most comprehensive, state-of-the-art methodology, is provided by Lee et al.[29,48] EPA Method 604, and the method by Ribick et al.[28] are also extensively used for routine monitoring of phenols.

Table 7

RESULTS OF EI, −VE CI, +VE CI MS DATA FOR PRIORITY POLLUTANT PHENOLS

Parameter	Figure 5		Figure 6		Figure 7	
	Scan no.	EI +ve ion	Scan no.	CI −ve ion	Scan no.	CI +ve ion
Phenol	106	94	—	—	105	95
2-Chlorophenol	109	128;130;92	107	142	108	129
2-Nitrophenol	140	139;109	135	139	135	140
2,4-Dimethylphenol	147	122;107	141	121;106;134	141	123
2,4-Dichlorophenol	152	162;98;126	146	162;142;176	146	163
4-Chloro-3-methyl-phenol	185	142;107	176	156;177;141	176	143
2,4,6-Trichlorophenol	202	196;97;132;160	191	196	192	199;163
2,4-Dinitrophenol	235	184;91;154;107	220	167;154;184	221	185
4-Nitrophenol	252	139;109;93	236	139	237	140
2-Methyl-4,6-dinitrophenol	277	198;105;121;168	258	181;198;168	259	199;227
Pentachlorophenol	355	266;264;165	336	230;196	337	267

The following methods provide an easy introduction into phenol analysis. The methods are intended as a starting point for the analyst on which he can build his skills.

Phenolics (Colorimetric-Automated 4-Aminoantipyrine)

I.   Scope and application
  A.   This automated colorimetric method is suitable for the determination of the concentration of phenolic compounds in unpolluted or slightly polluted natural waters. Phenolic compounds with a substituent in the para position usually do not produce color with 4-aminoantipyrine. However, the para substituents of the phenol, such as carboxyl, halogen, hydroxyl, methoxyl, or sulfonic acid group, are expelled. These phenolic compounds produce color with 4-aminoantipyrine. The working range is 1 to 30 μg/l using phenol as a standard.
  B.   The detection limit for this method is 1 μg/l phenol.

II.   Principle and theory
  A.   Phenolic materials are steam-distilled from the sample in an automated system under acid conditions. The distillate is reacted with 4-aminoantipyrine at pH 10 ± 0.2 in the presence of potassium persulfate to form a colored antipyrine dye, which is extracted into chloroform. Absorbance of the chloroform extract is measured at 480 nm.

III.   Interference
  A.   Steam-distillable nonphenolic compounds (e.g., formaldehyde), which will also react with 4-aminoantipyrine and give a higher apparent phenol concentration, interfere.
  B.   Oxidizing agents such as chlorine interfere with the color-forming reaction and can oxidize phenols in the sample. Excess oxidizing agents, detected by starch iodide test paper, should be removed after sampling by addition of ferrous sulfate or sodium arsenite.
  C.   Sulfur compounds interfere with the color-forming reaction. Interference from H_2S or SO_2 will be removed if the sample is preserved (see IV.A below).

IV.   Sampling procedure and storage
  A.   Samples should be taken in glass bottles with Teflon®-lined caps. Due to biochemical oxidation, preservatives are required to maintain the original phenol concentration of the sample. Immediately after sample collection, the following steps should be carried out.
       1.   Add copper sulfate to a concentration of 1.0 g $CuSO_4 5H_2O$ per liter of sample to inhibit bacterial degradation and also to precipitate mercaptans and sulfides which would otherwise interfere.
       2.   Acidify the sample with phosphoric acid (H_3PO_4) to a pH of less than 4 to ensure adequate copper ion activity and to liberate any dissolved hydrogen sulfide or sulfur dioxide.
       3.   After preservative addition, the sample should be stored cold (5 to 10°C) and in the dark. Analysis should be done within 24 h of sampling.

V.   Sample preparation
  A.   Sample aliquots either should be free from turbidity or should be filtered through a glass-fiber filter.

VI.   Apparatus
    A.   Technicon Auto Analyzer unit consisting of:
        1.   Industrial sampler (modified to eliminate aspiration of air while the probe is between sample cup and wash solution).
        2.   Industrial sampler (modified to eliminate aspiration of air while the probe is between sample cup and wash solution).
        3.   Manifold.
        Note: Glass and Teflon® transmission lines should be used wherever possible.
        4.   Proportioning pump, Carlo-Erba model 1512/20.
        5.   Heating bath.
        6.   Constant temperature circulating pump, Haake FJ, or equivalent.
        7.   Colorimeter equipped with 50-mm flow cell and 480-nm filters.
        8.   Range expander.
        9.   Recorder.

    B.   Distillation apparatus (see Section XI.B below) constructed of borosilicate glass. The inner heating tube is constructed of silica. The vertical section of the side arm, which is connected to the bottom of the main distillation tube, is wrapped with heating tape and warmed by the application of approximately 10 V AC through a variable transformer. Approximately 110 V AC is applied to the heating tube. All parts of the system carrying heated vapor are wrapped with asbestos insulation.
    C.   Variable transformers (two).
    D.   Extractor. The extractor consists of glass tubing, 3 mm I.D. and about 90 cm long, bent into a U-shape and filled with thin strips of Teflon® twisted together.
    E.   Separator. The separator is a T-piece of glass tubing, 3 mm I.D. Inside the arm through which the aqueous-air-solvent mixture is fed, a piece of Teflon® measuring 1.5 mm by 0.8 mm by 10 cm is inserted and held in place along the bottom of the tube with a spiral made of 26-gauge Chromel wire.
    F.   Displacement bottles (four). The displacement bottles are constructed from 500-ml wash bottles (Kontes K-331750 or equivalent) with modified inlet and outlet tubes as shown in Figure 5.

VII.   Reagents
    A.   Buffered 4-aminoantipyrine (4-AAP) solution: Dissolve 27 g potassium bicarbonate ($KHCO_3$), 27 g boric acid (H_3BO_3), and 35 g potassium hydroxide (KOH) in approximately 750 ml distilled water and dilute to 1 l. Dissolve 2.1 g 4-aminoantipyrine, $C_{11}H_{13}N_3O$, in this buffer solution. Prepare daily.
    B.   Potassium persulfate solution, 2.5%: Dissolve 25 g potassium persulfate, $K_2S_2O_8$, in distilled water and dilute to 1 l. Adjust the pH to approximately 11 with potassium hydroxide. Prepare daily.
    C.   Wash solution: Dissolve 1 g copper sulfate, $CuSO_4 \cdot 5H_2O$, in distilled water. Add 2 ml phosphoric acid, H_3PO_4, and dilute to 1 l with distilled water. Approximately 6 l of wash solution is required daily.
    D.   Phenol stock solution, 1000 mg/l: Dissolve 1.00 g phenol, C_6H_5OH, in distilled water (phenol-free) and dilute to 1 l. Preserve with copper sulfate solution and store in a dark bottle in the refrigerator. Prepare weekly.
    E.   Phenol intermediate solution, 10 mg/l: Dilute 10 ml of phenol stock solution (7.4) to 1 l with distilled water containing copper sulfate and phosphoric acid in the same proportions as in the wash solution (7.3).

F. Phenol standard solution, 0.10 mg/l: Dilute 10 ml of intermediate solution (7.5) to 1 l. Preserve with copper sulfate solution.

G. Phenol working standards: Prepare a series of working standards in the required concentration range by appropriate dilutions of the standard solution (7.6). Prepare daily. Preserve with copper sulfate solution.

VIII.   Procedure
A. Set the system in operation with distilled water in all lines and distillation heaters off.

B. When the flow of liquid through the distillation tube is steady, turn on the power to the column and adjust the transformers to give a distillation rate of about 80%.

C. On shutdown, procedures VIII.A and VIII.B should be followed in reverse order.

D. Run the standards and samples at 10/h using the manifold.

E. Check for and eliminate any air bubbles restricting the flow in the displacement bottles. Discard chloroform from the bottles daily.

IX.   Calculations
A. Prepare a calibration curve derived from the peak heights obtained with the standard solutions.

B. Determine the concentration of phenolics in the samples by comparing sample peak heights with the calibration curve.

X.   Precision and accuracy
A. In a single laboratory (National Water Quality Laboratory), the coefficients of variation at phenol levels of 2.6 μg/l and 3.6 μg/l phenol were ± 5.8% and ± 5.6%, respectively.

B. The percent recovery of a spike of 1 μg/l at the phenol level of 2.6 μg/l ranged from 60 to 120% with the mean of ten determinations being 94%.

XI.   References
A. Goulden, P.D., Brooksbank, P., and Day, M. B., Determination of submicrogram levels of phenol in water, *Anal. Chem.*, 45, 2430, 1973.

B. Goulden, P. D., personal communication.

C. *Standard Methods for the Examination of Water and Wastewater*, 13th ed, American Public Health Association, Washington, D.C., 1971.

Pentachlorophenol (Gas Chromatographic)

I.   Scope and application
A. This method is applicable to the determination of pentachlorophenol in natural and wastewaters. The method is not suitable for the analysis of pulp mill condensate due to the interferences of the chlorinated degradation products of lignin. As little as 0.01 μg/l pentachlorophenol can be quantitated.

II.   Principle and theory
A. Pentachlorophenol is extracted from an acidified aqueous sample with benzene. The PCP is then extracted from the benzene with a 0.1 *M* solution of potassium carbonate. This procedure extracts only acidic compounds such as halogenated phenols and substituted phenoxy acid herbicides. Neutral and basic compounds

(e.g., lindane, heptachlor, and aldrin) remain in the benzene fraction. Acetylation of the pentachlorophenol is accomplished by the addition of acetic anhydride to the carbonate solution. The pentachlorophenyl acetate is then extracted with 10.0 ml hexane. The acetylation and subsequent hexane extract involve only acidic alcohols, leaving the carboxyl acids in the carbonate solution. This method has the advantage of elimination of the herbicide interferences that occur in the diazomethane reaction.

III. Interferences
A. The chlorinated degradation products of lignin are interferences that pass through this method. In natural and wastewater, however, these are not significant. Acidification to pH less than 4.0 followed by the addition of copper sulfate inhibits bacterial degradation and liberates any dissolved hydrogen sulfide or sulfur dioxide. None of the 17 substituted phenols tested interfered with pentachlorophenol.

IV. Sampling procedure and storage
A. The samples are collected in clean 1-l glass whiskey bottles that have long necks. As alternative to the above acid preservation technique is to add a few sodium hydroxide pellets directly to the sample for preservation. The bottles are capped with clean aluminum foil and the regular cap. The preserved samples should be stored in the dark to avoid photodecomposition and at a temperature of 4°C.

V. Apparatus
A. Glassware:
1. Erlenmeyer flasks, 125 ml.
2. Separatory funnels, 250 ml.
3. Stirrers and Teflon® stirring bars.
4. 10 ml pipette.
5. Disposable pipettes, 2 ml.
6. Centrifuge tubes with stoppers, 15 ml.
7. All-glass distillation apparatus for acetic anhydride purification.
B. Gas chromatographic equipment:
1. Varian 2800 equipped with Ni-63 electron capture detectors or equivalent.
C. Columns:
1. Glass, 6 mm O.D., 1.8 m packed with 11% OV-17/QF-1 on Chromosorb Q, 80-100 mesh.
2. Glass, 6 mm O.D., 1.8 m packed with 4% OV-101 and 6% OV-210 on Chromosorb W, 80-100 mesh.
3. Glass, 6 mm O.D., 1.8 m packed with 4% OV-225 on high performance Chromosorb Q, 80-100 mesh.
D. Conditions:
Injection temperature — 210°C
Column temperature — 200°C
Detector temperature — 215 °C
N_2 flow — 30 ml/min
Chart speed — 5 min/in
Recorder — 1mV Honeywell

VI.   Reagents
    A.   Pentachlorophenol, Aldrich — 00%.
    B.   Acetone, Fisher washing grade.
    C.   Petroleum ether, Caledon pesticide grade.
    D.   Benzene, Fisher pesticide grade.
    E.   Hexane, Caledon pesticide grade.
    F.   Acetic anhydride, MCB — 99%, distilled three times.
    G.   0.1 M potassium carbonate, Analar, analytical reagent twice extracted with 100 ml of benzene by shaking in a separatory funnel.
    H.   NaOH Pellets and aristar sulfuric acid.
    I.   Purification of acetic anhydride.
        The purification of acetic anhydride is very important, since it exhibits an interfering peak in the determination of pentachlorophenyl acetate.
        1.   The following apparatus are used in the distillation:
            a.   Varian heating mantle.
            b.   60-cm Vigreux distillation column packed with prewashed glass beads or Raschig rings. The bottom of the column can be packed with the Raschig rings or loosely with precleaned glass wool. Wrap the column with glass wool during the distillation.
            c.   Water-cooled condenser and Claisen distillation head. In the first distillation, collect only the fraction above 135°C. In the second distillation, collect only the 138 to 140°C fraction. In the final distillation, with 400 ml of acetic anhydride discard the first and last 50 ml portions, collecting only the balance, 300 ml.

VII.   Procedure
    A.   The preserved sample is adjusted to pH 2 with sulfuric acid just before extraction, then 30 ml of benzene is added to 1 l of water and stirred for 45 min with a magnetic stirrer.
    B.   Remove the benzene contained in the neck of the bottle with a disposable pipette and transfer it to a separatory funnel.
    C.   Repeat the extraction with 30 ml of benzene and stir for 30 min. Transfer the benzene into the same separatory funnel.
    D.   Repeat the extraction a third time with 10 ml of benzene. Add distilled water until the benzene comes to the top of the bottle. If an emulsion forms add a few drops of saturated sodium sulfate solution, and agitate the organic layer. Transfer the benzene to the separatory funnel.
    E.   Extract the combined extract of benzene with 100 ml of 0.1 M potassium carbonate in 30-ml portions.
    F.   Transfer the aqueous extracts into a 125-ml Erlenmeyer flask after each extraction.
    G.   Prepare a series of three standards to cover the expected working range by fortifying 100 ml of 0.1 M potassium carbonate with pentachlorophenol in three 125 ml Erlenmeyer flasks.
    H.   Add 1 ml of acetic anhydride to each sample and standard.
    I.   Pipette 10 ml of hexane into each sample and standard.
    J.   Stir for 1 h at room temperature and at such a speed that the hexane comes almost to the bottom of the flask without touching it.
    K.   Make the appropriate dilutions or concentrations, if necessary, and analyze the hexane layer by gas liquid chromatography.

VIII. Calculations
A. The concentration of pentachlorophenol is determined by comparison of peak heights or areas of the samples with the peak heights or areas of the standards which were derivatized simultaneously. This can be done using calibration curves or the equation:

Pentachlorophenol

$$\mu g/l \ (ppb) = \frac{H_{sam}}{H_{std}} \times \frac{W_{std}}{V_{sam}} \times \frac{V_{sam\,ext}}{V_{std\,ext}}$$

$$\frac{V_{std\,inj}}{V_{sam\,inj}} \times 10^{-3}$$

where H_{sam} = peak height or area of sample; H_{std} = peak height or area of standard; W_{std} = weight (ng) or PCP in standard; V_{sam} = volume of sample (l); $V_{sam\,ext}$ = final volume of sample extract (ml); $V_{std\,ext}$ = final volume of standard extract (ml); $V_{std\,inj}$ = injection volume of standard (ml); and $V_{sam\,inj}$ = injection volume of sample (ml).

IX. Precision and accuracy
A. In one laboratory 0.175 μg/l ± 0.004 μg/l pentachlorophenol was measured at a 99% level of confidence. A level of 1.500 μg/l ± 0.01 μg/l was measured at a 95% level of confidence. At the 0.175 ppb level a relative error of 1.1% was found.

X. Confirmatory of identity
A. The confirmation of pentachlorophenol is based on multicolumn techniques. The derivative pentachlorophenyl acetate is very readily confirmed by GC/MS.

XI. Remarks
A. The extraction of PCP can also be performed using the XAD-7 resin according to Section XII.C.

XII. References
A. Chau, A. S. Y. and Coburn, J. A., Procedure for Analysis of Pentachlorophenol in Water and Wastewater, Water Quality Branch Report.
B. Chau, A. S. Y. and Coburn, J.A., *J. Assoc. Off. Anal. Chem.*, 57, 389, 1974.
C. Coburn, J. A. and Chau, A. S. Y., *J. Assoc. Off. Anal. Chem.*, 59, 862, 1976.
D. Bevenue, A. and Beckman, H., *Residue Rev.*, 19, 93, 1967.
E. Rudling, L., *Water Res.*, 4, 533, 1970.

REFERENCES

1. *Organic Chemistry, College Outline Series*, 6th ed., Barnes & Noble, New York.
2. **Bovey, R. W. and Young, A. L.**, *The Science of 2,4,5-T and Associated Phenoxy — Herbicides*, John Wiley & Sons, New York.

3. **Gilman, A. P., Douglas, U. M., Arbuckle-Sholtz, T., and Jamieson, J. W. S.,** *Chlorophenols and Their Impurities: A Health Hazard Evaluation,* Bureau of Chemical Hazards, Environmental Health Directorate, National Health and Welfare, Ottawa, Ontario, Canada, 1982.
4. **Pavlov, B. and Terentyev, T.,** *Organic Chemistry,* Gordon and Breach Science Publishers, New York, 1965.
5. **Jones, P. A.,** *Chlorophenols and Their Impurities in the Canadian Environment,* Environment Canada Econ. Tech. Rev. Report EPS-3-EC-81-2, p. 434.
6. *Nanogen Index,* Nanogens International, P.O. Box 487, Freedom, CA, 95019.
7. **Vogt, C. R., Kapila, S., and Manahan, S. E.,** Fused silica capillary column gas chromatography with tandem flame ionization. Photoionization detection for the characterization of *in situ* coal gasification by-products, *Int. J. Environ. Anal. Chem.,* 12, 27, 1982.
8. **Orazio, C. E., Kapka, S., and Manahan, S. E.,** High-performance liquid chromatographic determination of phenols as phenolates in a complex mixture, *J. Chromatogr.,* 262, 434, 1983.
9. **Bridle, T. R., Afghan, B. K., Sachdev, A., Campbell, H. W., Wilkinson, R. J., and Carron, J.,** The Formation and Fate of PCDD's and PCDF's during Chlorophenol Combustion.
10. National Research Council of Canada, NRCC No. 18578, adapted from an EPA document (reference not available), National Research Council of Canada, Montreal Road, Ottawa, Ontario, Canada, K1A OR6.
11. *CRC Crit. Rev. Toxicol.,* 7(1), 4, 1980.
12. **Jones, P. A.,** *1983 Supplement, Chlorophenols and Their Impurities in the Canadian Environment,* Economic and Technical Review Report, EPS 3-EC-81-2, Environment Canada, Environmental Protection Service, Ottawa, Ontario, Canada.
13. *Crit. Rev. Environ. Control,* 12(4), 310, 1982.
14. Environment Canada, CCB-In-4-80 Infonotes Substances in the List of Priority Chemicals 1979 — Chlorophenols. Place Vincent Massey Building, Hull, Quebec.
15. National Swedish Environment Protection Board, Wallenberg Laboratory, Stockholm, Phenolic compounds — analytical methods in relation to environmental aspects, in *Report of the Workshop of the Working Party 8, Specific Analytical Problems Concerted Action Analysis of Organic Micro Pollutants in Water,* OMP/33/82, COST 648 B15, Commission of the European Communities, Joint Research Centre, Dubendore, Switzerland, May 18—19, 1982, 36.
16. **Regnier, Z. and Watson, A. E. P.,** *The Development of Colorimetric and Fluorometric Automated Methods for Trace Phenol Estimation in Natural Waters,* Water Quality Division, Water Chemistry Subdivision, Editorial and Publications Division, Inland Waters Directorate, Environment Canada, Ottawa, Ontario K1A OE7, Canada.
17. National Research of Canada, *Chlorinated Phenols,* NRCC No. 18578, National Research Council of Canada, Montreal Road, Ottawa, Ontario, K1A OR6.
18. **Bevenue, A. and Beckman, H.,** Pentachlorophenol: a discussion of its properties and its occurrence as a residue in human and animal tissues, *Residue Rev.,* 19, 83, 1967.
19. **Rao, K. R.,** Ed., Pentachlorophenol; chemistry, pharmacology and environmental toxicology, in *Proceedings Symposium Pensacola, Fla.,* Plenum Press, New York, 1977, 402, 1978.
20. **West, K. F. and West, P. W.,** *Anal. Chem.,* 38, 68, 1966.
21. **Fichnolz, G. G., Nagel, A. E., and Huges, R. H.,** *Anal. Chem.,* 37, 863, 1965.
22. **Chau, A. S. Y. and Coburn, J. A.,** Determination of pentachlorophenol in natural and waste water, *J. Assoc. Off. Anal. Chem.,* 57(2), 389, 1974.
23. **Afghan, B. K., Belliveau, P. E., Larose, R. H., and Ryan, J. F.,** An improved method for determination of trace quantities of phenols in natural waters, *Anal. Chem. Acta,* 71, 355, 1974.
24. American Public Health Association, American Water Works Association and American Pollution Control Federation, *Standard Methods for the Examination of Water and Wastewater,* 13th ed., American Public Health Association, New York, 1971, 501.
25. **Kaplin, V. T. and Freuko, N. G.,** *Gig. Sanit.,* 26, 68, 1961.
26. **Afghan, B. K., Carron, J., Goulden, P. D., Lawrence, J., Leger, D., Onuska, F., Sherry, J., and Wilkinson, R.,** Recent Advances in Ultratrace Analysis of Dioxins and Related Halogenated Hydrocarbons. *Can. J. Chem.,* 65, 1086, 1987.
27. **Scott, B. F., Nagy, E., Hart, J., and Afghan, B. K.,** National Water Research Institute (Canada), reprinted from Industrial Research and Development, April, 1982.
28. **Ribick, M. A., Smith, L. M., Dubay, G. R., and Stalling, D. L.,** Applications and Results of Analytical Methods Used in Monitoring Environmental Contaminants, for Submission to ASTM, presented at The ASTM Committee E35 on Aquatic Toxicology Symposium, October, 1979, Chicago, IL, U.S. Department of Interior, U.S. Fish and Wildlife Service, Columbia National Fisheries Research Laboratory, Route No. 1, Columbia, MO, 65201.
29. **Lee, H. B., Weng, L., and Chau, A. S.,** Chemical derivatization analysis of pesticide residues. IX. Analysis of phenol and 21 chlorinated phenols in natural waters by formation of pentafluorobenzyl ether derivatives, *J. Assoc. Off. Anal. Chem.,* 67(6), 1086, 1984.

30. **Smith, B.,** Investigation of Reagents for the Qualitative Analysis of Phenols, Chalmers Tekniska Hogskolas Handlingar, Transactions of Chalmers University of Technology, Gothenburg, Sweden (avd. Kemi ocb Kemisk Teknologi 41) Nr 263, 1963.

31. **Smith, B., Persmark, U., and Edman, E.,** The use of tetracyanoethylene for the qualitative analysis of phenols, *Acta Chem. Scand.,* 17, 709, 1963.

32. **Houghton, G. U. and Pelly, R. G.,** A colorimetric method for the determination of traces of phenol in water, *Analyst,* 62, 117, 1937.

33. **Churacek, J. and Coupek, J.,** Phenols — liquid column chromatography, *J. Chromatogr. Library,* 3, 441, 1975.

34. **Tomlinson, E., Jefferies, T. M., and Riley, C. M.,** Ion-pairs high-performance liquid chromatography, *J. Chromatogr.,* 159, 315, 1978.

35. Waters Canada, 3688 Nashua Drive, Mississauga, Ontario, L4V 1M3, Canada.

36. **Ogan, K. and Katz, E.,** Liquid chromatographic separation of alklyphenols with fluorescence and ultraviolet detection, *Anal. Chem.,* 53, 160, 1981.

37. **Realini, P. A.,** Determination of priority pollutant phenols in water by HPLC, *J. Chromatogr. Sci.,* 19, 124, 1981.

38. **Hoffsommer, J. C., Glover, D. J., and Hazzard, C. Y.,** Quantitative analysis of polynitrophenols in water in the micro to nanogram range by reversed-phase ion-pair liquid chromatography, *J. Chromatogr.,* 195, 435, 1980.

39. **Wegman, R. R. C. and Wammes, J. I. J.,** Determination of nitrophenols in surface water samples by reversed-phase ion-pair high performance liquid chromatography, in *Report of the Workshop of the Working Party 8, Specific Analytical Problems Concerted Action Analysis of Organic Micro Pollutants In Water,* OMP/33/82, Cost 64B Bis, Commission of the European Communities, Joint Research Centre, Dubendore, Switzerland, May 18-19, 1982, 59.

40. Bioanalytical Systems Inc., 2701 Kent Avenue, West Lafayette, IN, 47906.

41. **Weissharr, D. E., Tallman, D. E., and Anderson, J. L.,** Kel-F-graphite composite electrode as an electrochemical detector for liquid chromatography and application to phenolic compounds, *Anal. Chem.,* 53, 1809, 1981.

42. **Shoup, R. E. and Mayer, G. S.,** Determination of environmental phenols by liquid chromatography/ electrochemistry, *Anal. Chem.,* 54, 1164, 1982.

43. **Armentrout, D. N., McLean, J. D., and Long, M. W.,** Trace determination of phenolic compounds in water by reverse phase liquid chromatography with electrochemical detection using a carbon-polyethylene tubular anode, *Anal. Chem.,* 51, 1039, 1979.

44. **Corcia, A. D., Samperi, R., Sebastiani, E., and Severini, C.,** Gas chromatographic analysis of water phenolic pollutants using acid-washed graphatised carbon black, *Chromatographia,* 14(2), 86, 1981.

45. **Mathew, J. and Elzerman, A. W.,** Gas-liquid chromatographic determination of some chlor- and nitro-phenols by direct acetylation in aqueous solution, *Anal. Lett.,* 14(A16), 1351, 1981.

46. **Lamparski, L. L. and Nestrick, T. J.,** Determination of trace phenols in water by gas chromatographic analysis of heptafluorobutyl derivatives, *J. Chromatogr.,* 156, 143, 1978.

47. **Lehtonen, M.,** Gas chromatographic determination of phenols as 2,4-dinitrophenyl ethers using glass capillary columns and an electron-capture detector, *J. Chromatogr.,* 202, 413, 1980.

48. **Lee, H. B. and Chau, A. S. Y.,** Analysis of pesticide residues by chemical derivatization. VII. Chromatographic properties of pentafluorobenzyl ether derivatives of thirty-two phenols, *J. Assoc. Off. Anal. Chem.,* 66(4), 1029, 1983.

49. *Master Analytical Scheme For The Analysis of Organic Compounds in Water,* Athens Environmental Research Laboratory Office of Research and Development, U.S. Environmental Protection Agency, Athens, GA, 30605.

50. **Keuh, D. W. and Dougherty, R. C.,** Pentachlorophenol in the Environment: Evidence for Its Origin from Commercial Pentachlorophenol by Negative Chemical Ionization Mass Spectrometry, submitted to Environmental Science and Technology, 1980. Department of Chemistry, Florida State University, Tallahassee, FL, 32306.

51. **Carron, J. M. and Afghan, B. K.,** unpublished work, National Water Quality Laboratory, 867 Lakeshore Road, Burlington, Ontario, Canada, L7R 4A6.

52. Radian Corporation, *Phenol in Methylene Chloride (FRN 21402 CRN 119; August 1978),* Analabs, Inc., 80 Republic Drive, New Haven, CT, 06473.

53. Supelco, Phenol STD, Mix, Supelco Canada Ltd., 46-220 Wyncroft Road, Oakville, Ontario, Canada, L6K 3V1.

54. **Goulden, P. D., Brooksbank, P., and Day, M. B.,** Determination of submicrogram levels of phenol in water, *Anal. Chem.,* 45, 2430, 1973. Adapted for WQB method. Phenolica (Colorimetric — Automated), *Analytical Methods Manual 1979,* Inland Waters Directorate, Water Quality Branch, Ottawa, Ontario, Canada.

Chapter 5

ANALYSIS OF CHLORINATED DIBENZO-*p*-DIOXINS AND DIBENZOFURANS IN THE AQUATIC ENVIRONMENT

R. E. Clement and H. M. Tosine

TABLE OF CONTENTS

I. INTRODUCTION

Considering their hydrophobic nature, it is perhaps surprising that the chlorinated dibenzo-*p*-dioxins (CDD) and chlorinated dibenzofurans (CDF) are of considerable importance to the aquatic environment. However, current techniques for isolating ultratrace quantities of organic compounds from very complex mixtures and performing quantitative analysis by gas chromatography-mass spectrometry are so effective that parts-per-trillion (ppt: picograms per gram) concentrations of these compounds can now be determined with some considerable degree of confidence. Determination of these substances in fish and other aquatic wildlife has shown that the impact and distribution of CDD/CDF in the aquatic ecosystem is considerable.

Concern with the CDD/CDF groups of chemicals even at such low concentrations is a result of considerable research into the toxicity of 2,3,7,8-tetrachlorodibenzo-*p*-dioxin (2,3,7,8-4CDD). Even though a dozen of the 135 CDD/CDF compounds are now considered to be very toxic, intense scientific, media, and public attention world-wide has been focused specifically on 2,3,7,8-4CDD. Only recently has a significant effort been put forth to examine the wide range of CDD and CDF present in the environment.

Attention towards 2,3,7,8-4CDD has resulted from several highly publicized environmental contamination episodes. These include an accident at a 2,4,5-trichlorophenol plant in Seveso, Italy;[2-7] spraying of the 2,3,7,8-4CDD-contaminated herbicide, Agent Orange, in Vietnam;[8-12] and the spraying of chlorinated dibenzo-*p*-dioxin contaminated waste oil in Missouri, which eventually led to the evacuation of the town of Times Beach.[13-16] A serious incident also occurred in 1968 in Japan, in which many people used cooking oil contaminated with a polychlorinated biphenyl (PCB) formulation called Kanechlor 400.[17] It was found that this PCB formulation contained significant amounts of chlorinated dibenzofurans.[18-20] Environmental effects relating to the CDD/CDF groups of compounds were noted as early as 1957, when millions of broiler chickens in the U.S. died from feed fats contaminated by toxic impurities.[21] The toxic impurities were isolated and later identified as various CDD congeners. A number of industrial accidents in addition to the one at Seveso, Italy, have occurred, starting as early as 1949.[22]

The analytical chemistry of CDD/CDF has progressed significantly over the past 20 years, from a state in which simply identifying the presence of unspecified CDD congeners at parts-per-million (ppm) concentrations was difficult, to the present where low parts-per-trillion concentrations of all of the most toxic CDD and CDF congeners can be determined with confidence in samples containing hundreds of other compounds at concentrations many thousands of times greater. This development was a necessary result of the intense public concern generated by the possible health impact of the CDDs and CDFs.

Because of the complex interrelationships which exist in the environment, a study of CDD/CDF in the aquatic system must consider a much wider base than water, sediments, and biota. Inputs to the aquatic system from ground water containing leachate from toxic waste dumps or landfills, industrial effluent, and long-range atmospheric transport are possible. Consequently, current analytical methods for CDD/CDF in stack emissions, soils, incinerator fly ash, and ambient air will be described here in addition to water, sediments, and biota. Although broad in scope, this review of CDD/CDF analytical methods is not intended to be comprehensive. Analysis of industrial products such as chlorophenols and herbicides; human blood, adipose tissue, and mother's milk; and PCB oils and soot from transformer fires, are not specifically covered. However, reference is made to published reports where treatment of these other areas is more complete.

A. Structure, Nomenclature, and Properties

The chlorinated dibenzo-p-dioxins (CDD) and chlorinated dibenzofurans (CDF) are groups of structurally related compounds, as shown in Figure 1. A total of 75 different CDD and 135 CDF compounds are possible, having from 1 to 8 chlorine substituents. Table 1 shows the number of CDD and CDF compounds according to degree of chlorination.

There is some confusion regarding the naming of these substances. The CDD or CDF molecules having the same number of chlorine substituents are isomers of each other. The 75 individual CDD compounds, regardless of degree of chlorination, are congeners. Likewise, the 135 CDF compounds are all congeners of one another. In this paper, the term congener group will be used to refer to a single group of CDD or CDF isomers. For example, there are 22 isomers of the tetrachlorinated dibenzo-p-dioxins (4CDD). The 4CDD congener group refers to the entire group of 22 isomers. Different CDD or CDF congener groups are often incorrectly called homologues.

Confusion can result when referring to the chlorinated dioxins as simply "dioxins". The term dioxin has at various times been used with reference to: all of the chlorinated dibenzo-p-dioxins, only the congeners with four or more chlorines, the tetrachlorinated dibenzo-p-dioxin congener group, and the single compound 2,3,7,8-tetrachlorodibenzo-p-dioxin (2,3,7,8-4CDD). In this paper, the abbreviated names shown in Table 1 will be used. Specific congeners will be identified by these abbreviations, and the numbers showing the positions of chlorine atoms on the basic structure as indicated in Figure 1 and in the above example (2,3,7,8-4CDD). Other reported acronyms for the tetrachlorinated dibenzo-p-dioxins include 4-DD, TCDD, T_4CDD and tetra-CDD. Many other methods of naming the other congener groups have also been used. The naming system proposed here (Table 1) was chosen because of its simplicity and unambiguous nature. A recent publication has called for the standardized use of nomenclature.[1]

Little information exists on the physical properties of the entire CDD/CDF group of compounds. Most studies have been concerned with the properties of only a few congeners. In many cases, those not equipped with laboratories specially designed to handle toxic substances are unwilling to work with the relatively large quantities of CDDs and CDFs needed for such investigations. Generally, the CDDs and CDFs are hydrophobic and lipo-

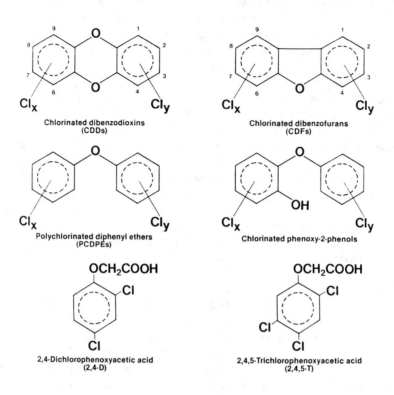

FIGURE 1.   Structure of chlorinated dibenzo-*p*-dioxins, chlorinated dibenzofurans, and related compounds.

philic. Higher chlorinated congeners have lower aqueous solubility than the lower chlorinated congeners. CDDs and CDFs with four or more chlorines have low vapor pressures, however, volatility from water and the upper layers of soil can be appreciable. The environmental transport and fate of CDDs/CDFs will be discussed later in more detail.

Table 2 is a summary of some physical/chemical properties reported in the literature for a few of the 210 CDD/CDF. The octanol-water partition coefficient (log K_{ow}) of a compound is an important property because it can be directly related to aqueous solubility, soil adsorption, and bioaccumulation factors.[23]

Reported values of log K_{ow} range from 6.15 to 7.02 for 2,3,7,8-4CDD, and from 7.20 to 8.70 for 1,3,6,8-4CDD. Aqueous solubilities for 2,3,7,8-4CDD in Table 2 vary from 7.91 to 200 parts-per-trillion (10^{-9} g/l). The large differences in reported solubilities and log K_{ow} values are partly due to differences in methodologies used in their determination, however, it is generally very difficult to work with these hydrophobic substances.[367] Schroy and co-workers[24] have assembled a compendium of reported properties for 2,3,7,8-4CDD which includes density, heat of formation, and heat of vaporization. Methods are needed to improve the accuracy with which physical/chemical properties are measured for CDD/CDF congeners, to enable the development of improved models concerning environmental transport and fate of these chemicals.

Table 1
NUMBER OF CDD AND CDF CONGENERS ACCORDING TO DEGREE OF CHLORINATION

Congener group	Abbreviated name	Formula	Number of isomers
Monochlorodibenzo-p-dioxin	1CDD	$C_{12}H_7O_2Cl_1$	2
Dichlorodibenzo-p-dioxin	2CDD	$C_{12}H_6O_2Cl_2$	10
Trichlorodibenzo-p-dioxin	3CDD	$C_{12}H_5O_2Cl_3$	14
Tetrachlorodibenzo-p-dioxin	4CDD	$C_{12}H_4O_2Cl_4$	22
Pentachlorodibenzo-p-dioxin	5CDD	$C_{12}H_3O_2Cl_5$	14
Hexachlorodibenzo-p-dioxin	6CDD	$C_{12}H_2O_2Cl_6$	10
Heptachlorodibenzo-p-dioxin	7CDD	$C_{12}H_1O_2Cl_7$	2
Octachlorodibenzo-p-dioxin	8CDD	$C_{12}H_0O_2Cl_8$	1
Total congeners			75
Monochlorodibenzofuran	1CDF	$C_{12}H_7OCl_1$	4
Dichlorodibenzofuran	2CDF	$C_{12}H_6OCl_2$	16
Trichlorodibenzofuran	3CDF	$C_{12}H_5OCl_3$	28
Tetrachlorodibenzofuran	4CDF	$C_{12}H_4OCl_4$	38
Pentachlorodibenzofuran	5CDF	$C_{12}H_3OCl_5$	28
Hexachlorodibenzofuran	6CDF	$C_{12}H_2OCl_6$	16
Heptachlorodibenzofuran	7CDF	$C_{12}H_1OCl_7$	4
Octachlorodibenzofuran	8CDF	$C_{12}H_0OCl_8$	1
Total congeners			135

Table 2
PHYSICAL-CHEMICAL PROPERTIES OF CDD/CDF REPORTED IN THE LITERATURE

Property	Congener	Values	Ref.
Log K_{ow}	2,3,7,8-4CDD	6.15	24
		6.64	25
		7.02	26
	1,3,6,8-4CDD	7.20	26
		8.70	27 (28,29)
	8CDD	11.76	27 (28,29)
	2,3,7,8-4CDF	5.82	26
Water solubility (20°C) (ng/l)	2,3,7,8-4CDD	7.91 ± 2.7	30
		19.3 ± 3.7	31
		200	32
	1,3,6,8-4CDD	320	27,33
	8CDD	0.4	27,33
Vapor pressure (atm.)	1,3,6,8-4CDD (20°C)	5.3×10^{-9}	27,34
	1,3,6,8-4CDD (100°C)	1.1×10^{-8}	27,34
	8CDD (20°C)	8.6×10^{-11}	27,34
	8CDD (100°C)	1.7×10^{-10}	27,34

B. Formation and Sources to the Environment

With the exception of small amounts of CDDs and CDFs synthesized for research purposes and to provide analytical standards, these compounds are not manufactured intentionally. There are two principal inputs of CDDs/CDFs to the environment: trace contaminants in industrial chemicals such as chlorinated phenols, phenoxy herbicides, and PCBs; and products of combustion from such sources as municipal incinerators, wood burning, chemical waste combustion, automotive emissions, pathological waste, and PCB-filled transformer fires.

Extensive work has been performed regarding the synthesis of CDDs,[35-46,158,215] Reaction of chlorobenzenes with catechol can form CDDs in high yield.[35] Nilsson produced CDDs by the cyclization of 2-phenoxyphenols (commonly called pre-dioxins).[36] This reaction can proceed by thermochemical or photochemical decomposition. During analysis by gas chromatography of the fully chlorinated 2-phenoxyphenol, the compound undergoes immediate ring closure to form 8CDD.[36] Lower chlorinated CDDs can be formed through dechlorination of higher chlorinated congeners through photolysis.[37,38] An important synthetic route for the formation of CDDs is by pyrolytic dimerization of ortho-chlorinated phenols or their salts.[35,39] Unexpected products may be obtained through intervention of the Smiles rearrangement.[40,41] All 22 of the 4CDD isomers have been prepared using controlled flow pyrolysis of di-, tri-, and tetrachlorophenols.[42] Recent studies have emphasized the synthesis and purification of individual CDD congeners for use as analytical standards.[43-46]

Although interest in the CDF congeners is more recent than for the CDDs, considerable efforts have been made to study the synthesis of various CDF congeners.[47-55,251,257,259,284] The principal routes of CDF synthesis have been discussed by Bell and Gara[50] and include pyrolysis of PCBs,[53,55] direct halogenation of the dibenzofuran structure, Pd-catalyzed cyclization of diphenyl ethers (DPE), and photochemical cyclization of DPE. Choudhry[49] also produced CDFs by photolysis of PCBs.

Sources to the environment of CDDs and CDFs have been discussed previously.[56-64] An excellent review of chemical sources is available in a U.S. Environmental Protection Agency report.[59] Of greatest concern in industrial chemicals are the CDDs formed during the manufacture of chlorophenols and their derivatives (2,4-D, 2,4,5-T). In most cases, manufacturing processes can be controlled to minimize the amounts of CDDs formed. Much greater inputs to the environment arise from the uncontrolled burning of these chemicals or products treated with these chemicals. For example, high levels of CDFs have been measured in the soot after fires involving PCB transformers.[65] A special committee of experts recently ranked municipal and industrial incineration sources as the major environmental inputs of CDDs.[66] This is supported by others.[56,57] The findings of trace levels of CDDs in virtually every type of combustion product they tested led Dow Chemical scientists to propose the "Trace Chemistries of Fire Hypothesis"[67] from which they concluded that CDDs are ubiquitous in the environment.

C. Human and Animal Health Effects

Many studies investigating the acute toxicity of 2,3,7,8-4CDD and other CDD/CDF congeners to animals have been performed. In the most sensitive species, the guinea pig, an LD_{50} of 0.6 µg/kg body weight was determined.[68] However, large differences in LD_{50} are observed in various species. In one study,[69] values of 2µg/kg (guinea pig), 70 µg/kg (rhesus monkeys), and 284 µg/kg (mice) were determined. Large differences in LD_{50} were also observed between various CDD congeners tested. For example, LD_{50} for guinea pigs were 31µg/kg, 125 µg/kg, and 30,000 µg/kg for 1,2,3,7,8,-5CDD, 1,2,4,7,8-5CDD, and 2,3,7-3CDD, respectively.[69] The most toxic CDDs and CDFs (with the exception of 8CDD/8CDF) are those in which the 2, 3, 7, and 8 positions are filled by chlorine atoms (Figure 1). The toxicity and observed effects are similar for CDDs and CDFs having the same chlorine substitution patterns.

Chronic and acute effects of 2,3,7,8-4CDD have been extensively studied. Several reviews of this work have been reported previously.[60-62,70-77] The common biological and toxicological responses observed in animal studies include: a wasting syndrome, manifested by progressive weight loss, decreased food consumption, and other effects; skin disorders, acneform eruption, or chloracne; impaired liver function; general suppression of the immune system; reproductive disorders; modulation of chemical carcinogenesis; and the induction of numerous enzymes. These wide-ranging effects are highly species dependent.

The major acute effect observed in humans is that of chloracne, which is generally observed after exposure to relatively high concentrations of CDDs/CDFs such as can occur from occupational or accidental exposure. Studies on the effects of these chemicals on humans are extremely difficult to evaluate as testing is limited to epidemiological investigations initiated after occupational or accidental exposure to CDDs/CDFs. Consequently, assessment of the degree of hazard to humans is generally based upon extrapolation from results of animal studies. One such approach used a set of weighting factors for CDDs/CDFs to calculate their total concentration in a sample in terms of 2,3,7,8-4CDD toxic equivalents.[78] In one assessment of risk to humans from CDD and CDF exposure, it was recommended that the daily intake of these substances, as 2,3,7,8-4CDD toxic equivalents, should not exceed 10 pg/kg body weight per day.[61] Since this total intake is the sum for all exposure from ambient air, water, food, and other exposure, analytical methods must be able to achieve ultratrace limits of detection for all these matrices.

II. ANALYTICAL METHODOLOGY

Because of the low detection limits generally required for the analysis of environmental samples for CDDs and CDFs, optimization of the entire analytical process is usually needed. This includes procedures for sampling, extraction, concentration, removal of coextractives, and detection by a sensitive and selective technique. Elimination of false positive and false negative results are both vital to adequately protect human health without causing undue alarm in the general public. The technique of gas chromatography-mass spectrometry is the only one that can provide the necessary instrumental sensitivity and selectivity from potential interferences. Strict quality control procedures using surrogate compounds to monitor the effectiveness of analytical procedures have been developed specifically for GC-MS determination of CDDs and CDFs in the environment. Other reviews of the analytical methodology for CDDs and CDFs have been reported.[62,79-86]

A. Sampling and Extraction

Sampling procedures and their effect on precision and accuracy of CDD/CDF determinations in environmental samples have not been extensively studied, especially for such matrices as water, sediments, and biota. More attention has been given to soils, and to sampling procedures for stationary sources such as municipal incinerator stacks. Procedures for sampling CDDs/CDFs in ambient air and large-volume water are currently under active investigation. Most of the work reported to date has employed grab-sampling techniques, with little or no replication. Although extensive quality control procedures are often documented for sample extraction and analysis, special quality control procedures used during sampling are seldom described.

Many factors for the design of a sampling program for CDDs/CDFs should be considered. Some of these are in the category of general "good laboratory practice" and are the same considerations that would be applied to any environmental survey. However, because of the increased effects of sample contamination, inhomogeneity, and analyte instability at ultratrace levels, the degree of care required when collecting samples for CDD and CDF determination is often greater than for other analytes. The principal factors include the following.

Use of data — A survey designed to locate a point source would not be the same as one to delineate cleanup boundaries after a spill or to determine background levels in the environment. Many other sampling/analytical consideration are affected by this factor.

Sampling location — Proximity to other potential sources of CDD/CDF must be considered. Chemical waste dumps, chemical industries, sites of former chemical manufacturing/

formulation plants, proximity to emissions from combustion operations (industrial and municipal) are potential sources of CDDs/CDFs.

Representative sample — This is often difficult to obtain in environmental surveys. Soils and sediments may contain stones, plants, or small animal life. Surface or ground water, or industrial effluents, may have appreciable quantities of particulates. Factors such as bioavailability and bioaccumulation affect the levels found in fish of differing ages, species, and sizes. Generally, estimates of bulk concentrations are made from very small aliquots of the whole, especially in environmental studies. Sample homogeneity is often poor in environmental work; but this may be improved by mixing or grinding, and by compositing many samples taken in several locations.

Statistical considerations — Provost[87] has stated that most environmental sampling studies are not amenable to classical statistical techniques. However, statistical methods can be employed to help determine the number of samples needed for a required precision.[87] Application of the techniques to surveys of CDD/CDF levels in environmental studies shows that many samples are required to achieve good precision with a high degree of confidence. Unfortunately, the cost of CDD/CDF analysis is often prohibitive in such surveys. As a minimum objective in this work, frequent replicates should be analyzed to provide an estimate of the variance in the total analytical procedure. Because of factors described earlier, obtaining true replicates is not necessarily a trivial matter.

Field blanks — Although lab blanks are routinely analyzed in CDD/CDF determinations, field blanks are often not taken. For analyses at ultratrace levels, such as parts-per-quadrillion (ppq: 10^{-15}g/g) determination of CDDs/CDFs in drinking water, the importance of field blanks increases.

Sample cross-contamination — Contamination is an important consideration during sampling, especially in the determination of concentrations at ppt levels or lower. At such concentrations, a small amount of contamination of low-level samples from high-level samples can completely change the CDD/CDF concentration profile that is determined. It is especially important to be aware of the proximity of possible CDD/CDF sources, and to either eliminate their influence on the sample being collected, or to test their possible effect by employing field blanks.

Sample preservation and storage — The factors are not as significant for CDDs/CDFs as for many other analytes. No preservatives are needed, and generally precautions are limited to protection of samples from UV radiation, and storage at room temperature or cooler. Storage time could be an important consideration for matrices such as water, where CDDs/CDFs could be adsorbed onto the sample container surface. However, such effects have not been well documented.

Required detection limits — Detection limits determine the sample sizes that are extracted, and should be defined before sampling is performed. Generally, low ppt concentrations of CDDs/CDFs can be determined in soils/sediments and biota using extraction sizes of about 20 g. Less than 1.0 ppt of individual CDD/CDF congeners can be detected in a 1.0 l aqueous sample or industrial effluent. Parts-per-quadrillion levels of CDDs and CDFs in drinking water require sample sizes of at least 10 l.

Although many extraction procedures have been reported for CDDs/CDFs in virtually every type of environmental matrix, the differences in these methods are often relatively minor. Aqueous samples are usually extracted by liquid/liquid procedures, while particulates are generally extracted by Soxhlet apparatus, sometimes after acid treatment. More variation exists in methods for the extraction of fish and other biota, primarily in the procedures used to break down the sample matrix before actual solvent extraction is performed.

Common to most methods is the use of stable isotope-labeled CCD or CDF congeners to monitor the efficiency of extraction, cleanup, and GC-MS analysis. A number of these ^{13}C-labeled or [37]Cl-labeled standards are now available commercially, and their use will be discussed in detail in a later section. Although many methods for 2,3,7,8-4CDD have been reported, while relatively few have been validated for the whole range of CDD and CDF congeners (tetra- to octachlorinated), the physical and chemical behavior of CDDs/CDFs is such that a sampling/extraction procedure designed for 2,3,7,8-4CDD will probably be sufficient for the determination of all CDDs/CDFs virtually without modification.

1. Sampling and Extraction of Water
a. Sampling Methods

The principal considerations in sampling water (surface waters, drinking water, groundwater, industrial effluent) for CDDs/CDFs are the detection limits required and the presence of particulates. Studies have shown that such hydrophobic materials as 2,3,7,8-4CDD are strongly bound to particulates in an aquatic environment.[88-90] In one investigation,[88] between 80 and 99% of 2,3,7,8-4CDD was removed from water by passing samples through a 1.2 micron glass fiber filter. Since different extraction methods are generally performed for aqueous and particulate matrices, water containing visible particulates should be filtered before extraction, and the filtered particulate extracted separately as described below. Unfortunately, no studies have been performed to date to determine whether CDDs/CDFs dissolved in water will also be removed by the filter itself, or by sorption onto particles trapped on the filter. Kurtz[91] has shown that chlorinated hydrocarbon pesticides in water can be separated from water by adsorption to cellulose triacetate membrane filters. Recoveries of organics ranged from 68 to 113%. Sampling by filtration methods has the advantages of convenience, simplicity, and ability to easily achieve very large concentration factors. It is likely that such sampling methods can be used for CDDs/CDFs. If it is desired to determine accurately the distribution of CDDs/CDFs between particulates and the aqueous phase, it may be better to separate these components by centrifugation rather than filtration.

Methods using sorbent materials are advantageous for aqueous samples because of their inherent ease of use and because sampling, separation from the aqueous phase, and analyte concentration are all performed in a single step. The use of sorbent materials to sample organic compounds in water has been discussed in a recent review.[92] Two different adsorbents have been tested for use in sampling water for CDDs/CDFs.[93,94] In one study,[93] octadecyl (C18) reverse-phase adsorbent was used to trap 2,3,7,8-4CDD from drinking water samples. A sampling cartridge was designed with a 9.0 cm. O.D. fiberglass filter to remove particulates before passing water through the C18 adsorbent. A second filter was placed immediately before the sorbent material, and was needed to maintain high flow rates (500 ml/min) during sampling 10 l of drinking water. CDDs were removed from the sorbent cartridge by elution using acetone. Filters were extracted by Soxhlet using benzene and combined with the acetone eluant, after solvent exchange into benzene. Recoveries of 2,3,7,8-4CDD exceeded 60%, using this system. The method of sample delivery to the cartridge, and the solvent in which isotopically labeled 2,3,7,8-4CDD was initially dissolved before spiking into water to determine recoveries, were found to affect the distribution of spiked 2,3,7,8-4CDD between the filter, container walls, and sorbent material. Another group[94] used XAD-2 sorbent to isolate CDDs and CDFs from drinking water. The sampling method was tested with 200 l of tap water, and used a spiking solution containing several CDD/CDF congeners in acetonitrile solvent. After sampling, cartridges were eluted with 300 ml of 15% acetone/hexane (v/v). Decreasing recoveries were observed for the higher chlorinated CDD/CDF congeners, and ranged from about 30 to 100% for water fortified at the 1 pg/l level.

A batch sampling procedure especially designed for testing raw and treated water from water treatment plants has been reported.[95,96] Important elements of this procedure included the following:

1. Presampling visit to treatment plants to ensure adequate sampling conditions, including installation of organically inert sample transfer lines.

2.   Sample containers solvent rinsed and proven clean before use by GC-MS analysis of bottle rinsings.
3.   Collection of subsamples for analysis of support chemistry parameters such as turbidity and pH.
4.   All samples collected in duplicate.
5.   Field blanks taken for both raw and treated samples.
6.   Samples refrigerated during shipping; strict time protocol for completion of extractions within 5 days of receipt at laboratory was followed.
7.   Validation criteria were developed for analytical data that included detection of CDD/CDF congeners in both replicate samples.

Other sampling methods for CDDs/CDFs in water have not been reported, however, a detailed guide to ground-water sampling has been prepared[97] that describes techniques that could be used for sampling water for CDDs/CDFs.

b. Extraction Methods

Most work for the determination of CDDs/CDFs in aqueous samples involves batch liquid-liquid extraction (LLE) of water using solvents such as dichloromethane or pentane. Because CDDs/CDFs are hydrophobic, any common organic solvent not miscible with water could be used. For 1.0 l sample sizes, extraction is easily accomplished using a separatory funnel, or by vigorous stirring of the sample and solvent in a volumetric flask. Emulsions can occur, especially when extracting surface or ground water, or industrial effluent. It is common practice to spike samples with isotopically labeled CDD/CDF surrogates before extraction.

The conditions affecting efficiency of LLE for determination of 2,3,7,8-4CDD have been studied.[98-100] The pH was found to have little effect on recovery of 2,3,7,8-4CDD from water by LLE using a separatory funnel. Average recoveries in replicate determinations at pH 2, 7, and 10 were over 90%. No difference in efficiency was found using either dichloromethane solvent or 15% dichloromethane in hexane.[99] However, a high spike level of 100 ng 2,3,7,8-4CDD in 1.0 l water samples was used in this work. Sample stability was also studied with respect to pH, storage times, and degree of water chlorination. Spiking levels of 1000 ng/l were used to test stability at pH 2, 7, and 10 up to 14 d in water containing a residual chlorine content of 10 ppm. Storage temperatures of 4 and 25°C were employed. Losses were observed in chlorinated water, regardless of temperature and pH. Recoveries of 30 to 50% were obtained after 14 d, compared to 66 to 80% for nonchlorinated samples. Further studies[99] showed that only 10% loss was obtained after 6 d storage.

Other methods of LLE using specialized apparatus have been reported.[101-104] Brenner et al.[101] used a Ludwig-perforator and *n*-hexane to extract the impinger water collected during sampling incinerator stack emissions. Peters et al.[102] used a Kontes continuous extraction device (Kontes No. K-583250), operated for 16 h, to extract 1.0 l samples spiked with 20 ng of ^{13}C-2,3,7,8-4CDD in acetone. In 8 grab samples, percent recoveries of 94 to 103 (average 98 ± 0.6) were obtained. A device for batch extraction of large (up to 200 l) volumes of water has been reported.[103] In this device, called the Aqueous Phase Liquid Extractor (APLE), water is centrifuged and held in a stainless-steel drum, in which dichloromethane is recirculated by pumping through a sprayer bar. Contaminants from 200 l samples can be isolated and concentrated in less than 2 h. In extraction studies using tap water, a mean recovery of 92% for 18 organochlorine contaminants was obtained for spikes ranging from 0.05 to 1.0 ppt. A method that shows promise for field extraction of large volumes of water for CDDs/CDFs is based on the mixer-settler principle, and has been described by Ahnhoff and Josefsson.[104] Advantages of this technique over the APLE are those of portability and use of solvents lighter than or heavier than water.

2. Extraction of Fish and Other Biota

Two principal difficulties in the analysis of biota for CDDs/CDFs are the destruction of the matrix to allow efficient solvent extraction of analytes, and removal of the large quantity of organic co-extractives before GC-MS analysis. Sample cleanup techniques are described later. Several procedures have been described for breaking down the sample matrix but most of them fall into three categories: acid or base digestion, followed by LLE; ultrasonic disruption of cells and solvent extraction; and solvent extraction with no special pretreatment.

a. Sample Pretreatment

The analytical results obtained for CDDs/CDFs in fish depend upon whether the whole fish is homogenized, or only a portion (i.e., edible fillet). One study[105] showed that the fatty portions of fish gave significantly higher 2,3,7,8-4CDD concentrations than the lean fillet. Grinding whole fish should give higher CDD/CDF concentrations, which may be advantageous in general environmental surveys designed to locate areas of concern. However, for protection of human health, fillets provide better samples because their concentrations of CDDs/CDFs more closely measure the levels that would actually be consumed. In addition, limited data have been presented that show the lean fillet gives values for 2,3,7,8-4CDD that are relatively independent of weight and length of fish, for a specific species and location.[105]

Many biota analysis procedures employ acid or base digestion to break down the sample matrix prior to extraction.[194] Treatment using concentrated base may result in decomposition of higher chlorinated CDD/CDF congeners.[106,107,195] Up to 60% decomposition of 8CDD at room temperature, and over 90% loss after refluxing, was found in one study[106] after base digestion. The possibility of forming 2,3,7,8-4CDD from precursor compounds by base hydrolysis has also been discussed.[108,188] For these reasons, acid digestion using HCl or H_2SO_4 is generally preferred over alkaline hydrolysis. However, base digestion may be acceptable if only lower chlorinated CDD/CDF congeners are to be determined. Losses are also possible by photodecomposition for pretreatment or extraction procedures that involve the use of alcohol.[109]

Other biota pretreatment procedures include freeze-drying[110] and blending with anhydrous sodium sulfate to obtain a free-flowing powder[111] which is then extracted. Much developmental work has been performed using the fish matrix, and in a round robin study of eight laboratories,[112] no significant differences in reported concentrations of 2,3,7,8-4CDD were observed due to the use of different digestion or extraction techniques.

b. Extraction Procedures

Once the sample matrix has been broken down, extraction can be performed using a wide range of solvent systems. Hexane is the most common solvent employed,[112] although acetonitrile,[113] dichloromethane,[111] chloroform/methanol,[114] acetone/hexane,[115] benzene/hexane,[116] and benzene/dichloromethane[117] have also been used. After tissue breakdown or digestion with concentrated acid or base, LLE by shaking with solvent is the most widely used extraction technique. Other methods have also been successfully employed. Direct solvent extraction of tissue after grinding using a homogenizer has been performed.[110,116] Homogenization apparatus are available that can perform virtually total disruption of biota samples, and therefore give good CDD/CDF recoveries. Freeze-drying animal tissue can produce sample in the form of a fine powder. Extraction methods have been reported whereby the powdered sample is packed into a glass column and the CDDs/CDFs are directly eluted with solvent.[111,118] Such procedures may be amenable to automation of the extraction/cleanup process. Soxhlet extraction has been used for plant material such as grain, dry leaves, and grass.[119] This may be an efficient extraction method for vegetal material because of evidence that since there is little uptake of CDDs/CDFs in plants, most of these compounds will be found on the outer surfaces.[125] Direct solvent extraction with methanol followed by dichloromethane after chopping up vegetal tissues gave average recoveries of 80% for 2,3,7,8-4CDD, with no digestion step.[122] One of the principal difficulties in extracting complex samples after pretreatment by acid or base is the formation of emulsions during LLE.[119,123-124] These can often be broken by addition of sodium sulfate,[123] sodium carbonate,[124] or methanol.[122] Methodology for CDDs/CDFs in human tissue is not within the scope of this report. However, some excellent reviews on this subject are available.[62,120-121]

3. Soils, Sediments, and Incinerator Fly Ash

Due to a number of environmental contamination episodes, notably at Seveso, Italy, herbicide spraying in Vietnam, and various locations in Missouri, U.S., an extensive effort in the sampling and analysis of soils has been made. More recently, incinerator fly ash has been of interest because it has been shown to contain significant concentrations of a large range of CDD and CDF congeners. Less work has been reported for the analysis of sediments, but methodology is generally similar to that used for soils. Little pretreatment is used for these sample types, compared to fish and other biota. Grinding is sometimes performed, as well as acid treatment before extraction — especially for fly ash and other carbonaceous sample types.

Sampling procedures have been studied in detail for soils.[5,7,126-130] Studies of areas highly contaminated by 2,3,7,8-4CDD required a rectangular grid to be mapped over the sampling area so a systematic determination of 2,3,7,8-4CDD concentration gradients could be performed.[5,7,126,129] Statistical considerations in preparing sample strategies for cleanup of soil contaminated with 2,3,7,8-4CDD have also been addressed.[137] Guarding against sample cross-contamination by careful cleaning of all sampling apparatus and containers is an important consideration.[130] In one investigation,[130] improved estimates of CDD/CDF contamination were obtained by establishing 2 m diameter circular plots at each sampling location, and collecting 15 to 20 soil plugs (2 cm diameter × 5 cm deep) in each plot. The soil plugs were then manually ground until all particles passed through a standard sieve (mesh size 0.08 in.) before subsampling for CDD/CDF determination. Such compositing methods are important in estimates of average concentrations over large areas, since significant variations can be observed in soil concentrations found in multiple samples taken from even a small area. In one study,[129] a 1-yd^2 area of a contaminated site was divided into 1-yd^2 areas, and soil samples were collected in the center of each square. Concentrations of 2,3,7,8-4CDD varied from 8.1 to 57 ppb, which was much greater than the intralaboratory variation for these analyses.

Special precautions have to be taken when depth sampling of soils or sediments to obtain depth concentration profiles, because of the possibility of contaminating lower layers with soil or sediment from upper layers. For depth profiling of soils, trenching seems to be the best sampling method.[127,129] In this procedure, a trench is first excavated to expose a vertical face of the sampling location. After carefully cleaning the exposed face, samples at various depths are taken from bottom to top by inserting horizontally core cylinders or other sample collection devices. Design of a special device for precise sampling of soils down to 1 mm thickness for 2,3,7,8-4CDD mobility studies has been reported.[128] Sediments are collected by scooping or shovelling random grabs, or by devices such as the Eckman dredge,[131] Ponar grab sampler,[126] or a box corer.[132,133] Suspended sediment can be sampled[134] by extending a sampling line under water to a desired point, and pumping water to a continuous flow separator that removes the suspended sediment by centrifugation.

Virtually no studies have been performed regarding sampling techniques for fly ash from municipal incinerators. It is not known whether the concentrations of CDDs/CDFs from a single grab sample of fly ash collected by electrostatic precipitators, which is a common collection procedure, gives concentrations representative of the bulk fly ash collected. Concentrations of total CDDs in daily samples of municipal incinerator fly ash collected over a 1-week period varied from 57 to 220 ppb in one study,[135] but other factors such as refuse composition and incinerator operating conditions may account for this variation. To obtain estimates of daily CDD/CDF concentrations on fly ash, it is recommended to take grabs at fixed intervals during the day (i.e., every 3 h) to obtain a composite sample which is ground, coned and quartered, and subsampled several times to obtain a final sample for extraction.[136]

a. Sample Extraction: Soils and Sediments

Many methods have been reported for the extraction of soils and sediments.[113,127,128,132,138-151] To date, only a few of these have been tested for the whole range of CDD/CDF congeners — most methods were developed specifically for the determination of 2,3,7,8-4CDD. Very little pre-extraction treatment is usually required. Soils are usually dried before extraction, either by air-drying,[142,144] low heat[143] (40°C), or most commonly, mixing with anhydrous sodium sulfate before extracting.[138,140-142,147] Soil is sometimes extracted directly without drying or other treatment.[127,128] Before extraction, virtually all current methodologies require the spiking of samples with isotopically labeled CDD or CDF surrogates to determine extraction/cleanup efficiencies. To remove large inhomogeneous sample components, such as twigs and stones, samples are sometimes sieved before extraction.[143,146] A few methods specify special pretreatment such as refluxing samples with potassium hydroxide/ethanol,[113,189] but these are generally not required.

Solvent extraction is performed by mechanical or manual shaking,[127,138-141,146,147,150] or by Soxhlet.[132,135,142-144,148,149] Manual shaking or "jar" extraction methods,[139,149] are preferred in studies where rapid analysis of many samples is required, such as on-site analysis by GC-MS-MS during cleanup operations.[140] One investigation compared "jar" extraction techniques to Soxhlet[149] and found that both methods were about equally efficient, and that rapid "jar" extraction was adequate for determination of 2,3,7,8,-4CDD in soils above 1 ppb. However, it has also been reported that Soxhlet extraction is better than stirring or "jar" extraction methods for the wide variety of soil types encountered. When performing rapid extraction methods it should be noted that the surrogate CDDs/CDFs used to validate these methods are only spiked on the surface of the matrix and, even after thorough mixing and aging, may not be as strongly adsorbed on the soil/sediment surface as native CDDs/CDFs. This is especially true for the higher chlorinated CDD/CDF congeners. Therefore, those wishing to employ "jar" extraction methods should validate their extraction efficiencies compared to Soxhlet extraction for each different type of soil or sediment matrix.[140]

A wide variety of solvents and solvent mixes have been used for extracting soils and sediments. The most common are hexane, dichloromethane, benzene and toluene. Samples are often pre-wetted or extracted with acetone or methanol to remove water and improve contact between the sample and extracting solvent.

b. Sample Extraction: Fly Ash

Soxhlet extraction is the preferred technique for recovering CDD/CDF from fly ash. Benzene and toluene are the usual extraction solvents employed. Extraction efficiencies for various solvents and sample pretreatment methods have been well studied.[152-155,157] In one investigation[154] using Soxhlet extraction with no sample pretreatment, benzene and toluene gave about equal extraction efficiencies, and were much superior to cyclohexane, dichloromethane, or methanol. An interesting observation was that although methanol extracted less than 2% of the total CDD/CDF compared to benzene or toluene, re-extraction of the same fly ash with benzene after an initial methanol extraction gave greater CDD/CDF recoveries than extraction with benzene only.

Acid treatment with HCl before Soxhlet extraction has been shown to increase CDD/CDF recoveries.[152,153,155] This is especially true for the higher chlorinated CDD/CDF congeners, which are more strongly adsorbed to carbonaceous particles in the fly ash. Acid extraction is needed to break down the sample matrix, which has been shown to dissolve in discrete steps with increasing acidity,[156] suggesting the presence of specific fractions within fly ash. It was also found that in the extraction of fly ash with various solvents, the absolute amount of CDD/CDF extracted changed with the solvent, rather than the number of extractions required for complete analyte recovery.[157] The authors postulated either the existence of occlusions within the fly ash matrix or that some CDD were occluded in matrices of a polymeric nature. Acid treatment before extraction should increase analyte recoveries in either situation. After acid treatment, 24-h extractions are sufficient to recover over 98% of total extractable CDDs/CDFs.[153]

Although benzene or toluene extraction by Soxhlet after acid treatment is now generally accepted as the most effective procedure for recovery of CDD/CDF from fly ash, several other solvent systems have been used. Rapid "jar" extraction[350] and ultrasonic extraction[152,159] techniques for CDD/CDF in fly ash have been reported, however, extraction efficiencies were found to be significantly lower than those obtained by Soxhlet extraction.

4. Stationary Sources, and Ambient Air

As stated in the Introduction, combustion processes represent one of the major sources of CDDs and CDFs to the environment. Atmospheric emissions can affect aquatic systems directly by discharge of scrubber water or particulate collected from abatement processes into receiving waters, or by long range atmospheric transport. The possibility of long range transport of organics was discussed by Eisenreich and co-workers,[161] who concluded that atmospheric deposition represents the principal source of PCBs to the Upper Great Lakes. Many methods of analysis of stack emissions for CDD/CDF have been reported, especially for municipal waste incinerators, but only those in common use will be discussed here. Although methods for CDDs in ambient air have been in use since the Seveso accident,[5] it has been only recently that ambient air methods have been studied in detail, especially with regard to the distribution in air between particulates and the vapor phase.

a. Sampling of Incinerator Stack Emissions

Several reviews and detailed descriptions of incinerator stack sampling techniques have been reported.[80,162-168] Evaluations of the collection efficiencies of some of these techniques have been performed.[169,170] These sampling techniques are the most complex that have been developed for any environmental matrix for the determination of CDDs/CDFs because of the complex nature of stack emissions; which consist of hot vapors and particulates, with the vapors consisting of primarily water and a complex mixture of vaporized organic compounds and inorganic gases, including HCl. It is common practice to collect the particulates by filtration before condensation of the vapor component by impingers or sorbent traps. Sample collected by the various components of the sampling system can then be analyzed separately to give an estimate of both particulate and vapor-phase components.

Key features of any of the sampling systems include the following: sampling must be isokinetic to obtain a representative sample;[165] the filter is contained in a heated area to prevent condensation of vapors; and sampling is performed by traversing the stack diameter with a sampling probe, so emissions across the whole stack profile can be sampled. Isokinetic sampling means that the sampling rate is such that vapors supplied to the nozzle of the sample train probe are extracted without altering the flow. Typical stack gas sampling volumes employed range from 1 to 20 m^3. Most designs are based on the U.S. Environmental Protection Agency Modified Method 5 train.[162,164,166,169] Collection of vapors is accomplished by a combination of ice-cooled impingers and/or various sorbent traps such as XAD-,[165,167,169] florisil,[171] or polyurethane foam plugs.[168] A comparison of commonly used stack sampling trains has been reported.[168]

The recovery of sample from sampling trains is a laborious process, and results in several subsamples if separate estimates of particulate and vapor-phase components are to be performed.[171,172] In one study,[172] seven different sample types were generated for each sampling event, of which five were analyzed separately. It is not known at this time whether the separate fractions collected are really representative of the vapor/particulate distributions of CDDs/CDFs in the emissions, because absorption/desorption may occur when stack gases move through the particulate filter. Extraction of the various fractions collected is generally performed using techniques already described. Particulates are extracted by Soxhlet apparatus after acid treatment, usually using benzene or toluene solvent, while the condensed aqueous portion can be extracted by LLE using a variety of solvents. CDDs/CDFs can be recovered from sorbent cartridges by direct elution[136] or Soxhlet extraction.[171]

b. Sampling of Ambient Air

Early methods of ambient air sampling collected only the suspended particulate matter by high-volume filtration methods.[5,173,174] Analysis was for 2,3,7,8-4CDD only. In one study,[174] no 2,3,7,8-4CDD losses were observed from particles during sampling, although detection limits were only in the ppb range. However, in studies using both a glass fiber filter and a sorbent cartridge,[179] significant amounts of PCBs and other chlorinated hydrocarbons were found on the sorbent cartridge. More recent ambient air methods use both a filter and sorbent cartridge design for high-volume air collection.[175-178] Polyurethane foam[75,176] is becoming increasingly popular as the medium used for collection of vapor-phase CDDs/CDFs due to its favorable flow characteristics and relative ease of use. In one study,[176] the tetra- and pentachlorinated CDD/CDF congeners were detected principally on the polyurethane foam sorbent cartridge, while hexa-, hepta-, and octachlorinated congeners were detected on the filter-collected particulates. Another cartridge design[177,178] employed silica gel contained in a removable extraction thimble in a Teflon® housing. Extraction of both filters and sorbent cartridges is usually by Soxhlet, with benzene and toluene as the most popular solvents. More ambient air work is needed to assess the possible contribution of CDDs/CDFs to toxic deposition, long range transport, and for monitoring the effect of emissions from combustion sources.

5. Concentration Techniques

After sample pretreatment and extraction, the CDDs and CDFs are in a matrix consisting of possibly thousands of coextractives in 100 to 300 ml of organic solvent. Concentration of this extract must be performed, often at several stages of the cleanup procedure. Solvent exchange is also frequently performed, depending upon the specific cleanup technique used. This may involve bringing the extract to dryness. It has been shown that significant analyte losses can occur using common evaporation techniques.[180,181] By allowing 100 μl of a concentrated fly ash extract in benzene to evaporate to dryness under ambient conditions (~ 40 h) and reconstituting by adding 100 μl benzene with ultrasonic agitation, about 10% of all CDDs (tetra- to octachlorinated) were lost in replicate tests.[180] In another investigation of 2,3,7,8-4CDD losses,[181] evaporation of toluene solutions to dryness resulted in recovery of only 47%; and only 2% of 2,3,7,8-4CDD was recovered when evaporation was performed at 100°C. However, evaporation of benzene or dichloromethane solutions at room temperatures could be performed to give 90% recovery. Nitrogen evaporation and vacuum evaporation techniques were all found to be effective for the concentration of 2,3,7,8-4CDD solutions.[181] Common preconcentration techniques for trace analysis of organic compounds have been reviewed.[182]

B. Removal of Interferences

A wide range of sample complexity exists from raw and treated drinking waters to particulate samples such as soils, sediments, and incinerator fly ash; and to extremely complex matrices such as fish and other biota. As the sample complexity increases, identification and quantitation of CDDs/CDFs becomes increasingly difficult due to the presence of a large number of organic compound coextractives, many of which are at concentrations thousands of times greater than concentrations of the CDDs/CDFs. The need for positive detection of CDDs/CDFs at ppt levels or lower dictates the application of highly specific and efficient cleanup techniques.

The extent of cleanup required for a specific application is also dependent upon the detection system employed. Very sophisticated high resolution GC-MS systems are available that can detect CDDs/CDFs in the presence of most other organic substances, however, these systems are very costly and are only available to relatively few laboratories. More extensive sample cleanup reduces the need for specialized equipment, however, there is a trade-off between cleanup efficiency and the sample preparation time needed. Even if not required for the detection system employed, extensive cleanup has the following additional benefits:

1.   Reduction in the possibility of false positives
2.   Improved analyte S:N by removing background
3.   Extended GC column life
4.   Less frequent MS ion source maintenance

Many cleanup procedures have been developed to isolate either the whole group of CDDs/CDFs or to isolate only specific congeners, such as 2,3,7,8,-4CDD, from the many coextractives present in environmental samples. The proliferation of cleanup schemes illustrates the wide variability of environmental samples and the many potential interferences in their analysis for CDDs/CDFs. Many of these cleanup schemes, however, have much in common. Although details on their use may differ, the general elements of sample extract cleanup include the following: bulk matrix removal by acid or base treatment of the solvent extract using aqueous wash or packed acid/base column elution, liquid-solid column chromatography to separate CDDs/CDFs from other chlorinated organics by elution on a variety of solid phases using solvents of varying polarity, and additional fractionation by HPLC for further cleanup or for toxic congener isolation from other CDDs/CDFs.

Two previous reviews[80,83] have described in detail several of the common cleanup methods employed. It is not the intention here to cover in detail all of the methods that have been reported. Rather, greater treatment is given to the individual elements of the cleanup, while only two of the most widely used total cleanup schemes are described in full.

1. Bulk Matrix Removal

It is sometimes difficult to separate this operation from the extraction/concentration procedure. Acid or base treatment is often used as an integral part of sample extraction, as described previously. For simple matrices such as water, bulk matrix removal can be accomplished simultaneously with sampling by passing water through sorbent cartridges capable of retaining the CDDs/CDFs. Significant efforts for bulk matrix removal must be made to separate lipids associated with biological matrices and to remove organic coextractives obtained during the analysis of complex soils/sediments. In addition to acid and base washing, concentrated organic solvent extracts may be processed by a number of liquid-solid chromatographic columns to isolate CDDs/CDFs from most other organics.

a. Magnesia/Celite 545

Following hexane extraction of crustaceans and fish, efficient cleanup was accomplished by elution of the concentrated extract through dry packed magnesium/Celite 545.[188] With a total loading of 8 g of lipids, this procedure was successful in removing 95% lipid, while retaining all spiked 4CDD congeners. Percent recoveries of [37]Cl-4CDD spikes ranged from 71 ± 12 (65 ppt spike) to 87 ± 21 (0.7 ppt spike). However, this method has not been shown to be effective for other CDD/CDF congener groups. Table 3 lists other cleanup schemes that incorporate the use of Celite 545.

b. Size Exclusion Chromatography

Size exclusion chromatography was found to be an effective method of removing lipids from fish extracts.[190] An initial base/acid treatment was used to remove most macromolecules. Size exclusion chromatography on a PMS 60S column further separated low molecular weight compounds from those ranging to 1000 Da. This technique was applied in combination with reversed-phase partitioning on a C-8 HPLC column followed by nonaqueous reversed-phase separation on C-18. HPLC methods are described below in greater detail.

c. Modified Silica Gel

To supplement or replace acid/base washing of the sample extract, many cleanup procedures employ silica-gel impregnated with concentrated H_2SO_4 or NaOH.[191] Lipids are oxidized and retained on the acid treated silica-gel (44% H_2So_4W/W) while CDD/CDF pass through, eluted with hexane solvent. This method has been used for biological samples,[191] municipal incinerator fly ash,[199] and soils.[147] Although very effective in removing hydrocarbons and easily oxidizable organics from sample extracts, it is easy to overload the capacity of cleanup columns used with extracts from heavily contaminated samples. In severe cases, the acid silica can actually form a plug, preventing sample flow through the column. For such samples a prewash with H_2SO_4 or multiple passes of the sample through fresh acid silica columns may be required.

2. Separation of Chlorinated Organics

A high hydrocarbon background can adversely affect CDD/CDF detection limits, while the presence of highly polar coextractives can interfere with chromatographic resolution and retention times, if cleanup is not performed. However, the most severe interferences for CDD/CDF determinations are from chlorinated organics giving ions having the same nominal masses as CDD/CDF congeners under the mass spectrometer conditions used for analysis. Possible interfering substances include: chlorinated pesticides such as DDT and DDE, some PCBs, polychlorinated naphthalenes (PCN), chlorinated diphenyl ethers (CDPE), and hydroxy and methoxy PCBs. These compounds are not removed by the procedures described above, and require additional cleanup steps.

a. Silica Gel and Florisil

Preliminary separation of chlorinated organic compounds can be performed using silica gel. CDDs are eluted with PCBs and PCNs from silica gel, but are separated from DDT and its analogs.[185] The retention times of these compounds increases with decreasing chlorine substitution,[185] which is opposite to the effect observed in gas chromatography. Detailed work by Leoni[192] using sequential elution of silica gel with increasingly polar solvents achieved only partial separation of DDT and DDE from PCBs. Varying results using silica gel can be obtained depending upon the degree of activation by thermal treatment that is performed.

Firestone[124] and O'Keefe et al.[188] found Florisil to be an efficient means of separating PCBs and DDE from 4CDDs for biological (bovine milk) samples. Use of a Florisil minicolumn was found to reduce PCBs to insignificant levels during cleanup of samples fortified with 4CDD. Florisil has also been applied to other biological samples such as chicken livers[195] to separate PCBs, DDE, and residual lipids from CDDs/CDFs. For the analysis of 2,3,7,8-4CDD, Florisil has been used with azulene as a visual marker compound.[203] PCBs are eluted just before azulene, while 2,3,7,8-4CDD elutes just after azulene using the solvent system described.

b. Alumina

In a recent worldwide survey on CDD/CDF analysis capability,[204] cleanup of sample extracts using alumina was identified as the most common cleanup method employed for all matrices surveyed (soil, water, fish, fly ash). Both acidic and basic alumina have been used.[186] Activation of the alumina by solvent washing and/or heating under inert gas flow is generally required, and the effectiveness of the cleanup is dependent upon performing the activation very reproducibly for each batch of samples. Activation affects the water content in the alumina and, therefore, the activity, since water molecules adsorb easily on the most energetic active sites.[205]

By sequential elution of alumina using solvents of increasing polarity, a series of chlorinated organics including PCBs, PCNs, chlorobenzenes, and polychlorinated terphenyls have been separated from CDD/CDF.[186] In another investigation,[187] a mixture of Aroclors® 1254, 1260, and 1262 was separated from di-, tri-, and tetrachlorinated dibenzo-*p*-dioxins using sequential elution of alumina with solvents of increasing polarity. Lamparski achieved 99.8% removal efficiency of PCBs from a solvent extract of trout fish using basic alumina.[191]

Alumina has also been used as a packing material for HPLC columns,[185] where a microparticulate alumina column (10 μm-AloxT) was used with a mobile phase of *n*-hexane to achieve a clear separation of all common PCB and PCN mixtures from the CDDs. It was also demonstrated that this type of HPLC system with a liquid chromatography-electron capture detector may be used as a screening method for CDDs in addition to performing effective separations of CDDs from other chloro-organics.[185] The CDFs were not analyzed in this work, although they coeluted with the CDDs. No real environmental samples were analyzed by this method. In another application,[206] alumina was used in thin-layer chromatography to separate 2,3,7,8-4CDD and 8CDD from Aroclors® 1221, 1254, 1260, and pesticides including p,p¹-DDT, p,p¹-DDD, p,p¹-DDE, and dieldrin.

c. Carbon

The use of powdered carbon as an adsorbent to isolate CDDs and CDFs in the presence of other chloro-organics is increasing in popularity.[204] The technique usually preferred is to use powdered carbon dispersed on the surface of shredded glass fiber filter as reported by Smith.[195] Using this adsorbent material, CDD/CDF were isolated from other coextractives from whole fish samples. Recoveries of CDD/CDF spiked at levels of 25 to 250 ppt ranged from 65 to 100% with a standard deviation of 5%. The sample loading on the carbon column when dispersed on glass fibers was reported to be three times greater than carbon dispersed on polyurethane foam.[111,195] A major advantage of carbon adsorbent, which traps molecules having planar structures while allowing others to be easily eluted, is that it is especially effective in isolating CDDs/CDFs in the presence of high levels of PCBs.[195]

The carbon-glass fiber adsorbent has been incorporated into a complex cleanup scheme described in detail below, that is effective in isolating CDDs/CDFs from a wide variety of environmental samples. A similar method used a packing consisting of a mixture of powdered activated carbon (Amoco PX21) and silica gel (1:25 W/W).[183] By this method, 2,3,7,8-4CDD could be isolated from fly ash and soil and quantified by GC-ECD. However, samples could not be analyzed by GC-MS because of excessive background noise. Carbon adsorbent has been used as part of cleanup schemes for the isolation of CDDs/CDFs from fish,[208,209] sediment,[150,209] and human adipose tissue.[211] Carbon adsorbent has also been used as part of a semiautomated cleanup method.[212]

3. High Performance Liquid Chromatography (HPLC)

HPLC has been used to analyze CDDs in pentachlorophenol[213] and environmental samples.[214] However, since GC-MS is generally required for determination of CDDs/CDFs in environmental samples at ppb to ppt levels, HPLC is more commonly used as a cleanup method.[42,114,191,196,197,215-222,270,273] Lamparski[191] used a reversed-phase chromatographic column (Zorbax ODS) with methanol isocratic mobile phase for fish cleanup that removed 99.999% of PCBs from sample extracts when combined with basic alumina cleanup. Isomer-specific determination of 2,3,7,8-4CDD could be performed when DDE was present in the sample at concentrations two million times greater than that of 2,3,7,8-4CDD.

Lamparski and Nestrick extended the use of HPLC for cleanup to the separation of all 22 isomers of 4CDD[42] and to the separate speciation of the 4CDD, 5CDD, 6CDD, 7CDD, and 8CDD congener groups.[196] The fractionation scheme combined reversed phase (ODS) with normal phase (silica) HPLC. This fractionation scheme provided distinct groups of isomers in each HPLC fraction collected that could be further separated and quantitated by GC-MS.

The use of HPLC for final extract purification prior to GC-MS determination of CDDs/CDFs was also reported by Ryan and Pilon.[114] A reversed-phase C8 HPLC column with acetonitrile mobile phase removed interferences and provided a separation of the tetra- to octachlorinated CDD/CDF congener groups. This procedure was applied to a wide variety of sample types including fish tissue. An activated carbon (Amoco PX-21) on silica-gel HPLC column[197] was used to cleanup samples heavily contaminated with PCBs. In many cases, the cleanup was effective enough to provide sample extracts suitable for analysis by GC/ECD. Tong also showed that quantitation of CDD/CDF could be performed with non-MS detection by employing a 2-step HPLC fractionation for extracts of municipal incinerator fly ash.[270,273] Advantages of HPLC cleanup are rapid fractionation, possibility of automated operation, and the ability to identify a wide range of organic compounds in samples, including CDDs/CDFs.[219] Since there is no single GC column that can resolve all of the different CDD/CDF congeners, fractionation of sample extracts using HPLC is often required for isomer-specific determinations.[42,191,215,216,218,221-223] When combined with other cleanup methods, HPLC techniques can provide CDD/CDF isomer-specific analysis capability even when using packed column GC-MS.[42,216] HPLC also can be used as an additional cleanup step for further cleanup of difficult samples when the usual cleanup used is unsuccessful.[222]

4. Complete Cleanup Methodologies

The cleanup methods described above for isolation of CDDs/CDFs in extracts from environmental samples are seldom used individually. Almost all cleanup methods reported for extracts derived from environmental samples are combinations of several of the above techniques. A National Research Council of Canada (NRCC) report[80] described in detail the various cleanup schemes reported up to 1981.

Most cleanups reviewed by NRCC employed an acid/base treatment for initial extraction and lipid saponification step. Whether the acid/base treatment was performed by washing the extract with an aqueous solution or eluting the solvent extract through a column packed with acid and/or base on silica gel seemed to be a personal preference; superior performance of one method over the other has not been demonstrated. Most methods also employed alumina column cleanup, often in conjunction with silica-gel or Florisil cleanup.

Advances made since 1981 have primarily included refinements to the original schemes to include CDD/CDF congener groups other than 4CDD, and determination of all 2,3,7,8-substituted CDD/CDF congeners in addition to 2,3,7,8-4CDD. The use of carbon adsorbent and HPLC in cleanup has also increased in recent years, although these methods had been introduced by 1981. Early work in the analysis of CDDs/CDFs in environmental samples heavily emphasized the determination of only 2,3,7,8-4CDD.

To show how complex cleanup schemes are developed from the individual elements described above, two complete cleanup methods are described in detail below. They were

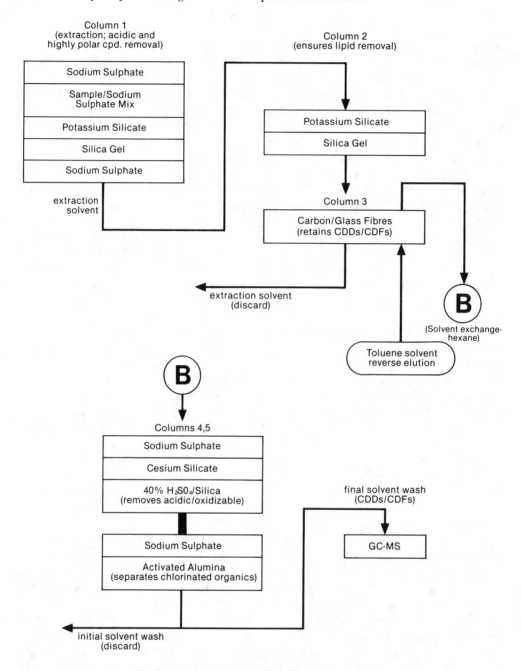

FIGURE 2.   Flow chart illustrating carbon fiber cleanup.

selected on the basis of proven effectiveness, ability to analyze a range of CDD/CDF congeners or congener groups, and to illustrate use of many of the individual cleanup elements described above. Other cleanup schemes are shown in Table 3.

a. Carbon/Glass Fiber Cleanup[198]
A schematic illustration of this cleanup is shown in Figure 2. The sample is blended with anhydrous sodium sulphate, dried and reblended to form a smooth flowing powder. The powdered sample is spiked with isotopically labeled CDD and/or CDF surrogate standards and solvent-extracted.

The extract is passed through a series of silica based adsorbents and a carbon/glass fiber adsorbent. The CDD/CDF as well as PCB, PCN are retained on the carbon fiber and are recovered by reverse elution with toluene. The toluene extract is applied to two tandem columns containing cesium or potassium silicate with sulfuric acid/silica gel followed by alumina. Elution from these two columns provide the final fractionation of CDD/CDF from the other chlorine containing organics. This scheme has been applied to a variety of species of freshwater fish,[117] snapping turtle fat, Baltic seal fat,[200] aquatic soils,[201] and soot from an office fire involving PCB.[202]

b. Dow Clean-Up[196]

Particulate samples are extracted by Soxhlet, followed by concentration of the extract to a small volume (1 to 3 ml). The bulk matrix components (lipids, fats, other oxidizable organics) are retained on the first of a sequence of columns and comprised of NaOH modified silica gel, H_2SO_4/silica gel. The chlorinated organics are collected, concentrated and passed through the second column: $AgNO_3$ on silica and alumina. The CDD fraction is free from the majority of PCB and other chlorinated interferences.

Fractionation of the CDD congeners is accomplished through use of reverse phase HPLC. The 6CDD, 7CDD, and 8CDD fractions are immediately subjected to capillary GC-MS for analysis. Further fractionation of the 4CDD extract, to provide isomer specific analysis of all 22 4CDD isomers, is accomplished through use of normal phase HPLC. This cleanup scheme has shown to be applicable to environmental particulate samples, biological samples, and fly ash samples.

C. Analysis by Gas Chromatography and Mass Spectrometry

GC-MS is universally accepted as the instrumental technique required for the analysis of CDDs and CDFs in environmental samples. If the cleanup procedure used is effective in removing the bulk of sample coextractives, then the application of GC-MS will be independent of the original sample matrix. Detection is then based solely upon the gas chromatograph column conditions, and mass spectrometer techniques that are employed.

A multitude of GC-MS techniques are available. Classification of these techniques can be by GC resolution (high resolution-HRGC, use of wall-coated open tubular columns; low resolution-LRGC, use of packed columns), method of MS ionization (electron impact or EIMS; positive ion chemical ionization-PCIMS: negative ion chemical ionization-NCIMS; atmospheric pressure ionization-APIMS); mass resolution (quadrupole MS-low resolution or LRMS; double focusing MS-high resolution or HRMS), or by multiple analyzer MS configurations (tandem mass spectrometry or MS-MS). The possible combinations of these techniques and others lead to a great many GC-MS configurations, however, only a few are in common use. The techniques of HRGC-LRMS and HRGC-HRMS, primarily employing EI ionization, are the most commonly used methods. NCIMS and GC-MS-MS have also been successfully applied to CDD/CDF determinations, but are in routine use by only a few groups. Generally, the choice of GC column for either total congener groups or isomer-specific analysis is the most important GC-MS variable.

1. Chromatographic Properties of CDDs/CDFs
a. Packed Column Analysis-LRGC

Packed columns are generally used for nonisomer-specific analysis of CDD/CDF. Table 3 summarizes the reported applications of packed columns to CDD/CDF determinations, listed according to the stationary phase used and matrix analyzed. The most widely used liquid phases in packed column work are SE-30, OV-3, OV-101, and a combination of OV-17/Poly S-179. An efficient column for packed column analysis of CDD/CDF congener groups was demonstrated by Eiceman et al.[255] A phase of Carbowax 20M on chromosorb

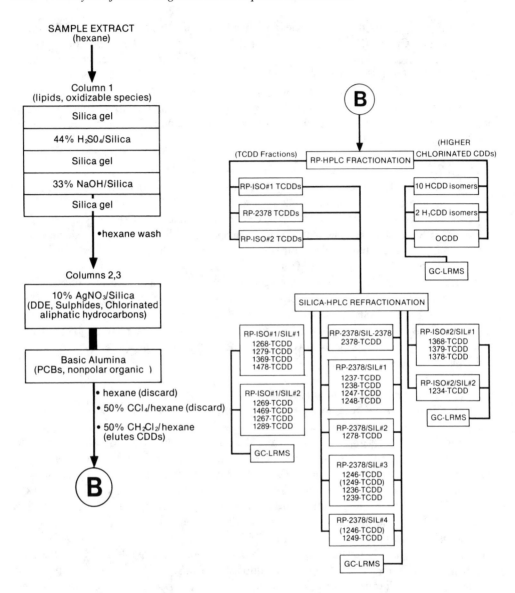

FIGURE 3.   Flow chart illustrating Dow cleanup.

W was exhaustively extracted by Soxhlet apparatus using methanol and then heat-treated to produce a very thin (0.2%) carbowax stationary phase. The tetra- to octachlorinated CDD congener groups were completely separated from each other using this column.

Isomer-specific 2,3,7,8-4CDD analysis in the presence of all 22-4CDD isomers using packed columns has been demonstrated by Lamparski and Nestrick.[216] Sample prefractionation by HPLC was required to perform this separation. Presumably, the use of a variety of packed columns with HPLC heart-cutting techniques could be used to separate all CDD/CDF congeners. However, the use of WCOT columns began to replace packed columns for CDD/CDF determinations during the period 1975 to 1980, when the commercial availability of WCOT columns greatly increased. A development of special significance was the manufacture of reproducible, stable WCOT columns of fused silica, made flexible by covering the fused silica with a polyimide coating. Advantages over packed columns include the following: lower bleed for GC-MS analysis; lower carrier flows needed, allowing direct

Table 3
APPLICATION OF PACKED COLUMN GC TO THE DETERMINATION
OF CDDs AND CDFs IN VARIOUS SAMPLES[a]

Stationary phase	Application	Congeners determined	Ref.
3% OV-1	Commercial pentachlorophenols	6CDD, 7CDD, 8CDD	236, 237, 242
	Commercial pentachlorophenols	tetra to octa CDD/CDF	241
	Commercial PCBs	2,3,7,8-4CDF, 2,3,4,7,8-5CDF	239
3% OV-3	Bovine milk	6CDD, 8CDD	107
	Biological extracts	2,3,7,8-4CDD	247
	Beef fat	2,3,7,8-4CDD	249
	Fish, water, soil	2,3,7,8-4CDD	243
	Bovine milk	2,3,7,8-4CDD	244
	Commercial herbicides	2,3,7,8-4CDD	245
	Beef adipose tissue	2,3,7,8-4CDD	267
3% OV-7	Metal surface swabs	2,3,7,8-4CDD	256
1% OV-17	Wood dust	CDDs, CDFs	268
	Reaction products of chlorophenoxyphenols	CDDs, CDFs	36
2% OV-17	Commercial herbicides	2,3,7,8-4CDD	233
	Foliage	4CDD	258
3% OV-17	Commercial herbicides	CDDs	269
3% OV-101	Treated wood	8CDD	260
5% OV-101	Chicken fat	CDDs	228
0.1% OV-105	Bovine milk	6CDD, 8CDD	107
3% OV-210	Biological tissue and wood	CDDs	333
	Commercial herbicides	4CDD	253
	Perchlorination	8CDD	230
3% OV-225	Commercial herbicides	2,3,7,8-4CDD	235
1.5% OV-17/2.0% QF-1	Pesticides	CDDs, 2,3,7,8-4CDD	229
1.5% OV-17/2% QF-1	PCBs	2,3,7,8-4CDF, 2,3,4,7,8-5CDF	239
3% OV-1/5% OV-210/ 6% OV-225	PCBs	2,3,7,8-4CDF, 2,3,4,7,8-5CDF	239
0.6% OV-17/0.4% Poly S-179	Coal combustion	CDDs	265
	Fish	2,3,7,8-4CDD	191
	Particulate samples	isomer-specific CDDs	196
	Synthesis	6CDDs	215
	Purified pentachlorophenol	6CDD, 7CDD, 8CDD	250
	Soils	2,3,7,8-4CDD	262
	Human milk	4CDD, 6CDD, 7CDD, 8CDD	264
3% SE-30	Pentachlorophenol	6CDD, 8CDD	240
	Commercial herbicides	1CDDs to 4CDDs	269
5% SE-30	Fish	2,3,7,8-4CDD	232
2.5% SE-52	Fats, oils, fatty acids	CDDs	189
1% Apiezon M	Chlorophenols	CDFs	238
1.5% Apiezon M	Human tissue	CDFs	254

Table 3 (continued)
APPLICATION OF PACKED COLUMN GC TO THE DETERMINATION
OF CDDs AND CDFs IN VARIOUS SAMPLES[a]

Stationary phase	Application	Congeners determined	Ref.
2.5% BMBT liquid crystal	Human milk	2,3,7,8-4CDD	266
1% Hi-Eff 8BP	Commercial herbicides	2,3,7,8-4CDD	233
0.2% Carbowax 20M	Incinerator fly ash	CDDs, CDFs	225, 252, 255
10% DC-200	PCBs	CDFs	19
3% Dexil 300	Effluents, soils, chemical wastes	4CDDs	263
	Cooking oils	CDFs	246
5% QF-1	Treated wood	8CDD	260
15% QF-1/10% DC-200	PCBs	CDFs	19
1.2% Silar 10C	Commercial gelatins	isomer-specific 6CDD, 7CDD, 8CDD	124
3% SP-2100	Cow's milk/blood	6CDDs, 7CDDs, 8CDD	194
5% SP-2401	Commercial PCBs	2,3,7,8-4CDF, 2,3,4,7,8-5CDF	239
5% UCW-98	Pesticides	CDDs, 2,3,7,8-4CDD	229
3% XE-60	Commercial chlorophenols	CDDs	227
	Corn oil	2,3,7,8-4CDD, 8CDD	226, 234
	Commercial pentachlorophenol	CDDs, CDFs	242

[a] For analysis conditions, consult original references.

connection with the MS ion source; greater sensitivity, better peak shapes, better isomer separation; and, fused silica column physical flexibility allows easier installation. For highly complex environmental samples, WCOT columns are essential to achieve the best detection limits, greatest isomer specificity, and maximum freedom from interfering co-extractives. Packed columns are still useful for simple mixtures, in screening applications, and as the initial column used for multidimensional gas chromatographic separations.[148,271,272] Using a 10-foot packed Silar 10C column followed by a combined DB-17/SP2330 capillary column interfaced to a high resolution mass spectrometer; PCB fluids, fly ash extract, and hexa-chlorophene were analyzed for 2,3,7,8-4CDD and 2,3,7,8-4CDF without preliminary cleanup.[271]

b. Wall-Coated Open Tubular (Capillary) Column Analysis-HRGC

Capillary columns used for the analysis of CDDs and CDFs can generally be divided into two classifications, relatively nonpolar phases such as OV-101 and SE-54 that separate CDD/CDF congeners into well-defined groups depending upon the number of chlorine atoms,[290] and more polar columns such as Silar 10C, SP-2330, SP-2340, and CP SIL-88 needed to perform isomer-specific determinations. Isomer-specific analysis can also be performed by a combination of HPLC prefractionation and capillary column GC-MS. No single column has yet been developed that will separate all 210 CDD/CDF congeners. However, because of the large difference in relative toxicity between CDD and CDF congeners having 2,3,7,8-

chlorine substitution and the other CDD/CDF congeners, for environmental applications it is only necessary to achieve isomer specificity for the following: 2,3,7,8-4CDD, 2,3,7,8-4CDF, 1,2,3,7,8-5CDD, 1,2,3,7,8-5CDF, 2,3,4,7,8-5CDF, 1,2,3,6,7,8-6CDD, 1,2,3,7,8,9-6CDD, 1,2,3,4,7,8,-6CDD, 1,2,3,6,7,8-6CDF, 1,2,3,7,8,9-6CDF, 1,2,3,4,7,8-6CDF and 2,3,4,6,7,8-6CDF. These five CDD and seven CDF congeners are often collectively referred to as the "dirty dozen".

Separation of discrete CDD/CDF congener groups, depending upon the degree of chlorination, has been demonstrated for OV-101,[274,284,288] OV-17,[274] SE-30,[286] and DB-5 (SE-54).[50,287,289,291] Retention data for 21 of 28 5CDF congeners using a 60-m DB-5 column were reported.[51] Retention indices for all 135 CDFs on a 30-m DB-5 were also determined.[291] A model was presented that described the relationship between the retention characteristics of CDFs and their structure.[291]

Many studies of the retention characteristics of the 4CDD congener group have been reported.[46,118,123,209,275-277,279,280,282] Isomer-specific determination of 2,3,7,8-4CDD can be performed using CP Sil,[46] Silar 10C,[209,276-278] OV-17,[209] SP-2330,[46,279] and SP-2331.[280] In one study, isomer-specific analysis of 2,3,7,8-4CDD was confirmed by determining its retention index on three GC columns[118] including methyl silicone, OV-17, and SE-54 phases. All 2,3,7,8-substituted CDD congeners can be uniquely resolved using Silar 10C[278] or SP 2330.[285] However, unique determination of all 2,3,7,8-substituted CDF congeners using a single column has not been reported. Studies of the 38 CDF congeners using SE-54 and SP 2330 columns showed that 2,3,7,8-4CDF could be resolved using both columns, but not on either individual column.[48,55] It was noted that a close eluting isomer (2,3,4,8-4CDF) was barely separated from 2,3,7,8-4CDF from both columns,[48] and could provide a problem in analysis. It was demonstrated that 2,3,4,7,8-5CDF, 1,2,3,6,7,8-6CDF, 1,2,3,7,8,9-6CDF, and 2,3,4,6,7,8-6CDF are uniquely separated using SP-2330,[50] while 1,2,3,4,7,8-6CDF and 1,2,3,7,8-5CDF could be determined using both SP-2330 and DB-5.[285] The toxic congener 2,3,7,8-4CDF could not be unambiguously determined using any single column, or combination of SP-2330, SP-2340, and DB-5.[50] Other columns that have been used for CDD analysis include SP 2100[281] and a polymeric liquid crystal column.[283] The liquid crystal column completely separated 2,3,7,8-4CDD from an Aroclor® mixture, and from most other 4CDD isomers.

2. Determination of CDD/CDFs by Electron Impact GC-MS

The two requirements of high sensitivity and high selectivity must be achieved to determine CDDs/CDFs in environmental samples. A detection system should be able to give a positive response to between 1 to 100 pg of any CDD or CDF congener. For the most toxic congeners (2,3,7,8-4CDD, 2,3,7,8-4CDF) an absolute sensitivity of 10 pg or less is desirable. High selectivity is required because potential interferences are often present in sample extracts even after extensive cleanup has been performed. A third requirement of high specificity to distinguish between the dirty dozen and other CDD/CDF congeners is needed for some applications, and is provided by techniques already described.

GC-MS is the only analytical method that can provide the high sensitivity and selectivity needed to determine concentrations of CDDs/CDFs in environmental samples. A variety of different GC-MS techniques are available for this work including positive ion electron impact (EIMS), positive (PCIMS) and negative (NCIMS) chemical ionization, atmospheric pressure chemical ionization (APIMS), high (HRMS) and low resolution (LRMS) mass spectrometry, and tandem mass spectrometry (MS-MS). In most reported applications to date EIMS was used. LRMS using the quadrupole analyzer is common, especially for applications involving determination of CDD/CDF congener groups, although extensive use of HRMS has also been employed.

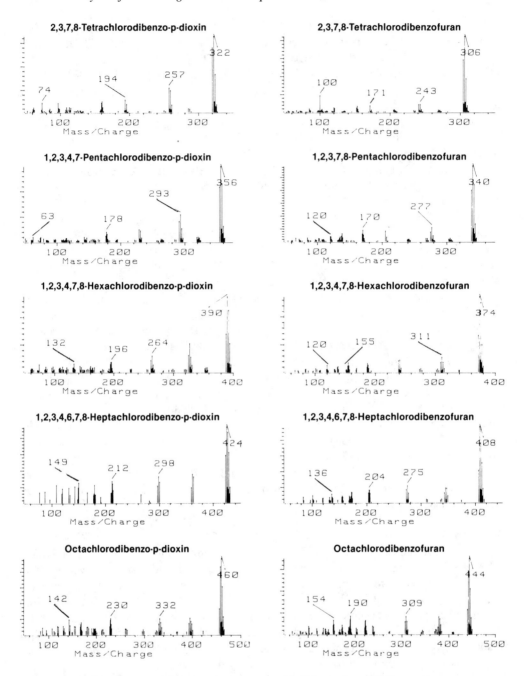

FIGURE 4.   Electron impact mass spectra at 70 eV for CDD and CDF congeners.

a. Characteristics of Electron Impact Mass Spectra

The fragmentation patterns of CDDs[35,39,285,292-297] and CDFs[51,285,289,297] have been discussed previously. Figure 4 shows the EI mass spectra for representative members of each CDD and CDF congener group containing four to eight chlorines. All the CDD mass spectra shown in Figure 4 are dominated by intense parent ions (base peak) and contain major fragment ions due to losses of CO and Cl and formation of $[M-COCl]^+$ and $[M-2COCl]^+$ ions. Strong doubly-charged molecular ions are also observed. Buser and Rappe[294] have shown that positional CDD isomers can be distinguished by their low mass ion region (<200

Table 4
CHARACTERISTIC CDD/CDF ION MASSES FOR GC-MS/SIM ANALYSIS

Congener group	Ion description	CDD masses		CDF masses		Relative abundance
		Nom.	Exact	Nom.	Exact	
Tetrachlorinated	M$^+$	320	319.8965	304	303.0939	77
	[M + 2]$^+$	322	321.8936	306	305.8987	100
	[M + 4]$^+$	324	323.8906	308	307.8957	49
	[M − COCl]$^+$	257	256.9328	241	240.9379	a
Pentachlorinated	M$^+$	354	353.8576	338	337.8627	62
	[M + 2]$^+$	356	355.8546	340	339.8597	100
	[M + 4]$^+$	358	357.8517	342	341.8568	65
	[M − COCl]$^+$	291	290.8938	275	274.8989	a
Hexachlorinated	M$^+$	388	387.8186	372	371.8237	51
	[M$^+$2]$^+$	390	389.8156	374	373.8207	100
	[M + 4]$^+$	392	391.8127	376	375.8186	81
	[M − COCl]$^+$	325	324.8548	309	308.8599	a
Heptachlorinated	M$^+$	422	421.7796	406	405.7847	44
	[M$^+$2]$^+$	424	423.7767	408	407.7818	100
	[M + 4]$^+$	426	425.7737	410	409.7788	98
	[M − COCl]$^+$	359	358.8259	343	342.8209	a
Octachlorinated	M$^+$	456	455.7407	440	439.7407	34
	[M + 2]$^+$	458	457.7377	442	441.7428	88
	[M + 4]$^+$	460	459.7348	444	443.7398	100
	[M + 6]$^+$	462	461.7318	446	445.7369	65
	[M − COCl]$^+$	393	392.7769	377	376.7820	a

a Depends upon instrument parameters. Generally, the COCl-loss ion for the base peak will be at a relative abundance of 20 to 35%.

amu), however, the differences in mass spectra between members of a congener group are small, and could not be used to distinguish between isomers in real environmental sample analysis. Isomers of CDF congener groups yield identical EI Spectra.[285] The EI spectra of CDFs have similar characteristics and fragmentation patterns as CDD congeners.

For analytical purposes, the important features of the spectra shown in Figure 4 are the characteristic isotope patterns in the molecular ion region. These are the typical patterns expected for chlorine-containing molecules due to the presence of the stable isotopes of chlorine,[35]Cl and [37]Cl. For trace CDD/CDF determination in environmental samples, it is not possible to obtain complete mass spectra. Identification of CDD/CDF congeners is then performed by the technique of selected ion monitoring, where only two or three characteristic ions of each congener group are monitored.

b. Analysis by HRGC-LRMS

Analysis of cleaned-up sample extracts using GC-MS is performed using the technique of selected ion monitoring (SIM). By choosing only a few characteristic ions for each congener instead of the complete mass spectrum, increased sensitivity of a factor of 100 or more can be obtained. However, the effect of interferences becomes more prominent as fewer ions are monitored, and it is important to follow strict criteria for the identification of CDD/CDF peaks. Table 4 lists the characteristic masses most often monitored for each CDD and CDF congener containing four or more chlorines. There is a trade-off between the number of ions monitored and sensitivity. At least two characteristic masses should be monitored for each congener group to provide some degree of confidence that the peaks detected are not interferences. For low resolution MS analysis, monitoring at least three ions per congener group is recommended. The use of GC-MS/SIM for the analysis of CDDs/

CDFs has been discussed previously.[297,298] This technique has also been called "mass frag-
mentometry" and "multiple ion detection". However, the term selected ion monitoring is
preferred because molecular ions as well as fragment ions are monitored, and either one or
a number of ions can be selected.

An important aspect of GC-MS/SIM determination of CDDs/CDFs is the addition of
isotopically labeled CDDs to samples before sample extraction and cleanup. Determination
of the percent recoveries of these surrogates can be used as indicators of the efficiency of
the entire analytical process. The ability to monitor isotopically labeled surrogates in the
same analysis as native CDDs/CDFs are measured is an important advantage of GC-MS
detection. Fully labeled ^{37}Cl and ^{13}C standards are available commercially, including ^{13}C
analogs of all of the most toxic congeners.[299] Methods for the determination of 2,3,7,8-
4CDD generally require the addition of either ^{37}Cl-2,3,7,8-4CDD or ^{13}C-2,3,7,8-4CDD
before extraction, followed by selective cleanup, GC-MS/SIM detection, and correction of
the native 2,3,7,8-4CDD concentration for recovery losses as determined by monitoring the
surrogate standard.[300] Sometimes both ^{37}Cl-2,3,7,8-4CDD and ^{13}C-2,3,7,8-4CDD are used
for this application. One standard is a surrogate to monitor recovery efficiency, while the
other is an internal standard used for quantitation. If separate determination of percent
recovery is not needed, then direct quantitation using the surrogate can be performed. By
assigning a response factor to the surrogate using its peak height or area determined by GC-
MS in the final sample extract, based upon the amount of surrogate added to the sample
before extraction, accurate quantitation can be performed with automatic recovery correction.
Care must be taken when using the ^{37}Cl-labeled surrogate, since its mass spectrum gives a
single molecular ion peak (no Cl-isotope cluster) and, therefore, its response factor is not
the same as that of either native or ^{13}C-labeled 2,3,7,8-4CDD. If external quantitation is
performed, an injection standard should also be employed to correct for between-injection
variations, especially for the split/splitless injection technique. Although high quality native
and labeled standards are available, a recent study has shown that extensive quality checks
of commercial standards should be part of an overall quality control program for CDD/CDF
analysis.[301] When standards of all the congeners or congener groups to be quantitated are
not available, it may be possible to interpolate response factors for the missing congeners.[305]
Best sensitivity for CDD analysis is obtained at a MS electron energy of 35 eV rather than
the commonly used 70 eV,[306] however, Chittim et al.[51] has shown that the highest sensitivity
for GC-EIMS analysis of 5CDFs is 70 eV.

A number of criteria must be satisfied for peaks detected by GC-MS/SIM analysis before
they are identified as CDD/CDF congeners. These include the following:

1. Correct capillary column GC retention time
2. Simultaneous response of all characteristic ions monitored
3. Correct peak area ratios for all characteristic ions monitored
4. Peaks detected must be present with signal:noise greater than 3:1

In the case of isomer-specific determinations, criterion (1) refers to the exact retention
time match between surrogate or external standards. For total congener group determinations,
the retention time of each peak must be within the known retention window of its corresponding congener group. This is determined by analysis of a mixture containing most
CDDs/CDFs, or at least the first and last eluting members of each congener group. Such a
mixture can be derived from extracts of environmental samples such as municipal incinerator
fly ash that contain most congeners.

Criterion (3) can refer to the correct theoretical ratios of ions from the molecular ion
cluster, which depend on the relative natural abundances of ^{35}Cl and ^{37}Cl and number of
chlorine substituents present. Generally, a deviation of ± 10 to 20% is allowed on theoretical

ratios. Alternatively, the allowed deviation can be taken as a fixed percentage from the isotope ratios determined from periodic analysis of standards. This latter approach is recommended since MS tuning procedures can affect ratios differing by even a few amu. Also, ratios of the [M-COCl]$^+$ or similar ion abundances are expected to be even more variable than ratios of ion abundances from members of the molecular ion cluster.

A number of additional considerations may be used in the interpretation of GC-MS/SIM data. Confidence of identification is increased when GC-MS analysis is preceded by an efficient and highly specific sample cleanup.[302] Analysis of a number of samples from the same matrix/source generally produces data that show the same relative distributions of isomers/congener groups. Therefore, samples producing patterns differing from those expected should be examined in greater detail for the possible presence of interferences. This similarity of CDD/CDF patterns is very well illustrated in the analysis of municipal incinerator emissions.[172] It is also advisable to check the results of LRMS determinations by occasional re-analysis using HRMS. This is especially important when new types of samples, or samples from a new geographical area are analyzed, since new interferences may be present that are not removed by the existing procedures. The difficulties of analyzing and validating CDD/CDF data, especially when determined near the detection limit, have been discussed by Crummett.[303,304]

c. Analysis by HRGC-HRMS

Although the initial cost of HRMS instrumentation is considerably greater than that of LRMS equipment, HRMS is extensively used for determination of CDD/CDF in environmental samples. Many chlorinated contaminants exist that are potential interferences in the GC-LRMS analysis of CDD/CDF in environmental samples. Several studies have been reported that investigated these contaminants.[142,295,307-311,313] In one investigation,[311] the effect of PCBs, polychlorinated naphthalenes (PCN), chlorinated diphenyl ethers (DPE), chlorinated methoxy and hydroxy-PCBs (MEO-PCB, HO-PCB), chlorinated methoxy/hydroxy diphenyl ethers, chlorinated benzyl phenyl ethers (BzPE) and chlorinated biphenylenes were studied. The potential interferences were added to samples which were then analyzed by HRGC-HRMS to evaluate their effect. By employing an efficient cleanup procedure that included activated carbon,[208] no interferences were observed. However, PCN at concentrations greater than 1 ppm and nonortho-substituted PCBs greater than several ppb could produce interferences that obscure low-level CDD/CDF signals. Table 5 is a summary of potential CDD/CDF interferences. Many of these interferences can be eliminated by HRMS at resolution 10,000 or better. Another advantage of HRMS is that lower background is generally obtained, which may give lower detection limits than can be obtained using LRMS.

The use of HRMS has been largely confined to the determination of 2,3,7,8-4CDD, or the group of 4CDD congeners.[113,315-320] This is principally due to the increased importance of 2,3,7,8-4CDD compared to other CDD/CDF congeners, but also because HRMS instrumentation is not as flexible as LRMS-quadrupole analyzers for rapid ion switching when a wide range of CDD/CDF congeners are determined. Modern HRMS instruments with laminated magnets and improved software now approach the flexibility of quadrupole systems for multiple ion detection, and it is possible to determine all CDDs/CDFs containing 4 to 8 chlorines at a resolution of 10,000 in a single analysis.[321] Thoma and Hutzinger[325] have shown that using HRMS can allow reduced sample cleanup for the quantitative determination of tetra- to octachlorinated CDDs/CDFs in extracts of municipal incinerator fly ash.

d. Other Mass Spectrometric Methods

Direct mass spectrometry — Direct analysis by mass spectrometry with no prior GC separation has been reported.[231,232,322-324] Baughman and Meselson[231] achieved a 1-pg detection limit for 2,3,7,8-4CDD using a double-focusing MS and direct sample inlet by

Table 5

**POTENTIAL INTERFERENCES IN GC-MS ANALYSIS OF
ENVIRONMENTAL SAMPLES FOR CDD/CDF**

Compound	Mass of interfacing ion	MS resolution	Ref.
Interferences for 4CDD Determination: Masses 257; 259; 320; 322			
Tetrachlorobiphenyls	256.9506	14,400	142
	258.9476	14,500	142
Pentachlorobiphenyls	256.9461	19,200	142
	258.9432	19,300	142
Hexachlorobiphenyls	256.9320	321,100	142
	258.9290	323,600	142
Heptachlorobiphenyls	256.9275	48,700	142
	258.9245	48,800	142
	321.8678	12,500	80
Octachlorobiphenyls	256.9134	13,200	142
	256.9089	10,800	142
Nonachlorobiphenyls	319.8521	7,200	80
	321.8491	7,300	80
Tetrachloromethoxybiphenyl	319.9329	8,900	80
	321.9299	8,900	80
Tetrachlorobenzylphenyl ether	319.9329	8,900	80
	321.9300	8,900	80
Pentachlorobenzylphenyl ether	319.9143	18,100	80
	321.9114	18,200	80
DDT	319.9321	9,100	80
	321.9292	9,100	80
DDE	319.9321	9,100	80
	321.9292	9,100	80
Tetrachloroxanthene	319.9143	18,100	80
	321.9114	18,200	80
Hydroxytetrachlorodibenzofuran	319.8966	cannot resolve	80
	321.8936	cannot resolve	80
Tetrachlorophenylbenzoquinone	319.8966	cannot resolve	80
	321.8936	cannot resolve	80
Interferences for CDD/CDF determination			
Polychlorinated naphthalenes (PCN)	CDF interference is separated by GC		285
Polychlorinated terphenyls (PCT)	CDF interference, removed in cleanup		285
Polychlorinated diphenyl ethers (DPE)	CDF interference, removed in cleanup		285,311
Chlorinated polynuclear aromatic compounds	CDF interference, separated by GC		285
Chlorinated methoxy-PCBs (MeO-PCB)	CDD and CDF interference		310,311
Chlorinated hydroxy-PCBs (HO-PCB)	CDF interference		311
Chlorinated methoxy-diphenyl-ethers (MeO-DPE)	CDD and CDF interference		311
Chlorinated hydroxy-diphenyl-ethers (HO-DPE)	CDD interference; can undergo ring closure in heated injection port		311—314
Chlorinated benzylphenyl ethers (BZPE)	CDD interference		311
Chlorinated biphenylenes	CDD interference		311

repeated scans over a narrow mass range and producing a single time-averaged scan. Sample cleanup was needed since 2,3,7,8-4CDD response was reduced in the presence of large sample sizes. They applied this technique to the analysis of fish.[232] However, the benefits of GC-MS selectivity and ease of quantitation far outweigh the advantages of rapid analysis by the direct MS method. Dougherty[323] and Kuehl et al.[324] employed direct-probe negative chemical ionization MS as a screening tool. They were able to detect tetra-, penta-, hexa-, and octa-CDDs in fish extracts, but no isomer speciation was possible. Reynolds et al.[322] described a high pressure ionization mass sprectrometer capable of detecting less than 10 fg of 2,3,7,8-4CDD, however, it was not used to analyze real environmental samples. Because of the high complexity of environmental samples, use of direct MS methods will probably be limited to special screening applications.

Chemical ionization mass spectrometry — Negative ion chemical ionization mass spectrometry (NCIMS)[282,289,326-334] and positive ion chemical ionization mass spectrometry (PCIMS)[289,332-335] have been used for CDD/CDF determination. Both NCIMS spectra[289,327,330-332,334] and PCIMS spectra[289,332,334] have been reported. The methane PCIMS spectra of CDDs and CDFs have as base peak the protonated molecular ion [MH]$^+$, while the methane NCIMS spectra of CDFs give [M]$^-$ base peaks. The methane NCIMS spectra of CDDs have as base peak [M-35]$^-$ for penta- to octachlorinated congeners. For tetrachlorinated CDD congeners, NCIMS may be used to differentiate between the isomers depending upon the presence or absence of peri-chlorines when oxygen is added to the MS ion source.[331] Since the only 4CDD isomer with no peri-chlorines is 2,3,7,8-4CDD, NCIMS can be selective for this isomer. Thus, while most oxygen-NCIMS spectra of CDDs and CDFs have a base peak of [M-19]$^-$, the oxygen-NCIMS spectrum of 2,3,7,8-4CDD has its base peak at m/z 176. Miles found that the oxygen NCIMS mass spectra of hexachlorinated CDDs exhibited a base peak of [M-Cl]$^-$[237] as reported by Hass et al.[332] However, this may have been caused by operation at higher MS source temperatures. Unfortunately, relatively few real applications of chemical ionization techniques to environmental sample analysis have been reported. Although use of NCIMS may be a valuable tool for isomer speciation, variable response factors for different members of a specific congener group make this technique difficult to apply to the quantitative analysis of samples where many isomers are present. A study by Rappe et al.[335] showed that response factors for both CDDs and CDFs were more variable for NCIMS than EIMS, and that while NCIMS response factors were greater than EIMS response factors for the CDFs studied, the response factors for CDDs by NCIMS were lower.

Atmospheric pressure ionization MS (APIMS) and tandem mass spectrometry (MS-MS) — A number of studies have used APIMS for the determination of 2,3,7,8-4CDD.[300,336-339,342] For the analysis of fish tissue,[336] use of APIMS allowed selective determination of 2,3,7,8-4CDD even in the presence of PCBs. APIMS was also shown to be highly specific for the 2:2 chloro-substituted 4CDD isomers.[338] In the analysis of 4CDD isomers,[340,341] APIMS gave a detection limit of 0.5 pg for 2,3,7,8-4CDF.[341] The use of APIMS for ultrasensitive determination of specific CDD/CDF isomers seems promising, but additional work on congeners other than 4CDD/4CDF and more extensive application to real environmental samples are needed.

MS-MS methods have also been used to analyze CDDs and CDFs.[140,343-352] Most studies have emphasized isomer-specific determination of 2,3,7,8-4CDD, but all CDD/CDF congeners from tetra to octachlorinated were determined in a few investigations.[350,351,353] Tetrachlorinated CDDs in fish extracts were determined by direct probe and API source in one study.[343] The M$^+$ ion of 4CDD (m/z 320) was allowed to pass through the first mass analyzer to a collision cell,[343] where further collisionally induced dissociation occurred, followed by monitoring the COCl-loss ion (m/z 63) using the second MS analyzer. Because of the dual-MS selection 4CDDs could be selectively determined in the presence of PCBs at 1,000 times greater concentration with no GC separation. However, for very complex environmental samples, GC separation before MS-MS detection is generally required. However, the increased specificity of MS-MS allows rapid elution using a short GC column and reduced sample cleanup,[140,349,350] compared to conventional GC-MS. The accuracy and precision of rapid GC-MS-MS determination of 2,3,7,8-4CDD in soil was found to be comparable to that of conventional GC-MS and lengthy sample cleanup.[140]

An advantage of such rapid analysis methods is that a mobile GC-MS-MS system can be transported to the site of emergency or cleanup operations to provide screening analyses within hours.[353] The number of environmental applications of GC-MS-MS reported to date is somewhat disappointing, considering the tremendous potential of this technique.

D. Bioanalytical Methods

There has been considerable interest in development of bioanalytical procedures for rapid screening determination of the toxic CDD/CDF congeners.[353-366] Radioimmunoassay techniques were used to determine 2,3,7,8-4CDD,[353,354] and 2,3,7,8-4CDF,[355] with the best detection limits approximately 20 pg. Other methods based on the induction of aryl hydrocarbon hydroxylase (AHH) activity,[356,357] and the cytosol receptor assay,[358] have been reported. Helder and Seinen used an early-life-stage toxicity test to screen for CDDs/CDFs.[365,366] A flat-cell bioassay was shown to have a detection limit of 3 to 30 ppt for the most toxic 2,3,7,8-4CDD and 2,3,7,8-4CDF congeners, respectively, compared to other CDD/CDF congeners tested that were over 100 times less sensitive to this test.[362-364]

These methods are of interest for large scale surveys or emergency response situations where many samples must be tested rapidly to delineate the areas of greatest contamination. More expensive and lengthy GC-MS procedures would then need to be applied to confirm the presence of CDDs/CDFs and to determine accurate concentrations. Bioassay methods are generally nonspecific for 2,3,7,8-4CDD, but indicate total biological activity, including other 2,3,7,8-substituted CDDs/CDFs. For some applications, this may be advantageous since a hazard assessment of CDDs/CDFs must consider all toxic congeners. Further, samples that exhibit high activity but low CDD/CDF concentrations should be examined for the presence of other toxic substances. Some bioassay tests are specific enough that reduced sample cleanup is not needed, although removal of other active compounds such as polyaromatic hydrocarbons may be required.

In spite of the promise of rapid and inexpensive screening analysis, application of bioanalytical techniques for CDDs/CDFs in the environment has not attained wide use. Although not as sensitive as rigorous sample preparation and GC-MS determination, the sensitivity of bioassay procedures is low enough to permit assessment of contaminated areas requiring cleanup. More work is needed to further improve sensitivity, correlate biological response with toxic CDD/CDF congener concentrations, and apply these methods to real environmental problems.

III. DISTRIBUTION AND FATE IN THE ENVIRONMENT

The single compound 2,3,7,8-4CDD is undoubtedly the most intensively studied with respect to its toxic properties and environmental distribution. Other CDD and CDF congeners, especially the most toxic 2,3,7,8-substituted compounds, are also under active investigation. A number of reviews have described the sources and environmental distribution of these substances and have been discussed earlier.[56-64,79,162,296,300,368-371] Although 2,3,7,8-4CDD is not always found in environmental samples, most environmental surveys have found that when low ppt detection limits are achieved and all CDD/CDF congeners are investigated, some congeners will be detected. There are enough data, therefore, to strongly suggest that taken as a group, the CDDs/CDFs are ubiquitous contaminants of the environment. These compounds are distributed via air, water, and soil, and are resistant to microbial degradation and photodecomposition. Although they have low vapor pressures, volatility can be important, and in aqueous systems bioaccumulation can be significant. Many investigations of the environmental transport and fate of 2,3,7,8-4CDD have been conducted. Although much less work in this area has been performed for the other CDD/CDF congeners, their behavior can often be approximated by studies performed using 2,3,7,8-4CDD, and applying corrections based upon differences in known physical parameters such as vapor pressure, water solubility, and log K_{ow}. In a recent review of the biological fate of organic priority pollutants in the aquatic environment,[373] 2,3,7,8-4CDD was rated as persistent, volatile, and accumulative.

Several studies of the fate of 4CDD congeners in model aquatic systems have been conducted.[89,90,374-378] The principal removal mechanisms from water are volatilization, bioaccumulation by aquatic organisms, and adsorption onto sediments.[374-377] Mill has described an environmental fate assessment of 4CDDs based upon previous studies and the physical properties of 2,3,7,8-4CDD.[379] Unfortunately, his assessment was limited by the lack of experimental data in this area. The affinity of 2,3,7,8-4CDD for suspended particulates in water was studied by Srinivasan and Fogler,[380] who found that hydroxy-aluminum-clay sorbents can be used as efficient removal agents. Webster et al.[381] found that the solubilities of CDDs in water are enhanced by the presence of dissolved humic matter. An hypothesis for the sorption of hydrophobic chemicals from water has been presented by Mackay and Powers.[382] More studies are needed to further elucidate the environmental fate of CDDs and CDFs in the aquatic environment.

A. Environmental Transport and Degradation

1. Soil Mobility

Primarily because of several environmental contamination episodes, many investigations of the mobility of 2,3,7,8-4CDD in soil have been conducted.[128,138,151,383-394] Generally, 2,3,7,8-4CDD is strongly bound to soil and is not easily leached.[393,394] Vertical penetration in soil at Seveso was found to occur primarily in the first 20 cm within 10 to 15 months.[383] Extrapolation back to initial surface concentrations indicated that 2,3,7,8-4CDD was not removed from the surface layer by other mechanisms. The vertical rate of 2,3,7,8-4CDD migration is affected by the presence of organic solvents,[128,157] such as might be present in chemical waste sites. Mobility is very slow if 2,3,7,8-4CDD is dispersed in saturated hydrocarbon solvent,[128,389] but is increased substantially with aromatic solvents and chloroform. The partitioning of 2,3,7,8-4CDD between soil solid-phase and soil water extracts was found to be dependent in part by the entire group of chlorinated organic compounds present.[151] Increasing soil organic matter decreased the volatilization flux of 2,3,7,8-4CDD.[395] In a study of the offsite transport of 2,3,7,8-4CDD from a herbicide disposal facility,[385] vaporization was found to be the major loss mechanism. Because 2,3,7,8-4CDD is primarily confined to the upper soil layers, water and wind erosion are also significant transport mechanisms.[384] A model of the movement of 2,3,7,8-4CDD in soil has been proposed which was based on a temperature-driven transport process.[386] Although vapor transport may not be the only process, the model which was based principally upon this mechanism showed good agreement with experimental data. Losses from soil and soil mobility can be expected to occur at slower rates for the higher chlorinated CDD/CDF, although few studies to date have been performed for any CDD/CDF congeners other than the 4CDDs.

2. Plant Uptake

A few studies have examined the possibility of uptake and accumulation of 2,3,7,8-4CDD by plants.[125,369,392,396-397] Wipf et al.[125,395] found a lack of significant plant uptake. In one study low amounts of 2,3,7,8-4CDD were found in fruit,[125] and 95% of the amount detected was on the peel. The authors concluded that surface dust rather than plant uptake accounted for amounts found. Studies on oats and soybeans also failed to detect any significant plant uptake.[369] Young found levels of 2,3,7,8-4CDD in plant roots were the same as the levels in soil where plants were grown,[392] and postulated a "passive" mechanism of plant uptake. Kenaga and Norris[398] concluded from these studies that plants do not bioconcentrate 2,3,7,8-4CDD from soil. In more recent studies,[389, 396] it was concluded that significant adsorption of 2,3,7,8-4CDD by plant roots can occur. However, the above-ground portions of plants become contaminated principally through evaporation of 2,3,7,8-4CDD from soil.[389,396]

3. Long Range Atmospheric Transport

Atmospheric transport of organics is an important means of environmental distribution of contaminants. A recent report states that 20 to 25% of the pollutant load to the Great Lakes is from toxic deposition.[399] Studies of atmospheric deposition of chlorinated hydrocarbons to the Mediterranean found levels were correlated with the movements of air masses,[400] and that dry deposition was more important than wet deposition for these compounds. Dry deposition was also found to be more significant than wet deposition by factors of 1.5 to 5.0 for deposition of trace organics such as chlorinated pesticides and PCBs to the Great Lakes ecosystem.[161]

Little work has been done concerning atmospheric deposition of the CDDs/CDFs. Evidence of atmospheric transport of these substances is principally from studies reported by Hites and co-workers in which sediment cores were analyzed.[132,160,401-405] Even in remote areas such as Lake Superior,[403] CDDs and CDFs were detected. By radiodating cross-sections of sediment cores, concentration profiles were determined for CDDs/CDFs that showed distinct increases in CDD/CDF levels at about 1940, which suggests that the incineration of chlorinated organic compounds is a principal source of CDD/CDF input to the environment.[402,404] Congener profiles of the CDDs/CDFs detected in sediment cores were the same for sediments from the Great Lakes,[132,160,402-404] Swiss lake sediments,[401] and urban air particulates from St. Louis and Washington, D.C.[404] Very low levels of lower chlorinated CDDs/CDFs were found, while 8CDD was the predominant congener detected.

Little is known of the fate of CDDs/CDFs during atmospheric transport. Townsend has postulated that the CDD isomer and congener distributions initially reflect the chemical conditions of the combustion source,[406] however, upon emission a chemical equilibrium takes place that leads to constant CDD ratios that reflect the chemical conditions of the atmosphere. No direct experimental evidence of this phenomenon has yet been presented.

4. Photodegradation and Microbial Decomposition

Although 2,3,7,8-4CDD and other CDDs/CDFs are considered to be persistent in the environment, photodegradation and microbial decomposition have been observed. Many studies of the photoreactivity of 2,3,7,8-4CDD have been reported,[407-416] and results of these investigations have been reviewed previously.[59,60,408] Rapid photolysis of CDDs and CDFs can be observed in the presence of hydrogen donors including alcohols, ethers, hydrocarbons, and natural products. The three requirements for photolysis are: dissolution in a light-transmitting film, presence of an organic hydrogen donor, and ultraviolet light of the appropriate wavelength. The principal degradation pathway is dechlorination to the lower chlorinated congeners. Attempts have been made to accelerate cleanup of contaminated areas by photochemical action by addition of solubilizing agents and hydrogen donors to contaminated soils.[345,410,416] Although practical limitations may prevent the use of photodecomposition as a cleanup measure for large-scale contaminated areas, it may be an effective means of treating localized areas of contamination such as interior wall surfaces or for lab-scale cleanup.[149] While photodecomposition of CDDs and CDFs clearly occurs, it is still not known precisely how significant a role this mechanism plays in real environmental settings.

Only a few investigations of microbial decomposition of 2,3,7,8-4CDD have been reported.[417-423] No data have been reported to date concerning microbial decomposition of higher chlorinated CDDs/CDFs or for 4CDD isomers other than 2,3,7,8-4CDD. Although a few strains have shown some limited success in the slow degradation of 2,3,7,8-4CDD, it must be concluded so far that microbes cannot contribute to rapid decontamination of 2,3,7,8-4CDD in soil.

B. Bioaccumulation in Aquatic Organisms

Isensee has shown that bioaccumulation factor (BCFs) for 2,3,7,8-4CDD of up to 7000 times the concentration in water are possible.[424] Total amounts accumulated were directly related to aqueous concentrations, and equilibrium was reached in 7 to 15 days. Adams predicted a BCF of 7900 for 2,3,7,8-4CDD to fathead minnows,[425] and a time to 90% of steady state of 48 days. In a study of six CDD congeners,[426] BCFs in rainbow trout and fathead minnows ranged from 85 to 2278 (trout) and 514 to 5703 (minnows). The six congeners investigated did not have the same order of BCF in the two species tested. Low BCFs from 800 to 2500 were found by Gobas et al.[427] for male and female guppies. Five CDD congeners and 8CDF were tested, but none were 2,3,7,8-substituted. Recent investigations have been concerned with the uptake by fish of CDDs/CDFs from municipal incinerator fly ash.[207,210,430-432] Fly ash has been shown to contain a range of CDD and CDF congeners that can be leached into water at trace levels.[428,429] O'Keefe et al.[430] found that 2,3,7,8-4CDD and 2,3,7,8-4CDF were not accumulated by goldfish exposed to fly ash that contained both of these compounds. However, Kuehl et al.[207,431] studied bioaccumulation of 4CDD isomers from fly ash, and bioaccumulation of all CDDs/CDFs from fly ash,[432] and found that the toxic congeners were preferentially retained by fish.[210] The depuration half-life of 2,3,7,8-4CDD from carp was found to be 300 to 325 days,[432] and the rate of depuration increased with increasing degree of chlorination. Kuehl et al.[314] also found selective accumulation of the toxic congeners in carp from contaminated river sediment.

IV. FUTURE RESEARCH NEEDS

Although the analysis of environmental samples for CDDs and CDFs is at an advanced level, there is still considerable room for improvement. The existing data base consists almost entirely of 2,3,7,8-4CDD concentrations, although in many cases the higher chlorinated CDDs or CDFs are more significant in terms of environmental impact. An expanded data base to include much more data on all CDD/CDF congeners is needed for a proper assessment of their importance. Improved analytical methods that can give more results with increased confidence and at much lower cost are required to develop such a data base. Further, cost effective methods for rapid, routine determination of all toxic CDD/CDF congeners are needed. Techniques such as GC-MS-MS that may be able to perform these analyses with reduced sample preparation should be investigated with much more vigor than has been done to date. Such techniques, combined with automated cleanup and automated data processing, can provide the necessary data base. Another possibility is to refine our knowledge of CDD/CDF structure/biological activity relationships so truly quantitative bioanalytical methods can be developed.

One area that should be addressed is the proliferation of methodologies and approaches to CDD/CDF analysis, that has resulted in a very large data base of variable quality data. It is not necessary or even desirable for all groups to employ identical techniques, however, concensus is needed on what specific elements are needed to produce data of a specified quality. For example, what surrogates should be used, at what levels should they be spiked, and how should recovery data be used to determine data quality and in data calculation? Data should be reported with more complete statements of the quality control employed such as use of field/lab blanks, replicates, surrogate recoveries for individual samples, and so forth. More interlaboratory comparisons and round-robin investigations are needed. Real environmental sample standard reference materials with known concentrations of a range of CDDs and CDFs are not currently available. New data interpretation methods are required to handle the quantity of information on CDDs/CDFs being generated. Pattern recognition methods such as described by Stalling et al.[224,248] will allow scientists to extract the maximum amount of information from their data.

Perhaps the most important advances in the CDD/CDF field must be made in the area of hazard/risk assessment. Especially needed are better means of relating chronic and acute toxic effects in laboratory animal experiments to humans. When more is known in this area, it will be possible to set realistic and meaningful guidelines for allowable CDD/CDF levels in the various environmental compartments. This, in turn, will greatly assist the development and standardization of analytical methodologies, by providing clearly defined target values required for detection limits, accuracy and precision. This development is needed soon, because analogs to CDDs/CDFs are present in the environment, and they may also pose a hazard to human health. Buser has discussed the brominated/chlorinated dibenzo-*p*-dioxins and dibenzofurans,[261] of which there are over 400 tetra-, penta-, and hexahalogenated compounds having 2,3,7,8-substitution and, therefore, of suspected high toxicity.

REFERENCES

1. **Clement, R. E.,** Reporting chlorinated dibenzo-*p*-dioxin and dibenzofuran data in scientific publications, *Chemosphere,* 15, 1157, 1986.
2. **Fortunati, G. U.,** The Seveso accident, *Chemosphere,* 14, 729, 1985.
3. **Sambeth, J.,** The Seveso accident, *Chemosphere,* 12, 681, 1983.
4. **Whiteside, T.,** *The Pendulum and the Toxic Cloud,* Yale University Press, London, 1979.
5. **Pocchiari, F., di Domenico, A., Silano, V., and Zapponi, G.,** Environmental impact of the accidental release of tetrachlorodibenzo-*p*-dioxin (TCDD) at Seveso (Italy), in *Accidental Exposure to Dioxins — Human Health Aspects,* Coulston, F. and Pocchiari, F., Eds., Academic Press, New York, 1983, 5.
6. **Gough, M.,** *Dioxin, Agent Orange — the Facts,* Plenum Press, New York, 1986, chap. 9.
7. **Monteriolo, S. C., di Domenico, A., Silano, V., Viviano, G., and Zapponi, G.,** 2,3,7,8-TCDD levels and distribution in the environment at Seveso after the ICMESA accident on July 19th, 1976, in *Chlorinated Dioxins and Related Compounds — Impact on the Environment,* Hutzinger, O., Frei, R. W., Merian, E., and Pocchiari, F., Eds., Pergamon Press, Oxford, 1982, 127.
8. **Gough, M.,** *Dioxin, Agent Orange — the Facts,* Plenum Press, New York, 1986, 43.
9. **JRB Associates,** Review of Literature on Herbicides, Including Phenoxy Herbicides and Associated Dioxins, Vol. 1, Analysis of Literature, report prepared for U.S. Veterans Adminstration, VA Contract No. V101(93) P-823, McLean, VA, 1981.
10. **JRB Associates,** Review of Literature on Herbicides, Including Phenoxy Herbicides and Associated Dioxins, Vol. II, Annotated Bibliography, report prepared for U.S. Veterans Administration, VA Contract No. V101(93) P-823, McLean, VA, 1981.
11. **Clement Associates,** Review of Literature on Herbicides, Including Phenoxy Herbicides and Associated Dioxins, Vol. IV, Annotated Bibliography of Recent Literature on Health Effects, report prepared for U.S. Veterans Administration, VA Contract No. V101(93) P953, Arlington, VA, 1984.
12. **Young, A. L., Kang, H. K., and Shepard B. M.,** Rationale and description of the federally sponsored epidemiological research in the United States on the phenoxy herbicides and chlorinated dioxin contaminants, in *Chlorinated Dioxins and Dibenzofurans in the Total Environment, Vol. II,* Keith, L. H., Rappe, C., and Choudhary, G., Eds., Butterworth Publishers, Stoneham, MA, 1985, 155.
13. **Gough, M.,** *Dioxin, Agent Orange — the Facts,* Plenum Press, New York, 1986, chap. 7.
14. **Yanders, A. F., Kapila, S., and Schreiber, R. J., Jr.,** Dioxin: field research opportunities at Times Beach, Missouri, in *Chlorinated Dioxins and Dibenzofurans in Perspective,* Rappe, C., Choudhary, G., and Keith, L. H., Eds., Lewis Publishers, Chelsea, MI, 1986, 237.
15. **Yanders, A. F.,** The Missouri dioxin episode, *Chemosphere,* 15, 1571, 1986.
16. **Kloepfer, R. D.,** 2,3,7,8-TCDD contamination in Missouri, *Chemosphere,* 14, 739, 1984.
17. **Reggiani, G.,** An overview on the health effects of halogenated dioxins and related compounds — the Yusho and Taiwan episodes, in *Accidental Exposure to Dioxins — Human Health Aspects,* Coulston, F. and Pocchiari, F., Eds., Academic Press, New York, 1983, 39.
18. **Nagayama, J., Kuratsune, M., and Masuda, Y.,** Determination of chlorinated dibenzofurans in kanechlors and yusho oil, *Bull. Environ. Contam. Toxicol.,* 15, 9, 1976.
19. **Roach, J. A. G. and Pomerantz, I. H.,** Finding of chlorinated dibenzofurans in a Japanese polychlorinated biphenyl sample, *Bull. Environ. Contam. Toxicol.,* 12, 338, 1974.
20. **Bowes, G. W., Mulvihill, M. J., Simoneit, B. R. T., and Risebrough, R. W.,** Identification of chlorinated dibenzofurans in American polychlorinated biphenyls, *Nature,* 256, 305, 1975.

21. **Firestone, D.,** Etiology of chick edema disease, *Environ. Health Perspect.,* 5, 59, 1973.
22. *Chem. Eng. News,* 61, 44, 1983.
23. **Wyman, W. J.,** Octanol-water partition coefficient, in *Handbook of Chemical Property Estimation Methods,* Lyman, W. J., Reehl, W. F., and Rosenblatt, D. H., Eds., McGraw-Hill, New York, 1982.
24. **Schroy, J. M., Hileman, F. D., and Cheng, S. C.,** Physical/chemical properties of 2,3,7,8-4CDD, *Chemosphere,* 14, 877, 1985.
25. **Marple, L., Berridge, B., and Throop, L.,** Measurement of the water-octanol partition coefficient of 2,3,7,8-tetrachlorodibenzo-*p*-dioxin, *Environ. Sci. Technol.,* 20, 397, 1986.
26. **Burkhard, L. P. and Kuehl, D. W.,** *N*-Octanol/water partition coefficients by reverse phase liquid chromatography/mass spectrometry for eight tetrachlorinated planar molecules, *Chemosphere,* 15, 163, 1986.
27. **Webster, G. R. B., Kriesen, K. J., Sarna, L. P., and Muir, D. C. G.,** Environmental fate modelling of chlorodioxins: determination of physical constants, *Chemosphere,* 14, 609, 1985.
28. **Webster, G. R. B., Sarna, L. P., and Muir, D. C. G.,** Octanol-water partition coefficient of 1,3,6,8-TCDD and OCDD by reverse phase HPLC, in *Chlorinated Dioxins and Dibenzofurans in the Total Environment,* Vol. II, Keith, L. H., Rappe, C., and Choudhary, G., Eds., Butterworths, Stoneham, MA, 1985, 79.
29. **Sarna, L. P., Hodge, P. E., and Webster, G. R. B.,** Octanol-water partition coefficients of chlorinated dioxins and dibenzofurans by reversed-phase HPLC using several C_{18} columns, *Chemosphere,* 13, 975, 1984.
30. **Adams, W. J. and Blaine, K. M.,** A water solubility determination of 2,3,7,8-TCDD, *Chemosphere,* 15, 1397, 1986.
31. **Marple, L., Brunck, R., and Throop, L.,** Water solubility of 2,3,7,8-tetrachlorodibenzo-*p*-dioxin, *Environ. Sci. Technol.,* 20, 180, 1986.
32. **Crummett, W. B. and Stehl, R. H.,** Determination of chlorinated dibenzo-*p*-dioxins and dibenzofurans in various materials, *Environ. Health Perspect.,* 5, 15, 1973.
33. **Friesen, K. J., Sarna, L. P., and Webster, G. R. B.,** Aqueous solubility of polychlorinated dibenzo-*p*-dioxins determined by high pressure liquid chromatography, *Chemosphere,* 14, 1267, 1985.
34. **Rordorf, B. F., Sarna, L. P., and Webster, G. R. B.,** Vapor pressure determination for several poly-chlorodioxins by two gas saturation methods, *Chemosphere,* 15, 2073, 1986.
35. **Pohland, A. E. and Yang, G. C.,** Preparation and characterization of chlorinated dibenzo-*p*-dioxins, *J. Agric. Food Chem.,* 20, 1093, 1972.
36. **Nilsson, C. A., Andersson, K., Rappe, C., and Westermark, S. O.,** Chromatographic evidence of the formation of chlorodioxins from chloro-2-phenoxyphenols, *J. Chromatogr.,* 96, 137, 1974.
37. **Buser, H. R.,** Formation and identification of tetra- and pentachlorodibenzo-*p*-dioxins from photolysis of two isomeric hexachlorodibenzo-*p*-dioxins, *Chemosphere,* 8, 251, 1979.
38. **Buser, H. R.,** Preparation of qualitative standard mixtures of polychlorinated dibenzo-*p*-dioxins and di-benzofurans by ultraviolet and gamma-irradiation of the octachloro compounds, *J. Chromatogr.,* 129, 303, 1976.
39. **Langer, H. G., Brady, T. P., and Briggs, P. R.,** Formation of dibenzodioxins and other condensation products from chlorinated phenols and derivatives, *Environ. Health Perspect.,* 5, 3, 1973.
40. **Gray, A. P., Cepa, S. P., and Cantrell, J. S.,** Intervention of the Smiles rearrangement in synthesis of dibenzo-*p*-dioxins, 1,2,3,6,7,8- and 1,2,3,7,8,9-hexachlorodibenzo-*p*-dioxins (HCDD), *Tetrahedron Lett.,* 33, 2873, 1975.
41. **Kende, A. S. and DeCamp, M. R.,** Smiles rearrangements in the synthesis of hexachlorodibenzo-*p*-dioxins, *Tetrahedron Lett.,* 33, 2877, 1975.
42. **Nestrick, T. J., Lamparski, L. L., and Stehl, R. H.,** Synthesis and identification of the 22 tetrachloro-rodibenzo-*p*-dioxin isomers by high performance liquid chromatography and gas chromatography, *Anal. Chem.,* 51, 2273, 1979.
43. **Dobbs, A. J., Jappy, J., and Wadham, A. E.,** Synthesis and separation of polychlorodibenzo-*p*-dioxins (PCDDs), *Chemosphere,* 12, 481, 1983.
44. **Taylor, M. L., Tiernan, T. O., Ramalingam, B., Wagel, D. J., Garrett, J. H., Solch, J. G., and Ferguson, G. L.,** Synthesis, isolation, and characterization of the tetrachlorinated dibenzo-*p*-dioxins and other related compounds, in *Chlorinated Dioxins and Dibenzofurans in the Total Environment,* Vol. II, Keith, L. H., Rappe, C., and Choudhary, G., Eds., Butterworth, Stoneham, MA, 1985, 17.
45. **Ap Simon, J. W., Collier, T. L., and Venayak, N. D.,** Synthesis of polychlorodibenzo-*p*-dioxins as analytical and toxicological standards, *Chemosphere,* 14, 881, 1985.
46. **Gelbaum, L. T., Patterson, D. G., and Groce, D. F.,** Preparation of dioxin standards for chemical analysis, in *Chlorinated Dioxins and Dibenzofurans in Perspective,* Rappe, C., Choudhary, G., and Keith, L. H., Eds., Lewis Publishers, Chelsea, MI, 1986, 479.

47. **Bell, R. A.,** Synthesis of uniformly labelled ¹³C-polychlorinated dibenzofurans, in *Chlorinated Dioxins and Dibenzofurans in the Total Environment,* Choudhary, G., Keith, L. H., and Rappe, C., Eds., Butterworth, Woburn, MA, 1983, 59.

48. **Mazer, T., Hileman, F. D., Noble, R. W., Hale, M. D., and Brooks, J. J.,** Characterization of tetrachlorodibenzofurans, in *Chlorinated Dioxins and Dibenzofurans in the Total Environment,* Choudhary, G., Keith, L. H., and Rappe, C., Eds., Butterworth, Woburn, MA, 1983, 23.

49. **Choudhry, G. G., van den Broecke, J. A., and Hutzinger, O.,** Formation of polychlorodibenzofurans (PCDFs) by the photolyses of polychlorobenzenes (PCBzs) in aqueous acetonitrile containing phenols, *Chemosphere,* 12, 487, 1983.

50. **Bell, R. A. and Gara, A.,** Synthesis and characterization of the isomers of polychlorinated dibenzofurans, tetra- through octa-, in *Chlorinated Dioxins and Dibenzofurans in the Total Environment,* Vol. II, Keith, L. H., Rappe, C., and Choudhary, G., Eds., Butterworth, Woburn, MA, 1985, 3.

51. **Chittim, B. G., Madge, J. A., and Safe, S. H.,** Pentachlorodibenzofurans: synthesis, capillary gas chromatography and gas chromatographic/mass spectrometric characteristics, *Chemosphere,* 15, 1931, 1986.

52. **Humppi, T.,** Synthesis of polychlorinated phenoxyphenols (PCPP), phenoxyanisoles (PCPA), dibenzo-*p*-dioxins (PCDD), dibenzofurans (PCDF) and diphenyl ethers (PCDE), *Chemosphere,* 15, 2003, 1986.

53. **Buser, H. R. and Rappe, C.,** Formation of polychlorinated dibenzofurans (PCDFs) from the pyrolysis of individual PCB isomers, *Chemosphere,* 8, 157, 1979.

54. **Safe, S. and Safe, L.,** Synthesis and characterization of twenty-two purified polychlorinated dibenzofuran congeners, *J. Agric. Food Chem.,* 32, 68, 1984.

55. **Mazer, T., Hileman, F. D., Noble, R. W., and Brooks, J. J.,** Synthesis of the 38 tetrachlorodibenzofuran isomers and identification by capillary column gas chromatography/mass spectrometry, *Anal. Chem.,* 55, 104, 1983.

56. **Hutzinger, O., Berg, M. V. D., Olie, K., Opperhuizen, A., and Safe, S.,** Dioxins and furans in the environment: evaluating toxicological risk from different sources by multi-criteria analysis, in *Dioxins in the Environment,* Kamrin, M. A. and Rogers, P. W., Eds., Hemisphere Publishing, New York, 1985, 9.

57. **Weerasinghe, N. C. A. and Gross, M. L.,** Origins of polychlorodibenzo-*p*-dioxins (PCDD) and polychlorodibenzofurans (PCDF) in the environment, in *Dioxins in the Environment,* Kamrin, M. A. and Rogers, P. W., Eds., Hemisphere Publishing, New York, 1985, 133.

58. **Hutzinger, O., Blumich, M. J., van den Berg, M., and Olie, K.,** Sources and fate of PCDDs and PCDFs: an overview, *Chemosphere,* 14, 581, 1985.

59. **Esposito, M. P., Tiernan, T. O., and Dryden, F. E.,** Dioxins, EPA-600/2-80-197, U.S. Environmental Protection Agency, Cincinnati, OH, 1980.

60. *Polychlorinated Dibenzo-p-dioxins:* Criteria for their Effects on Man and His Environment, NRCC 18574, National Research Council of Canada, Ottawa, 1981.

61. **Birmingham, B., Clement, R., Harding, D., Pearson, R., Rokosh, D., Smithies, W., Szakolcai, A., Hanna Thorpe, B., Tosine, H., and Wells, D.,** Chlorinated dioxins and dibenzofurans in Ontario — analysing and controlling the risks: development of scientific criteria document leading to multi-media standards for polychlorinated dibenzo-*p*-dioxins (PCDDs) and polychlorinated dibenzofurans (PCDFs), *Chemosphere,* 15, 1835, 1986.

62. Ontario Ministry of the Environment, Scientific Criteria Document for Standard Development, No. 4-84. Polychlorinated Dibenzo-*p*-dioxins (PCDDs) and Polychlorinated Dibenzofurans (PCDFs), September 1985.

63. **Sheffield, A.,** Sources and releases of PCDD's and PCDF's to the Canadian Environment, *Chemosphere,* 14, 811, 1985.

64. **Sheffield, A.,** Polychlorinated Dibenzo-*p*-dioxins (PCDDs) and Polychlorinated Dibenzofurans (PCDFs): Sources and Releases, Environment Canada Report EPS 5/HA/2, July 1985.

65. **desRosiers, P. E. and Lee, A.,** PCB fires: correlation of chlorobenzene isomer and PCB homolog contents of PCB fluids with PCDD and PCDF contents of soot, *Chemosphere,* 15, 1313, 1986.

66. Health and Welfare Canada, Report of the Ministers' Expert Advisory Committee on Dioxins, November 1983.

67. **Crummett, W. B.,** Environmental chlorinated dioxins from combustion — the trace chemistries of fire hypothesis, in *Chlorinated Dioxins and Related Compounds — Impact on the Environment,* Hutzinger, O., Frei, R. W., Merian, E., and Pocchiari, F., Eds., Pergamon Press, Oxford, 1982, 253.

68. **Schwetz, B. A., Norris, J. M., Sparschu, G. L., Rowe, U. K., Gehring, P. J., Emerson, J. L., and Gerbig, C. G.,** Toxicology of chlorinated dibenzo-*p*-dioxins, *Environ. Health Perspect.,* 5, 87, 1973.

69. **McConnell, E. E. and Moore, J. A.,** The toxicopathology of TCDD, in *Dioxin: Toxicological and Chemical Aspects,* Cattabeni, F., Cavallaro, A., and Galli, G., Eds., SP Medical Science Books, London, 1978, 137.

70. **Poland, A. and Knutson, J. C.,** 2,3,7,8-Tetrachlorodibenzo-*p*-dioxin and related halogenated aromatic hydrocarbons: examination of the mechanisms of toxicity, *Annu. Rev. Pharmacol. Toxicol.,* 22, 517, 1982.

71. **Kociba, R. J. and Schwetz, B. A.,** Toxicity of 2,3,7,8-tetrachlorodibenzo-*p*-dioxin (TCDD), *Drug Metab. Rev.,* 13, 387, 1982.

72. **Neal, R. A.,** Biological effects of 2,3,7,8-tetrachlorodibenzo-*p*-dioxin in experimental animals, in *Public Health Risks of the Dioxins,* Lowrance, W. W., Ed., William Kaufmann, Inc., Los Altos, CA, 1984, 15.

73. **Kociba, R. J.,** Summary and critique of rodent carcinogenicity studies of chlorinated dibenzo-*p*-dioxins, in *Public Health Risks of the Dioxins,* Lowrance, W. W., Ed., William Kaufmann, Inc., Los Altos, CA, 1984, 77.

74. **Mattison, D. R., Nightingale, M. S., and Silbergeld, E. K.,** Reproductive toxicity of tetrachlorodibenzo-*p*-dioxin, in *Public Health Risks of the Dioxins,* Lowrance, W. W., Ed., William Kaufmann, Inc., Los Altos, CA, 1984, 217.

75. **McNulty, W. P.,** Fetocidal and teratogenic actions of TCDD, in *Public Health Risks of the Dioxins,* Lowrance, W. W., Ed., William Kaufmann, Inc., Los Altos, CA, 1984, 245.

76. **Dean, J. H. and Lauer, L. D.,** Immunological effects following exposure to 2,3,7,8-tetrachlorodibenzo-*p*-dioxin: a review, in *Public Health Risks of the Dioxins,* Lowrance, W. W., Ed., William Kaufmann, Inc., Los Altos, CA, 1984, 275.

77. **Smuckler, E. A.,** Biological effects of dioxins and other halogenated polycyclics, in *Dioxins in the Environment,* Kamrin, M. A. and Rogers, P. W., Eds., Hemisphere Publishing, New York, 1985, 215.

78. **Barnes, D. G, Bellin, J., and Cleverly, D.,** Interim procedures for estimating risks associated with exposures to mixtures of chlorinated dibenzodioxins and dibenzofurans (CDDs and CDFs), *Chemosphere,* 15, 1895, 1986.

79. **Mitchell, M. F., McLeod, H. A., and Roberts, J. R.,** Polychlorinated Dibenzofurans: Criteria for their Effects on Humans and the Environment, National Research Council of Canada Report No. NRCC 22846, Ottawa, Canada, 1984.

80. Polychlorinated Dibenzo-*p*-dioxins: Limitations to the Current Analytical Techniques, NRCC 18576, National Research Council of Canada, Ottawa, Canada, 1981.

81. **Crummett, W. B.,** Status of analytical systems for the determination of PCDDs and PCDFs, *Chemosphere,* 12, 429, 1983.

82. **Karasek, F. W. and Onuska, F. I.,** Trace analysis of the dioxins, *Anal. Chem.,* 54, 309A, 1982.

83. **Tiernan, T. O.,** Analytical chemistry of polychlorinated dibenzo-*p*-dioxins and dibenzofurans: a review of the current status, in *Chlorinated Dioxins and Dibenzofurans in the Total Environment,* Choudhary, G., Keith, L. H., and Rappe, C., Eds., Butterworth, Woburn, MA, 1983, 211.

84. **Rappe, C.,** Analysis of polychlorinated dioxins and furans, *Environ. Sci. Technol.,* 18, 78A, 1984.

85. **Crummett, W. B., Nestrick, T. J., and Lamparski, L. L.,** Analytical methodology for the determination of PCDDs in environmental samples: an overview and critique, in *Dioxins in the Environment,* Kamrin, M. A. and Rodgers, P. W., Eds., Hemisphere Publishing, New York, 1985, 57.

86. **Webster, G. R. B., Olie, K., and Hutzinger, O.,** Polychlorodibenzo-*p*-dioxins and polychlorodibenzofurans, in *Mass Spectrometry in Environmental Sciences,* Karasek, F. W., Hutzinger, O., and Safe, S., Ed., Plenum Press, New York, 1985, 257.

87. **Provost, L. P.,** Statistical methods in environmental sampling, in *Environmental Sampling for Hazardous Wastes,* ACS Symposium Series, 267, Schweiter, G. E. and Santolucito, J. A., Eds., American Chemical Society, Washington, D.C., 1984, 79.

88. **Lamparski, L. L., Nestrick, T. J., Frawley, N. N., Hummel, R. A., Kocher, C. W., Mahle, N. H., McCoy, J. W., Miller, D. L., Peters, T. L., Pillepich, J. L., Smith, W. E., and Tobey, S. W.,** Perspectives of a large scale environmental survey for chlorinated dioxins: water analysis, *Chemosphere,* 15, 1445, 1986.

89. **Isensee, A. R. and Jones, G. E.,** Distribution of 2,3,7,8-tetrachlorodibenzo-*p*-dioxin (TCDD) in aquatic model ecosystem, *Environ. Sci. Tech.,* 9, 668, 1975.

90. **Ward, C. T. and Matsumura, F.,** Fate of 2,3,7,8-tetrachlorodibenzo-*p*-dioxin (TCDD) in a model aquatic environment, *Arch. Environ. Contam. Toxicol.,* 7, 349, 1978.

91. **Kurtz, D. A.,** Analysis of water for chlorinated hydrocarbon pesticides and PCB's by membrane filters, in *Trace Organic Analysis: A New Frontier in Analytical Chemistry,* Hertz, H. S. and Chesler, S. N., Eds., Proceedings of the 9th Materials Research Symposium, April 10-13, 1978, Gaithersburg, MD, NBS Special Publication 519, April, 1979, 19.

92. **Wegman, R. C. and Melis, P. H. A. M.,** Organic pollutants in water, *CRC Crit. Rev. Anal. Chem.,* 16, 281, 1985.

93. **O'Keefe, P., Meyer, C., Smith, R., Hilker, D., Aldous, K., and Wilson, L.,** Reverse-phase adsorbent cartridge for trapping dioxins in drinking water, *Chemosphere,* 15, 1127, 1986.

94. **LeBel, G. L., Williams, D. T., and Lau, B. P.-Y.,** Evaluation of XAD-2 resin cartridge for concentration/isolation of chlorinated dibenzo-*p*-dioxins and furans from drinking water at the parts-per-quadrillion level, in *Chlorinated Dioxins and Dibenzofurans in Perspective,* Rappe, C., Choudhary, G., and Keith, L. H., Eds., Lewis Publishers, Chelsea, MI, 1986, 329.

95. **Tosine, H. M., Clement, R. E., and Hunsinger, R.,** A survey of selected drinking water supplies in Ontario for chlorinated dibenzo-*p*-dioxins and chlorinated dibenzufurans, Ontario Ministry of the Environment Report.

96. **Tosine, H. M., Clement, R. E., and Hunsinger, R.,** Methodology and Data for Survey of Raw/Treated Drinking Water for Chlorinated Dioxins and Furans, presented at the 189th National Meeting of the American Chemical Society, Miami Beach, FL, September 1985.

97. **Barcelona, M. J., Gibb, J. P., Helfrich, J. A., and Garske, E. E.,** Practical Guide for Ground-Water Sampling, EPA/600/2-85/104, September 1985.

98. **Hileman, G. D., Kirk, D. E., Mazer, T., Snyder, A. D., Warner, B. J., and McMillin, C. R.,** EPA (Environmental Protection Agency) Method Study 26, Method 613, 2,3,7,8-Tetrachlorodibenzo-*p*-dioxin, EPA-600/4-84-037, May, 1984.

99. **Wong, A. S., Orbanosky, M. W., Taylor, P. A., McMillin, C. R., Noble, R. W., Wood, D., Longbottom, J. E., Foerst, D. L., and Wesselman, R. J.,** Determination of 2,3,7,8-tetrachlorodibenzo-*p*-dioxin in industrial and municipal wastewaters, method 613: development and detection limits, in *Chlorinated Dioxins and Dibenzofurans in the Total Environment,* Choudhary, G., Keith, L. H., and Rappe, C., Eds., Butterworth Publishers, Woburn, MA, 1983, 165.

100. **McMillin, C. R., Hileman, F. D., Kirk, D. E., Mazer, T., Warner, B. J., Wesselman, R. J., and Longbottom, J. E.,** Determination of 2,3,7,8-tetrachlorodibenzo-*p*-dioxin in industrial and municipal wastewaters, method 613: performance evaluation and preliminary method study results, in *Chlorinated Dioxins and Dibenzofurans in the Total Environment,* Choudhary, G., Keith, L. H., and Rappe, C., Eds., Butterworth Publishers, Woburn, MA, 1983, 181.

101. **Brenner, K. S., Dorn, I. H., and Herrmann, K.,** Dioxin analysis in stack emissions, slags, and the wash water circuit during high-temperature incineration of chlorine-containing industrial wastes, II, *Chemosphere,* 15, 1193, 1986.

102. **Peters, T. L., Nestrick, T. J., and Lamparski, L. L.,** The determination of 2,3,7,8-tetrachlorodibenzo-*p*-dioxin in treated wastewater, *Water Res.,* 18, 1021, 1984.

103. **McCrea, R. C. and Fischer, J. D.,** Design and Testing of an Aqueous Phase Liquid-Liquid Extractor (APLE) for the Determination of Organochlorine Contaminants, Environment Canada, Inland Waters Directorate report.

104. **Ahnhoff, M. and Josefsson, B.,** Apparatus for in-site solvent extraction of non-polar organic compounds in sea and river water, *Anal. Chem.,* 48, 1268, 1976.

105. **Tosine, H., Smillie, D., and Rees, G. A. V.,** Comparative monitoring and analytical methodology for 2,3,7,8-TCDD in fish, in *Human and Environmental Risks of Chlorinated Dioxins and Related Compounds,* Tucker, R. E., Young, A. L., and Gray, A. P., Eds., Plenum Press, New York, 1983, 127.

106. **Albro, P. W. and Corbett, B. J.,** Extraction and clean-up of animal tissues for subsequent determination of mixtures of chlorinated dibenzo-*p*-dioxins and dibenzofurans, *Chemosphere,* 7, 381, 1977.

107. **Lamparski, L. L., Mahle, N. H., and Shadoff, L. A.,** Determination of pentachlorophenol, hexachlorodibenzo-*p*-dioxin, and octachlorodibenzo-*p*-dioxin in bovine milk, *J. Agric. Food Chem.,* 26, 1113, 1978.

108. **O'Keefe, P. W.,** A neutral cleanup procedure for TCDD residues in environmental samples, in *Dioxin, Toxicological and Chemical Aspects,* Cattabeni, F., Cavallaro, A., and Galli, G., Eds., SP Medical Science Books, New York, 1978, 13.

109. **Crosby, D. G., Wong, A. S., Plimmer, J. R., and Woolson, E. A.,** Photodecomposition of chlorinated dibenzo-*p*-dioxins, *Science,* 173, 748, 1971.

110. **O'Keefe, P., Meyer, C., Hilker, D., Aldous, K., Jelus-Tyror, B., Dillon, K., Donnelly, R., Horn, E., and Sloan, R.,** Analysis of 2,3,7,8-tetrachlorodibenzo-*p*-dioxin in great lakes fish, *Chemosphere,* 12, 325, 1983.

111. **Huckins, J. N., Stalling, D. L., and Smith, W. A.,** Foam-charcoal chromatography for analysis of polychlorinated dibenzodioxins in herbicide orange, *J. Assoc. Off. Anal. Chem.,* 61, 32, 1978.

112. **Ryan, J. J., Pilon, J. C., Conacher, H. B. S., and Firestone, D.,** Interlaboratory study on determination of 2,3,7,8-tetrachlorodibenzo-*p*-dioxin in fish, *J. Assoc. Off. Anal. Chem.,* 66, 700, 1983.

113. **Harless, R. L., Oswald, E. O., Wilkinson, M. K., Dupuy, A. E., McDaniel, D. O., and Tai, H.,** Sample preparation and gas chromatography-mass spectrometry determination of 2,3,7,8-tetrachlorodibenzo-*p*-dioxin, *Anal. Chem.,* 52, 1239, 1980.

114. **Ryan, J. J. and Pilon, J. C.,** High performance liquid chromatography in the analysis of chlorinated dibenzodioxins and dibenzofurans in chicken liver and wood shaving samples, *J. Chromatogr.,* 197, 171, 1980.

115. **Bowes, G. W., Simoneit, B. R., Burlingame, A. L., deLappe, B. W., and Risebrough, R. W.,** The search for chlorinated dibenzofurans and chlorinated dibenzo-*p*-dioxins in wildlife populations showing elevated levels of embryonic death, *Environ. Health Perspect.,* 5, 191, 1973.

116. **O'Keefe, P., Hilker, D., Meyer, C., Aldous, K., Shane, L., Donnelly, R., Smith, R., Sloan, R., and Horn, E.,** Tetrachlorodibenzo-*p*-dioxins and tetrachlorodibenzofurans in Atlantic coast striped bass and in selected Hudson River fish, waterfowl and sediments, *Chemosphere,* 13, 849, 1984.

117. **Stalling, D. L., Smith, L. M., Petty, J. D., Hogan, J. W., Johnson, J. L., Rappe, C., and Buser, H. R.,** Residues of polychlorinated dibenzo-*p*-dioxins and dibenzofurans in Laurentian Great Lakes fish, in *Human and Environmental Risks of Chlorinated Dioxins and Related Compounds,* Tucker, R. E., Young, A. L., and Gray, A. P., Eds., Plenum Press, New York, 1983, 221.

118. **Norstrom, R. J., Hallett, D. J., Simon, M., and Mulvihill, M. J.,** Analysis of Great Lakes herring gull eggs for tetrachlorodibenzo-*p*-dioxins, in *Chlorinated Dioxins and Related Compounds: Impact on the Environment,* Hutzinger, O., Frei, R. W., Merian, E., and Pocchiari, F., Eds., Pergamon Press, Oxford, 1982, 173.

119. **Hummel, R. A.,** Clean-up techniques for the determination of parts per trillion residue levels of 2,3,7,8-tetrachlorodibenzo-*p*-dioxin (TCDD), *J. Agric. Food Chem.,* 25, 1049, 1977.

120. **Albro, P. W., Crummett, W. B., Dupuy, A. E., Jr., Gross, M. L., Hanson, M., Harless, R. L., Hileman, F. D., Hilker, D., Jason, C., Johnson, J. L., Lamparski, L. L., Lau, B. P.-Y., McDaniel, D. D., Meehan, J. L., Nestrick, T. J., Nygren, M., O'Keefe, P., Peters, T. L., Rappe, C., Ryan, J. J., Smith, L. M., Stalling, D. L., Weerasinghe, N. C. A., and Wendling, J. M.,** Methods for the quantitative determination of multiple, specific polychlorinated dibenzo-*p*-dioxin and dibenzofuran isomers in human adipose tissue in the parts-per-trillion range. An interlaboratory study, *Anal. Chem.,* 57, 2717, 1985.

121. **Stanley, J. S., Going, J. E., Redford, D. P., Kutz, F. W., and Young, A. L.,** A survey of analytical methods for measurement of polychlorinated dibenzo-*p*-dioxins (PCDD) and polychlorinated dibenzofurans (PCDF) in human adipose tissues, in *Chlorinated Dioxins and Dibenzofurans in the Total Environment,* Vol. II, Keith, L. H., Rappe, C., and Choudhary, G., Eds., Butterworth Publishers, Stoneham, MA, 1985, 181.

122. **Leoni, A., Fichtner, C., Frare, G., Balasso, A., Mauri, C., and Facchetti, S.,** A neutral clean-up procedure combined with a high resolution GC-MS technique for the detection of the 2,3,7,8-TCDD in various vegetal tissues, *Chemosphere,* 12, 493, 1983.

123. **Buser, H. R.,** Determination of 2,3,7,8-tetrachlorodibenzo-*p*-dioxin in environmental samples by high resolution gas chromatography and low resolution mass spectrometry, *Anal. Chem.,* 49, 918, 1977.

124. **Firestone, D.,** Determination of polychlorodibenzo-*p*-dioxins and polychlorodibenzofurans in commercial gelatins by gas liquid chromatography, *J. Agric. Food Chem.,* 25, 1274, 1977.

125. **Wipf, H.-K., Homberger, E., Neuner, N., Ranalder, U. B., Vetter, W., and Vuillemier, J. P.,** TCDD-levels in soil and plant samples from the Seveso area, in *Chlorinated Dioxins and Related Compounds: Impact on the Environment,* Hutzinger, O., Frei, R. W., Merian, E., and Pocchiari, F., Eds., Pergamon Press, Oxford, 1982, 115.

126. **Harrison, D. D. and Crews, R. C.,** A field study of soil and biological specimens from a herbicide storage and aerial test staging site following long-term contamination with TCDD, in *Human and Environmental Risks of Chlorinated Dioxins and Related Compounds,* Tucker, R. E., Young, A. L., and Gray, A. P., Eds., Plenum Press, New York, 1983, 323.

127. **di Domenico, A., Merli, F., Boniforti, L., Camoni, I., Di Muccio, A., Taggi, F., Vergori, L., Colli, G., Elli, G., Gorni, A., Grassi, P., Invernizzi, G., Jemma, A., Luciani, L., Cattabeni, F., De Angelis, L., Galli, G., Chiabrando, C., and Fanelli, R.,** Analytical techniques for 2,3,7,8-tetrachlorodibenzo-*p*-dioxin detection in environmental samples after the industrial accident at Seveso, *Anal. Chem.,* 51, 735, 1979.

128. **Palausky, J., Harwood, J. J., Chevenger, T. E., Kapila, S., and Yanders, A. F.,** Disposition of tetrachlorodibenzo-*p*-dioxin in soil, in *Chlorinated Dioxins and Dibenzofurans in Perspective,* Rappe, C., Choudhary, G., and Keith, L. H., Eds., Lewis Publishers, Chelsea, MI, 1986, 211.

129. **Harris, D. J.,** 2,3,7,8-Tetrachlorodibenzo-*p*-dioxin sampling methods, in *Environmental Sampling for Hazardous Wastes,* ACS Symposium Series 276, Schweitzer, G. E. and Santolucito, J. A., Eds., American Chemical Society, Washington, D.C., 1984, 27.

130. **McLaughlin, D. L. and Pearson, R. G.,** Concentrations of PCDD and PCDF in Soil from the Vicinity of the SWARU Incinerator, Hamilton, Ontario Ministry of the Environment Report, No. ARB-013-85-Phyto, September 1985.

131. **Elder, V. A., Proctor, B. L., and Hites, R. A.,** Organic compounds near dumpsites in Niagara Falls, New York, *Biomed. Mass Spectrom.,* 8, 409, 1981.

132. **Czuczwa, J. M. and Hites, R. A.,** Environmental fate of combustion generated polychlorinated dioxins and furans, *Environ. Sci. Technol.,* 18, 444, 1984.

133. **Onuska, F. I., Mudroch, A., and Terry, K. A.,** Identification and determination of trace organic substances in sediment cores from the Western Basin of Lake Ontario, *J. Great Lakes Res.,* 9, 169, 1983.

134. **Kuntz, K. W. and Warry, N. D.,** Chlorinated organic contaminants in water and suspended sediments of the lower Niagara River, *J. Great Lakes Res.,* 9, 241, 1983.

135. **Clement, R. E., Viau, A. C., and Karasek, F. W.,** Daily variations in composition of extractable organic compounds in fly-ash from municipal waste incineration, *Int. J. Environ. Anal. Chem.,* 17, 257, 1984.

136. **Ozvacic, V., Wong, G., Tosine, H., Clement, R., Osborne, J., and Thorndyke, S.,** Determination of Chlorinated Dibenzo-*p*-dioxins, Chlorinated Dibenzofurans, Chlorinated Biphenyls, Chlorobenzenes and Chlorophenols in Air Emission and Other Process Streams at SWARU in Hamilton, Report No. ARB-02-84-ETRD, Ontario Ministry of the Environment, Toronto, Ontario, 1984.

137. **Exner, J. H., Keffer, W. D., Gilbert, R. O., and Kinnison, R. R.,** A sampling strategy for remedial action at hazardous waste sites: cleanup of soil contaminated by tetrachlorodibenzo-*p*-dioxin, in *Chlorinated Dioxins and Dibenzofurans in Perspective,* Rappe, C., Choudhary, G., and Keith, L. H., Eds., Lewis Publishers, Chelsea, MI, 1986, 139.

138. **Freeman, R. A., Schroy, J. M., Hileman, F. D., and Noble, R. W.,** Environmental mobility of 2,3,7,8-TCDD and companion chemicals in a roadway soil matrix, in *Chlorinated Dioxins and Dibenzofurans in Perspective,* Rappe, C., Choudhary, G., and Keith, L. H., Eds., Lewis Publishers, Chelsea, MI, 1986, 171.

139. **Viswanathan, T. S. and Kloepfer, R. D.,** The presence of hexachloroxanthene at Missouri dioxin sites, in *Chlorinated Dioxins and Dibenzofurans in Perspective,* Rappe, C., Choudhary, G., Keith, L. H., Eds., Lewis Publishers, Chelsea, MI, 1986, 201.

140. **Smith, J. S., Ben Hur, D., Urban, M. J., Kleopfer, R. D., Kirchmer, C. J., Smith, W. A., and Viswanathan, T. S.,** Comparison of a new rapid extraction GC/MS/MS and the contract laboratory program GC/MS methodologies for the analysis of 2,3,7,8-tetrachlorodibenzo-*p*-dioxin, in *Chlorinated Dioxins and Dibenzofurans in Perspective,* Rappe, C., Choudhary, G., and Keith, L. H., Eds., Lewis Publishers, Chelsea, MI, 1986, 367.

141. **Donnelly, J. R., Vonnahme, T. L., Hedin, C. M., and Niederhut, W. J.,** Evaluation of RCRA method 8280 for analysis of dioxins and dibenzofurans, in *Chlorinated Dioxins and Dibenzofurans in Perspective,* Rappe, C., Choudhary, G., and Keith, L. H., Eds., Lewis Publishers, Chelsea, MI, 1986, 399.

142. **Donnelly, J. R., Sovocool, G. W., Tondeur, Y., Billets, S., and Mitchum, R. K.,** Some analytical considerations relating to the development of a high-resolution mass spectrometric dioxin analytical protocol, in *Chlorinated Dioxins and Dibenzofurans in Perspective,* Rappe, C., Choudhary, G., and Keith, L. H., Eds., Lewis Publishers, Chelsea, MI, 1986, 381.

143. **Anderson, J. W., Herman, G. H., Thielen, D. R., and Weston, A. F.,** Method verification for determination of tetrachlorodibenzodioxin in soil, *Chemosphere,* 14, 1115, 1985.

144. **Kitunen, V. H., Valo, R. J., and Salkinoja-Salonen, M. S.,** Contamination of soil around wood-preserving facilities by polychlorinated aromatic compounds, *Environ. Sci. Technol.,* 21, 96, 1986.

145. **Balasso, A., Fichtner, C., Frare, G., Leoni, A., Mauri, C., and Facchetti, S.,** An efficient method for the quantitative determination of 2,3,7,8-TCDD at ppt level in soil samples, *Chemosphere,* 12, 499, 1983.

146. **Kloepfer, R. D. and Kirchner, C. J.,** Quality assurance plan for 2,3,7,8-tetrachlorodibenzo-*p*-dioxin monitoring in Missouri, in *Chlorinated Dioxins and Dibenzofurans in the Total Environment,* Vol. II, Keith, L. H., Rappe, C., and Choudhary, G., Eds., Butterworth Publishers, Stoneham, MA, 1985, 355.

147. **Kloepfer, R. D., Yue, K. T., and Bunn, W. W.,** Determination of 2,3,7,8-tetrachlorodibenzo-*p*-dioxin in soil, in *Chlorinated Dioxins and Dibenzofurans in the Total Environment,* Vol. II, Keith, L. H., Rappe, C., Choudhary, G., Butterworth Publishers, Stoneham, MA, 1985, 367.

148. **Wegman, R. C. C., Freudenthal, J., de Korte, G. A. L., Groenemeijer, G. S., and Japenga, J.,** A modified clean-up procedure for the determination of PCDDs in soil samples, *Chemosphere,* 15, 1107, 1986.

149. **Solch, J. G., Ferguson, G. L., Tiernan, T. O., Van Ness, B. F., Garrett, J. H., Wagel, D. J., and Taylor, M. L.,** Analytical methodology for determination of 2,3,7,8-tetrachlorodibenzo-*p*-dioxin in soils, in *Chlorinated Dioxins and Dibenzofurans in the Total Environment,* Vol. II, Keith, L. H., Rappe, C., and Choudhary, G., Eds., Butterworth Publishers, Stoneham, MA, 1985, 377.

150. **Smith, R. M., O'Keefe, P. W., Aldous, K. M., Hilker, D. R., and O'Brien, J. E.,** 2,3,7,8-tetra-chlorodibenzo-*p*-dioxin in sediment samples from Love Canal storm sewers and creeks, *Environ. Sci. Technol.,* 17, 6, 1983.

151. **Jackson, D. R., Roulier, M. H., Grotta, H. M., Rust, S. W., and Warner, J. S.,** Solubility of 2,3,7,8-TCDD in contaminated soils, in *Chlorinated Dioxins and Dibenzofurans in Perspective,* Rappe, C., Choudhary, G., and Keith, L. H., Eds., Lewis Publishers, Chelsea, MI, 1986, 185.

152. **Lustenhouwer, J. W. A., Olie, K., and Hutzinger, O.,** Chlorinated dibenzo-*p*-dioxins and related compounds in incinerator effluents: a review of measurements and mechanisms of formation, *Chemosphere,* 9, 501, 1980.

153. **Kooke, R. M. M., Lustenhouwer, J. W. A., Olie, K., and Hutzinger, O.,** Extraction efficiencies of polychlorinated dibenzo-*p*-dioxins and polychlorinated dibenzofurans from fly ash, *Anal. Chem.,* 53, 461, 1981.

154. **Clement, R. E., Viau, A. C., and Karasek, F. W.,** Comparison of extraction efficiencies for toxic organic compounds on fly ash, *Can. J. Chem.,* 62, 2629, 1984.

155. **Stieglitz, L., Zwick, G., and Roth, W.,** Investigation of different treatment techniques for PCDD/PCDF in fly ash, *Chemosphere,* 15, 1135, 1986.
156. **Green, J.,** Determination of acid-base and solubility behaviour of lignite fly ash by selective dissolution in mineral acids, *Anal. Chem.,* 50, 1975, 1978.
157. **Morselli, L., Trifiro, F., and Vittori, B.,** Analysis of the effluents of an urban-solid refuse incinerator: study of methods of extraction and analysis for the quantiative determination of polychlorodibenzo-*p*-dioxins, *Ann. Chim.,* 557, 1981.
158. **Aniline, O.,** Preparation of chlorodibenzo-*p*-dioxins for toxicological evaluation, in *Chlorodioxins — Origin and Fate,* Blair, E. H., Ed.,*Am. Chem. Soc. Adv. Chem. Ser.,* 120, 126, 1973.
159. **Eiceman, G. A., Viau, A. C., and Karasek, F. W.,** Ultrasonic extraction of polychlorinated dibenzo-*p*-dioxins and other organic compounds from fly ash from municipal incinerators, *Anal. Chem.,* 52, 1492, 1980.
160. **Czuczwa, J. M. and Hites, R. A.,** Airborne dioxins and dibenzofurans: sources and fates, *Environ. Sci. Technol.,* 20, 195, 1986.
161. **Eisenreich, S. J., Looney, B. B., and Thornton, J. D.,** Airborne organic contaminants in the Great Lakes and ecosystem, *Environ. Sci. Technol.,* 15, 30, 1981.
162. American Society of Mechanical Engineers, *Study on State-of-the-Art of Dioxin from Combustion Sources,* ASME Report, Arthur D. Little, New York, 1981.
163. **Junk, G. A. and Jerome, B. A.,** Sampling methods for organic compounds in stacks, *Am. Lab.,* December 16, 1983.
164. **Stanley, J. S., Haile, C. L., Small, A. M., and Olson, E. P.,** Sampling and Analysis Procedures for Assessing Organic Emissions from Stationary Combustion Sources in Exposure Evaluation Division Studies, EPA Methods Manual EPA-560-5-82-014, Office of Toxic Substances, Washington, D.C., January 1982.
165. **Ozvacic, V.,** A review of stack sampling methodology for PCDDs/PCDFs, *Chemosphere,* 15, 1173, 1986.
166. **Velzy, C. O.,** ASME standard sampling and analysis methods for dioxins/furans, *Chemosphere,* 15, 1179, 1986.
167. **Hagenmaier, H., Kraft, M., Jager, W., Mayer, U., Lutzke, K., and Siegel, D.,** Comparison of various sampling methods for PCDDs and PCDFs in stack gas, *Chemosphere,* 15, 1187, 1986.
168. **Brenner, K. S.,** PU-foam-plug-technique and extractive co-distillation (Bleidner apparatus), versatile tools for stack emission sampling and sample preparation, *Chemosphere,* 15, 1917, 1986.
169. **Cooke, M., DeRoos, F., Rising, B., Jackson, M. D., Johnson, L. D., and Merrill, R. G., Jr.,** Dioxin sampling from hot stack gas using source assessment sampling system and modified method 5 trains — an evaluation, presented at Ninth Annual Research Symposium on Land Disposal, Incineration and Treatment of Hazardous Waste, Ft. Mitchell, KY, March 1983.
170. **Tiernan, T. O., Taylor, M. L., Van Ness, G. F., Garrett, J. H., Solch, J. G., Deis, D. A., and Guinivan, T. L.,** Laboratory evaluation of the collection efficiency of a stack sampling train for trapping airborne tetrachlorodibenzo-*p*-dioxin, tetrachlorodibenzofuran, and pentachlorophenol present in flue gases, in *Chlorinated Dioxins and Dibenzofurans in the Total Environment,* Vol. II, Keith, L. H., Rappe, C., and Choudhary, G., Eds., Butterworth Publishers, Stoneham, MA, 1985, 525.
171. **Clement, R. E., Tosine, H. M., Osborne, J., Ozvacic, V., and Wong, G.,** Levels of chlorinated organics in a municipal incinerator, in *Chlorinated Dioxins and Dibenzofurans in the Total Environment,* Vol. II, Keith, L. H., Rappe, C., and Choudhary, G., Eds., Butterworth Publishers, Stoneham, MA, 1985, 489.
172. **Taylor, M. L., Tiernan, T. O., Garrett, J. H., Van Ness, G. F., and Solch, J. G.,** Assessments of incineration processes as sources of supertoxic chlorinated hydrocarbons: concentrations of polychlorinated dibenzo-*p*-dioxins/dibenzofurans and possible precursor compounds in incinerator effluents, in *Chlorinated Dioxins and Dibenzofurans in the Total Environment,* Choudhary, G., Keith, L. H., and Rappe, C., Eds., Butterworth Publishers, Woburn, MA, 1983, 125.
173. **Harvan, D. J., Hass, J. R., Schroeder, J. L., and Corbett, B. J.,** Detection of tetrachlorodibenzodioxins in air filter samples, *Anal. Chem.,* 53, 1755, 1981.
174. **di Domenico, A., Silano, V., Viviano, G., and Zapponi, G.,** Accidental release of 2,3,7,8-tetrachlorodibenzo-*p*-dioxin (TCDD) at Seveso, Italy. VI. TCDD levels in atmospheric particles, *Ecotoxicol. Environ. Safety,* 4, 346, 1980.
175. **Hunt, G. T.,** Ambient monitoring of semi-volatile organics using high volume sorbent samplers — critical quality control features and other considerations in sample collection and network design, paper presented at the 1986 EPA/APCA Symposium on Measurement of Toxic Air Pollutants, Raleigh, NC, April 28-30, 1986.
176. **Oehme, M., Mano, S., Mikalsen, A., and Kirschner, P.,** Quantitative method for the determination of femtogram amounts of polychlorinated dibenzo-*p*-dioxins and dibenzofurans in outdoor air, *Chemosphere,* 15, 607,1986.
177. **Smith, R. M., O'Keefe, P. W., Hilker, D. R., and Aldous, K. M.,** Determination of picogram per cubic meter concentrations of tetra- and pentachlorinated dibenzofurans and dibenzo-*p*-dioxins in indoor air by high-resolution gas chromatography/high resolution mass spectrometry, *Anal. Chem.,* 58, 2414, 1986.

178. **O'Keefe, P. W., Silkworth, J. B., Gierthy, J. F., Smith, R. M., DeCaprio, A. P., Turner, J. N., Eadon, G., Hilker, D. R., Aldous, K. M., Kaminsky, L. S., and Collins, D. N.,** Chemical and biological investigations of a transformer accident at Binghamton, NY, *Environ. Health Perspect.,* 60, 201, 1985.

179. **Billings, W. N. and Bidleman, T. F.,** High volume collection of chlorinated hydrocarbons in urban air using three solid adsorbents, *Atmos. Environ.,* 17, 383, 1983.

180. **Bowers, W. D., Parsons, M. L., Clement, R. E., and Karasek, F. W.,** Component loss during evaporation-reconstitution of organic environmental samples for gas chromatographic analysis, *J. Chromatogr.,* 207, 203, 1981.

181. **O'Keefe, P., Meyer, C., and Dillon, K.,** Comparison of concentration techniques for 2,3,7,8-tetrachlorodibenzo-*p*-dioxin, *Anal. Chem.,* 54, 2623, 1982.

182. **Karasek, F. W., Clement, R. E., and Sweetman, J. A.,** Preconcentration for trace analysis of organic compounds, *Anal. Chem.,* 53, 1050A, 1981.

183. **Cutié, S. S.,** Recovery efficiency of 2,3,7,8-tetrachlorodibenzo-*p*-dioxin from active carbon and other particulates, *Anal. Chim. Acta,* 123, 25, 1981.

184. **Stalling, D. L., Tindle, R. C., and Johnson, J. L.,** Cleanup of pesticide and polychlorinated biphenyl residues in fish extracts by gel permeation chromatography, *J. Assoc. Off. Anal. Chem.,* 55, 32, 1972.

185. **Dolphin, R. J. and Willmott, F. W.,** Separation of chlorinated dibenzo-*p*-dioxins from chlorinated congeners, *J. Chromatogr.,* 149, 161, 1978.

186. **Albro, P. W. and Parker, C. E.,** General approach to the fractionation and class determination of complex mixtures of chlorinated aromatic compounds, *J. Chromatogr.,* 197, 155, 1980.

187. **Porter, M. L. and Burke, J. A.,** Separation of three chlorodibenzo-*p*-dioxins from some polychlorinated biphenyls by chromatography on an aluminum oxide column, *J. Assoc. Off. Anal. Chem.,* 54, 1426, 1971.

188. **O'Keefe, P. W., Meselson, M. S., and Baughman, R. W.,** Neutral cleanup procedure for 2,3,7,8-tetrachlorodibenzo-*p*-dioxin residues in bovine fat and milk, *J. Assoc. Off. Anal. Chem.,* 61, 621, 1978.

189. **Ress, J. R., Higginbotham, G. R., and Firestone, D.,** Methodology for chlorinated aromatics in fats, oils, and fatty acids, *J. Assoc. Off. Anal. Chem.,* 53, 628, 1970.

190. **Niemann, R. A., Brumley, W. C., Firestone, D., and Sphon, J. A.,** Analysis of fish for 2,3,7,8-tetrachlorodibenzo-*p*-dioxin by electron capture capillary gas chromatography, *Anal. Chem.,* 55, 1497, 1983.

191. **Lamparski, L. L., Nestrick, T. J., and Stehl, R. H.,** Determination of part-per-trillion concentrations of 2,3,7,8-tetrachlorodibenzo-*p*-dioxin in fish, *Anal. Chem.,* 51, 1453, 1979.

192. **Leoni, V.,** The separation of fifty pesticides and related compounds and polychlorobiphenyls into four groups by silica gel microcolumn chromatography, *J. Chromatogr.,* 62, 63, 1971.

193. **Armour, J. A. and Burke, J. A.** Method for separating polychlorinated biphenyls from DDT and its analogs, *J. Assoc. Off. Anal. Chem.,* 53, 761, 1970.

194. **Firestone, D., Clower, M., Jr., Borsetti, A. P., Teske, R. H., and Long, P. E.,** Polychlorodibenzo-*p*-dioxin and pentachlorophenol residues in milk and blood of cows fed technical pentachlorophenol, *J. Agric. Food Chem.,* 27, 1171, 1979.

195. **Smith, L. M.,** Carbon dispersed in glass fibers as an adsorbent for contaminant enrichment and fractionation, *Anal. Chem.,* 53, 2152, 1981.

196. **Lamparski, L. L. and Nestrick, T. J.,** Determination of tetra-, hexa-, hepta-, and octachlorodibenzo-*p*-dioxin isomers in particulate samples at parts per trillion levels, *Anal. Chem.,* 52, 2045, 1980.

197. **Korfmacher, W. A., Rushing, L. G., Nestorick, D. M., Thompson, H. C., Jr., Mitchum, R. K., and Kominsky, J. R.,** Analysis of dust and surface swab samples for octachlorodibenzo-*p*-dioxin and heptachlorodibenzo-*p*-dioxins by fused silica capillary GC with EC detection, *J. High Res. Chromatogr. Chromatogr. Commun.,* 8, 12, 1985.

198. **Smith, L. M., Stalling, D. L., and Johnson, J. L.,** Determination of part-per-trillion levels of polychlorinated dibenzofurans and dioxins in environmental samples, *Anal. Chem.,* 56, 1830, 1984.

199. **Nestrick, T. J. and Lamparski, L. L.,** Isomer-specific determination of chlorinated dioxins for assessment of formation and potential environmental emission from wood combustion, *Anal. Chem.,* 54, 2292, 1982.

200. **Rappe, C., Buser, H. R., Stalling, D. L., Smith, L. M., and Dougherty, R. C.,** Identification of polychlorinated dibenzofurans in environmental samples, *Nature,* 292, 524, 1981.

201. **Petty, J. D., Smith, L. M., Bergqvist, P.-A., Johnson, J. L., Stalling, D. L., and Rappe, C.,** Composition of polychlorinated dibenzofuran and dibenzo-*p*-dioxin residues in sediments of the Hudson and Housatonic rivers, in *Chlorinated Dioxins and Dibenzofurans in the Total Environment,* Choudhary, G., Keith, L. H., and Rappe, C., Eds., Butterworth Publishers, Woburn, Ma, 1983, 203.

202. **O'Keefe, P., Smith, R. M., Aldous, K. M., Hilker, D. R., and Jelus-Tyror, B. L.,** Analysis for 2,3,7,8-tetrachlorodibenzofuran and 2,3,7,8-tetrachlorodibenzo-*p*-dioxin in a soot sample from a transformer explosion in Binghamton, New York, *Chemosphere,* 11, 715, 1982.

203. **Nowicki, H. G., Kieda, C. A., Current, V., and Schaefers, T. H.,** Column chromatography fractionation of complex waste water sample extracts for measurement of ppt levels of 2,3,7,8-tetrachlorodibenzo-*p*-dioxin *J. High Resolut. Chromatogr. Chromatogr. Commun.,* 4, 178, 1981.

195

204. **Clement, R. E. and Lennox, S. A.,** Worldwide survey on chlorinated dibenzo-*p*-dioxin (CDD) and dibenzofuran (CDF) analysis capability, *Chemosphere,* 15, 1941, 1986.
205. **Combellas, C. and Drochon, B.,** Influence of water in adsorption chromatography on alumina, *Anal. Lett.,* 16(A20), 1647, 1983.
206. **Williams, D. T. and Blanchfield, B. J.,** Thin-layer chromatographic separation of two chlorodibenzo-*p*-dioxins from some polychlorinated biphenyls and organochlorine pesticides, *J. Assoc. Off. Anal. Chem.,* 54, 1429, 1971.
207. **Kuehl, D. W., Cook, P. M., and Batterman, A. R.,** Studies on the bioavailability of 2,3,7,8-TCDD from municipal incinerator flyash to freshwater fish, *Chemosphere,* 14, 871, 1985.
208. **Stalling, D. L., Petty, J. D., Smith, L. M., Rappe, C., and Buser, H. R.,** Isolation and analysis of polychlorinated dibenzofurans in aquatic samples, in *Chlorinated Dioxins and Related Compounds: Impact on the Environment,* Hutzinger, O., Frei, R. W., Merian, E., and Pocchiari, F., Eds., Pergamon Press, Oxford, 1982, 77.
209. **Lawrence, J., Onuska, F., Wilkinson, R., and Afghan, B. K.,** Methods research: determination of dioxins in fish and sediment, *Chemosphere,* 15, 1085, 1986.
210. **Kuehl, D. W., Cook, P. M., Batterman, A. R., and Butterworth, B. C.,** Isomer dependent bioavailability of polychlorinated dibenzo-*p*-dioxins and dibenzofurans from municipal incinerator fly ash to carp, *Chemosphere,* 16, 657, 1987.
211. **Lapeza, C. R., Jr., Patterson, D. G., Jr., and Liddle, J. A.,** Automated apparatus for the extraction and enrichment of 2,3,7,8-tetrachlorodibenzo-*p*-dioxin in human adipose tissue, *Anal. Chem.,* 58, 713, 1986.
212. **O'Keefe, P. W., Smith, R. M., Hilker, D. R., Aldous, K. M., and Gilday, W.,** A semiautomated cleanup method for polychlorinated dibenzo-*p*-dioxins and polychlorinated dibenzofurans in environmental samples, in *Chlorinated Dioxins and Dibenzofurans in the Total Environment,* Vol. II, Keith, L. H., Rappe, C., and Choudhary, G., Eds., Butterworth Publishers, Stoneham, MA, 1985, 111.
213. **Pfeiffer, C. D.,** Determination of chlorinated dibenzo-*p*-dioxins in pentachlorophenol by liquid chromatography, *J. Chromatogr. Sci.,* 14, 386, 1976.
214. **McGinnis, G. D., Ervin, H. E., Amlicke, B. C., and Ingram, L. L., Jr.,** Analysis of low-levels of pentachlorophenol and dioxins by high-performance liquid chromatography, *J. Am. Wood Preserv. Assoc.,* 233, 1979.
215. **Lamparski, L. L. and Nestrick, T. J.,** Synthesis and identification of the 10 hexachlorodibenzo-*p*-dioxin isomers by high performance liquid and packed column gas chromatography, *Chemosphere,* 10, 3, 1981.
216. **Lamparski, L. L. and Nestrick, T. J.,** The isomer-specific determination of tetrachlorodibenzo-*p*-dioxin at part per trillion concentrations, in *Chlorinated Dioxins and Related Compounds: Impact on the Environment,* Hutzinger, O., Frei, R. W., Merian, E., and Pocchiari, F., Eds., Pergamon Press, Oxford, 1982, 1.
217. **Smith, R. M., Hilker, D. R., O'Keefe, P. W., Aldous, K. M., Meyer, C. M., Kumar, S. N., and Jelus-Tyror, B. M.,** Determination of tetrachlorodibenzo-*p*-dioxins and tetrachlorodibenzofurans in environmental samples by high performance liquid chromatography, capillary gas chromatography and high resolution mass spectrometry, in *Human and Environmental Risks of Chlorinated Dioxins and Related Compounds,* Tucker, R. E., Young, A. L., and Gray, A. P., Eds., Plenum Press, New York, 1983, 73.
218. **Nestrick, T. J., Lamparski, L. L., Shadoff, L. A., and Peters, T. L.,** Methodology and preliminary results for the isomer-specific determination of TCDDs and higher chlorinated dibenzo-*p*-dioxins in chimney particulates from wood-fueled domestic furnaces located in eastern, central and western regions of the United States, in *Human and Environmental Risks of Chlorinated Dioxins and Related Compounds,* Tucker, R. E., Young, A. L., and Gray, A. P., Eds., Plenum Press, New York, 1983, 95.
219. **Tong, H. Y., Shore, D. L., Karasek, F. W., Helland, P., and Jellum, E.,** Identification of organic compounds obtained from incineration of municipal waste by high-performance liquid chromatographic fractionation and gas chromatography-mass spectrometry, *J. Chromatogr.,* 285, 423, 1984.
220. **Barnhart, E. R., Ashley, D. L., Reddy, V. V., and Patterson, D. G.,** Reversed-phase liquid chromatography of chloro-derivatives of dioxin, pyrene, and biphenylene on C-18 and pyrene columns, *J. High Resolut. Chromatogr. Chromatogr. Commun.,* 9, 528, 1986.
221. **Wagel, D., Tiernan, T. O., Taylor, M. L., Ramalingam, B., Garrett, J. H., and Solch, J. G.,** Development of high-performance liquid chromatographic methods for separation and automated collection of tetrachlorodibenzo-*p*-dioxin isomers, in *Chlorinated Dioxins and Dibenzofurans in Perspective,* Rappe, C., Choudhary, G., and Keith, L. H., Eds., Lewis Publishers, Chelsea, MI, 1986, 305.
222. **Swerev, M. and Ballschmiter, K.,** HPLC and heart cutting techniques of polychlorinated dibenzofurans and polychlorinated dibenzodioxins, *Chemosphere,* 15, 1123, 1986.
223. **O'Keefe, P. W., Smith, R., Meyer, C., Hilker, D., Aldous, K., and Jelus-Tyror, B.,** Modification of a high performance liquid chromatographic-gas chromatographic procedure for separation of the 22 tetrachlorodibenzo-*p*-dioxin isomers, *J. Chromatogr.,* 242, 305, 1982.

224. **Stalling, D. L., Norstrom, R. J., Smith, L. M., and Simon, M.,** Patterns of PCDD, PCDF, and PCB contamination of Great Lakes fish and birds and their characterization by principal component analysis, *Chemosphere,* 14, 627, 1985.

225. **Eiceman, G. A., Clement, R. E., and Karasek, F. W.,** Variations in concentrations of organic compounds including polychlorinated dibenzo-*p*-dioxins and polynuclear aromatic hydrocarbons in fly ash from a municipal incinerator, *Anal. Chem.,* 53, 955, 1981.

226. **Williams, D. T. and Blanchfield, B. J.,** Screening method for the detection of chlorodibenzo-*p*-dioxins in the presence of chlorobiphenyls, chloronaphthalenes, and chlorodibenzofurans, *J. Assoc. Off. Anal. Chem.,* 55, 93, 1972.

227. **Firestone, D., Ress, J., Brown, N. L., Barron, R. P., and Damico, J. N.,** Determination of polychlorodibenzo-*p*-dioxins and related compounds in commercial chlorophenols, *J. Assoc. Off. Anal. Chem.,* 55, 85, 1972.

228. **Metcalfe, L. D.,** Proposed source of chick edema factor, *J. Assoc. Off. Anal. Chem.,* 55, 542, 1972.

229. **Woolson, E. A., Thomas, R. F., and Ensor, P. D.,** Survey of polychlorodibenzo-*p*-dioxin content in selected pesticides, *J. Agric. Food Chem.,* 20, 351, 1972.

230. **Hutzinger, O., Safe, S., and Zitko, V.,** Analysis of chlorinated aromatic hydrocarbons by exhaustive chlorination. Qualitative and structural aspects of the perchloro-derivatives of biphenyl, naphthalene, terphenyl, dibenzofuran, dibenzodioxin and DDE, *Int. J. Environ. Anal. Chem.,* 2, 95, 1972.

231. **Baughman, R. and Meselson, M.,** An improved analysis for tetrachlorodibenzo-*p*-dioxins, in *Chlorodioxins — Origin and Fate,* Blair, E. H., Ed., Adv. Chem. Ser. 120, American Chemical Society, Washington, D.C., 1973, 92.

232. **Baughman, R. and Meselson, M.,** An analytical method for detecting TCDD (Dioxin): levels in samples from Vietnam, *Environ. Health Perspect.,* 5, 27, 1973.

233. **Elvidge, D. A.,** The gas chromatographic determination of 2,3,7,8-tetrachlorodibenzo-*p*-dioxin in 2,4,5-trichlorophenoxyacetic acid ("2,3,4-T"), 2,4,5-T ethylhexyl ester, formulations of 2,4,5-T esters and 2,4,5,-trichlorophenol, *Analyst,* 96, 721, 1971.

234. **Williams, D. T. and Blanchfield, B. J.,** An improved screening method for chlorodibenzo-*p*-dioxins, *J. Assoc. Off. Anal. Chem.,* 55, 1358, 1972.

235. **Buser, H. R. and Bosshardt, H. P.,** Determination of 2,3,7,8-tetrachlorodibenzo-1,4-dioxin at parts per billion levels in technical grade 2,4,5-trichlorophenoxy acetic acid, in 2,4,5-T alkyl ester and 2,4,5-T amine salt herbicide formulations by quadrupole mass fragmentography, *J. Chromatogr.,* 90, 71, 1974.

236. **Villanueva, E. C., Jennings, R. Q., Burse, V. W., and Kimbrough, R. D.,** A comparison of analytical methods for chlorodibenzo-*p*-dioxins in pentachlorophenol, *J. Agric. Food Chem.,* 23, 1089, 1975.

237. **Villanueva, E. C., Burse, V. W., and Jennings, R. W.,** Chlorodibenzo-*p*-dioxin contamination of two commercially available pentachlorophenols, *J. Agric. Food Chem.,* 21, 739, 1973.

238. **Nilsson, C. A. and Renberg, L.,** Further studies on impurities in chlorophenols, *J. Chromatogr.,* 89, 325, 1974.

239. **Bowes, G. W., Mulvihill, M. J., DeCamp, M. R., and Kende, A. S.,** Gas chromatographic characteristics of authentic chlorinated dibenzofurans; identification of two isomers in American and Japanese polychlorinated biphenyls, *J. Agric. Food Chem.,* 23, 1222, 1975.

240. **Blaser, W. W., Bredeweg, R. A., Shadoff, L. A., and Stehl, R. H.,** Determination of chlorinated dibenzo-*p*-dioxins in pentachlorophenol by gas chromatography-mass spectrometry, *Anal. Chem.,* 48, 984, 1976.

241. **Buser, H. R. and Bosshardt, H. P.** Determination of polychlorinated dibenzo-*p*-dioxins and dibenzofurans in commercial pentachlorophenols by combined gas chromatography-mass spectrometry, *J. Assoc. Off. Anal. Chem.,* 59, 562, 1976.

242. **Mieure, J. P., Hicks, O., Kaley, R. G., and Michael, P. R.,** Determination of trace amounts of chlorodibenzo-*p*-dioxins and chlorodibenzofurans in technical grade pentachlorophenol, *J. Chrom. Sci.,* 15, 275, 1977.

243. **Shadoff, L. A., Hummel, R. A., Lamparski, L. L., and Davidson, J. H.,** A search for 2,3,7,8-tetrachlorodibenzo-*p*-dioxin (TCDD) in an environment exposed annually to 2,4,5-trichlorophenoxy-acetic acid ester (2,4,5-T) herbicides, *Bull. Environ. Contam. Toxicol.,* 18, 478, 1977.

244. **Mahle, N. H., Higgins, H. S., and Getzendaner, M. E.,** Search for the presence of 2, 3, 7, 8-tetrachlorodibenzo-*p*-dioxin in bovine milk, *Bull. Environ. Contam. Toxicol.,* 18, 123, 1977.

245. **Ramstad, T., Mahle, N. H., and Matalon, K.,** Automated cleanup of herbicides by adsorption chromatography for the determination of 2,3,7,8-tetrachlorodibenzo-*p*-dioxin, *Anal. Chem.,* 49, 386, 1977.

246. **Morita, M., Nakagawa, J., Akiyama, K., Mimura, S., and Isono, N.,** Detailed examination of polychlorinated dibenzofurans in PCB preparations and Kanomi Yusho oil, *Bull. Environ. Contam. Toxicol.,* 18, 67, 1977.

247. **Shadoff, L. A. and Hummel, R. A.,** Determination of 2,3,7,8-tetrachlorodibenzo-*p*-dioxin in biological extracts by gas chromatography mass spectrometry, *Biomed. Mass Spectrom.,* 5, 7, 1978.

248. **Stalling, D. L., Peterman, P. H., Smith, L. M., Norstrom, R. J., and Simon, M.,** Use of pattern recognition in the evaluation of PCDD and PCDF residue data from GC/MS analyses, *Chemosphere,* 15, 1435, 1986.

249. **Kocher, C. W., Mahle, N. H., Hummel, R. A., Shadoff, L. A., and Getzendaner, M. E.,** A search for the presence of 2,3,7,8-tetrachlorodibenzo-*p*-dioxin in beef fat, *Bull. Environ. Contam. Toxicol.,* 19, 229, 1978.

250. **Pfeiffer, C. D., Nestrick, T. J., and Kocher, C. W.,** Determination of chlorinated dibenzo-*p*-dioxins in purified pentachlorophenol by liquid chromatography, *Anal. Chem.,* 50, 800, 1978.

251. **Buser, H. R.,** Formation of polychlorinated dibenzofurans (PCDFs) and dibenzo-*p*-dioxins (PCDDs) from the pyrolysis of chlorobenzenes, *Chemosphere,* 8, 415, 1979.

252. **Thulp, M. Th. M. and Hutzinger, O.,** Rat metabolism of polychlorinated dibenzo-*p*-dioxins, *Chemosphere,* 7, 761, 1978.

253. **Shadoff, L. A., Blaser, W. W., Kocher, C. W., and Fravel, H. G.,** Chlorinated benzyl phenyl ethers: possible interference in the determination of chlorinated dibenzo-*p*-dioxins in 2,4,5-trichlorophenol and its derivatives, *Anal. Chem.,* 50, 1586, 1978.

254. **Kuroki, H. and Musada, Y.,** Determination of polychlorinated dibenzofuran isomers retained in patients with Yusho, *Chemosphere,* 7, 771, 1978.

255. **Eiceman, G. A., Clement, R. E., and Karasek, R. W.,** Analysis of fly ash from municipal incinerators for trace organic compounds, *Anal. Chem.,* 51, 2343, 1979.

256. **Erk, S., Taylor, M. L., and Tiernan, T. O.,** Determination of 2,3,7,8-tetrachlorodibenzo-*p*-dioxin residues on metal surfaces, *Chemosphere,* 8, 7, 1979.

257. **Norstrom, A., Chaudhary, S. K., Albro, P. W., and McKinney, J. D.,** Synthesis of chlorinated dibenzofurans and chlorinated amino-dibenzofurans from the corresponding diphenyl ethers and nitro-diphenylethers, *Chemosphere,* 8, 331, 1979.

258. **Sundstrom, G., Jensen, S., Jansson, B., and Erne, K.,** Chlorinated phenoxyacetic acid derivatives and tetrachlorodibenzo-*p*-dioxin in foliage after application of 2,4,5-trichlorophenoxyacetic acid esters, *Arch. Environ. Contam. Toxicol.,* 8, 441, 1979.

259. **Gara, A., Nilsson, C.-A., Anderson, K., and Rappe, C.,** Synthesis of higher chlorinated dibenzofurans, *Chemosphere,* 8, 405, 1979.

260. **Dobbs, A. J. and Grant, C.,** Octachlorodibenzo-*p*-dioxin in wood treatment materials and treated wood, *Chemosphere,* 10, 1185, 1981.

261. **Buser, H.-R.,** Brominated and brominated/chlorinated dibenzodioxins and dibenzofurans: potential environmental contaminants, *Chemosphere,* 16, 713, 1987.

262. **Chess, E. K. and Gross, M. L.,** Determination of tetrachlorodioxins by mass spectrometric metastable decomposition monitoring, *Anal. Chem.,* 52, 2057, 1980.

263. **Van Ness, G. F., Solch, J. G., Taylor, M. L., and Tiernan, T. O.,** Tetrachlorodibenzo-*p*-dioxins in chemical wastes, aqueous effluents, and soils, *Chemosphere,* 9, 553, 1980.

264. **Langhorst, M. L. and Shadoff, L. A.,** Determination of parts-per-trillion concentrations of tetra-, hexa-, hepta- and octa-chlorodibenzo-*p*-dioxins in human milk samples, *Anal. Chem.,* 52, 2037, 1980.

265. **Mahle, N. H. and Whiting, L. F.,** The formation of chlorodibenzo-*p*-dioxins by air oxidation and chlorination of bituminous coal, *Chemosphere,* 9, 693, 1980.

266. **Shadoff, L. A.,** The determination of 2,3,7,8-tetrachlorodibenzo-*p*-dioxin in human milk, *Am. Chem. Soc. Symp. Ser.,* 136, 1980, 277.

267. **Gross, M. L., Sun, T., Lyon, P. A., Wojinski, S. F., Hilker, D. R., Dupuy, A. E., Jr., and Heath, R. G.,** Method validation for the determination of tetrachlorodibenzodioxin at the low parts-per-trillion level, *Anal. Chem.,* 53, 1902, 1981.

268. **Levin, J.-O. and Nilsson, C.-A.,** Chromatographic determination of polychlorinated phenols, phenoxyphenols, dibenzofurans and dibenzodioxins in wood-dust from worker environments, *Chemosphere,* 6, 443, 1977.

269. **Cochrane, W. P., Singh, J., Miles, W., and Wakeford, B.,** Determination of chlorinated dibenzo-*p*-dioxin contaminants in 2,4-D products by gas chromatography/mass spectrometric techniques, *J. Chromatogr.,* 217, 289, 1981.

270. **Tong, H. Y. and Karasek, F. W.,** Comparison of quantitation of polychlorinated dibenzodioxins and polychlorinated dibenzofurans in complex environmental samples by high resolution gas chromatography with flame ionization, electron capture and mass spectrometric detection, *Chemosphere,* 15, 1141, 1986.

271. **Ligon, W. V., Jr. and May, R. J.,** Determination of selected chlorodibenzofurans and chlorodibenzodioxins using two-dimensional gas chromatography/mass spectrometry, *Anal. Chem.,* 58, 558, 1986.

272. **Ligon, W. V., Jr. and May, R. J.,** Isomer specific analysis of selected chlorodibenzofurans, *J. Chromatogr.,* 294, 87, 1984.

273. **Tong, H. Y., Shore, D. L., and Karasek, F. W.,** Isolation of polychlorinated dibenzodioxins and polychlorinated dibenzofurans from a complex organic mixture by two-step liquid chromatographic fractionation for quantitative analysis, *Anal. Chem.,* 56, 2442, 1984.

274. **Buser, H.-R.,** High-resolution by gas chromatography of polychlorinated dibenzo-*p*-dioxins and dibenzofurans, *Anal. Chem.,* 48, 1553, 1976.
275. **Buser, H.-R.,** Analysis of TCDD's by gas chromatography-mass spectrometry using glass capillary columns, in *Dioxin, Toxicological and Chemical Aspects,* Cattabeni, F., Cavallero, A., and Galli, G., Eds., SP Medical Science Books, New York, 1978, 27.
276. **Buser, H.-R. and Rappe, C.,** High-resolution gas chromatography of the 22 tetrachlorodibenzo-*p*-dioxin isomers, *Anal. Chem.,* 52, 2257, 1980.
277. **Buser, H.-R. and Rappe, C.,** High-resolution gas chromatography of the 22 tetrachlorodibenzo-*p*-dioxin (TCDD) isomers, in *Chlorinated Dioxins and Related Compounds: Impact on the Environment,* Hutzinger, O., Frei, R. W., Merian, E., and Pocchiari, F., Eds., Pergamon Press, Oxford, 1982, 15.
278. **Buser, H.-R. and Rappe, C.,** Isomer-specific separation of 2,3,7,8-substituted polychlorinated dibenzo-*p*-dioxins by high-resolution gas chromatography/mass spectrometry, *Anal. Chem.,* 56, 442, 1984.
279. **SUPELCO, Inc.,** Rapid separation of 2,3,7,8-TCDD from other TCDD isomers, SUPELCO GC Bulletin 793C, 1983.
280. **SUPELCO, Inc.,** New capillary column for EPA analysis of 2,3,7,8-TCDD, *Supelco Reporter,* 4(4), 1985.
281. **Korfmacher, W. A. and Mitchum, R. K.,** Relative retention times of 39 polychlorinated dibenzo-*p*-dioxins using SP 2100 fused silica capillary chromatography, *J. High Resolut. Chromatogr. Chromatogr. Commun.,* 3, 681, 1982.
282. **Oehme, M. and Kirschmer, P.,** Isomer-selective determination of tetrachlorodibenzo-*p*-dioxins using hydroxyl negative ion chemical ionization mass spectrometry combined with high-resolution gas chromatography, *Anal. Chem.,* 56, 2754, 1984.
283. **Naikwadi, K. P. and Karasek, F. W.,** Gas chromatographic separation of 2,3,7,8-tetrachlorodibenzo-*p*-dioxin from polychlorinated biphenyls and tetrachlorodibenzo-*p*-dioxin isomers using a polymeric liquid crystal capillary column, *J. Chromatogr.,* 396, 203, 1986.
284. **Kuroki, H., Haraguchi, K., and Masuda, Y.,** Synthesis of polychlorinated dibenzofuran isomers and their gas chromatographic profiles, *Chemosphere,* 13, 561, 1984.
285. **Buser, H.-R., Rappe, C., and Bergqvist, P.-A.,** Analysis of polychlorinated dibenzofurans, dioxins and related compounds in environmental samples, *Environ. Health Perspect.,* 60, 293, 1985.
286. **Harless, R. L. and Lewis, R. G.,** Quantitative determination of 2,3,7,8-tetrachlorodibenzo-*p*-dioxin residues by gas chromatography/mass spectrometry, in *Chlorinated Dioxins and Related Compounds: Impact on the Environment,* Hutzinger, O., Frei, R. W., Merian, E., and Pocchiari, F., Eds., Pergamon Press, Oxford, 1982, 25.
287. **Chiu, C., Thomas, R. S., Lockwood, J., Li, K., Halman, R., and Lao, R. C.,** Polychlorinated hydrocarbons from power plants, wood burning and municipal incinerators, *Chemosphere,* 12, 607, 1983.
288. **Tiernan, T. O., Taylor, M. L., Garrett, J. H., Van Ness, G. F., Solch, J. G., Deis, D. A., and Wagel, D. J.,** Chlorodibenzodioxins, chlorodibenzofurans and related compounds in the effluents from combustion processes, *Chemosphere,* 12, 595, 1983.
289. **Fung, D., Boyd, R. K., Safe, S., and Chittim, B. G.,** Gas chromatographic/mass spectrometric analysis of specific isomers of polychlorodibenzofurans, *Biomed. Mass Spectrom.,* 12, 247, 1985.
290. **Lau, B. P.-Y., Sun, W.-F., and Ryan, J. J.,** Complication of using hydrogen as the GC carrier gas in chlorinated dibenzofuran and dibenzodioxin GC/MS analysis, *Chemosphere,* 14, 799, 1985.
291. **Hale, M. D., Hileman, F. D., Mazer, T., Shell, T. L., Noble, R. W., and Brooks, J. J.,** Mathematical modeling of temperature programmed capillary gas chromatographic retention indexes for polychlorinated dibenzofurans, *Anal. Chem.,* 57, 640, 1985.
292. **Buu-Hoi, N. P., Saint-Ruf, G., and Mangane, M.,** The fragmentation of dibenzo-*p*-dioxin and its derivatives under electron impact, *J. Heterocycl. Chem.,* 9, 691, 1972.
293. **Curley, A., Jennings, R. W., Burse, V. W., and Villanueva, E. C.,** Electron impact spectra of several aromatic systems, in *Mass Spectrometry and NMR Spectroscopy in Pesticide Chemistry,* Hague, R. and Biros, F. J., Eds., Plenum Press, New York, 1974, 71.
294. **Buser, H. R. and Rappe, C.,** Identification of substitution patterns in polychlorinated dibenzo-*p*-dioxins (PCDDs) by mass spectrometry, *Chemosphere,* 7, 199, 1978.
295. **Meyer, C. M., O'Keefe, P. W., Briggs, R. G., and Hilker, D. R.,** Differences in the chromatographic and mass spectral properties of 3,3′, 5,5′-tetrachlorodiphenoquinone and 2,3,7,8-tetrachlorodibenzo-*p*-dioxin, *Biomed. Environ. Mass Spectrom.,* 13, 47, 1986.
296. **Mahle, N. H. and Shadoff, L. A.,** The mass spectrometry of chlorinated dibenzo-*p*-dioxins, *Biomed. Mass Spectrom.,* 9, 45, 1982.
297. **Buser, H. R. and Rappe, C.,** Analysis of polychlorinated dibenzo-*p*-dioxins and dibenzofurans in chlorinated phenols by mass fragmentography, *J. Chromatogr.,* 107, 295, 1975.
298. **Holmstedt, B.,** Mass fragmentography of TCDD and related compounds, in *Dioxin: Toxicological and Chemical Aspects,* Cattabeni, F., Cavallaro, A., and Galli, G., Eds., SP Medical Science Books, New York, 1978, 13.

299. **Cambridge Isotope Laboratories, Inc.,** 20 Commerce Way, Woburn, MA, 01801.

300. **Cairns, T., Fishbein, L., and Mitchum, R. K.,** Review of the dioxin problem. Mass spectrometric analyses of tetrachlorodioxins in environmental samples, *Biomed. Mass Spectrom.,* 7, 484, 1980.

301. **Clement, R. E., Alfieri, A., and Bobbie, B.,** Investigation of the quality of commercial chlorinated dibenzo-*p*-dioxin (CDD) standards, *Chemosphere,* 15, 1947, 1986.

302. **Hummel, R. A. and Shadoff, L. A.,** Specificity of low resolution gas chromatography-low resolution mass spectrometry for the detection of tetrachlorodibenzo-*p*-dioxin in environmental samples, *Anal. Chem.,* 52, 191, 1980.

303. **Crummett, W. B.,** Fundamental problems related to validation of analytical data elaborated on the example of TCDD, *Toxicol. Environ. Chem. Rev.,* 3, 61, 1979.

304. **Crummett, W. B.,** The problem of measurements near the limit of detection, *Ann. N.Y. Acad. Sci.,* 320, 43, 1979.

305. **Viau, A. C. and Karasek, F. W.,** Quantitative aspects in the determination of polychlorinated dibenzo-*p*-dioxins by high resolution gas chromatography-mass spectrometry, *J. Chromatogr.,* 270, 235, 1983.

306. **Clement, R. E., Tosine, H. M., and Alfieri, A.,** Response factor studies for analysis of polychlorinated dibenzo-*p*-dioxins by GC-EIMS, presented at the 32nd Annual Conference on Mass Spectrometry and Allied Topics, San Antonio, TX, May 29, 1984.

307. **Kuehl, D. W., Durhan, E., Butterworth, B. C., and Linn, D.,** Tetrachloro-9H-carbazole, a previously unrecognized contaminant in sediments of the Buffalo River, *J. Great Lakes Res.,* 10, 210, 1984.

308. **Weeren, R. D. and Asshauer, J.,** Problems and results of trace analysis of 2,3,7,8-tetrachlorodibenzo-*p*-dioxin in 2,4,5-trichlorophenoxyacetic acid and its esters, *J. Assoc. Off. Anal. Chem.,* 68, 917, 1985.

309. **Patterson, D. G., Holler, J. S., Groce, D. F., Alexander, L. R., Lapeza, C. R., O'Connor, R. C., and Liddle, J. A.,** Control of interferences in the analysis of human adipose tissue to 2,3,7,8-tetrachlorodibenzo-*p*-dioxin (TCDD), *Environ. Toxicol. Chem.,* 5, 355, 1986.

310. **Phillipson, D. W. and Puma, B. J.,** Identification of chlorinated methoxybiphenyls as contaminants in fish as potential interferences in the determination of chlorinated dibenzo-*p*-dioxins, *Anal. Chem.,* 52, 2328, 1980.

311. **Smith, L. M. and Johnson, J. L.,** Evaluation of interferences from seven series of polychlorinated aromatic compounds in an analytical method for polychlorinated dibenzofurans and dibenzo-*p*-dioxins in environmental samples, in *Chlorinated Dioxins and Dibenzofurans in the Total Environment,* Choudhary, G., Keith, L. H., and Rappe, C., Eds., Butterworth Publishers, Woburn, MA, 1983, 321.

312. **Jensen, S. and Renberg, L.,** Contaminants in pentachlorophenol: chlorinated dioxins and predioxins (chlorinated hydroxy-diphenyl ethers), *Ambio,* 1(4), 1, 1972.

313. **Rappe, C. and Nilsson, C.-A.,** An artifact in the gas chromatographic determination of impurities in pentachlorophenol, *J. Chromatogr.,* 67, 247, 1972.

314. **Kuehl, D. W., Cook, P. M., Batterman, A. R., Lotherback, D., and Butterworth, B. C.,** Bioavailability of polychlorinated dibenzo-*p*-dioxins and dibenzofurans from contaminated Wisconsin River sediment to carp, *Chemosphere,* 16, 667, 1987.

315. **Freudenthal, J.,** The quantitative determination of TCDD with different mass spectrometric methods, in *Dioxin: Toxicological and Chemical Aspects,* Cattabeni, F., Cavallaro, A., and Galli, G., Eds., SP Medical Science Books, New York, 1978, 43.

316. **Harless, R. L. and Oswald, E. O.,** Low- and high-resolution gas-chromatography-mass spectrometry (GC-MS) method of analysis for the presence of 2,3,7,8-tetrachlorodibenzo-*p*-dioxin (TCDD) in environmental samples, in *Dioxin: Toxicological and Chemical Aspects,* Cattabeni, F., Cavallaro, A., and Galli, G., Eds., SP Medical Science Books, New York, 1978, 51.

317. **Chapman, J. R., Warburton, G. A., Ryan, P. A., and Hazelby, D.,** Application of the MS80 and a new high resolution selected ion monitoring routine to the determination of trace levels of TCDD, *Biomed. Mass Spectrom.,* 7, 597, 1980.

318. **Millington, D. S., Parr, V. C., and Hall, K.,** New techniques in selected ion detection, *Ann. Chim.,* 69, 629, 1979.

319. **Kuehl, D. W., Butterworth, B. C., and Johnson, K. L.,** Supplemental quality assurance criteria for high-resolution gas chromatography/high-resolution mass spectrometric determination of 2,3,7,8-tetrachlorodibenzo-*p*-dioxin in biological tissue, *Anal. Chem.,* 58, 1598, 1986.

320. **Stanley, J. S. and Sack, T. M.,** Protocol for the Analysis of 2,3,7,8-Tetrachlorodibenzo-*p*-Dioxin by High-Resolution Gas Chromatography/High-Resolution Mass Spectrometry, EPA 600/4-86-004, January 1986.

321. **Taguchi, V. Y., Reiner, E. J., Hallas, B., and Wang, D. T.,** High-resolution selected ion monitoring of polychlorinated dibenzo-*p*-dioxins (PCDD) and dibenzofurans (PCDF) using non-PFK lockmasses, presented at the 35th ASMS Conference on Mass Spectrometry and Allied Topics, Denver, CO, May 28, 1987.

322. **Reynolds, W. D., Mitchum, R. K., Newton, J., Bystroff, R. I., Pomernacki, C., Brand, H. A., and Siegel, M. W.,** An ultra-sensitive mass spectrometer system for quantitative biological studies, *Chem. Instrum.,* 8, 63, 1977.

323. **Dougherty, R. C.,** Negative chemical ionization mass spectrometry: applications in environmental analytical chemistry, *Biomed. Mass Spectrom.,* 8, 283, 1981.

324. **Kuehl, D. W., Dougherty, R. C., Tondeur, Y., Stalling, D. L., Smith, L. M., and Rappe, C.,** Negative chemical ionization studies of polychlorinated dibenzo-*p*-dioxins, dibenzofurans and naphthalenes in environmental samples, in *Environmental Health Chemistry,* McKinney, J. D., Ed., Ann Arbor Science, Ann Arbor, MI, 1981, 245.

325. **Thoma, H. and Hutzinger, O.,** Analysis of polychlorodibenzodioxins (PCDD) and polychlorodibenzofurans (PCDF) in fly ash by high resolution mass spectrometry, *Chemosphere,* 15, 2115, 1986.

326. **Hunt, D. F., Harvey, T. M., and Russel, J. W.,** Oxygen as a reagent gas for the analysis of 2,3,7,8-tetrachlorodibenzo-*p*-dioxin by negative ion chemical ionization mass spectrometry, *J. Chem. Soc. Chem. Commun.,* 151, 1975.

327. **Miles, W. F., Gurprasad, N. P., and Malis, G. P.,** Isomer-specific determination of hexachlorodibenzo-*p*-dioxins by oxygen negative chemical ionization mass spectrometry, gas chromatography, and high-pressure liquid chromatography, *Anal. Chem.,* 57, 1133, 1985.

328. **Larameé, J. A., Arbogast, B. C. and Deinzer, M. L.,** Electron capture negative ion chemical ionization mass spectrometry of 1,2,3,4-tetrachlorodibenzo-*p*-dioxin, *Anal. Chem.,* 58, 2907, 1986.

329. **Crow, F. W., Bjorseth, A., Knapp, K. T., and Bennett, R.,** Determination of polyhalogenated hydrocarbons by glass capillary gas chromatography-negative ion chemical ionization mass spectrometry, *Anal. Chem.,* 53, 619, 1981.

330. **Cavallaro, A., Bandi, G., Invernizzi, G., Luciani, L., Mongini, E., and Gorni, G.,** Negative ion chemical ionization MS as a structure tool in the determination of small amounts of PCDD and PCDF, in *Chlorinated Dioxins and Related Compounds: Impact on the Environment,* Hutzinger, O., Frei, R. W., Merian, E., and Pocchiari, F., Eds., Pergamon Press, Oxford, 1982, 55.

331. **Hass, J. R. and Friesen, M. D.,** Qualitative and quantitative methods for dioxin analysis, *Ann. N.Y. Acad. Sci.,* 320, 28, 1979.

332. **Hass, J. R., Friesen, M. D. and Hoffman, M. K.,** The mass spectrometry of polychlorinated dibenzo-*p*-dioxins, *Org. Mass Spectrom.,* 14, 9, 1979.

333. **Hass, J. R., Friesen, M. D., Harvan, D. J., and Parker, C. E.,** Determination of polychlorinated dibenzo-*p*-dioxins in biological samples by negative chemical ionization mass spectrometry, *Anal. Chem.,* 50, 1474, 1978.

334. **Hass, J. R., Friesen, M. D., and Hoffman, M. K.,** Recent mass spectrometric techniques for the analysis of environmental contaminants, in *Environmental Health Chemistry,* McKinney, J. D., Ed., Ann Arbor Science, Ann Arbor, MI, 1981, 219.

335. **Rappe, C., Marklund, S., Nygren, M., and Gara, A.,** Parameters for identification and confirmation in trace analyses of polychlorinated dibenzo-*p*-dioxins and dibenzofurans, in *Chlorinated Dioxins and Dibenzofurans in the Total Environment,* Choudhary, G., Keith, L. H., and Rappe, C., Eds., Butterworth Publishers, Woburn, MA, 1983, 259.

336. **Mitchum, R. K., Moler, G. F., and Korfmacher, W. A.,** Combined capillary gas chromatography/atmospheric pressure negative chemical ionization/mass spectrometry for the determination of 2,3,7,8-tetrachlorodibenzo-*p*- dioxin in tissue, *Anal. Chem.,* 52, 2278, 1980.

337. **Korfmacher, W. A., Rowland, K. R., Mitchum, R. K., Daly, J. J., McDaniel, R. C., and Plummer, M. V.,** Analysis of snake tissue and snake eggs for 2,3,7,8-tetrachlorodibenzo-*p*-dioxin via fused silica GC combined with atmospheric pressure ionization MS, *Chemosphere,* 13, 1229, 1984.

338. **Mitchum, R. K., Korfmacher, W. A., Moler, G. F., and Stalling, D. L.,** Capillary gas chromatography/atmospheric pressure negative chemical ionization mass spectrometry of the 22 isomeric tetrachlorodibenzo-*p*-dioxins, *Anal. Chem.,* 54, 719, 1982.

339. **Korfmacher, W. A., Moler, G. F., Delongchamp, R. R., Mitchum, R. K., and Harless, R. L.,** Validation study for the gas chromatographic-atmospheric pressure ionization-mass spectrometric method for the isomer-specific determination of 2,3,7,8-tetrachlorodibenzo-*p*-dioxin, *Chemosphere,* 13, 669, 1984.

340. **Korfmacher, W. A. and Mitchum, R. K.,** Separation of three tetrachlorodibenzofuran isomers via fused silica GC combined with API-MS, *J. High Resolut. Chromatogr. Chromatogr. Commun.,* 2, 294, 1981.

341. **Korfmacher, W. A., Mitchum, R. K., Hileman, F. D., and Mazer, T.,** Analysis of 2,3,7,8-tetrachlorodibenzofuran by fused silica GC combined with atmospheric pressure ionization MS, *Chemosphere,* 12, 1243, 1983.

342. **Mitchum, R. K., Korfmacher, W. A., and Moler, G. F.,** Validation study for the gas chromatography/atmospheric pressure ionization/mass spectrometry method for isomer-specific determination of 2,3,7,8-tetrachlorodibenzo-*p*-dioxin, in *Chlorinated Dioxins and Dibenzofurans in the Total Environment,* Choudhary, G., Keith, L. H., and Rappe, C., Eds., Butterworth Publishers, Woburn, MA, 1983, 273.

343. **French, J. B., Davidson, W. R., Reid, N. M., and Buckley, J. A.,** Trace monitoring by tandem mass spectrometry, in *Tandem Mass Spectrometry,* McLafferty, F. W., Ed., John Wiley & Sons, New York, 1983, 353.

344. **Bursey, M. M. and Hass, J. R.,** Tandem mass spectrometry for environmental problems, in *Tandem Mass Spectrometry,* McLafferty, F. W., Ed., John Wiley & Sons, New York, 1983, 465.

345. **Tou, J. C., Zakett, D., and Caldecourt, V. J.,** Tandem mass spectrometry of industrial chemicals, in *Tandem Mass Spectrometry,* McLafferty, F. W., Ed., John Wiley & Sons, New York, 1983, 435.

346. **Shushan, B., Fulford, J. E., Thomson, B. A., Davidson, W. R., Danylewych, L. M., Ngo, A., Nacson, S., and Tanner, S. D.,** Recent applications of triple quadrupole mass spectrometry to trace chemical analysis, *Int. J. Mass Spectrom. Ion Phys.,* 46, 225, 1983.

347. **Clement, R. E., Bobbie, B., and Taguchi, V.,** Comparison of instrumental methods for chlorinated dibenzo-*p*-dioxin (CDD) determination — interim results of a round-robin study involving GC-MS, MS-MS, and high resolution MS, *Chemosphere,* 15, 1147, 1986.

348. **Voyksner, R. D., Hass, J. R., Sovocool, G. W., and Bursey, M. M.,** Comparison of gas chromatography/high resolution mass spectrometry and mass spectrometry/mass spectrometry for detection of polychlorinated biphenyls and tetrachlorodibenzofuran, *Anal. Chem.,* 55, 744, 1983.

349. **Sakuma, T., Davidson, W. R., Lane, D. A., Thomson, B. A., Fulford, J. E., and Quan, E. S. K.,** The rapid analysis of gaseous PAH and other combustion related compounds in hot gas streams by APCI/MS and APCI/MS/MS, in *Polynuclear Aromatic Hydrocarbons: Chemical Analysis and Biological Fate,* Cooke, M. and Dennis, A. J., Eds., Battelle Press, Columbus, OH, 1981, 179.

350. **Shushan, B., Tanner, S. D., Clement, R. E., and Bobbie, B.,** The rapid determination of PCDDs and PCDFs in municipal waste incinerator fly ash by one-step "jar" extraction followed by low resolution capillary gas chromatography-tandem mass spectrometry ("Flash" GC/MS/MS), *Chemosphere,* 14, 843, 1985.

351. **Sakuma, T., Gurprasad, N., Tanner, S. D., Ngo, A., Davidson, W. R., McLeod, H. A., Lau, B. P.-Y., and Ryan, J. J.,** The application of rapid gas chromatography-tandem mass spectrometry in the analysis of complex samples for chlorinated dioxins and furans, in *Chlorinated Dioxins and Dibenzofurans in the Total Environment,* Vol. II, Keith, L. H., Rappe, C., and Choudhary, G., Eds., Butterworth Publishers, Stoneham, MA, 1985, 139.

352. **Shushan, B., Ngo, A., Ozvacic, V., Wong, G., DeBrou, G., Bobbie, B., Clement, R. E., Chittim, B., and Thorndyke, S.,** The development of techniques for the rapid screening of municipal refuse incinerator fly-ash and stack gas for PCDD/PCDF using a transportable tandem mass spectrometer (GC/MS/MS) system, *Proceedings of the 7th Annual Technology Transfer Conference, Vol. A: Air Quality Research,* Ontario Ministry of Environment, Toronto, Ontario, 1986, 45.

353. **Albro, D. W., Luster, M. I., Chae, K., Chaudhary, S. K., Clark, G., Lawson, L. D., Corbett, J. T., and McKinney, J. D.,** A radioimmunoassay for chlorinated dibenzo-*p*-dioxins, *Toxicol. Appl Pharmacol.,* 50, 137, 1979.

354. **McKinney, J., Albro, P., Luster, M., Corbett, B., Schroeder, J., and Lawson, L.,** Development and reliability of a radioimmunoassay for 2,3,7,8-tetrachlorodibenzo-*p*-dioxin, in *Chlorinated Dioxins and Related Compounds: Impact on the Environment,* Hutzinger, O., Frei, R. W., Merian, E., and Pocchiari, F., Eds., Pergamon Press, Oxford, 1982, 67.

355. **Luster, M. I., Albro, P. W., Chae, K., Lawson, L. D., Corbett, J. T., and McKinney, J. D.,** Radioimmunoassay for quantitation of 2,3,7,8-tetrachlorodibenzofuran, *Anal. Chem.,* 52, 1497, 1980.

356. **Bradlaw, J. A. and Casterline, J. L., Jr.,** Induction of enzyme activity in cell culture: a rapid screen for detection of planar polychlorinated organic compounds, *J. Assoc. Off. Anal. Chem.,* 62, 904, 1979.

357. **Safe, S.,** Determination of 2,3,7,8-TCDD toxic equivalent factors (TEFs): support for the use of the in-vitro AHH induction assay, *Chemosphere,* 16, 791, 1987.

358. **Hutzinger, O., Olie, K., Lustenhouwer, J. W. A., Okey, A. B., Bandiera, S., and Safe, S.,** Polychlorinated dibenzo-*p*-dioxins and dibenzofurans: a bioanalytical approach, *Chemosphere,* 10, 19, 1981.

359. **Sawyer, T., Bandiera, S., Safe, S., Hutzinger, O., and Olie, K.,** Bioanalysis of polychlorinated dibenzofuran and dibenzo-*p*-dioxin mixtures in fly ash, *Chemosphere,* 12, 529, 1983.

360. **Safe, S., Mason, G., Farrell, K., Keys, B., Piskorska-Pliszczynska, J., Madge, J. A., and Chittim, B.,** Validation of in vitro bioassays for 2,3,7,8-TCDD toxic equivalents, presented at the 6th International Symposium on Chlorinated Dioxins and Related Compounds, Fukuoka, Japan, September, 16—19, 1986, *Chemosphere,* in press.

361. **Stanker, L., Watkins, B., Vanderlaan, M., and Budde, W. L.,** Development of an immunoassay for chlorinated dioxins based on a monoclonal antibody and an enzyme linked immunosorbent assay (ELISA), presented at the 6th International Symposium on Chlorinated Dioxins and Related Compounds, Fukuoka, Japan, September 16—19, 1986, *Chemosphere,* in press.

362. **Gierthy, J. F. and Crane, D.,** Development of in vitro bioassays for chlorinated dioxins and dibenzofurans, in *Chlorinated Dioxins and Dibenzofurans in the Total Environment,* Vol. II, Keith, L. H., Rappe, C., and Choudhary, G., Eds., Butterworth Publishers, Stoneham, MA, 1985, 267.

363. **Gierthy, J. F., Crane, D., and Frenkel, G. D.,** Application of an in vitro keratinization assay to extracts of soot from a fire in a PCB-containing transformer, *Fund. Appl. Pharmacol.,* 4, 1036, 1984.

364. **Gierthy, J. F. and Crane, D.,** Development and application of an in vitro bioassay for dioxinlike compounds, in *Chlorinated Dioxins and Dibenzofurans in Perspective,* Rappe, C., Choudhary, G., and Keith, L. H., Eds., Lewis Publishers, Chelsea, MI, 1986, 269.

365. **Helder, Th. and Seinen, W.,** Standardization and application of an E.L.S.-bioassay for PDDDs and PCDFs, *Chemosphere,* 14, 183, 1985.

366. **Helder, Th. and Seinen, W.,** Relative toxicities of some CDDs and CDFs and toxic potentials of incineration products, assessed by the E.L.S.-bioassay, *Chemosphere,* 15, 1165, 1986.

367. **Bharath, A., Mallard, C., Orr, D., Ozburn, G., and Smith, A.,** Problems in determining the water solubility of organic compounds, *Bull. Environ. Contam. Toxicol.,* 33, 133, 1984.

368. **Barnes, D. G.,** Possible consequences of sharing an environment with dioxins, in *Accidental Exposure to Dioxins: Human Health Aspects,* Coulston, F. and Pocchiari, F., Eds., Academic Press, New York, 1983, 259.

369. **Kearney, P. C., Isensee, A. R., Helling, C. S., Woolson, E. A., and Plimmer, J. R.,** Environmental significance of chlorodioxins, in *Chlorodioxins — Origin and Fate,* Blair, E. H., Ed., ACS Adv. Chem. Ser. 120, American Chemical Society, Washington, D.C., 1973, 105.

370. **Laycock, D. E.,** Dioxins: assembling the pieces, *Can. Chem. News,* p. 7, November/December 1986.

371. **Josephson, J.,** Chlorinated dioxins and furans in the environment, *Environ. Sci. Technol.,* 17, 124A, 1983.

372. **Patterson, D. G., Alexander, L. R., Gelbaum, L. T., O'Connor, R. C., Maggio, U., and Needham, L. L.,** Synthesis and relative response factors for the 22 tetrachlorodibenzo-*p*-dioxins (TCDDs) by electron-impact ionization mass spectrometry, *Chemosphere,* 15, 1601, 1986.

373. **Richards, D. J. and Shieh, W. K.,** Biological fate of organic priority pollutants in the aquatic environment, *Wat. Res.,* 20, 1077, 1986.

374. **Corbet, R. L., Muir, D. C. G., and Webster, G. R. B.,** Fate of 1,3,6,8-T$_4$CDD in an outdoor aquatic system, *Chemosphere,* 12, 523, 1983.

375. **Tsushimoto, G., Matsumura, F., and Sago, R,.** Fate of 2,3,7,8-tetrachlorodibenzo-*p*-dioxin (TCDD) in an outdoor pond and in model aquatic ecosystems, *Environ. Toxicol. Chem.,* 1, 61, 1982.

376. **Marcheterre, L., Webster, G. R. B., Muir, D. C. G., and Grift, N. P.,** Fate of ^{14}C-octachlorodibenzo-*p*-dioxin in artificial outdoor ponds, *Chemosphere,* 14, 835, 1985.

377. **Muir, D. C. G., Townsend, B. E., and Webster, G. R. B.,** Bioavailability of ^{14}C-1,3,6,8-tetrachlorodibenzo-*p*-dioxin and ^{14}C-octachlorodibenzo-*p*-dioxin to aquatic insects in sediment and water, in *Chlorinated Dioxins and Dibenzofurans in the Total Environment,* Vol. II, Keith, L. H., Rappe, C., and Choudhary, G., Eds., Butterworth Publishers, Stoneham, MA, 1985, 89.

378. **Yochim, R. S., Isensee, A. R., and Jone, G. E.** Distribution and toxicity of TCDD and 2,4,5-T in an aquatic model ecosystem, *Chemosphere,* 7, 215, 1978.

379. **Mill, T.,** Prediction of environmental fate of tetrachlorodibenzodioxin, in *Dioxins in the Environment,* Kamrin, M. A. and Rodgers, P. W., Eds., Hemisphere Publishing, New York, 1985, 173.

380. **Srinivasan, K. R. and Fogler, H. S.,** Binding of OCDD, 2,3,7,8-TCDD and HCB to clay-based sorbents, in *Chlorinated Dioxins and Dibenzofurans in Perspective,* Rappe, C., Choudhary, G., and Keith, L. H., Eds., Lewis Publishers, Chelsea, MI, 1986, 531.

381. **Webster, G. R. B., Muldrew, D. H., Graham, J. J., Sarna, L. P., and Muir, D. C. G.,** Dissolved organic matter mediated aquatic transport of chlorinated dioxins, *Chemosphere,* 15, 1379, 1986.

382. **Mackay, D. and Powers, B.,** Sorption of hydrophobic chemicals from water: a hypothesis for the mechanism of the particle concentration effect, *Chemosphere,* 16, 745, 1987.

383. **Belli, G., Bressi, G., Calligarich, E., Cerlesi, S., and Ratti, S. P.,** Analysis of the TCDD-distribution as a function of the underground depth for data taken in 1977 and 1979 in Zone A at Seveso (Italy), in *Chlorinated Dioxins and Related Compounds: Impact on the Environment,* Hutzinger, O., Frei, R. W., Merian, E., and Pocchiari, F., Eds., Pergamon Press, Oxford, 1982, 137.

384. **Young, A. L., Cairney, W. J., and Thalken, C. E.,** Persistence, movement and decontamination studies of TCDD in storage sites massively contaminated with phenoxy herbicides, *Chemosphere,* 12, 713, 1983.

385. **Thibodeaux, L. J.,** Offsite transport of 2,3,7,8-tetrachlorodibenzo-*p*-dioxin from a production disposal facility, in *Chlorinated Dioxins and Dibenzofurans in the Total Environment,* Choudhary, G., Keith, L. H., and Rappe, C., Eds., Butterworth Publishers, Woburn, MA, 1983, 75.

386. **Freeman, R. A. and Schroy, J. M.,** Environmental mobility of TCDD, *Chemosphere,* 14, 873, 1985.

387. **Walters, R. W., Guiseppi-Elie, A., Rao, M. M., and Means, J. C.,** Desorption of 2,3,7,8-TCDD from soils into water-/methanol and methanol liquid phases, in *Chlorinated Dioxins and Dibenzofurans in Perspective,* Rappe, C., Choudhary, G., and Keith, L. H., Eds., Lewis Publishers, Chelsea , MI, 1986, 157.

388. **Young, A. L. and Cockerham, L. G.,** Fate of TCDD in field ecosystems — assessment and significance for human exposures, in *Dioxins in the Environment,* Kamrin, M. A. and Rodgers, P. W., Eds., Hemisphere Publishing, New York, 1985, 153.

389. **Palausky, J., Kapila, S., Manahan, S. E., Yanders, A. F., Malhotra, R. K., and Clevenger, T. E.,** Studies on vapor phase transport and role of dispersing medium on mobility of 2,3,7,8-TCDD in soil, *Chemosphere,* 15, 1389, 1986.

390. **Kearney, P. C., Woolson, E. A., Isensee, A. R., and Helling, C. S.,** Tetrachlorodibenzodioxin in the environment: sources, fate and decontamination, *Environ. Health Perspec.,* 5, 273, 1973.

391. **di Domenico, A., Silano, V., Viviano, G., and Zapponi, G.,** Accidental release of 2,3,7,8-tetrachlorodibenzo-*p*-dioxin (TCDD) at Seveso, Italy. IV. Vertical distribution in soil, *Ecotoxicol. Environ. Safety,* 4, 327, 1980.

392. **Young, A. L.,** Long-term studies on the persistence and movement of TCDD in a natural ecosystem, in *Human and Environmental Risks of Chlorinated Dioxins and Related Compounds,* Tucker, R. E., Young, A. L., and Gray, A. P., Eds., Plenum Press, New York, 1983, 173.

393. **Young, A. L., Thalken, C. E., Arnold, E. L., Cupello, J. M., and Cockerham, L. G.,** Fate of 2,3,7,8-Tetrachlorodibenzo-*p*-dioxin (TCDD) in the Environment: Summary and Decontamination Recommendations, U.S. Air Force Academy, Dept. of Chem. Biol. Sci., USAFA-TR-76-18, 1976.

394. **Helling, C. S., Isensee, A. R., Woolson, E. A., Ensor, P. D. J., Jones, G. E., Plimmer, J. R., and Kearney, P. C.,** Chlorodioxins in pesticides, soils, and plants, *J. Environ. Qual.,* 2, 171, 1973.

395. **Wipf, H. K., Homberger, E., Neuner, N., and Schenker, F.,** Field trials on photodegradation of TCDD on vegetation after spraying with vegetable oil, in *Dioxin: Toxicological Aspects,* Cattabeni, F., Cavallaro, A., and Galli, G., Eds. SP Medical Science Books, London, 1978, 201.

396. **Facchetti, S., Balasso, A., Fichtner, C., Frare, G., Leoni, A., Mauri, C., and Vasconi, M.,** Studies on the absorption of TCDD by plant species, in *Chlorinated Dioxins and Dibenzofurans in Perspective,* Rappe, C., Choudhary, G., and Keith, L. H., Eds., Lewis Publishers, Chelsea, MI, 1986, 225.

397. **Facchetti, S., Balasso, A., Fichtner, C., Frare, G., Leoni, A., Mauri, C., and Vasconi, M.,** Studies on the absorption of TCDD by some plant species, *Chemosphere,* 15, 1387, 1986.

398. **Kenaga, E. E. and Norris, L. A.,** Environmental toxicity of TCDD, in *Human and Environmental Risks of Chlorinated Dioxins and Related Compounds,* Tucker, R. E., Young, A. L., and Gray, A. P., Eds., Plenum Press, New York, 1983, 277.

399. *Chem. Eng. News,* March 9, 1987, 11.

400. **Villaneuve, J.-P. and Cattini, C.,** Input of chlorinated hydrocarbons through dry and wet deposition to the Western Mediterranean, *Chemosphere,* 15, 115, 1986.

401. **Czuczwa, J. M., Niessen, F., and Hites, R. A.,** Historical record of polychlorinated dibenzo-*p*-dioxins and dibenzofurans in Swiss lake sediments, *Chemosphere,* 14, 1175, 1985.

402. **Czuczwa, J. M. and Hites, R. A.,** Historical record of polychlorinated dioxins and furans in Lake Huron sediments, in *Chlorinated Dioxins and Dibenzofurans in the Total Environment,* Vol. II, Keith, L. H., Rappe, C., and Choudhary, G., Eds., Butterworth Publishers, Stoneham, MA, 1985, 59.

403. **Czuczwa, J. M, McVeety, B. D., and Hites, R. A.,** Polychlorinated dibenzodioxins and dibenzofurans in sediments from Siskiwit Lake, Isle Royale, *Chemosphere,* 14, 623, 1985.

404. **Czuczwa, J. M. and Hites, R. A.,** Sources and fate of PCDD and PCDF, *Chemosphere,* 15, 1417, 1986.

405. **Czuczwa, J. M. and Hites, R. A.,** Dioxins and dibenzofurans in air, soil and water, in *Dioxins in the Environment,* Kamrin, M. A. and Rodgers, P. W., Eds., Hemisphere Publishing, New York, 1985, 85.

406. **Townsend, D. I.,** The use of dioxin isomer group ratios to identify sources and define background levels of dioxins in the environment. A review, update, and extension of the present theory, in *Chlorinated Dioxins and Related Compounds: Impact on the Environment,* Hutzinger, O., Frei, R. W., Merian, E., and Pocchiari, F., Eds., Pergamon Press, Oxford, 1982, 265.

407. **Plimmer, J. R., Klingebiel, U. I., Corsby, D. G., and Wong, A. S.,** Photochemistry of dibenzo-*p*-dioxins, in *Chlorodioxins — Origin and Fate,* Blair, E. H., Ed., ACS Adv. Chem. Ser. 120, American Chemical Society, Washington, D.C., 1973, 44.

408. **Choudhary, G. G. and Hutzinger, O.,** Photochemical formation and degradation of polychlorinated dibenzofurans and dibenzo-*p*-dioxins, *Residue Rev.,* 84, 113, 1982.

409. **Crosby, D. G. and Wong, A. S.,** Environmental degradation of 2,3,7,8-tetrachlorodibenzo-*p*-dioxin (TCDD), *Science,* 195, 1337, 1977.

410. **Botre, C., Memoli, A., and Alhaique, F.,** TCDD solubilization and photodecomposition in aqueous solutions, *Environ. Sci. Technol.,* 12, 335, 1978.

411. **Stehl, R. H., Papenfuss, R. R., Bredeweg, R. A., and Roberts, R. W.,** The stability of pentachlorophenol and chlorinated dioxins to sunlight, heat, and combustion, in *Chlorodioxins — Origin and Fate,* Blair, E. H., Ed., ACS Adv. Chem. Ser., 120, American Chemical Society, Washington, D.C., 1973, 119.

412. **Liberti, A., Brocco, D., Allegrini, I., and Bertoni, G.,** Field photodegradation of TCDD by ultra-violet radiations, in *Dioxin: Toxicological and Chemical Aspects,* Cattabeni, F., Cavallaro, A., and Galli, G., Eds., SP Medical Science Books, New York, 1978, 195.

413. **Wong, A. S. and Crosby, D. G.,** Decontamination of 2,3,7,8-tetrachlorodibenzo-*p*-dioxin (TCDD) by photochemical action, in *Dioxin: Toxicological and Chemical Aspects,* Cattabeni, F., Cavallaro, A., and Galli, G., Eds., SP Medical Science Books, New York, 1978, 185.

414. **Choudhry, G. G. and Webster, G. R. B.,** Quantum yields for the photodecomposition of polychlorinated dibenzo-*p*-dioxins (PCDDs) in water-acetonitrile solution, *Chemosphere,* 14, 893, 1985.
415. **Barbeni, M., Pramauro, E., and Pelizzetti, E.,** Photochemical degradation of chlorinated dioxins, biphenyls, phenols and benzene on semiconductor dispersion, *Chemosphere,* 15, 1913, 1986.
416. **Crosby, D. G.,** Methods of photochemical degradation of halogenated dioxins in view of environmental reclamation, in *Accidental Exposure to Dioxins: Human Health Aspects,* Coulston, F. and Pocchiari, F., Eds., Academic Press, New York, 1983, 149.
417. **Kearney, P. C., Woolson, E. A., and Ellington, C. P., Jr.,** Persistence and metabolism of chlorodioxins in soils, *Environ. Sci. Technol.,* 6, 1017, 1972.
418. **Matsumura, F. and Benezet, H. J.,** Studies on the bioaccumulation and microbial degradation of 2,3,7,8-tetrachlorodibenzo-*p*-dioxin, *Environ. Health Perspect.,* 5, 253, 1973.
419. **Klecka, G. M. and Gibson, D. T.,** Metabolism of dibenzo-*p*-dioxin and chlorinated dibenzo-para-dioxins by a *Beijerinckia* species, *Appl. Environ. Microbiol.,* 39, 288, 1980.
420. **Hutter, R. and Philippi, M.,** Studies on microbial metabolism of TCDD under laboratory conditions, in *Chlorinated Dioxins and Related Compounds: Impact on the Environment,* Hutzinger, O., Frei, R. W., Merian, E., and Pocchiari, F., Eds., Pergamon Press, Oxford, 1982, 87.
421. **Camoni, I., Di Muccio, A., Pontecorvo, D., Taggi, F., and Vergori, L.,** Laboratory investigation for the microbiological degradation of 2,3,7,8-tetrachlorodibenzo-*p*-dioxin in soil by addition of organic compost, in *Chlorinated Dioxins and Related Compounds: Impact on the Environment,* Hutzinger, O., Frei, R. W., Merian, E., and Pocchiari, F., Eds., Pergamon Press, Oxford, 1982, 95.
422. **Matsumura, F., Quensen, J., and Tsushimoto, G.,** Microbial degradation of TCDD in a model ecosystem, in *Accidental Exposure to Dioxins: Human Health Aspects,* Coulston, F. and Pocchiari, F., Eds., Academic Press, New York, 1983, 105.
423. **Matsumura, F., Quensen, J., and Tsushimoto, G.,** Microbial degradation of TCDD in a model ecosystem, in *Human and Environmental Risks of Chlorinated Dioxins and Related Compounds,* Tucker, R. E., Young, A. L., and Gray, A. P., Eds., Plenum Press, New York, 1983, 191.
424. **Isensee, A. R.,** Bioaccumulation of 2,3,7,8-tetrachlorodibenzo-para-dioxin, *Ecol. Bull.* (Stockholm), 27, 255, 1978.
425. **Adams, W. J., DeGraeve, G. M., Sabourin, T. D., Cooney, J. D., and Mosher, G. M.,** Toxicity and bioconcentration of 2,3,7,8-4CDD to fathead minnows *(Pimephales promelas), Chemosphere,* 15, 1503, 1986.
426. **Muir, D. C. G., Marshall, W. K., and Webster, G. R. B.,** Bioconcentration of PCDDs by fish: effects of molecular structure and water chemistry, *Chemosphere,* 14, 829, 1985.
427. **Gobas, F. A. P. C., Shiu, W. Y., Mackay, D., and Opperhuizen, A.,** Bioaccumulation of PCDDs and OCDF in fish after aqueous and dietary exposure, *Chemosphere,* 15, 1985, 1986.
428. **Karasek, F. W., Charbonneau, G. M., Reuel, G. J., and Tong, H. Y.,** Determination of organic compounds leached from municipal incinerator fly ash by water at different pH levels, *Anal. Chem.,* 59, 1027, 1987.
429. **Carsch, S., Thoma, H., ahd Hutzinger, O.,** Leaching of polychlorinated dibenzo-*p*-dioxins and polychlorinated dibenzofurans from municipal waste incinerator fly ash by water and organic solvents, *Chemosphere,* 15, 1927, 1986.
430. **O'Keefe, P. W., Hilker, D. R., Smith, R. M., Aldous, K. M., Donnelly, R. J., Long, D., and Pope, D. H.,** Nonaccumulation of chlorinated dioxins and furans by goldfish exposed to contaminated sediment and flyash, *Bull. Environ. Contam. Toxicol.,* 36, 452, 1986.
431. **Kuehl, D. W., Cook, P. M., Batterman, A. R., and Lothenbach, D. B.,** Bioavailability of 2,3,7,8-tetrachlorodibenzo-*p*-dioxin from municipal incinerator fly ash to freshwater fish, *Chemosphere,* 14, 427, 1985.
432. **Kuehl, D. W., Cook, P. M., and Batterman, A. R.,** Uptake and depuration studies of PCDDs and PCDFs in freshwater fish, *Chemosphere,* 15, 2023, 1986.

Chapter 6

ANALYSIS OF POLYCYCLIC AROMATIC HYDROCARBONS IN ENVIRONMENTAL SAMPLES

Francis I. Onuska

TABLE OF CONTENTS

I. INTRODUCTION

Polycyclic aromatic hydrocarbons (PAH) represent a class of man-made environmental pollutants to which man has almost constantly been exposed. The forest fires throughout the ages represented a history of environmental pollution.

However, it was not until coal replaced wood as a fuel that cancer was recognized in chimney sweepers and which eventually led to the establishment of the carcinogenic nature of some of the PAHs. But not all PAHs are man made. Minerals made of pure PAHs and their analogs occur, especially with mercury ores. Pendletonite is pure coronene. Indralite is a mixture of angularly anneled PAH such as chrysene and picene along with alkylated analogs and nitrogen and sulfur analogs. Interestingly enough, over 100 components are present in indrialite, none appear to be carcinogens.[1] The source of the polynuclear hydrocarbon minerals appears to be the same as that of the man-made pollutants, they are formed by pyrolysis of organic materials at moderate temperatures. The interest in detecting PAH mixtures in trace quantities in air, water, sediments, and biota is to trace their origins and assess the biological risks of their presence.

A. Nomenclature and Structures

In general, any organic compound containing two or more aromatic rings may be considered to be a polycyclic organic matter (POM). In this group of chemicals a diverse range of chemical compounds can be classified such as polycyclic aromatic hydrocarbons, nitrogen-containing PAHs (azaarenes), sulfur-containing PAHs (thiaarenes), carbonyl and dicarbonyl arenes, and arene hydroxides and oxides. These compounds containing a heteroatom in their structures are known as heterocyclic aromatic compounds (HAC). Limited attention will be given to these groups of contaminants since the diversity of these compounds within each of these classes dramatically increase when substituents such as alkyl, amino, cyano, imino, and nitro, are present on the ring. The accepted nomenclature of polycyclic aromatic hydrocarbons is that adopted by the International Union of Pure and Applied Chemistry (IUPAC). In general, the rules that are applied to determine the orientation and peripheral numbering system assigned are given as follows:

1. The maximum number of ring structures is located in a horizontal row.
2. As many ring structures as possible are above and to the right of a horizontal row.
3. If more than one orientation fulfill these requirements, the one with the minimum number of ring structures at the lower left is chosen.
4. The carbon atoms are numbered in a clockwise direction starting with the atom in the most clockwise position of the uppermost ring structure or the uppermost ring which is furthest to the right and is not engaged in ring fusion.

Isomers are distinguished by lettering the peripheral sides of the molecule. Letters start with ''a'' for the side between carbon atoms 1 and 2 and then continuing clockwise around the molecule.

Phenanthrene

There are many exceptions to above given rules. The peripheral numbering system for heteroaromatic compounds is governed by the same principles as for the PAHs but with the additional clause saying that where there is a choice of orientation, low numbers are assigned to the heteroatoms in the molecule. If a choice still remains, the lowest number is given to oxygen in preference to sulfur and to sulfur in preference to nitrogen.

Quinoline   Carbazole   Benz c acridine
                                       (3,4-benzacridine)

The IUPAC nomenclature for selected PAHs and HAC compounds, along with alternate names, molecular formulas, formula weights, and structures are given in Table 1.

B. Physical-Chemical Properties

Polycyclic aromatic hydrocarbons and heterocyclic aromatic compounds are usually crystalline solid materials having high melting points and low vapor pressures, low water solubilities, and they are usually exhibiting some color.

Because of their relatively low vapor pressures, low water solubilities, and their aromaticity, they exhibit a strong adsorption affinity for surfaces. It is considered that PAHs and HACs are associated with suspended particulate matter in both the atmospheric and aquatic environments. Some physicochemical data are given in Tables 2 and 3.

Although polycyclic aromatic hydrocarbons comprise a diverse range of organic compounds, some general trends in physical-chemical properties are apparent. As the molecular weight of PAHs increases, their vapor pressure enhances saturation vapor, concentration decreases. At elevated temperatures, some PAHs, especially the lower molecular weight compounds, exist particularly in the vapor phase. Loss of material due to volatilization during source sampling at elevated temperatures and during ambient atmospheric sampling for relatively long periods have been reported.[2]

The published values for the solubilities of PAHs indicate a wide variation, reflecting in part the differences in preparation and determination of PAH solutions. In general, some trends can be observed. They can be summarized as follows:

- Water solubility of PAHs decreases, as their molecular weight increases.
- Molecules with a linear arrangement of fused aromatic rings, such as naphthalene and anthracene, are usually less soluble than that of PAHs, having angular or pericondensed structures such as chrysene and phenanthrene.
- Alkyl substitution of the aromatic ring decreases water solubility.

Temperature has a significant influence on the solubility of PAHs, as can be demonstrated in the case of water solubility of anthracene. Its solubility increases from 12.7 ± 0.4 μg/l to 55.7 ± 0.7 μg/l when the temperature is increased from 5 to 29°C.[3]

PAHs undergo chemical reactions characteristic of organic aromatic molecules, including 1,2- and 1,4-additions and additional eliminations and electrophilic substitutions.[4] The type of reaction that occurs, as well as the rate of reaction, is dependent upon the molecular structure of the particular PAH in addition to the reagent and medium.

In the presence of light and oxygen PAHs readily undergo photooxidation. The photooxidation of PAHs in both air and water, as well as other environmentally significant removal mechanisms will be discussed later.

PAHs absorb radiation at various wavelengths. Absorption of incident radiation results in activation of the ground state molecule to an excited singlet state which, may react with a second ground state molecule to produce a photoproduct or undergo a transition to an excited triplet state. Reaction of the excited triplet state molecule with ground state molecular oxygen produces excited singlet state PAH molecule to yield oxygenated products.[5] The primary products of photooxidation are endoperoxides. They can undergo further dealkylation and ring cleavage during photolysis. One-electron oxidations occur, producing radical cations. Being unstable, radical cations react with water molecules or some other nucleophile to produce diols, quinones, aldehydes, and dimers.

C. Bioaccumulation and Biodegradation

Evidence that PAHs accumulate in the tissues of aquatic organisms above those levels found in water is documented in data obtained from field studies. Freshwater and marine organisms exposed to PAHs from point sources, such as petroleum drilling activities, oil spills, or chronic leakage, accumulated PAHs at the $\mu g/g$ level.[6-12] Tissue level in biota from remote or relatively unpolluted areas are found in the ng/g range.[13-16] The bioconcentration factor for PAHs is about 1000 times its concentration in water.[17] Similar studies were conducted on the distribution of benzo[a]pyrene levels in water, sediment, plants, plankton, mollusks, and fish from water.[18] Benzo[a]pyrene was found to bioaccumulate in sediment and biota at 100 to 10,000 times the level found in water. An important phenomenon of the environmental fate of PAHs is the extent to which they are accumulated in aquatic organisms. Of major concern is the possibility that PAHs accumulate at tissue levels which are toxic.   In biota, uptake processes for PAHs are related to their hydrophobic character. The hydrophobicity of PAHs is the driving force for partitioning of PAHs between aqueous and lipid phases.[19-20] The extent to which a compound is bioaccumulated by aquatic organisms, relative to the surrounding water is called the bioconcentration factor (BCF) and it reflects an equilibrium between the competing processes of uptake against depuration and metabolism. This type of measurement can be made by exposing organisms for a period long enough to allow residue detection at equilibrium state.

PAHs are sparingly soluble in water and are found mostly adsorbed to particulates. However, an exchange equilibrium exists which means that some portion, even of the high molecular weight compounds dissolves and is present in solution. Since octanol-water partitioning provides a good estimate of particulate-water partitioning, the low molecular weight PAHs with low K_{ow} would be less firmly adsorbed and would be in higher concentration in water than to higher molecular weight compounds with higher K_{ow}.

D. Metabolism, Retention, and Toxicity

Some aquatic invertebrate and vertebrate species possess enzymatic systems which are capable of metabolizing PAHs. A mixed function oxidase (MFO) system, similar to that in mammals has been found in a number of aquatic species. These enzymes, known as aryl hydrocarbon hydroxylases, are believed to be responsible for virtually all oxidative primary products of aromatic hydrocarbon metabolism.[21] The primary metabolism of PAHs results in the formation of phenol, quinone and *trans*-dihydrodiol derivatives. These metabolites are more water-soluble than the parent compounds, thus water-lipid partitioning does not favor their accumulation in tissue lipids. They would be eliminated from animal tissues faster than native compounds. There is limited information available on the types of PAH metabolites formed by aquatic organisms. The pattern of dihydrodiol, quinone, and phenol derivatives produced by fish appears to be similar to metabolite profiles formed in mammalian systems.[21]

As far as retention of PAHs is of concern, it is of interest to note that studies with both fish[22] and birds[23] known to contain relatively high MFO activity levels depurate PAHs quite rapidly when high doses of PAHs are administered.

Table 1

LIST OF DICYCLIC AND POLYCYCLIC AROMATIC HYDROCARBONS, THEIR STRUCTURE, MOLECULAR WEIGHT, MELTING AND BOILING POINTS, AND THEIR CHARACTERISTIC MASS SPECTRA

Structure	IUPAC nomenclature	Molecular weight	Melting point (°C)	Boiling point (°C)[760]	Characteristic ions (m/z)
	Indan	118.18	−51	178	117,118,91
	Indene	116.16	−2	183	116,115,58
	Naphthalene	128.19	81	218	128,129,127
	2-Methylnaphthalene	142.20	35	241	142,141,115
	1-Methylnaphthalene	142.20	−22	245	
	Biphenyl	154.21	71	255	154,153,152
	2-Ethylnaphthalene	156.29	−7	258	141,156,115

Table 1 (continued)
LIST OF DICYCLIC AND POLYCYCLIC AROMATIC HYDROCARBONS, THEIR STRUCTURE, MOLECULAR WEIGHT, MELTING AND BOILING POINTS, AND THEIR CHARACTERISTIC MASS SPECTRA

Structure	IUPAC nomenclature	Molecular weight	Melting point (°C)	Boiling point (°C)^{760}	Characteristic ions (m/z)
	1-Ethylnaphthalene	156.23	− 14	259	
	2,6-Dimethylnaphthalene	156.23	110	262	156,141,155
	2,7-Dimethylnaphthalene	156.23	97	262	
	1,7-Dimethylnaphthalene	156.23		263	
	1,3-Dimethylnaphthalene	156.23		265	
	1,6-Dimethylnaphthalene	156.23		266	
	2,3-Dimethylnaphthalene	156.23	105	268	

211

1,4-Dimethylnaphthalene	156.23	8	268	
4-Methylbiphenyl	168.24	50	268	
1,5-Dimethylnaphthalene	156.23	80	269	
Azulene	128.19	100	270 d	
1,2-Dimethylnaphthalene	156.23	−4	271	
Acenaphthylene	152.21	93	~270 par d	152,151,153
3-Methylbiphenyl	168.24	5	273	
3,5-Dimethylbiphenyl	182.27		275	
Acenaphthene	154.21	96	279	154,153,152
1,3,7-Trimethylnaphthalene	170.25	14	280	

Table 1 (continued)
LIST OF DICYCLIC AND POLYCYCLIC AROMATIC HYDROCARBONS, THEIR STRUCTURE, MOLECULAR WEIGHT, MELTING AND BOILING POINTS, AND THEIR CHARACTERISTIC MASS SPECTRA

Structure	IUPAC nomenclature	Molecular weight	Melting point (°C)	Boiling point (°C)^{760}	Characteristic ions (m/z)
	2,3,5-Trimethylnaphthalene	170.25	25	285	
	2,3,6-Trimethylnaphthalene	170.25	101	286	
	Fluorene	166.23	117	294	166,165,167
	9-Methylfluorene	180.25	47		180,165,179
	4-Methylfluorene	180.25			
	3-Methylfluorene	180.25	85	316	
	2-Methylfluorene	180.25	104	318	

Compound				
1-Methylfluorene	180.25		~318	
1-Phenylnaphthalene	204.28	~45	334	
Phenanthrene	178.24	101	338	178,179,176
Anthracene	178.24	216	340	178,89,176
3-Methylphenanthrene	192.26	65	352	
2-Methylphenanthrene	192.26		355	
9-Methylphenanthrene	192.26	92	355	192,191,189
2-Methylanthracene	192.26	209	359 sub	192,191,165
4,5-Methylenephenanthrene	190.24	116	359	
4-Methylphenanthrene	192.26			

Table 1 (continued)
LIST OF DICYCLIC AND POLYCYCLIC AROMATIC HYDROCARBONS, THEIR STRUCTURE, MOLECULAR WEIGHT, MELTING AND BOILING POINTS, AND THEIR CHARACTERISTIC MASS SPECTRA

Structure	IUPAC nomenclature	Molecular weight	Melting point (°C)	Boiling point (°C)^{760}	Characteristic ions (m/z)
	1-Methylphenanthrene	192.26	123	359	
	2-Phenylnaphthalene	204.28	104	360	
	1-Methylanthracene	192.26	86	363	
	3,6-Dimethylphenanthrene	206.29		363	
	2,7-Dimethylanthracene	206.29	241	~370	206,205,160
	2,6-Dimethylanthracene	206.29	250	~370	
	2,3-Dimethylanthracene	206.29	252		

Fluoranthene	202.26	111	383	202,101,100
9,10-Dimethylanthracene	206.29	183		
Pyrene	202.26	156	393	202,101,100
2,7-Dimethylpyrene	230.32		396	
Benzo[b]fluorene	216.29	209	402	216,215,217
Benzo[c]fluorene	216.29		406	
Benzo[a]fluorene	216.29	190	407	
2-Methylpyrene	216.29		410	216,215,213
1-Methylpyrene	216.29		410	
4-Methylpyrene	216.29		410	

Table 1 (continued)

LIST OF DICYCLIC AND POLYCYCLIC AROMATIC HYDROCARBONS, THEIR STRUCTURE, MOLECULAR WEIGHT, MELTING AND BOILING POINTS, AND THEIR CHARACTERISTIC MASS SPECTRA

Structure	IUPAC nomenclature	Molecular weight	Melting point (°C)	Boiling point (°C)[760]	Characteristic ions (m/z)
	Benzo[ghi]fluoranthene	226.28		432	252,253,125
	Benzo[c]phenanthrene	228.30	68		
	Benzo[a]anthracene	228.30	162	435 sub	228,229,114
	Triphenylene	228.30	199	439	228,226,113
	Chrysene	228.30	256	441	228,226,229
	6-Methylchrysene	242.32			
	1-Methylchrysene	242.32	257		

Name	MW	mp	bp	Wavelengths
Naphthacene	228.30	257	450 sub	228,229,114
2,2′-Dinaphthyl	254.34	188	452[753] sub	228,229,126
Benzo[b]fluoranthene	252.32	168	481	252,253,125
Benzo[j]fluoranthene	252.32	166	~480	
Benzo[k]fluoranthene	252.32	217	481	252,253,125
Benzo[e]pyrene	252.32	179	493	252,126,253
Benzo[a]pyrene	252.32	177	496	252,253,125
Perylene	252.32	278		252,253,124
3-Methylcholanthrene	268.38	180		
Indeno[1,2,3-cd]pyrene	276.34			276,138,277

Table 1 (continued)

LIST OF DICYCLIC AND POLYCYCLIC AROMATIC HYDROCARBONS, THEIR STRUCTURE, MOLECULAR WEIGHT, MELTING AND BOILING POINTS, AND THEIR CHARACTERISTIC MASS SPECTRA

Structure	IUPAC nomenclature	Molecular weight	Melting point (°C)	Boiling point (°C)[760]	Characteristic ions (m/z)
	Dibenz[a,c]anthracene	278.36	205		
	Dibenz[a,h]anthracene	278.36	270		278,139,279
	Dibenz[a,i]anthracene	278.36	264		
	Dibenz[a,j]anthracene	278.36	198		
	Benzo[b]chrysene	278.36	294		278,276,279
	Picene	278.36	368	519	278,276,139
	Benzo[ghi]perylene	276.34	278		276,138,274

Anthanthrene	276.34			300,150,149
Coronene	300.36	439 cor	525?	
Dibenzo[a,e]pyrene	302.38	234		

Note: d = decomposes; par d = partly decomposes; sub = sublimes.

Table 2
PHYSICAL PROPERTIES OF POLYCYCLIC AROMATIC HYDROCARBONS

Compound	Density	Vapor pressure at 25°C	Sat. conc.(ng/m³)	Vapor equilibrium −10°C	Vapor 30°C
Fluorene	1.203	—	—	—	—
Anthracene	1.25	$2.6.10^{-5}$	$1.9.10^7$	—	—
Phenanthrene	1.79	$9.1.10^{-5}$	$6.5.10^7$	—	—
Fluoranthene	1.252	—	—	—	—
Pyrene	1.271	$9.1.10^{-8}$	$7.4.10^4$	5.810^2	$1.4.10^5$
Benzo[a]anthracene	—	$1.5.10^{-8}$	$1.3.10^3$	3.4	$2.8.10^3$
Chrysene	1.274	—	—	—	—
Benzo[k]fluoranthene	—	$1.3.10^{-11}$	$1.3.10^1$	$1.3.10^{-2}$	$3.0.10^1$
Benzo[a]pyrene	1.351	$7.3.10^{-10}$	$7.5.10^1$	$1.5.10^{-1}$	$1.6.10^2$
Benzo[e]pyrene	—	$7.4.10^{-10}$	$7.5.10^1$	$1.5.10^{-1}$	$1.6.10^2$
Perylene	1.35	—	—	—	—
Benzo[ghi]perylene	—	$1.3.10^{-11}$	1.5	$1.8.10^{-3}$	3.4
Dibenz[ghi,pqr]chrysene	1.377	$2.0.10^{-13}$	$2.0.10^{-2}$	$1.8.10^{-6}$	$5.8.10^{-2}$

Table 3
SOLUBILITY AND OCTANOL-
WATER PARTITION COEFFICIENTS
OF PAHs AT 25°C

Compound	Solubility (µg/l)	K_{ow}
Fluorene	800	—
Anthracene	59	4.5
Phenanthrene	435	4.46
2-Methylanthracene	21.3	4.77
9-Methylphenanthrene	261.0	4.77
1-Methylphenanthrene	269.0	4.77
Fluoranthene	260.0	5.03
Pyrene	133.0	4.98
9,10-Dimethylanthracene	56.0	5.13
Benzo[a]fluorene	45.0	5.34
Benzo[b]fluorene	29.6	5.34
Benzo[a]anthracene	11.0	5.63
Naphthacene	1.0	5.65
Chrysene	1.9	5.63
Triphenylene	43.0	5.63
7,12-Dimethylbenz[a]anthracene	1.5	6.36
Benzo[b]fluoranthene	2.4	6.21
Benzo[j]fluoranthene	2.4	6.21
Cholanthene	2.0	6.28
Benzo[a]pyrene	3.8	6.04
Benzo[e]pyrene	2.4	6.21
Perylene	2.4	6.21
Dibenzo[a,h]fluorene	0.8	6.57
Dibenzo[a,g]fluorene	0.8	6.57
Dibenzo[a,c]fluorene	0.8	6.57
3-Methylcholanthene	0.7	6.64
Dibenz[a,j]anthracene	0.4	6.86
Benzo[ghi]fluoranthene	0.5	6.78
Benzo[ghi]perylene	0.3	6.78
Coronene	0.14	7.36

Note: Salinity 32% @22°C

Table 4
RELATIVE CARCINOGENICITY INDEX OF
SOME PAHs

Compound	Carcinogenicity Index
Benzo[a]anthracene	+
7,12-Dimethylbenz[a]anthracene	+ + + +
Dibenz[a,j]anthracene	+
Dibenz[a,h]anthracene	+ + +
Dibenz[a,c]anthracene	+
Benzo[c]phenanthrene	+ + +
Dibenzo[a,g]fluorene	+
Dibenzo[a,h]fluorene	UC
Dibenzo[a,c]fluorene	+
Benzo[b]fluoranthene	+ +
Benzo[j]fluoranthene	+ +
Benzo[j]aceanthrylene	+ +
3-Methylcholanthrene	+ + + +
Benzo[a]pyrene	+ + +
Dibenzo[a,l]pyrene	UC
Dibenzo[a,h]pyrene	+ + +
Dibenzo[a,i]pyrene	+ + +
Indeno[1,2,3-cd]pyrene	+
Chrysene	UC
Dibenzo[b,def]chrysene	+ +
Dibenzo[def,p]chrysene	+

Note: UC = unknown.

Studies with petroleum indicate that the naphthalenes and phenolic compounds, not the multiple ring PAHs are mainly responsible for the acute toxicity of oils.[24] The PAHs vary substantially in their toxicity to marine environment. It appears that the toxicity increases with increasing molecular size until the 4- and 5-ring molecules are reached. These compounds are not found to be toxic within the limits of their solubility. Aside from short-term toxicity to aquatic organisms, the greatest concern about PAHs is their carcinogenic potential. The general relationship of structure to carcinogenic activity seems to favor the 4-, 5-, and 6-membered ring PAHs rather than smaller (3-ring structures) or larger (7-ring structure) such as coronene. In addition, the angular, rather than linear or highly condensed structures such as perylene, ananthene, etc., possess higher activity. Details can be found elsewhere.[25] Some relative carcinogenicity of PAHs are given in Table 4.

II. GENERAL ANALYTICAL PROCEDURES

A. Sampling and Storage
1. Water

Samples should be collected in glass bottles with plastic screw caps having precleaned Teflon liners. Since only submicrogram quantities of PAHs are usually present in most samples, all bottles must be scrupulously cleaned.

Generally, PAHs have been shown to be associated with suspended matter in waters, particularly in rivers or lakes that exhibit a relatively high turbidity. PAHs are adsorbed by particulate matter and, consequently, filtered samples report lower results. Thus, most analyses should be carried out using unfiltered water samples if possible, or examined separately on the filtered and particulate matter fractions. The size of the sample taken may vary between 1 and 4 l. Samples should be stored at approximately 4°C, but if several days elapse before the analysis is performed, partial extraction is carried out by the addition of 10 ml of nanogen quality benzene.

The highest concentrations of dissolved and particulate organic matter in the waters are found in the surface films. When organic pollutants, such as PAHs are present, they tend to accumulate in this surface film, particularly if they are either nonpolar compounds or show a surface activity. A major problem in the sampling of surface films is the inclusion of water with the film. In the ideal sampler, only the film of organic molecules, perhaps a few molecular layers in thickness, floating on the water surface, would be collected. The analytical results could then be expressed either in terms of surface area sampled or volume taken. In practice, all of the samplers collect some portion of the surface layer of water. Each sampler collects a different slice of the surface layer; thus results expressed in weights per unit volume are only comparable when the samples were taken with the same sampler. The available information on these subjects has been reviewed by Wangersky.[37]

The initial problem when attempting to evaluate PAHs in water is the concentrations of other contaminants in comparison to the amount of PAHs present.[27] Their high molecular weight and lack of polar substituents give a low solubility for water. Although literature claims most compounds to be "insoluble in water", values in the order of 7 to 12 µg/l are quoted by the *Merck Index*.[28] Bohm et al.[29] claimed that a wide variety of organic compounds, including alkylbenzene sulfonate, influence the solubility of PAHs in water.

PAHs present in the aqueous phase may be transferred to organic solvent by exhaustive extraction, performed at ambient temperature; light and air may alter or oxidize some of the PAHs. An important point, which has influenced the final assay procedure can be achieved by transferring PAHs from the aqueous phase by a variety of solvents. The amount of other organic matter removed depends on the type of an organic solvent used. Balgaires et al.[30] and Monkman et al.[31] confirmed this work using PAH contaminated samples concluding that it was possible to classify their extract into two groups:

1.   Iso-octane-cyclohexane
2.   Benzene-chloroform-acetone-methanol

As a consequence, they reported that the benzene extract contained all the compounds extractable by cyclohexane with some additional material not extracted by the latter. They recommended that cyclohexane be used as solvent when measuring for PAHs and made a further conclusion that, unlike benzene, cyclohexane is a useful solvent in the presence of UV-light. Most analytical laboratories use single or multiple step extractions of water samples either with benzene, cyclohexane, *n*-hexane, or *n*-pentane.[32] The extracts are concentrated and analyzed by gas chromatography, liquid, and thin-layer chromatography or gas chromatography-mass spectrometry. Recoveries have shown to be inconsistent and vary especially for higher fused ring PAHs.[33]

Therefore, it may be necessary to consider cleaning and preconcentrating the samples by complementary coupled liquid column chromatography or gel permeation chromatography techniques, which have a maximum specificity for the three to seven condensed ring PAHs. Recoveries for both phenanthrene and pyrene were greater than 90%.

2. Sediment Samples

Sediment samples have usually been collected by grab sampling techniques[34] and in some instances, especially modified samples have been used to collect the upper few centimeters of bottom sediments.[35] The initial sample size may vary as much as 10 to 100 g. Sediment samples are usually air dried and ground, then extracted while moist, or sonicated while moist employing a variety of solvents.[36]

Gieger and Blumer[36] proposed an extraction procedure for PAHs applicable to sediments, fossil fuels, and other environmental samples. The samples are Soxhlet-extracted with methanol for 24 h, then benzene is added and the extraction continues for a further 24 h. The hydrocarbons are then partitioned from the benzene-methanol extract into *n*-pentane. After removal of any extracted sulfur, the solution is subjected to gel permeation chromatography on a Sephadex LH-20 column. The fractions are then chromatographed on alumina-silica gel, to separate the PAHs from the saturated hydrocarbons and olefins. The remaining PAHs are purified by forming an adduct with 2,4,7-trinitro-9-fluorenone. The uncomplexed materials are then washed from the adduct with *n*-pentane. The PAHs are recovered after the adduct is split by percolation through a silica gel column and examined by gas chromatography-mass spectrometry. A modified version of this procedure was used by LaFlamme et al.[37] After the methanol/benzene Soxhlet extraction step, the remaining water is extracted with 5 vol hexane, followed by 1 vol methylene chloride. The extract is dried, combined, and redissolved in cyclohexane-nitromethane (1:1) and extracted with 5 vol nitromethane. The nitromethane is removed at 40°C under vacuum and subjected to silicic acid (2 g) column chromatography with *n*-hexane as the eluant from the PAHs. The hexane is removed and the eluate dissolved in methylene chloride and analyzed by GC/MS.

Reported concentrations of PAHs in sediments and soils on a global scale have ranged from not detectable in Nevada soil and the Amazon River to over 10,000 ppb in the Charles River, Boston, MA.

3. Fish and Biota Samples

Samples of wildlife species have included fish, birds, their eggs, oysters, and biota most of which are hand collected or netted. If some cases, dead species are analyzed in relation to direct environmental contamination such as oil spills.

The solid samples are usually wrapped in a precleaned foil and placed in containers for freezing and shipment. Samples are homogenized in a suitable blender with a desiccant (usually sodium sulfate) or sometimes an abrasive material such as sand. Selected tissues from some birds are homogenized and freeze-dried prior to extraction.

The determination of PAHs in fish, meat, poultry, and yeasts can be performed with *n*-hexane. The protein-rich samples such as fish and meat require a saponification with potassium hydroxide before the extraction step, in order to recover all the PAHs and the extracts concentrated by liquid-liquid extraction.[39] Systems such as methanol-water-cyclohexane, *N,N*-dimethylformamide-water-cyclohexane and column chromatography employing Sephadex LH-20 can be used. Polycyclic aromatic hydrocarbons in oysters were extracted by Onuska et al.[40] using the following procedure: 5 g of the sample (dried lipids) was extracted in a Soxhlet extractor with 200 ml cyclohexane during a 12-h period. The cyclohexane was then evaporated to between 1 to 2 ml using a rotary evaporator. Cleanup and fractionation of the extracts were performed on a silica gel column (2.5 × 30 cm). The blank samples were spiked with 9-phenyl anthracene as an internal standard and eluted from the column with 2500 ml of isooctane and 1500 ml of benzene. The benzene fraction was analyzed for PAHs by HRGC/MS. Similar procedures were applied to the determination of PAHs in mollusca *Cipanopaludina chinensis*.[40]

B. Preconcentration and Cleanup Procedures

The concentration of PAHs in water is too low to permit direct utilization of many of the modern analytical instruments. Concentration by a factor of a hundred or more is necessary in many instances. Furthermore, the water and many ions may interfere with many of the analytical procedures. Separation of the organic components from water and biota therefore accomplishes two purposes: (1) it removes interfering substances; and (2) at the same time concentrates enough PAHs to make analysis possible.

It is not surprising that considerable effort has been put into methods of separation and concentration. Even after organic compounds including PAHs have been concentrated and separated from a matrix such as water or lipids, the resulting mixture may be too complex for the analytical method. An analyst can follow any of a number of fractionation schemes, based perhaps on functional groups or acidity/basicity properties of various components present in the extract. Alternatively, if he wishes to measure only one class of compounds, he may try to design a concentration method which will be specific for the class of interest, thus achieving concentration and fractionation in one step. Both of these approaches have been followed for PAHs with some success.

1. Preconcentration

The simplest approach to the collection and separation of PAHs in water, sediment, and biota samples is to use some physical or chemical means of removing one fraction from solution or suspension. The techniques vary from simple filtration to collect particulate matter, to chemical methods, such as solvent extraction and coprecipitation. With each of these methods, the analyst must know the efficiency of collection and exactly which fraction is being collected. Very often the fraction is defined by the method of collection. Two methods which purport to collect the same fraction may in fact be sampling very different systems.

2. Centrifugation

A method for removing particulate matter from water samples which is not limited in volume sampled and which suffers less from problems of overlapping is continuous-flow centrifugation. Separation into density classes can be achieved by choice of speed of centrifugation. The continuous centrifugation was used to collect total suspended matter in the marine environment.[43] The drawback seems to be the separation by density, rather than the more usual separation by particle size.

3. Solvent Extraction

Solvent extraction is a most common technique for concentrating a particular fraction of organic contaminants, the fraction concentrated being determined by the choice of solvent. The obvious limitation on the method is that related to the solvent choice; the solvent should have only limited solubility in water, which limits the components removed to the less polar compounds. The solvent must be purified carefully, since the amounts of the various organic compounds collected from water or sediment samples will be about as large as the trace impurities in the solvents. Because of the relatively long sampling periods required for the processing of each sample, these methods must be used to characterize the organic compounds at a few selected stations and depths, rather than in large surveys. Once the separation into the organic solvent has been accomplished, any of a number of techniques of fractionation and analysis can be applied.

Ahnoff and Josefsson have described an *in situ* apparatus for solvent extraction.[44] This apparatus is immersed at the sampling depth, and water is pumped through a series of extraction chambers. Since the sampling apparatus is battery-powered, the unit may be immersed and kept under water for up to 48 h.

Wu and Suffet described an apparatus based on a Teflon® helix liquid-liquid extractor.[45] The extractor was optimized for the removal of organophosphorus pesticides, with an efficiency of about 80%.

Solvent extraction has proven to be most useful when applied to the water, sediment, and fish for which there exist an analytical method of great sensitivity. In general, solvent extraction is an excellent method for the concentration of specific compounds, chiefly nonpolar from various matrices. Special precautions must be taken to prevent contamination from trace materials in the solvents used. When coupled to modern separation and detection methods, they may offer the simplest and most direct approach to the measurement of PAHs. Because of the great variety of PAHs present in environmental samples, a true efficiency of extraction may be impossible to obtain. Working efficiencies, using model compounds, may be our only approach in trying to make the PAH-analyses truly quantitative.

4. Fractionation

Many different bases for the fractionation of PAHs have been used.[46] Prefractionation of organic compounds into discrete chemical classes is performed by adsorption column chromatography using small quantities of neutral aluminum oxide and silicic acid. The principal chemical classes investigated are aliphatic hydrocarbons, PAHs, polycyclic aromatic sulfur heterocycles, nitrogen-containing PAHs, and hydroxyl-containing PAHs. The nitrogen-containing PAHs are further separated into secondary nitrogen PAHs and tertiary nitrogen PAHs to facilitate their identification.[47]

5. Removal of Lipids

Gel permeation chromatography is being increasingly employed for separation and removal of lipids.[48] The high efficiency columns now available have increased the speed of analysis and provide very good separations of lipids and molecules with small size differences. Calibration curves using molar volumes, carbon number, or molecular weight may be used to obtain the desired information for characterization of the components. Gel permeation chromatography is effective in separating lipids and other emulsion-forming acids in fish samples. Generally, GPC is superior to liquid-liquid extraction procedures. Many packings are compatible with a wide range of organic solvents and they do not require regeneration. They have a relatively high capacity for lipids and allow the residues to be recovered nearly quantitatively.

a. Florisil

The most common adsorbent used in cleanup procedures for environmental samples is florisil. The pesticide grade adsorbent known as ''PR grade'' brand is available from a number of distributors. Florisil column chromatography is used for the cleanup of hydrocarbons from organochlorine pesticides, organophosphates as well as triazines, carbonates, and others, by elution with solvents of increasing polarity.[49]

b. Silica Gel

Silica gel is a widely used adsorbent for cleanup. It is commercially available in a wide range of polarity. Presently, chemically bonded stationary phases on silica gel are used as column packing materials for reversed-phase high performance chromatography. These materials consist of organic functional groups, e.g., phenyl-, ethyl-, octyl-, octadecyl-, and cyano- compounds, chemically bonded to a base support silica.[50,51]

c. Activated Carbon

Two different types of activated carbon are employed in adsorption chromatography: (1) graphitized (nonpolar) carbon, which is prepared by high temperature activation; and (2) oxidized (polar) carbon, which is prepared by low-temperature oxidation.[52]

The first type exhibits little selectivity for different types of compounds and adsorption is governed largely by the size of the sample molecules. The second type selectively adsorbs polar molecules in preference to larger nonpolar molecules.

6. Removal of Basic Nitrogen-Containing PAHs

There are a number of difficulties associated with the isolation and chemical class separation of PAHs from nitrogen-containing polycyclic aromatic hydrocarbons. It is difficult to obtain pure amino-PAHs using conventional adsorption and gel permeation chromatography due to the coelution of amino-PAHs and other specific nitrogen-containing PAHs.[53] Details of the column chromatographic separation method used to isolate aromatic hydrocarbons, neutral PAHs and nitrogen-containing compounds are given by Later et al.[54] Prefractionation of samples into discrete chemical classes is performed on small alumina columns (neutral aluminum oxide and silicic acid).

7. Removal of Sulfur-Containing PAHs

Poirier and Smiley[55] described a procedure for the separation and identification or organic sulfur-containing compounds from petroleum and oils. The method permits the concentration of the sulfur-containing compounds, although some aromatic hydrocarbons have been found in the sulfur compound concentrate. The procedure employs liquid-solid chromatography, gas chromatography, and mass spectrometry. The sulfur-containing PAHs were concentrated in the aromatic fraction by chromatography on a silica gel column. The 28 sulfides and sulfur-containing PAHs were identified by mass spectrometry.

Fused silica open tubular columns coated with SE-52, medium polarity (Superox 20M) and liquid crystal stationary phases and their mixtures were evaluated for the separation of isomeric PAHs containing sulfur heterocycles. Although columns containing Superox or mixtures of Superox in SE-52 were able to resolve all 3-ring isomers, no single column could resolve all of either the 4-ring or 5-ring isomers. On the other hand, the compounds that were unresolved on pure SE-52 could be resolved on a 50% BBBT-liquid crystal phase in SE-52, making possible the positive identification of these compounds.[56]

III. ANALYTICAL TECHNIQUES

The ultimate goal in analytical techniques employed for polycyclic aromatic hydrocarbon analyses is a single component identification and quantitation. However, due to their relative chemical similarity, PAHs mixed with other organic contaminants are rather difficult to separate, identify, and quantitate. One of the most effective means of achieving separation of a complex group of compounds is chromatography.

Since Tswett first discovered and applied this technique to separate plant pigments, it has undergone significant development, and today is the most prevalent separation method. Because of the extremely low PAH levels likely to be encountered, and the retardation of the extraneous organic matter, it is necessary to separate PAHs not only from each other, but from other matrices and co-contaminants. To accomplish these goals, the separation techniques of column chromatography, paper chromatography, thin layer chromatography, high performance liquid chromatography, gas chromatography, and supercritical fluid chromatography have been suggested. Some of them will be discussed below.

A. Thin-Layer Chromatography

Thin-layer chromatography (TLC) has been used for separation of PAHs but has proven useful for quantitative analysis of PAHs in recent years. Solvent extracts of environmental water samples contain substantial quantities of organic compounds other than PAHs. These compounds do not affect analysis of PAHs by TLC and fluorescence procedures but make gas chromatographic analyses of PAHs much more complex. TLC-methods for the separation of PAHs have been reviewed by Herod and James.[57] However, TLC is limited, since the separation conditions cannot be adequately controlled. Quantitative TLC is described by Daisey and Leyko.[58] PAHs and aliphatic hydrocarbons are separated from each other and from other classes of compounds. The nonpolar, aliphatic hydrocarbons remain at the start when the reverse phase TLC system is used. Aza-arene compounds and other more polar components move with the solvent system to the top portion of the TLC plate while the

Table 5
R_f AND R_b VALUES OF SOME PAHs[58]

Compound	R_f	R_b
Triphenylene	0.16	0.8
Benzo[a]pyrene	0.23	1.0
Anthanthrene	0.31	1.4
Chrysene	0.36	1.6
Benz[a]anthracene	0.57	2.5
Dibenzo[a,h]anthracene	0.59	2.6
Perylene	0.58	2.55
7,12-Dimethylbenz[a]anthracene	0.66	2.9
Benzo[e]pyrene	0.70	3.0
Benzo[ghi]perylene	0.70	3.0
Fluoranthene	0.73	3.2
Pyrene	0.73	3.2
Benzoanthrone	0.85	3.7
Benzo[h]quinoline	0.88	3.8
Benzo[f]quinoline	0.89	3.9
Benzo[c]quinoline	0.92	4.0
Acridine	0.96	4.2
Carbazole	0.97	4.25

Note: Support Material: Silica Gel + 20 percent ace-
tylated cellulose. Solvent: 1-propanol-acetone-
water (2:1:1 v/v/v). Sandwich Chamber.

PAHs spread through the plate with R_f 0.15 to 0.80 and can be subdivided into three PAH fractions. R_f and R_B values of some PAHs are given in Table 5.

Earlier TLC techniques for the separation of PAHs were reviewed by many authors.[59-61] TLC has been used in the separation of PAHs both as a preliminary or intermediate steps or as a quantitative estimation by means of UV absorption, fluorescence spectrometry, and densitometry. Measurements can be performed either directly on the thin-layer plate, or after removal of the appropriate spot of interest and subsequent extraction and filtration of a solid phase. The former method is more convenient but generally less accurate than the latter. A summary of some TLC references is shown in Table 6. A method is described by Kraft et al.[62] for the determination of PAHs in the diluted exhaust gas of diesel vehicles. Sampling is done by drawing off proportional streams from the dilution tunnel. The particulates deposited on filters are sublimed, and the sublimate is purified and prefractionated on silica gel. Further separation and quantitation of the PAHs is performed by two dimensional TLC in conjunction with *in situ* fluorescence spectrometry. Results and experimental data on the distribution of the emitted PAHs between particulate matter and the corresponding gas phase in diluted exhaust were presented. A modified Langmuir adsorption model is used to explain the effects of dilution ratio and sample temperature in the dilution tunnel. Comparison of the emission data for PAHs obtained from diluted and undiluted exhaust shows good agreement.

B. Column Chromatography

Column chromatography is perhaps the most widely used separation technique for poly-cyclic aromatic hydrocarbons. A large variety of adsorbents such as alumina, silica gel, cellulose acetate, Floridine, and synthetic gels have been used for separation. Adsorbents with uniform particle size, e.g., 80-100, 100-120, or +200 mesh and column diameter-to-length ratio of 1.2 to 1.5 are suggested. Maximum separation can be achieved by slowly increasing the polarity of the eluting solvents. The hydrocarbons are eluted from the column

Table 6
THIN LAYER CHROMATOGRAPHY — SELECTED REFERENCES

Stationary layer	Eluting medium	Details	Ref.
Alumina/acetyl cellulose/ 2% CaSO$_4$·2H$_2$O	n-Hexane/benzene(4:1) methanol/ether/water	Two dimensional densitometry	T-1
Silica/alumina/acetyl cellulose	MeOH/toluene/water	Two dimensional	T-2
Silica gel	Cyclohexane/benzene	Sep. PAH from HC	T-3
Alumina/acetyl cellulose	Pentane/ether; EtOH/toluene/water	Two dimensional	T-4
Silica/acetyl cellulose	1-PrOH/acetone/water	One-step; 18 PAH	T-5
Alumina (1st step) Silica/acetyl cellulose		Class separation PAH separation	T-6
C-18 Silica	MeOH/CH$_3$CN/water	Densitometry	T-7
Acetylated cellulose	MeOH/H$_2$O; MeOH/ ether/water	HPTLC	T-7

References:

T-1. **Basu, D. K. and Saxena, J.**, Monitoring PAHs in water. II. Extraction and recovery of six representative compounds with polyurethane foam, *Environ. Sci. Technol.*, 12, 791, 1978.

T-2. **Borneff, J. and Korte, H.**, Carcinogenic substances in water and soil. XXVI. A routine method for the determination of PAHs in water, *Arch. Hyg. Bakt.*, 153, 220, 1969.

T-3. **Brocco, D., Cantuti, V., and Cartoni, G. P.**, Determination of PAHs in atmospheric dust by a combination of TLC and gas chromatography, *J. Chromatogr.*, 49, 66, 1970.

T-4. **Chatot, G., Dangy-Caye, R., and Fontanges, R.**, Atmospheric PAHs. II. Application of chromatographic coupling in gas phase to the determination of PAHs, *J. Chromatogr.*, 72, 202, 1972.

T-5. **Daisey, J. M. and Leyko, M. A.**, TLC method for determination of PAHs and aliphatic hydrocarbons in airborne particulate matter, *Anal. Chem.*, 51, 24, 1975.

T-6. **Pierce, R. C. and Katz, M.**, Determination of atmospheric isomeric PAHs by TLC and fluorescence spectrometry, *Anal. Chem.*, 47, 1743, 1975.

T-7. **Poole, C. F., Butler, H. T., Coddens, M. E., Khatib, S., and van Dervennet, R.**, *J. Chromatogr.*, 302, 149, 1984.

in the following order: aliphatics, olefins, benzene derivatives, naphthalene derivatives, dibenzofuran fraction, anthracene fraction, chrysene fraction, benzopyrene, and coronene fractions.[63]

Alumina columns are often used and alumina is avilable in various grades, pH, and activity. For the separation of neutral PAH fractions, neutral alumina is generally used. The sample-to-adsorbent ratio varies between 1:100 and 1:1000. Elution is performed on deactivated, neutral alumina (13.7% water) by gradually increasing the polarity of the solvent system. Diethyl ether-pentane was used by Sawicki,[63] cyclohexane-ether by Cleary[64], cyclohexane-benzene[63] and benzene-methanol by Brass et al.[65] The main disadvantage with alumina column separation is that it is very time consuming, sometimes may cause decomposition of important components on the column and good reproducibility of fractionation is very difficult to achieve.

As a next choice, Davidson grade 12 silica gel or equivalent material of 100 to 200 mesh was found suitable for chromatography of PAHs.[63] The sample size-to-adsorbent ratio varies between 1:50 and 1:500. Some researchers use silica gel containing between 3 to 5% moisture content, while others prefer dried silica gel. The elution of adsorbed components is performed using benzene or hexane-benzene.[66] It should be recognized that silica gel is normally used in combination with other adsorbents or other separation methods and is rarely used as the sole separation adsorbents in the column chromatography. Chromatography on silica gel

Table 7
COLUMN CHROMATOGRAPHY - SELECTED REFERENCES

Column Packing	Eluent	Details	Ref.
Silica gel	*n*-Hexane	Benzopyrene fraction	C-1
Alumina	Toluene	Benzopyrene and benzo[k]fluoranthene	C-2
Alumina	*n*-Pentane/ether	12 PAHs up to coronene	C-3
Sephadex LH-20	Isopropanol	Benzo[a]pyrene	C-4
Acetylated cellulose	EtOH/toluene/H_2O	Benzo[a]pyrene	C-5
Alumina	*n*-Pentane-CH_2Cl_2	Detection: vis; UV; MS	C-6
Phenyl bonded silica	MeOH/water (70 + 30)sep.	4-Membered rings	C-7
C-18 uBond	Acetonitrile/H_2O	Reversed phase; UV det.	C-8

References:

C-1. **Davies, H. J., Lee, L. A., and Davidson, T. R.,** Fluorometric determination of benzo[a]pyrene in cigarette smoke condensate, *Anal. Chem.,* 38, 1752, 1966.

C-2. **Dubois, L., Zdrojewski, A., Baker, C., and Monkman, J. L.,** *Air Poll. Control Assoc.,* 17, 818, 1967.

C-3. **Sawicki, E.,** Tentative method of analysis for PAHs content of atmospheric matter, *Health Lab. Sci.,* 7, 31, 1970.

C-4. **Glader, R.,** The determination of carcinogenic PAHs in automobile exhaust gases by column chromatography, *Chromatographia,* 3, 236, 1972.

C-5. **Ohnishi, A., Endo, Y., Maeda, K., and Uehara, M.,** *Nippon Sembai Kosha Chuo Kenkyusho Kenkyu Hokoku,* 114, 231, 1972.

C-6. **Giger, W. and Blumer, M.,** PAHs in the environment: isolation and characterization by chromatography, UV and MS, *Anal. Chem.,* 46, 1663, 1974.

C-7. **Jinno, K. and Okamoto, M.,** Molecular-shape recognition of PAHs in reversed phase liquid chromatography, *Chromatographia,* 18, 495, 1984.

C-8. **May, W. E. and Wise, S. A.,** LC-determination of PAHs in air particulate extracts, *Anal. Chem.,* 56, 225, 1984.

suffers from all the disadvantages normally encountered with alumina. The relatively slow flow rates of mobile phases through silica gel bed causes additional problems.

Griest et al.[67] reported cellulose acetate use as a column material and applied it to the purification of PAHs. Cellulose acetate has been used also in TLC as a mixture with neutral alumina and in liquid chromatography.[68]

Gel filtration was first applied for the separation of PAHs by Wilk et al.[69] Other researchers confirmed its usefulness for separation of PAHs.[70] The separation is best achieved using Sephadex® LH-20 column eluted with 2-propanol as a solvent at a flow rate of 6 to 7 ml/min. Claimed recoveries were close to 100%. Although gel filtration is free from most of the drawbacks of other column chromatographic techniques, its main disadvantage is time. It requires up to 3 days for PAHs separation. Summary of column and gel filtration references is given in Table 7.

C. Liquid and High Performance Liquid Chromatography

Liquid chromatography (LC) and high performance/pressure liquid chromatography (HPLC) are recent techniques of column chromatography. The most important advancements made possible by HPLC are the development of high pressure pumps to deliver a mobile phase rapidly through a column and the development of special solid adsorbents that may be chemically bonded with a liquid phase. Very small particle sizes of the solid phase (between 10 to 1 μm O.D.) are highly efficient in terms of the generated number of theoretical plates.

Sorrell et al.[71] employed acetonitrile-water (70 + 30) as a mobile phase using 1 ml/min flow rate. Each of the three fractions was run separately and eluent was monitored by UV absorbance at 254 and 280 nm in conjunction with fluorescence measurements (excitation

Table 8
RATIO OF ABSORBANCE RESPONSES OF
VARIOUS PAHs

Compound	Ratio 280/254 nm	Fluorescence[a] UV
Phenanthrene	0.17	0.20
Fluoranthene	1.56	0.61
1-Methyl phenanthrene	0.17	0.58
2-Methyl phenanthrene	0.16	0.15
Pyrene	0.44	0.35
1-Methyl pyrene	1.00	0.85
Chrysene	0.25	0.41
Benzo[a]anthracene	1.58	0.53
Perylene	0.03	2.38
Benzo[a]pyrene	0.84	1.08
Benzo[b]fluoranthene	0.54	0.92
Benzo[k]fluoranthene	0.56	2.94
Dibenzo[ah]anthracene	6.55	0.38
Benzo[ghi]perylene	1.57	0.44
Indeno[1,2,3-cd]pyrene	0.47	0.35

[a] Fluorescence — excitation 250 nm; emission 340 kV; 1.0 μA; UV 254 nm; 0.01 AUFS.

wavelength of 250 nm and emission at 340 nm.). A second injection of the 50% methylene chloride fraction was monitored using UV absorbance at 254 and 340 nm, and the fluorescence excitation at 286 nm and a filtered emission of 418 kV. The lower percentage of acetonitrile in water increased peak spreading in addition to lengthening the time for analysis. The ratio of responses for the standard PAHs were calculated using the formula:

$$\text{Ratio of response} = \text{Peak height @ 280 nm/Peak Height @ 254 nm}$$

Although this technique cannot eliminate the possibility of the incorrect identification for a particular PAH having the same retention time, it would indicate an incorrect response ratio and the presence of another compound with different adsorption properties. In combination with the fluorescence response, some of the uncertainty in the identification or rejection may be eliminated. The ratios of some absorbance responses are given in Table 8.

The use of HPLC in oil spill identification has been evaluated in terms of analysis time and reliability of data obtained. The aromatic fraction was analyzed by reverse-phase chromatography on a 3 μm packing with detection at 210 and 287 nm, in less than 20 min. The profiles of various crude oils were differentiated by applying simple statistical parameters, namely the linear regression coefficients and the Euclidian distances. The effect of environmental weathering on the samples has also been investigated.[72]

Some HPLC references are provided in Table 9.

D. Gas Chromatography

Similar chemical structures of polycyclic aromatic hydrocarbons and their relatively high boiling points require high resolution chromatographic separation and thermally stable phases as emphasized by Grob.[73] From the available stationary phases, SE-30, OV-1, SE-52 are thermally stable but barely adequate for the use with packed columns. In general, the temperature required to separate some of the high boiling PAHs must reach 400°C which is not practical for packed column separations. In gas chromatography two choices of a

Table 9

HIGH PERFORMANCE LIQUID CHROMATOGRAPHY-
SELECTED REFERENCES

Column packing	Mobile phase	Detection	Ref.
Cellulose acetate	EtOH/Me$_2$Cl$_2$	UV; isocratic	HC-1
Spherosil XOA-400	Isooctane	Fluorescence	HC-2
2,4,7-trinitrofluoren-one silica	n-Heptane	UV; isocratic	HC-3
C-18 Silica	MeOH/water	UV; gradient	HC-4
Polyvinyl pyrolidone	2-Propanol	Fluorescence	HC-5
C-18 u-Bondapak	Acetonitrile/H$_2$O	Fluorescence/gradient	HC-6
u-Bondapak-amine	n-Hexane	UV; gradient	HC-6
C-18 u-Bondapak	AcCN/water	dye-laser	HC-7
Radialpak C-18	AcCN/water	Fluorescence	HC-8
Spherisorb silica	Hexane/CH$_2$Cl$_2$	UV; gradient	HC-9

References:

HC-1. **Klimish, H. J.,** Determination of PAHs separation of benzopyrenes by HPLC, *Anal. Chem.,* 45, 1960, 1973.

HC-2. **Strubert, W.,** Separation of fused ring aromatics on silica gel. Detection of nanogram or ppb amounts, *Chromatographia,* 4, 205, 1973.

HC-3. **Hagenmeier, H., Feierbend, R., and Faeger, W.,** HPLC of PAHs in water, *Wasser Abwasser Forsch.,* 10, 99, 1977.

HC-4. **Gold, A.,** Carbon black adsorbents: separation and identification of a carcinogen and some oxygenated polyaromatics, *Anal. Chem.,* 47, 1469, 1975.

HC-5. **Goldstein, G.,** Separation of PAHs on crosslinked polyvinyl pyrrolidone, *J. Chromatogr.,* 129, 61, 1976.

HC-6. **Wise, S. A., Chester, S. N., Hilpert, L. R., and May, W. E.,** Chemically bonded aminosilane stationary phase for the separation of PAH compounds, *Anal. Chem.,* 49, 2306, 1977.

HC-7. **Furuta, N. and Otsuki, A.,** Time-resolved fluorometry in detection of ultratrace PAHs in lake water by HPLC, *Anal. Chem.,* 55, 2407, 1983.

HC-8. **Kagi, R., Alexander, R., and Cumbers, M.,** PAHs in rock oysters: a baseline study, *Int. J. Environ. Anal. Chem.,* 22, 135, 1985.

HC-9. **Tong, H. Y. and Karasek, F. W.,** Quantitation of PAHs in diesel exhaust particulates by HPLC fractionation and HRGC, *Anal. Chem.,* 56, 2129, 1984.

column selection are available depending on the separation required and conditions used to achieve adequate separation of PAHs:

1. Packed columns
2. Open tubular columns
   a. Wall-coated open tubular columns
   b. Support coated open tubular columns

1. Packed Columns

In the packed column, the chromatographic process is limited by the slowness of diffusion of the sample molecules around the support particles and within their pores. Packed columns, however, are considerably less expensive, they can be repacked many times, they are durable, and they are more forgiving of poor injection techniques than even recently introduced wide-bore open tubular columns. On the other hand, wide-bore columns provide higher resolution with fewer adsorption problems than packed columns.[73] Because of this, a smaller selection of bonded and crosslinked phases for wide-bore columns will provide better separation in a shorter time than variety of liquid phases used in packed-column gas chromatography. Transfer of a separation from a packed column to a wide-bore open tubular column will result in a faster separation because of the higher linear velocity, with better quantitation and lower detection limits because of the higher column efficiency (400 to 700 theoretical plates per meter). Finally, wide-bore open tubular columns do not require a gas chromatograph specially dedicated for open tubular columns operation. In that respect, they provide a perfect means to familiarize with capillary gas chromatography.

A number of packing materials have been used for packing glass or metallic columns. The most common one is Chromosorb W. From the most popular liquid phases that have been employed for PAHs analyses are OV-1,[74] SE-52,[75] Apiezon L,[76] Dexsil 300,[77,78] Dexsil-400,[79] Dexsil-410,[80] and some mixed phases.[81,82] Also nematic liquid crystals were employed as suitable phases for separating PAHs,[83] However, a column bleeding at higher temperature than 185°C remains a problem when *N,N'-bis-(p*-benzylidene) α,α'-bi-*p*-toluidine is used as a liquid crystal phase.

2. Open Tubular Columns

Open tubular columns (OTC) or wall-coated open tubular columns (WCOT) represent a viable technique today in environmental organic trace analyses of micropollutants. Many applications have been reviewed by Onuska and Karasek.[84] The efficiency of the gas chromatographic column, its separation power, depends on many variables, and is a complex function of the diffusion of the sample components in the gaseous and liquid phases. The liquid stationary phase is distributed in the form of a very thin film on the inner wall of the capillary tubing. Thus, the diffusion of the sample molecules in and out of this phase is much easier than in the case of the packed column, where contact is not so intimate as it is in the nonporous capillary.

If one would summarize the advantages of open tubular columns over the packed columns, it can be stated that they provide much better separation in equal or shorter time or, if the same resolution is satisfactory, this can be achieved much faster in a much shorter time period. It is not our intention to compile a complete introduction to capillary gas chromatography but it is important to realize that one of the significant variables in choosing open tubular column, is the thickness of the stationary phase film. The best performance will be obtained when film thickness is between 0.17 to 0.35 μm for the analysis of PAHs.

Cantuti et al.[85] compared several liquid phases and found that SE-52 provided the most effective separation of PAHs. Later, Lee et al.[86,87] separated over 150 PAH components containing from 2 to 6 rings on a 12-m long SE-52 WCOT column. Onuska et al.[88] also demonstrated excellent separation of PAHs using SE-52 in environmental samples.

The determination of differently substituted PAHs, nitro-PAHs, carbazoles, keto-PAHs and azaarenes in aerosol samples is published by Stray et al.[89] Liquid carbon dioxide extraction is used to minimize the loss of reactive compounds. HPLC on chemically activated silica is employed to prefractionate the samples into subfractions with a minimum of overlap among different PAH-classes. Both, electron capture detection and negative ion chemical ionization combined with OTC-separation are used for identification and quantitative analyses.

Wright[90] described complex organic mixtures, such as coal liquefaction and oil shale products that are comprised of thousands of individual components. Even HRGC does not provide sufficient resolution to allow accurate identification and quantitation of many components of interest. The concept of dual OTCs combines the different resolving characteristics of two OTCs coated with different stationary phases into a single chromatographic run. In this set-up, both columns are connected to the same injector. Analysis of complex mixtures, in this fashion can confirm the identification and quantitation of components on two columns of different polarity with little increase of the analysis time. This technique can provide a means of obtaining quantitative data for individual components which are known to coelute on any one column and can alert analysts to caution that would be undetected by HRGC on a single OTC.

A modified cleanup of PAHs is described by Spitzer and Danneker.[91] n-Alkanes are separated from the PAHs and are eluted in one fraction. The distribution of 17 PAHs and 4 of their alkyl derivatives, 2 sulfur-containing heteroaromatic compounds and 2 azaarenes among the different fractions is reported. Azaarenes containing three rings are separated from PAHs and sulfur-containing compounds. OTCs coated with OV-215 and OV-25 were used in the separation of isomeric azaarenes and in PAH profile analysis, respectively.

Moser and Arm[92] described separation of 3- to 6-ring PAHs by OTC-high resolution chromatography using a liquid crystal as a stationary liquid phase. A deactivation method, which is compactible with N,N'-bis-p-butoxybenzylidene)-a,a'-bi-p-toluidine (BBBT), which has a nematic mesophase in the range 188 to 303°C is used. The column length and the film thickness of the stationary phase have been adjusted in such a way that PAHs can be separated in a short time.

PAHs adsorbed on fly ash from an incinerator were studied by Tausch and Stehlik.[93] Following Soxhlet extraction and preconcentration, analysis is performed by HRGC and HRGC/MS using high temperature glass OTCs coated with OV-1 and SE-54. Identification of individual compounds is based on retention data, relative detector responses, and mass spectra. Among other PAHs, series of gradually chlorinated PAHs containing four to seven rings were detected. For unequivocal identification, appropriate reference compounds were synthesized.

Some of the more significant references dealing with PAH analyses are summarized in Table 10.

E. Mass Spectrometry and HRGC/MS

The use of mass spectrometry (MS) and HRGC/MS to identify and quantify PAHs in environmental samples is escalating with the number of publications reporting PAH-analysis each year. Water and sediment samples contaminated with petroleum products continue to receive most of the attention. However, the major components of bunker oil dissolved in seawater were found to be acetophenone, phenol, cresols, and 2-phenyl-2-propanol besides naphthalene and its methyl homologues.[94] Brown et al.[95] have identified benzene and na-phthalene-related compounds, indanes, biphenyls, and higher PAHs using methylene chloride extraction of estuarine waters contaminated with No. 2 fuel oil. Dispersed nonvolatile hydrocarbons (bp >235°C, MW >190) in open ocean waters were measured using infrared spectrometry, UV, gas chromatography, and mass spectrometry. Fractions of the tetra-chloromethane extract were eluted from a silica gel column and vaporized into a mass spectrometer through a frit. The nonvolatile hydrocarbons were found to be more concentrated in the surface waters than in samples collected at greater depths (10 m). A study by Walker et al.[96] of oil-contaminated sediments using mass spectrometry, revealed that benzene is the most effective solvent for extraction. The mass spectrometry provided data indicating that the extracted material contained also weathered hdyrocarbon components. Concentration of saturated hydrocarbons decreased with depth, but some aromatics were detected in larger concentrations at greater depth. In a surface scrap sample of an estuarine tidal mud from the mouth of the Usk River, England, the quantity of hydrocarbons (1230 mg/l) was unusually high but it fit a pattern of weathered crude oil.[97] Other types of contaminants identified were PAHs (140 mg/l), phthalate esters (60 mg/l) and fatty alcohols (58 mg/l).

Extracts of periwinkles (sea snails) taken near a wharf at St. Andrews and Passamaguddy Bay from New Brunswick, Canada, contained PAHs that were identified by mass spectro-metry, UV, and fluorescence spectra by Zitko.[98]

Gibson et al.[99] reported oxidation of benzo(a)pyrene and benzo(a)anthracene by a mutant

Table 10
HIGH RESOLUTION GAS CHROMATOGRAPHY-
SELECTED REFERENCES

Stationary phase	Column	Prog. temp. (°C)	Detector	Ref.
SE-52	WCOT-glass	100—300	FID	GC-1
Versamid 900	WCOT-glass	88—201	FID	GC-2
OV-101	WCOT-SS	230 and 270	FID	GC-3
OV-101	WCOT-glass	100—265	FID	GC-4
SE-52	WCOT-glass	60—240	FID/MS	GC-5
SE-52	WCOT-glass	60—230	FID	GC-6
SE-52	WCOT-glass	60—240	GC/MS	GC-7
SE-54	WCOT-glass	100—220	FID	GC-7
SE-52 crosslinked	WCOT-FS	45—300	FTIR	GC-8
SE-52	WCOT-glass	50—250	FID	GC-9

References:

GC-1.   **Cantuti, V., Cartoni, G. P., Liberti, A., and Torri, A. G.,** Improved evaluation of PAHs in atmospheric dust by GC, *J. Chromatogr.,* 17, 60, 1965.

GC-2.   **Brocco, D., Cimmino, A., and Possanzini, M.,** Determination of aza-heterocyclic compounds in atmospheric dust by a combination of TLC and GC, *J. Chromatogr.,* 84, 371, 1973.

GC-3.   **Grimmer, G.,** Die quantitative Bestimmung von PAH mit der Kapillar Gaschromatographie, *Erdoel Kohle,* 24, 676, 1971.

GC-4.   **Doran, T. and McTaggart, N. G.,** The combined use of high efficiency HPLC and capillary GC for the determination of PAH in automotive exhaust condensates, *J. Chromatogr. Sci.,* 12, 715, 1974.

GC-5.   **Lee, M. L., Novotny, M., and Bartle, K. D.,** GC/MS and NMR determination of PAHs in airborne particulates, *Anal. Chem.,* 48, 1566, 1976.

GC-6.   **Onuska, F. I., Wolkoff, A. W., Comba, M., Larose, R. H., Novotny, M., and Lee, M. L.,** Gas chromatographic analysis of PAH in shellfish on short-WCOT glass capillary columns, *Anal. Lett.,* 9, 451, 1976.

GC-7.   **Borwitzky, H. and Schomburg, G.,** Separation and identification of PAHs in coal tar by using HRGC and GC/MS, *J. Chromatogr.,* 170, 99, 1979.

GC-8.   **Chiu, K. S., Biemann, K., Krishnan, K., and Hill, S. L.,** Structural characterization of PAHs by combined GC/MS and GC/FTIR, *Anal. Chem.,* 56, 1610, 1984.

GC-9.   **Lee, M. L., Vassilaros, D. L., White, C. M., and Novotny, M.,** Retention indices for programmed temperature capillary GC of PAHs, *Anal. Chem.,* 51, 768, 1979.

strain of bacterial species isolated from a polluted stream. Both, electron impact ionization mass spectra and chemical ionization mass spectra were used to characterize the major dihydrodiols formed for each compound. Benzo(a)pyrene produced *cis*-9,10-dihydroxy-9,10-dihydrobenzo(a) pyrene and benzo(a)anthracene produced *cis*-1,2-dihydroxy-1,2-dihydro-benzo(a) anthracene. The authors claimed that these products were not formed from arene oxide precursors, since the *trans*-isomers would have been produced.

Raw anthracene oil and four hydrogenation product mixtures were analyzed by GC/MS, HRGC, and nuclear magnetic resonance spectrometry by Schepple et al.[100]

When the samples were evaporated directly into the ion source using a catalyst, mass spectra showed highly condensed ring systems, especially when mild hydrogenation conditions were employed. Trace amounts of PAHs were found in all fractions of ten high boiling crude oil distillates, although the crude oils were from different geological sources.[101]

Mass spectrometry represents a selective means for determining the identity and specificity of PAHs. Relatively abundant molecular ions are produced under electron impact ionization and fragmentation can be even more suppressed by lowering an electron energy from 70 eV to 20 eV. Lower electron voltages provide mass spectra containing mainly singly-charged, intact molecular ions.[102] One problem encountered with electron impact ionization mass spectra of PAH-isomers is that almost identical mass spectra are produced, making their identification sometimes impossible. Chemical ionization methods can be used to distinguish among many isomeric PAHs as recommended by Hites and Dubay.[103] These authors established that mass spectra of PAHs exhibit characteristic ratios of the protonated molecular ion to the molecular ion when argon containing between 5 to 10% methane is employed as the reactant gas.

The application of high resolution mass spectrometry to the direct analysis of PAHs derived from coal and petroleum has been reviewed.[104] Aczel et al.[105] have analyzed coal and petroleum products and devised a data system capable to determine various mixtures containing up to 2900 components in 58 different types of compounds.

Review of some mass spectrometric methods is given in Table 11.

F. Photoluminescence Analysis

This technique has become well-established as a selective and sensitive method for PAH analysis and it has been found to be as sensitive and selective and much cheaper than a mass spectrometric technique. Since photoluminescence spectrometry is a nondestructive analytical method, individual fractions can be collected and subjected to further analysis. The selectivity of fluorescence analysis for PAHs has been refined by characterizing PAH-isomer at two wavelengths.[106]

PAHs have been quantitatively analyzed without chromatographic separation by means of a selective excitation and/or oxygen quenching method.[107] Sawicki et al.[108,109] reported detection limits of PAHs between 3 and 200 ng, depending upon the particular compound and a limit of detection for benzo(a)pyrene of 3 ng after a TLC separation.

Kunte[110] has reported a procedure for fluorometric analysis of PAHs extracted from a TLC plate. She investigated the efficiency of their removal from the plate. Detection limits down to 1 ng were reported and this procedure has been used in environmental analyses by Borneff et al.[111] and Reichert et al.[112]

Low-temperature fluorescence ($-197°C$) and phosphorescence spectra of PAHs have been used in their analysis. This technique offers a possibility of resolution of PAHs in a mixture with a very high sensitivity.[113-115] Fluorescence line narrowing spectrometry of PAHs at μg/kg in organic glasses employing lasers was studied by Brown et al.[116] Using miniaturized instrumentation, Hatano et al.[117] separated PAHs using a capillary column HPLC system. After separation components passed through a microcell having 3 μl volume and were detected at picogram quantities by means of spectrofluorimetry.

Several PAHs were detected by their characteristic X-ray excited optical luminescence without isolation of individual components present in environmental samples.[118] Luminescence analysis references are summarized in Table 12.

Table 11
GC/MS AND MASS SPECTROMETRY —
SELECTED REFERENCES

Technique	Detailed description	Ref.
High resolution; low volt.	Qualitative and quantitative	MS-1
High resolution	Acquisition	MS-2
Low resolution-SIM	Mass fragmentography	MS-3
Low resolution	Chemical ionization	MS-4
High resolution	Matrix analysis	MS-5
High resolution	Direct sample insertion	MS-6
High resolution-FT	Laser desorption to m/z 700	MS-7
Low resolution-SIM	HRGC/MS; quantitation	MS-8

References:

MS-1. **Aczel, T., Allan, D. E., Harding, J. H., and Knipp, E. A.,** Techniques for quantitative HRMS analyses of complex hydrocarbon mixtures, *Anal. Chem.,* 42, 341, 1970.

MS-2. **Aczel, T.,** Anwendung der Hochauflosenden Mass-spektrometrie bei der Analyse von Erdoel und Kohlenderivaten, *Erdoel Kohle,* 26, 27, 1973.

MS-3. **Matsushima, H. and Takahisa, H.,** PAHs in the environment: determination of PAHs in water by mass fragmentography, *Bunseki Kagaku,* 24, 505, 1975.

MS-4. **Simonsick, W. J. and Hites, R. A.,** Analysis of isomeric PAHs by charge-exchange CI-MS, *Anal. Chem.,* 56, 2749, 1984.

MS-5. **Stauffer, J. L., Levins, P. L., and Oberholtzer, J. E.,** HRMS matrix analysis of environmental samples, in *Carcinogenesis,* Vol. 3, Jones, P. W. and Freudenthal, R. I., Eds., Raven Press, New York, 89.

MS-6. **Klempier, N. and Binder, H.,** Determination of PAHs in soot by MS with direct sample insertion, *Anal. Chem.,* 55, 2104, 1983.

MS-7. **Tan, Y. L.,** Rapid simple sample preparation technique for analyzing PAHs in sediments, *J. Chromatogr.,* 176, 319, 1979.

MS-8. **Wilkins, C. L., Weil, D. A., Yang, C. L. C., and Ijames, C. F.,** High mass analysis by laser desorption FTMS, *Anal. Chem.,* 57, 520, 1985.

MS-9. **Onuska, F. I., Mudroch, A., and Terry, K. A.,** Identification and determination of trace organic substances in sediment cores from the western basin of Lake Ontario, *J. Great Lakes Res.,* 9, 169, 1983.

Table 12
SELECTED REFERENCES OF LUMINESCENCE
ANALYSIS OF PAHs

Method	Details	Ref.
Fluorescence	Water and soil, TLC	LU-1
Low temperature luminescence	Mixture analysis	LU-2
Fluorescence	Corrected spectra	LU-3
Luminescence-Spolskii effect	Low temperature of 77°K	LU-4
Luminescence-Spolskii effect	Low temperature identification	LU-5
Luminescence	X-ray excited spectra	LU-6
Fluorescence	Leaves, blooms, phytoplankton	LU-7
Fluorescence and UV	Exe. 300 nm — emission 420 nm	LU-8
UV-resonance Raman	Coal liquids	LU-9
C-13 NMR	Asphaltenes	LU-10

References:

LU-1. **Kunte, H.,** Carcinogenic substances in water and soil. XVIII. Determination of PAHs by means of mixed TLC and by fluorescence measurements, *Arch. Hyg. Bakt.,* 151, 193, 1967.

LU-2. **Hood, L. V. S. and Winefordner, J. D.,** TLC and low temperature luminescence measurement of mixtures of carcinogens, *Anal Chim. Acta,* 42, 199, 1968.

LU-3. **Porro, T. J., Anacreon, R. E., Flandreau, P. S., and Fagerson, I. S.,** Corrected fluorescence spectra of PAHs and their principal applications, *J. Assoc. Off. Anal. Chem.,* 56, 607, 1973.

LU-4. **Kirkright, G. F. and Lima, L. C. G.,** The detection and determination of PAH by luminescence spectrometry utilizing the Spolskii effect at 77°K, *Analyst,* 99, 338, 1974.

LU-5. **Colmsjo, A. and Stenberg, U.** Identification of PAH by Spolskii low temperature fluorescence, *Anal. Chem.,* 51, 145, 1979.

LU-6. **Woo, Ch. S., D'Silva, A. P., Fassel, U. A., and Oestreich, G. J.,** PAHs in coal-identification by their X-ray excited optical luminescence, *Env. Sci. Technol.,* 12, 173, 1975.

LU-7. **Hellmann, H.,** Fluorimetic determination of PAH in leaves, blooms and phytoplankton, *Fresenius Z. Anal. Chem.,* 287, 148, 1977.

LU-8. **Black, J. J., Halt, T. F., and Black, P. J.,** A novel integrative technique by locating and monitoring PAH discharges to the aquatic environment, *Environ. Sci. Technol.,* 16, 247, 1982.

LU-9. **Johnson, C. R. and Asher, S. A.,** A new selective technique for characterization of PAHs in complex samples: UV resonance Raman spectrometry of coal liquids, *Anal. Chem.,* 56, 2258, 1984.

LU-10. **Murphy, P. D. and Gerstein, B. C.,** Determination of chemical functionality in asphaltenes by high resolution solid state C-13, NMR, *Anal. Chem.,* 54, 522, 1982.

REFERENCES

1. **Rigandy, J. and Khleshey, S. P.,** *IUPAC Nomenclature of Organic Chemistry, Section A, Hydrocarbons,* Pergamon Press, Oxford, 1979, 1.
2. **Lao, R. C. and Thomas, R. S.,** The volatility of PAH and possible losses in ambient sampling in polycyclic aromatic hydrocarbons, in *Chemistry and Biological Effects,* Bjorseth, A. and Dennis, A. J., Eds., Batelle Press, Columbus, OH, 1979, 829.

3. **May, W. E., Wasik, S. P., and Freeman, D. H.,** Determination of the aqueous solubility of PAHs by a coupled column liquid chromatographic technique, *Anal. Chem.,* 50, 175, 1978.

4. **Tipson, R. S.,** Review on Oxidation of PAH, National Bureau of Standards Report No. 8363, U.S. Government Printing Office, Washington, D.C., 1964.

5. Particulate Polycyclic Organic Matter, U.S. National Academy of Sciences, Washington, D.C., 1972.

6. **Di Salvo, L. H., Guard, H. E., and Hunter, L.,** Tissue hydrocarbon burden of mussels as potential monitor of environmental hydrocarbon insult, *Environ. Sci. Technol.,* 9, 247, 1975.

7. **Grahl-Nielsen, O., Staveland, J. T., and Wilhelmsen, S.,** Aromatic hydrocarbons in benthic organisms from coastal areas polluted by Iranian crude oil, *J. Fish. Res. Board Can.,* 35, 615, 1978.

8. **Makie, P. R., Hardy, R., Whittle, K. J., Bruce, C., and McGill, A. S.,** The tissue hydrocarbon burden of mussels from various sites around the Scottish coast, in *Polycyclic Aromatic Hydrocarbons: Chemical and Biological Effects,* Battelle Press, Columbus, OH, 1980.

9. **Murray, H. E., Ray, L. E., and Giam, C. S.,** Analysis of marine sediment, water and biota for selected organic pollutants, *Chemosphere,* 10, 1327, 1981.

10. **Knutzen, J. and Shortland, B.,** PAHs in some algae and invertebrates from moderately polluted parts of Norway, *Water Res.,* 16, 421, 1981.

11. **Black, J. J., Hart, T. E., and Evans, E.,** HPLC studies of PAH pollution in a Michigan trout stream, in *Polycyclic Aromatic Hydrocarbons: Chemical Analysis and Biological Fate,* Cooke, M. and Dennis, A. J., Eds., Battelle Press, Columbus, OH, 1980, 343.

12. **Brink, R. H.,** Biodegradation of organic chemicals in the environment, in *Environmental Health Chemistry: The Chemistry of Environmental Agents of Potential Human Hazards,* McKinney, J. D., Ed., Ann Arbor Science Publishers, Ann Arbor, MI, 1981, 75.

13. **Cerniglia, C. E. and Gibson, D. T.,** Aromatic hydrocarbons: Degradation by bacteria and fungi, in *Oil Sand, Oil Shale Chemistry,* Stranaz, O. P. and Hour, E. M., Eds., Verlag Chemie International, New York, 1978, 191.

14. **Calder, J. A. and Lader, J. H.,** Effect of dissolved aromatic hydrocarbons on the growth of marine bacteria in batch culture, *Appl. Environ. Microbiol.,* 32, 95, 1976.

15. **Sirota, G. R., Uthe, J. F., Sreedharan, A., Matheson, S., Musial, T., and Hamilton, K.,** PAHs in lobsters *(Homarus americanus)* and sediments in the vicinity of coking facility, in Proceedings of the 7th Symposium on PAHs, Battelle, Columbus, OH, 1982.

16. **Mix, M. C., Schaffer, R. L., Hemingway, S. J.,** PAHs in bay mussels *(Mytilus edulis)* from Oregon, in *Phyletic Approaches to Cancer,* Dawe, C. J., Harshbarger, J. C., Kondo, S., Sugimura, T., Takayama, S., Eds., Japan Science Society Press, Tokyo, 1981, 167.

17. **Southworth, G. R., Beanchamp, J. J., and Schmieder, P. K.,** Bioaccumulation potential of PAHs in *Daphia pulex, Water Res.,* 12, 973, 1978.

18. **Il'nitsky, A. P., Lembik, J. L., Solenova, L. G., and Shabad, L. M.,** On the distribution of benzo [a] pyrene among different objects of the aqueous medium in water bodies, *Cancer Detect. Prev.,* 2, 471, 1979.

19. **Bruggeman, W. A., van der Steen, J., and Hutzinger, O.,** Reversed phase TLC of PAHs and PCBs. Relationship with hydrophobicity as measured by aqueous solubility and octanol-water partition coefficient, *J. Chromatogr.,* 238, 335, 1982.

20. **Stegeman, J. J.,** PAHs and their metabolites in the marine environment, in *Polycyclic Aromatic Hydrocarbons and Cancer,* Vol. 3, Gelboin, H. V. and Ts'O, P. O., Eds., Academic Press, New York, 1981, 59.

21. **Malins, D. C.,** Metabolism of aromatic hydrocarbons in marine organisms, *Ann. N.Y. Acad. Sci.,* 298, 482, 1977.

22. **Anderson, J. W., Neff, J. M., Cox, B. A., Tatem, H. E., and Hightower, G. M.,** Characteristics of dispersion and water soluble extracts of refined oils and their toxicity to estuarine crustaceans and fish, *Mar. Biol.,* 27, 75, 1974.

23. **Ott, F. S., Harris, R. P., and O'Hara, S. C. M.,** Acute and sublethal toxicity of naphthalene and three methylated derivatives to the estuarine copepod *Eurytemora affinis, Mar. Environ. Res.,* 1, 49, 1978.

24. **James, M. O. and Bend, J. R.,** PAH-induction of cytochrome P-450 dependent mixed-function oxidases in marine fish, *Toxicol. Appl. Pharmacol.,* 54, 117, 1980.

25. *Health Assessment Document for Polycyclic Organic Matter,* U.S. Environmental Protection Agency Office of Research and Development, Washington, D.C., 1979.

26. Method development and monitoring of PAHs in selected U.S. waters, EPA-600/1-77-052, U.S. Department of Commerce, PB-276 635 NTIS, prepared for Health Effects Research Lab, Cincinnati, OH.

27. **Clark, R. C., Jr. and Findley, J. S.,** *Proceedings of the Joint Conference on Preservation and Control of Oil Spills,* American Petroleum Institute, Washington, D.C., 1973, 161.

28. *The Merck Index,* 8th ed., Merck, Rahway, NJ, 1968.

29. **Boehm-Goessl, T. and Krueger, R.,** Solubilization of benzopyrene by alkylbenzene sulfonate acids, *Kolliod Z. Z. Polym.,* 206, 65, 1965.

30. **Balgaires, E. and Claeys, C.,** Essai de determination de la pollution des atmospheres par les fumees. *Incidence Anthracene Rev. Med. Miniere,* 34, 35, 1957.
31. **Monkman, J. L., Moore, G. E., and Katz, M.,** Analysis of PAH in particulate pollutants, *J. Am. Indust. Hyg.,* 23, 489, 1963.
32. **Acheson, M. A., Harrison, R. M., Perry, R., and Wellings, R. A.,** Factors affecting the extraction and analysis of PAH in water, *Water Res.,* 10, 207, 1976.
33. **Hoffman, D. and Wyder, E. L.,** Short term determination of carcinogenic aromatic hydrocarbons, *Anal. Chem.,* 32, 295, 1960.
34. **Goulden, P. D.,** *Environmental Pollution Analysis,* Heyden & Sons, London, 1978.
35. **LaFlame, R. E. and Hites, R. A.,** The global distribution of PAH in recent sediments, *Geochim. Cosmochim. Acta,* 42, 289, 1978.
36. **Gieger, W. and Schaffner, C.,** Determination of PAH in the environment by HRGC, *Anal. Chem.,* 50, 243, 1978.
37. **Wangersky, P. J.,** Particulate organic carbon in the Atlantic and Pacific Oceans, *Deep-Sea Res.,* 23, 457, 1976.
38. **Farrington, J. W., Tripp, B. W., Teal, J. M., Mille, G., Tjessem, K., Davis, A. C., Livramento, J. B., Hayward, N. A., and Frew, N. M.,** Biogeochemistry of aromatic hydrocarbons in the benthos of microcosm, in *Chemistry and Analysis of Hydrocarbons in the Environment,* Albaig, J., Frei, R. W., and Merian, E., Eds., Gordon and Breach Science Publishers, New York, 1983, 191.
39. **Grimmer, G.,** Analysis, data reporting and profile concept, presented at OECD-workshop on PAHs, Paris, 1981.
40. **Onuska, F. I., Wolkoff, A. W., Comba, M., Larose, R. H., Novotny, M., and Lee, M. L.,** Gas-chromatographic analysis of PAHs in shellfish on short-WCOT glass capillary columns, *Anal. Lett.,* 9, 451, 1976.
41. **Kalas, L., Mudroch, A., and Onuska, F. I.,** Bioaccumulation of arenes and organochlorine pollutants by *Cipangopaludina chinensis* (Mollusca:Gastropoda) from ponds in the Royal Botanical Gardens, Hamilton, Ontario, in *Hydrocarbons and Halogenated Hydrocarbons in the Aquatic Environment,* Afghan, B. K. and McKay, D., Eds., Plenum Press, New York, 1978.
42. **Webb, R. G.,** Isolating organic water pollutant, XAD resins, urethane foams, solvent extraction, PB Report No. 245647, U.S. National Technology Information Service, 1975.
43. **Jacobs, M. B. M. and Ewing, M.,** Suspended particulate matter: concentration in the major oceans, *Science,* 163, 380, 1969.
44. **Ahnoff, M. and Josefsson, B.,** Apparatus for in-situ solvent extraction of non-polar organic compounds in sea and river water, *Anal. Chem.,* 48, 1268, 1976.
45. **Wu, C. and Suffet, I. H.,** Design and optimization of a Teflon helix continuous liquid-liquid extraction apparatus and its application for the analysis of organophosphate pesticides in water, *Anal. Chem.,* 49, 231, 1977.
46. **Ogan, K. et al.,** *J. Chromatogr. Sci.,* 16, 517, 1978.
47. **Later, D. W., Lee, M. L., Bartle, K. D., Kong, R. C., and Vassilaros, D. L.,** Chemical class separation and characterization of organic compounds in synthetic fuels, *Anal. Chem.,* 53, 1612, 1981.
48. **Stalling, D. L., Smith, L. M., and Petty, J. D.,** Approaches to Comprehensive Analyses of Persistent Halogenated Environmental Contaminants, ASTM-Special Technological Publication No. 686, p302, 1979.
49. *Analytical Methods Manual, 1979,* Inland Waters Directorate, Water Quality Branch, Environment Canada, Part 5, Naquadat 18165.
50. **Amateis, P. G. and Taylor, L. T.,** Retention parameters of aromatic polar species via reversed-phase HPLC with application to coal liquefaction products, *Chromatographia,* 17(8), 431, 1983.
51. **Rayss, J., Dawidowicz, A., Suprynowicz, Z., and Buszewski, B.,** A study of the properties of octadecyl phases bonded to controlled-porosity glasses, *Chromatographia,* 17(8), 437, 1983.
52. **Kiselev, A. V. and Yashin, J. Ja.,** *Gas-Adsorption Chromatography,* Nauka, Moscow, 1967.
53. **Vassilaros, D. L., Stoker, P. W., Booth, G. M., and Lee, M. L.,** Capillary GC-determination of PAHs in vertebrate fish tissue, *Anal. Chem.,* 54, 106, 1982.
54. **Later, D. W., Lee, M. L., Bartle, K. D., Kong, R. C., and Vassilaros, D. L.,** Chemical class separation and characterization of organic compounds in synthetic fuels, *Anal. Chem.,* 53, 1612, 1981.
55. **Poirier, M. A. and Smiley, G. T.,** A novel method for separation of sulphur compounds in naphtha and middle distillate fractions of lloydminster heavy oil by GC/MS, *J. Chromatogr. Sci.,* 22, 304, 1984.
56. **Kong, R. C., Lee, M. L., Tominaga, Y., Pratap, R., Iwao, M., Castle, R. N., and Wise, S. A.,** Capillary column gas chromatographic resolution of isomeric polycyclic aromatic sulfur heterocycles in a coal liquid, *J. Chromatogr. Sci.,* 20, 502, 1982.
57. **Herod, A. A. and James, R. G.,** A review of methods for the estimation of PAHs with particular reference to coke oven emissions, *J. Inst. Fuel,* 51, 164, 1978.

58. **Daisey, J. M. and Leyko, M. A.**, TLC-Method for determination of PAH and aliphatic hydrocarbons in airborne particulate matter, *Anal. Chem.*, 51, 24, 1979.

59. **Van Langermeersch, A.**, *Chémie Analytique*, 50, 3, 1968.

60. **Halot, D.**, Hydrocarbures et pollution atmospherique, *Talanta*, 17, 729, 1970.

61. **Schaad, R. E.**, Chromatography of polycyclic aromatic hydrocarbons, *Chromatogr. Rev.*, 13, 61, 1970.

62. **Kraft, J., Hartung, A., Lies, K. H., and Schulze, J.**, Two dimensional TLC of substituted PAH in aerosols with negative ion mass spectrometry, *J. High Res. Chromatogr. Comm.*, 5, 489, 1982.

63. **Sawicki, E.**, Tentative method of analysis for PAH content of atmospheric matter, *Health Lab. Sci.*, 7, 31, 1970.

64. **Cleary, G. J.**, Discrete separation of PAH in airborne particulates using very long alumina columns, *J. Chromatogr.*, 9, 204, 1962.

65. **Brass, H. J., Elbert, W. C., Feige, M. A., Glick, E. M., and Lington, A. W.**, Black River Survey: Analysis for Hexane Organic Extractables and PAH, U.S. Steel Lorain, OH, Office of Enforcement and General Council, US-EPA, Cinncinati, OH, October 1974.

66. **Liberti, A., Morozzi, G., and Zoccolillo, L.**, Comparative determination of PAH in atmospheric dust by GLC and spectrophotometry, *Ann. Chim.*, 65, 573, 1975.

67. **Griest, W. H., Kubota, H., and Guerin, M. R.**, Resolution of PAH by packed column chromatography, *Anal. Lett.*, 8(12), 949, 1975.

68. **Klimish, H. J.**, Determination of PAH. Separation of benzopyrenes by HPLC, *Anal. Chem.*, 48, 1960, 1973.

69. **Wilk, M., Rochlitz, J., and Beude, K. D.**, Column chromatography of PAH on lipophilic Sephadex® LH-20, *J. Chromatogr.*, 24, 414, 1966.

70. **Novotny, M., Lee, M. L., and Bartle, K. D.**, The methods for fractionation, analytical separation and identification of PAHs in complex mixtures, *J. Chromatogr. Sci.*, 12, 606, 1974.

71. **Sorrell, R. K., Dressman, R. C., and McFarren, E. J.**, HPLC for the measurement of PAH in water, in *Proceedings of the Water Quality Technology Conference*, Kansas City, Mo, December 1977.

72. **Grimalt, J.**, Albaiges, oil spill identification by HPLC, *J. High Res. Chromatogr. Chromatogr. Commun.*, 5, 255, 1982.

73. **Grob, K. and Grob, K., Jr.**, On-column injection onto glass capillary columns, *J. Chromatogr.*, 151, 311, 1978.

74. **Lao, R. C., Thomas, R. S., and Monkman, J. L.**, Computerized GC/MS analysis of PAH in environmental samples, *J. Chromatogr.*, 112, 681, 1975.

75. **Chatot, G., Dangy-Caye, R., and Fontanges, R.**, Atmospheric PAHs. II. Application of chromatographic coupling in gas phase and determination of PAHs, *J. Chromatogr.*, 72, 202, 1972.

76. **DeMaio, L. and Corn, M.**, Chromatographic analysis of PAHs with packed columns. Application to air pollution studies, *Anal. Chem.*, 38, 131, 1966.

77. **Smith, J. D., Hauser, J. Y., and Bagg, J.**, PAHs in sediments of the Great Barrier Reef Region, Australia, *Marine Poll. Bull.*, 16, 110, 1985.

78. **Schulte, K. A., Larsen, D. J., Holnung, R. W., and Crable, J. W.**, Analytical methods used in a study of coke oven effluent, *J. Am. Ind. Hyg. Assoc.*, 36, 131, 1975.

79. **Lao, R. C., Oja, H., Thomas, R. S., and Monkman, J. L.**, Assessment of environmental problems using the correlation of GC and quadrupole mass spectrometry, *Sci. Total Environ.*, 2, 223, 1973.

80. **Lysyuk, L. S. and Korol, A. N.**, Selectivity of stationary phases for the resolution of PAHs, *Chromatographia*, 10, 712, 1977.

81. **Bhatia, K.**, Gas-chromatographic determination of PAHs, *Anal. Chem.*, 43, 609, 1971.

82. **Lane, D. A., Moe, H. K., and Katz, M.**, Analysis of PAH, some heterocyclics and aliphatics with a single gas chromatographic column, *Anal. Chem.*, 45, 1776, 1973.

83. **Janini, G. M., Muschik, J. A., Schroer, J. A., and Zielinski, W. L.**, Gas-liquid chromatographic evaluation and GC/MS application of a new high temperature liquid crystal stationary phases for PAH separation, *Anal. Chem.*, 48, 1879, 1976.

84. **Onuska, F. I. and Karasek, F. W.**, *Open Tubular Column Gas Chromatography in Environmental Sciences*, Plenum Press, New York, 1984.

85. **Cantuti, V., Cartoni, G. P., Liberti, A., and Torri, A. G.**, Improved evaluation of PAH in atmospheric dust by GC, *J. Chromatogr.*, 17, 60, 1965.

86. **Lee, M. L., Novotny, M., and Bartle, K. D.**, GC/MS and NMR determination of PAHs in airborne particulates, *Anal. Chem.*, 48, 1566, 1976.

87. **Lee, M. L., Vasillaros, D. L., White, C. M., and Novotny, M.**, Retention indices for programmed-temperature (GC)^2 of PAH, *Anal. Chem.*, 51, 768, 1979.

88. **Onuska, F. I., Wolkoff, A. W., Comba, M. E., Larose, R. H., Novotny, M., and Lee, M. L.**, Gas-chromatographic analysis of PAH in shellfish on short-WCOT glass capillary columns, *Anal. Lett.*, 9, 451, 1976.

89. **Stray, H., Mano, S., Mikalsen, A., and Oehme, M.,** Selective determination of substituted PAHs in aerosols with negative ion mass spectrometry, *J. High Res. Chromatogr. Chromatogr. Comm.,* 7, 74, 1984.

90. **Wright, C. W.,** Simultaneous quantitative analysis using dual capillary columns of different polarities, *J. High Res. Chromatogr. Chromatogr. Comm.,* 7, 83, 1984.

91. **Spitzer, T. and Danneker, W.,** Cleanup of PAH and 3-ring azaarenes and their GC-analysis on whisker-WCOT columns, *J. High Res. Chromatogr. Chromatogr. Comm.,* 7, 301, 1984.

92. **Moser, H. and Arm, H.,** Separation of 3- to 6-ring PAHs by capillary gas chromatography with a liquid crystal as stationary phase, *J. High Res. Chromatogr. Chromatogr. Comm.,* 7, 637, 1984.

93. **Tausch, H. and Stehlik, G.,** Analysis of PAH in the fly ash of an incineration plant for radioactive waste, *J. High Res. Chromatogr. Chromatogr. Comm.,* 8, 524, 1985.

94. **Guard, H. E., Hunter, L., and DiSalvo, L. H.,** Identification and potential biological effects of the major components in the seawater extract of a bunker fuel, *Bull. Environ. Contam. Toxicol.,* 14, 395, 1975.

95. **Brown, R. A., Elliott, J. L., Kelliher, J. M., and Searl, T. D.,** *Adv. Chem. Ser.,* 147, 172, 1975.

96. **Walker, J. D., Colwell, R. R., Hamming, M. C., and Ford, H. T.,** Extraction of petroleum hydrocarbons from oil contaminated sediments, *Bull. Environ. Contam. Toxicol.,* 13, 245, 1976.

97. **May, W. E., Chester, S. N., Cram, S. P., Gump, B. H., Hertz, H. S., Enagonio, D. P., and Dyszel, S. M.,** Chromatographic analysis of hydrocarbons in marine sediments and seawater, *J. Chromatogr. Sci.,* 13, 535, 1975.

98. **Zitko, V.,** Aromatic hydrocarbons in aquatic fauna, *Bull. Environ. Contam. Toxicol.,* 14, 621, 1975.

99. **Gibson, D. T., Mahadevan, V., Jerina, D. M., Yagi, H., and Yeh, H. J. C.,** Oxidation of the carcinogens benzo (a) pyrene and benzo (a) anthracene to dihydrodiols by a bacterium, *Science,* 189, 295, 1975.

100. **Schepple, S. E., Greenwood, G. J., and Crynes, B. L.,** Proceedings of the 23rd ASMS Conference, Houston, Tx, 1975.

101. **Nagy, B., and Colombo, U.,** *Fundamental aspects of petroleum geochemistry,* Elsevier Amsterdam, 1967.

102. **Jewell, D. M., Roberts, R. G., and Davis, B. E.,** Systematic approach to the study of aromatic hydrocarbons in heavy distillates and residues by adsorption chromatography, *Anal. Chem.,* 44, 2318, 1972.

103. **Hites, R. A. and Dubay, G. R.,.** Charge-exchange chemical ionization of PAHs, in *Carcinogenesis, Volume 3, Polycyclic Aromatic Hydrocarbons,* Jones, P. W. and Freudenthal, R. I., Eds., Raven Press, New York, 1978, 85.

104. **Aczel, T.,** *Rev. Anal. Chem.,* 1, 226, 1972.

105. **Aczel, T., Allan, D. E., Harding, J. H., and Knapp, E. A.,** Computer techniques for quantitative high resolution mass spectral analyses of complex hydrocarbon mixtures, *Anal. Chem.,* 42, 341, 1970.

106. **Hellmann, H.,** Fluorescence spectra of biogenic PAHs, *Fresenius Z. Anal. Chem.,* 278, 263, 1976.

107. **Heinrich, G. and Guesten, H.,** Fluorescence spectroscopic determination of carcinogenic PAHs in the atmosphere, *Fresenius Z. Anal. Chem.,* 278, 257, 1976.

108. **Sawicki, E., Carey, R. C., Dooley, A. E., Giscard, J. B., Monkman, J. L., Neligan, R. E., and Ripperton, L. A.,** Tentative method of microanalysis for 3,4-benzopyrene in airborne particulates and source effluents, *Health Lab. Sci.,* 7, 56, 1970.

109. **Sawicki, E., Stanley, T. W., Elbert, W. C., Meeker, J., and McPherson, S.,** Comparison of methods for the determination of 3,4-benzopyrene in particulates from urban and other atmospheres, *Atmos. Environ.,* 1, 131, 1967.

110. **Kunte, H.,** Carcinogenic substances in water and soil. XVII. Determination of PAHs by means of mixed TLC and by fluorescence measurements, *Arch. Hyg. Bakt.,* 151, 193, 1967.

111. **Borneff, J. and Korte, H.,** Carcinogenic substances in water and soil. XXVI. A routine method for the determination of PAHs in water, *Arch. Hyg. Bakt.,* 153, 220, 1969.

112. **Reichert, J., Kunte, H., Engelhardt, K., and Borneff, J.,** Carcinogenic substances in water and soil. XXVII. Further studies on the elimination from waste water of carcinogenic PAHs, *Arch. Hyg. Bakt.,* 155, 18, 1971.

113. **Hood, L. V. S. and Winefordner, J. D.,** TLC and low temperature luminiscence measurement of mixtures of carcinogens, *Anal. Chim. Acta,* 42, 199, 1968.

114. **Kirkbright, G. F. and Lima, L. C. G.,** The detection and determination of PAH by luminescence spectrometry utilizing Spolski effect at 77°K, *Analyst,* 99, 338, 1974.

115. **Colmsjo, A. and Stenberg, U.,** Identification of PAH by Spolski low temperature fluorescence, *Anal. Chem.,* 51, 149, 1979.

116. **Brown, J. C., Elelson, M. C., and Small, G. J.,** Fluorescence line narrowing spectroscopy in organic glasses containing ppb levels of PAHs, *Anal. Chem.,* 50, 1394, 1978.

117. **Hatano, H., Yamamoto, Y., Saito, M., Mochida, E., and Watanabe, S.,** High speed liquid chromatograph with a flow-spectrofluorimetric detector and the ultramicro determination of aromatic compounds, *J. Chromatogr.,* 83, 373, 1973.

118. **Woo, Ching S., D'Silva, A. P., Fassel, V. A., and Oestreich, G. J.,** PAHs in coal identification by their X-ray excited optical luminescence, *Environ. Sci. Technol.,* 12, 173, 1973.

Chapter 7

PHTHALATE ESTERS IN THE AQUATIC ENVIRONMENT

J. Kohli, J. F. Ryan, and B. K. Afghan

TABLE OF CONTENTS

I. INTRODUCTION

A. Identity and Sources of Phthalic Acid Esters

Phthalate esters or phthalic acid esters (PAEs) are terms used interchangeably to describe the esters of ortho-phthalic acid. They can be prepared by reacting phthalic acid with a specific alcohol as shown in Figure 1.

Industrially, PAEs are synthesized from phthalic anhydride, which is obtained by oxidation of either naphthalene or ortho-xylene in the presence of vanadium pentoxide.[1] Esterification of the anhydride and the appropriate alcohol is effected in the presence of sulfuric acid or *p*-toluene sulfonic acid as catalysts, or else noncatalytically at high temperature. Commercial preparations may well contain, in addition to the indicated diester, the monoester, phthalic acid, phthalic anhydride, and the alcohol.[2] Thus, commercially produced PAEs are not pure compounds, although varying degrees of purity can be approached.

In most common usage are those PAEs in which the two alcohol moieties are identical; however, mixed esters can also be produced. Abbreviations of common PAEs are given in the Appendix; abbreviated forms are used extensively in this text.

While PAEs refer to the ortho form of benzenedicarboxylic acid, meta and para forms occur and are termed isophthalate and terephthalate esters respectively. Compared with phthalate esters, they have some similarities and many differences in properties as well as end uses, toxicology and so forth. There are some indications[2] that certain organisms can and do synthesize PAEs during their normal metabolic activities. However, authorities have not reached any concensus on the relative importance of this source.[3,4] Probably most environmental PAEs result from human production and dispensation of phthalate esters themselves or of materials containing PAEs.

Human production and use of PAEs are for the most part dependent on their intrinsic qualities as plasticizers. More information on their properties in this regard is given in the following sections, along with a description of other uses, toxicological characteristics, and environmental dynamics. A discussion of the literature on analytical aspects relating to PAEs in aquatic media follows this overview.

Although PAEs are found in the atmosphere and in land biosphere compartments, the following is largely limited to aquatic media.

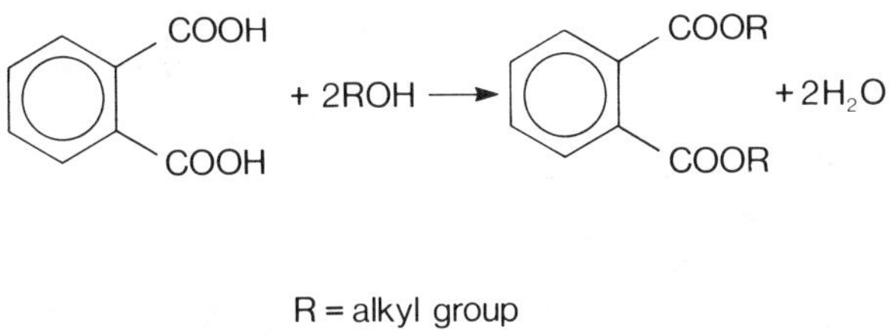

R = alkyl group

FIGURE 1.   Reaction producing phthalate esters.

B. Physical and Chemical Properties

Table 1 gives a summary of some of the key physical and chemical properties of the more common PAEs. In general, they have low vapor pressures, low water solubilities, and high boiling points. They are relatively colorless, odorless, and viscous liquids under ambient conditions.[2] Although PAEs are often treated as a single category of compounds in terms of toxicology and environmental releases/dynamics, there are significant differences among PAEs in physical and chemical properties (Table 1). Consequently this places importance on analytical methods for discriminating among the various PAEs.

Several properties are common to plasticizers, of which PAEs are paramount in importance. They must have low volatility — both innate as measured by vapor pressure, as well as when incorporated into a polymer. They must have low water and organic solvent extractability and low migration tendencies while incorporated in the polymer. They must form a homogenous mixture with the resin. The PAEs demonstrate ranges in these properties while acting as plasticizers. As the length of the alcohol side chains increases[5-7] PAE volatility decreases and the volatility also increases with the added branching of these side chains.[3,5]

When incorporated into a polymer the volatilities of PAEs are approximately proportional to their vapor pressures. Volatility is usually expressed in terms of percent weight loss at a specified temperature. There are significant differences among the PAEs (Table 2) because of this characteristic. PAEs are soluble in many common organic solvents but are poorly soluble in water, the water solubility being greater for PAEs with shorter side chains.[2] Many hydrophobic organics such as PAEs can be brought into solution via adsorption to external polar groups of various colloidal molecules. Thus, fulvic acids, found commonly in fresh water and soils,[8] have been shown to form water soluble complexes with dialkylphthalates.[8] Fulvic acids appear to complex (probably in a nonchemical manner of binding) different PAEs to varying degrees and the process is somewhat pH-dependent. For example, Matsuda and Schnitzer[9] found that fulvic acid complexation was 25% less at pH 7 than at pH 2.45. In contrast, Boehm and Quinn[10] found that organic matter in seawater had no effect on DnBP solubility. Very little is known about the actual chemical and physical form(s) taken on by PAEs under varying ambient or contaminated aquatic conditions.[2]

Chemical hydrolysis of simple esters to carboxylic acid and alcohol in the presence of catalytic quantities of an acid or base may occur.[7] Acidic hydrolysis of esters is usually reversible whereas alkaline hydrolysis is irreversible and alkaline hydrolysis of PAEs usually occurs step-wise, first to the monoester and then to the carboxylic or carboxylate anion.

Diester hydrolysis occurs somewhat faster than monoester hydrolysis, but reaction rates are further functions of side chain configuration.[11] Apparently, PAEs are less easily hydrolyzed than isophthalates or terephthalates.[11,12] Due to the reportedly slow hydrolysis rates, these reactions are expected to be relatively insignificant in PAE degradation in the aquatic environment.[2]

Although there is very little information regarding photolytic decomposition of PAEs, the process does not seem to be a major degradation pathway. For example, Gledhill et al.[13] reported no significant decrease in BBP concentrations in water after 28 days in samples exposed to sunlight as opposed to controls retained in the dark.

Similarly, little information is available on the thermal stability of PAEs. Pierce et al.[2] cited a study[14] indicating that PAEs were volatilized but not degraded during oxidative thermal decomposition at 400°C. They added that low temperature incineration and open burning of discarded plastics containing PAEs, may be an important secondary source of atmospheric and aquatic contamination. In contrast, Peakall[4] suggested that up to 66% of PAEs would be destroyed during low temperature incineration or open burning in dumps.

PAEs may also migrate from the host plastic or other material, to solids in contact with them. This has been demonstrated using dry powders such as silica and soil. This behavior increases with higher plasticizer content in the material and depends further on the type of polymer matrix, and the type of PAE.[7] The importance of humic materials (such as fulvic acids) in mobilizing PAEs in this manner is suspected but has not been rigorously defined.[8,15]

C. Production and Uses

Annual world production of PAEs (based on 1977 levels)[16] is in the range of 1.4×10^9 to 1.8×10^9 kg, of which the U.S. is responsible for 30 to 40%.[17] Although an accurate estimate of the quantitative end uses for PAEs is not readily available, as much as or more than 90% is directed to plasticizer uses.[18] Plasticizers include a number of other categories of chemicals, but PAEs are the most extensively used. A plasticizer is a material incorporated into a plastic to lend workability and distensibility and is physically dispersed within the polymer chain matrix to decrease the interaction (attractant) forces of adjacent chains, thereby facilitating slippage movement of adjacent molecular chains.

PAEs are used as plasticizing agents for resins and polymers — especially for polyvinyl chloride (PVC). They serve the same function in cellulosics and some types of elastomers. These and other end uses for PAEs in Canada were reviewed by Leah.[18] Table 3 provides a summary of this information. It is reiterated that the PAEs identified may not be pure, due to the methods of production.

The most widely used PAE in North America appears to be DEHP; in 1977 it accounted for 32% of the total U.S. PAE production.[16] Its isomer, DOP, has frequently been synonymously referred to and doubtless receives considerable use as well. Other common PAEs include BBP, DiDP, and DHnNDP.[2] Together, DEHP, DHnNDP, BBP, and DiDP accounted for 80% of the Canadian market in the late 1970s.[18]

Biological production (metabolic by-products) of PAEs has been noted. There appear to be no numerical estimates of the magnitude of this production category, although Mathur[19] has suggested a method for determining this contribution using $^{14}C/^{12}C$ ratios, keeping in mind that man-made PAEs are synthesized from fossil organics depleted in ^{14}C.

D. Regulatory Limits for PAEs

Table 4 provides a summary of various regulatory criteria, guidelines and objectives set for various PAEs in environmental milieux, based on human toxicology or toxicity to other organisms. The list must be considered incomplete and conditions for implementation of these levels are variable.

II. ENVIRONMENTAL CONSIDERATIONS

A. Environmental Concerns

Phthalic acid esters have received attention as potential environmental contaminants for several reasons. PAEs have been produced in increasing quantities since the 1940s[24] and few restrictions have been placed on them regarding acceptable levels in industrial wastes, liquid effluents, or work-place environments. Large quantities of PAEs undeniably reach the environment, both as pure compounds/mixtures or within the host resins in plastics. In the latter case, volatilization or leaching may permit further distribution of PAEs into various

Table 1
PHYSICAL AND CHEMICAL PROPERTIES OF SOME PHTHALATE ESTERS

Short Cain

Compound	Molecular weight	Specific gravity (at 25°C)	Melting point (°C)	Boiling point (°C) (at 760 mmHg)	Vapor pressure (mmHg)	λ_{max} MeOH (nm)	Log ϵ	Water solubility (g/100 g at 20°C)	Solvent solubility	Log $P_{o/w}$[a] (estimate)
Dimethylphthalate (DMP)	194.2	1.188—1.192	0.0 (freezes)	147(10 mm) 282	1.1×10^{-3}	225 274	3.92 3.10	0.5	Insoluble in petroleum oils; soluble in all other common organic solvents and oils	1.5
Diethylphthalate (DEP)	222.3	1.115—1.119	−40.5	158(10 mm) 296	4.0×10^{-4}	225 274	3.92 3.10	0.1	Soluble in all common organic solvents and oils	1.8
Di-n-butylphthalate (DnBP)	278.4	1.044—1.048	−35	192(10 mm) 340	1.4×10^{-5}	225 274	3.90 3.09	0.45 (25°C)	Soluble in all common organic solvents and oils	2.2
Di-iso-butylphthalate (DiBP)	278.4	1.040 (20/4)	−50	327	—	—	—	0.01	—	—
n-Butylbenzylphthalate (nBBP)	298.4	—	—	—	—	—	—	—	—	—
Dicyclohexylphthalate (DCHP)	330.4	1.29	63.0	231(10 mm)	4.5×10^{-7}	225 274	3.97 3.26	<0.01(25°C)	Soluble in acetone n-butanol, carbon tetrachloride, cyclohexane, ethyl acetate, ether, toluene	3—4

Table 1 (continued)
PHYSICAL AND CHEMICAL PROPERTIES OF SOME PHTHALATE ESTERS

Compound	Molecular weight	Specific gravity (at 25°C)	Melting point (°C)	Boiling point (°C) (at 760 mmHg)	Vapor pressure (mmHg)	Absorption Spectra λmax MeOH (nm)	Log ε	Water solubility (g/100 g at 20°C)	Solvent solubility	Log $P_{o/w}$ [a] (estimate)
					Short Cain					
n-Butyl-n-octyl-phthalate (nBnOP)	334.5	1.001	−50	340(740 mm)	—	—	—	—	—	—
n-Butyl-n-decyl-phthalate (nBnDP)	362.6	—	—	—	—	—	—	—	—	—
Di-n-octylphthalate (DnOP)	390.6	0.978 (20/4)	−25	220	<0.2(150°C)	—	—	—	—	3—4
Di-(2-ethylhexyl)-phthalate (DEHP)	390.6	0.980—0.985	−50	236(10 mm) 387(5 mm)	1.6×10^{-7}	—	—	<0.005(25°C)	Miscible with most common solvents	3—4
Di-iso-octyl-phthalate (DiOP)	390.6	0.981 (20/4)	−4	239(5 mm)	1.0(200°C)	—	—	—	—	3—4
n-Octyl-n-decyl-phthalate (nOnDP)	418.7	0.970	−28	250(5 mm)	—	—	—	—	—	3—4
Di-iso-decyl-phthalate (DiDP)	446.7	0.963—0.969	−48	256(10 mm)	3.5×10^{-9}	—	—	<0.005(25°C)	Limited solubility in glycerine, glycols and some amines. Soluble in most other organic liquids	3—4
Ditridecylphthalate (DTDP)	530.9	1.054	−37	235(3.5 mm)	—	—	—	—	—	3—4

[a]   Log of partition co-efficient (octanol/water)

Modified from Pierce, R. C. et al., National Research Council of Canada, Publication NRCC No. 17583, 1980. With permission.

Table 2
WEIGHT LOSS OF PLASTICIZED PVC RESIN DUE TO PHTHALATE ESTER VOLATILITY AND LIQUID EXTRACTION

Compound	Plasticizer concentration (phr)	Volatility[b] (%)	% Weight loss resulting from extraction[a]						
			Distilled water[c]	1% Soap solution[c]	1% Detergent solution[c]	Mineral oil[c]	iso-Octane[d]	Di-iso-butylene[d]	iso-Octane + toluene[d] (70:30)
Di-n-butylphthalate	54	7.20	0.50	1.00	0.90	1.05	1.85	2.10	14.5
Di-iso-butylphthalate	62.5	9.30	0.50	1.30	1.80	1.50	3.50	6.30	14.0
Di(2-ethylhexyl)phthalate	62.8	0.35	0.10	0.30	0.55	1.50	20.3	23.0	16.3
Di-iso-octylphthalate	64	0.40	0.05	0.40	0.50	1.30	20.2	23.4	16.4
Di-n-nonylphthalate	69.5	0.35	0.05	0.10	0.15	1.90	28.9	29.2	21.0
Di-iso-decylphthalate	66.5	0.30	0.15	0.15	0.40	7.05	28.5	29.3	20.8
Ditridecylphthalate	74.5	0.30	0.20	0.05	0.60	20.0	33.0	33.8	32.6

a   Samples 7 cm × 5 cm × 0.01 cm. Solvents neither stirred nor renewed.
b   Samples 7 cm × 5 cm × 0.01 cm in a circulating-air oven at 85°C.
c   240 h at 23°C.
d   24 h at 23°C.

From Pierce, R. C. et al., National Research Council of Canada, Publication NRCC No., 17583, 1980. With permission.

Table 3
END USES OF PAEs

Use categories	Major PAEs in use
Plasticizers in PVC resins	
Vinyl flooring	
Homogeneous vinyl tile	DOP
Vinyl asbestos tile	BBP, BOP, DOP,DiOP, benzoates
Vinyl chloride carpet backing	—
Vinyl roll goods (foamed or unfoamed)	—
Sheeting and film	DOP, DiOP
Wire and cable insulation	DTDP, nOnDP, DIDP
Plastisols	
"Spread" or "doctor blade" coatings on textiles or papers	—
Dipping of heated forms (e.g., dish drainers)	—
Slush moldings (e.g., footwear, toys)	—
Plasticizers in other materials	
Adhesives	
Polyvinyl acetate adhesives	—
Nitrocellulose film coating	DBP, DCHP
Nonplasticizer uses	
Insect repellent formulations	DEP
Defoaming agents in paper/paperboard manufacture	DMP
In production of denatured ethyl alcohol for use in cosmetics, rubbing alcohol, liquid soap, detergents	DEP
Inks for decorative packaging	DBP
In lacquers for industrial flooring and applying phosphor to fluorescent lamps	DOP, DBP
As cooling agents in military propellants	DEP, DBP
As a carrier for peroxide catalysts in the production of polyester fiberglass-reinforced plastics	DMP
As additives in industrial and lubricating oils	—
As carriers in pesticide formulations	—
As PCB substitutes in electrical capacitors	—

Note: — = no data provided on PAEs used.

From Leah, T. D., Environmental Contaminants Inventory Study No. 4, Report Series 47, Inland Waters Directorate, Ontario Region, Water Planning and Management Branch, Burlington Ontario, Fisheries and Environment Canada, Ottawa, 1977. With permission.

environmental compartments. Control on the quantities of PAEs reaching the environment via waste disposal are notoriously difficult to implement.

Concerns are amplified by the frequently very short useful lifetimes of many plastic and other products containing PAEs. Although poorly water soluble, PAEs are readily sequestered, absorbed, or adsorbed by organic residues and solid surfaces in aqueous environments. The long term ecological consequences of PAE deposition and accumulation have not been fully explored. Man's use of aquatic resources brings him directly and indirectly into contact with PAEs; the toxicological ramifications in this regard have also not been fully explored.[24]

Table 4
REGULATORY CRITERIA, GUIDELINES, AND OBJECTIVES FOR PAE LEVELS

Nature of stipulation	Substance	Value μg/l	Target organism	Ref.
Water quality criteria	PAEs	3	Freshwater aquatic life	20
Water quality criteria	PAEs	0.3	Fish and fish food supply	21
Recommended water quality objectives	DBP	4.0	Aquatic life	22
	DEHP	0.6	Aquatic life	22
	Other PAEs	0.2	Aquatic life	22
Discharge multimedia environmental goal(DMEG)[a]	PAEs	1.5	Aquatic	23
Discharge multimedia environmental goal (DMEG)	PAEs	100,000	Aquatic life	23
Discharge multimedia environmental goal (DMEG)	PAEs	100,000	Human health	23
Discharge multimedia environmental goal (DMEG)	DEHP	75,000	Human health	23
Discharge multimedia environmental goal (DMEG)	BBP	2,130,000	Human health	23
Ambient multimedia environmental goal (AMEG)[b]	PAEs	70.3	Aquatic life	23
Ambient multimedia environmental goal (AMEG)	BBP	50,000	Aquatic life	23
Ambient multimedia environmental goal (AMEG)	PAEs	70	Human health	23
Ambient multimedia environmental goal (AMEG)	DEHP	70	Human health	23
Ambient multimedia environmental goal (AMEG)	BBP	1,264	Human health	23
Drinking water criteria	DBP	34,000	Human health	21
Drinking water criteria	DEHP	15,000	Human health	21
Drinking water criteria	DEP	350,000	Human health	21
Drinking water criteria	DMP	313,000	Human health	21
Toxicity based criteria[c]	PAEs	940/3.0	Aquatic life	20

[a] (DMEG): The Discharge Multimedia Environmental Goal represents the "concentration of pollutants in undiluted (effluent) streams that will not adversely affect those...ecological systems exposed for short periods of time".

[b] (AMEG): The Ambient Multimedia Environmental Goal represents the "concentration of pollutants in emission streams, which after dispersion, will not cause the level of contamination in the ambient media (water) to exceed a safe continuous exposure concentration".

[c] First value indicates acute toxicity level to freshwater aquatic life; second value indicates chronic toxicity level to freshwater aquatic life.

Enviromental concerns about PAEs hinge on the relatively high and increasing rates of PAE release to the environment, on their demonstrated (although frequently low) toxicity to various organisms, and on their persistence.

B. Releases of PAEs to the Environment

Two major categories of releases of PAEs to the environment have been described:[2,18] (1) release from production and processing activities — including losses during PAE production, plasticizer and resin compounding, fabrication of PVC into products, and during production of adhesives, plastisols, coatings, etc., and (2) release from use and disposal of materials incorporating PAEs — including plasticizer losses during the useful lifetimes of the products or during incineration or landfill.

Table 5
ESTIMATED PAE RELEASE RATES TO THE
ENVIRONMENT BY INDUSTRIAL SOURCE AND
ACTIVITY

Activity	Industry loss rate (%)	Industry contribution to total production/ processing releases (%)
PAE production	0.1—0.3	4.0—5.3
PVC compounding/processing		
Flooring	1.0—5.0	8.3—18.6
Sheet and film	0.1—0.3	1.5—2.1
Wire and cable	0.1—0.2	0.6—0.7
Plastisols	0.1	71.3—80.7
Other vinyl extrusion and molding	0.1	0.1—0.3
Other		
Adhesives	2.0	0.3—0.6
Cellulose film coatings	0.1	0.3—0.6
Miscellaneous	1.0	0.2—0.5

From Leah, T. D., Environmental Contaminants Inventory Study No. 4, Report Series 47, Inland Waters Directorate, Ontario Region, Water Planning and Management Branch, Burlington, Ontario, Fisheries and Environment Canada, Ottawa, 1977. With permission.

1. Release from Production and Processing Activities

Based primarily on the pattern of PAE production/processing in Canada, Leah[18] estimated that approximately 2.0 to 4.5% of the total supply of PAEs was lost as gaseous and liquid effluents during this stage. Comparing all production and processing losses, plastisol processing contributed approximately 75% of the loss, while flooring product production contributed over 8%, and PAE production itself contributed over 4%. Losses from seven other subcategories were lower. Releases to the environment from production and processing are characteristically concentrated in industrial areas.

Table 5 gives estimates of releases from production and processing activities. It is not known whether these are typical of worldwide industrial loss rates either by industrial activity or by industrial contribution as a whole.

2. Release from Use and Disposal Activities

Quantitative data concerning the loss of PAEs during article use and disposal are not available. There are various methods whereby estimates can and have been derived, but most suffer from lack of data regarding one or more of the following:

- Average or range in PAE loss rates during the lifetime of a product.
- Longevity of specific products prior to disposal.
- Consumer use and disposal practices (including demographic and socioeconomic controlling factors bearing on quantities and length of useful life).
- Quantities of specific types of wastes flowing to various types of waste disposal facilities (landfill, burning/landfill, incineration, waste reclamation, etc.).
- Relative degradation efficiencies of PAEs in incinerators and dump fires; variations among incinerators in emission factors.

- Rates of decomposition, migration, and solubilization of specific PAEs in landfill situations or in underwater or surface situations.

- Interpretation of relative contributions toward disposal by industry, commercial, institutional and other sectors.

In addition to releases from the more traditionally recognized sources of plastics, adhesives, and so forth, PAEs have been detected in effluents from nylon, chemical, and textile plants, as well as pulp and paper mills and oil refineries.[2]

C. Environmental Levels

Phthalate esters are widely distributed in the ecosystem due to their ubiquitous use as plasticizer compounds. Weschler[140] identified phthalate esters as a significant component of Arctic aerosol. The concentrations of DEP, DnBP, and DEHP ranged from 0.1 to 20 $ng.m^{-3}$. In addition, n-alkanes and polydimethylsiloxanes were also identified in the cyclohexane and methylene chloride extracts of the Arctic aerosols. Giam et al.[141] and other workers have also reported the presence of phthalate esters in atmospheric particulate matters collected at remote sites. The concentration of DEHP and DBP were found to be higher, compared to those of DDTs and PCBs, both in the vapor phase and particulate matter in the atmosphere of the Gulf of Mexico. However, the levels of phthalate esters found were one or two orders magnitude lower than those reported for the continental and urban air.

It is difficult to infer probable background levels for PAEs in various aquatic environmental compartments such as freshwater, saltwater, sediments, and biota. PAEs, like many other anthropogenic chemical compounds, have been detected in many parts of the world at levels which cannot be said to reflect background concentrations. The added problem of PAE contamination during the analytical stage appears to have plagued many studies,[2] so that such results cannot always be fully credited. Table 6 provides a list of such results, giving environmental levels, location, and other pertinent data. Included are reported values which are clearly indicative of situations having PAE enrichment, as well as those where little or no PAE input has occurred.

Levels of PAEs in surface water are generally nondetectible or low because of limited solubility, biological degradation and physical or chemical breakdown or sequestering by solids. Typical levels are below or in the vicinity of the low µg/l range. Shackleford and Keith[25] ranked the PAEs most commonly reported (in water) in the recent literature, in decreasing order of frequency as follows: DEP, DOP, DBP, DEHP, BBP, and DnBP. The authors consider that their ranking does not accurately reflect the current relative prevalence of PAEs in aqueous milieux. From the authors' experience, DEHP and DBP are the most common PAE contaminants.

Most sediment samples examined for PAEs contained higher concentrations than did corresponding water samples,[2] ranging from nondetectible or less than 1.0 µg/kg up to 218,000 µg/kg. This is largely thought to occur due to low PAE solubility in water and the tendency for organic particulate matter to adsorb PAEs prior to their deposition as sediment or at the sediment/water interface.[2] Schwartz et al.[26] determined that most PAE was associated with sediment particles smaller than 16 µm in diameter. Aquatic biota contain PAE residues in the low mg/kg range (mostly based on whole organism body mass).

D. Toxicity and Physiological Effects

1. Aquatic Fauna

Although the acute toxicity of PAEs to mammals is low, there is evidence that aquatic organisms concentrate PAEs and are more vulnerable to PAEs than warmblooded animals.[52] Disturbances of the reproduction and growth of aquatic species have been demonstrated in water containing low ppb levels of PAEs.[53]

Table 6
ENVIRONMENTAL LEVELS OF PAEs IN AQUATIC MILIEUX[2,22]

Type of water and location	Compound	Concentration (μg/l)	Ref.
Freshwater			
Surface water: Black Bay, Lake Superior, Ontario	DEHP	300	27
Surface water: Hammond Bay, Lake Huron, MI	DnBP	0.04	27
Surface water: Lake Huron, MI	DEHP	5	27
Surface water: Lake Ontario (lake-wide)	Phthalate	2—5.0	28
Surface water: Lake Erie (lake-wide)	Phthalate	0.7—6	28
Surface water: Lake Huron (lake-wide)	Phthalate	0.8—3	28
Surface water: Lake Superior	Phthalate	≤0.1	28
Surface water: Lake Michigan (2 of 13 sites)	BBP	2—4	29
Surface water: Lake Michigan (3 of 13 sites)	DBP	1—14	29
Surface water: Lake Michigan (10 of 13 sites)	DEHP	1—137	29
Surface water: Nipigon Bay, Lake Superior (near pulp and paper mill outfall)	DOP	0.1—2	30
Surface water: St Clair River	DBP	1	31
Surface water: St. Clair River	DEHP	1.6—4.6	31
Surface water: Missouri River, McBain, MO	DEHP	4.9	27
Surface water: Missouri River, McBain, MO	DnBP	0.09	27
Surface water: Charles River, Boston, MA	Phthalate	0.88—1.9	32
Surface water: Delaware River, Philadelphia, PA (winter)	DOP	3—5	33
Surface water: Delaware River, Philadelphia, PA (summer)	DOP	0.06—2	33
Surface water: Delaware River, Philadelphia, PA (winter)	BBP	0.4—1	33
Surface water: Delaware River, Philadelphia, PA (summer)	BBP	9.3	33
Surface water: Delaware River, Philadelphia, PA (winter)	DMTP	0.06	33
Surface water: Delaware River, Philadelphia, PA (winter)	DMTP	nd	33
Surface water: Netherlands	DEHP	0.5—4	26
Surface water: Netherlands	DBP	0—1.5	26
Surface water: Tama River, Tokyo, Japan	DEHP	0.5—4.4	34
Surface water: Tama River, Tokyo, Japan	DnBP	0.4—5.6	34
Finished drinking water: 30 Ontario treatment plants	Phthalate	tr—9.0	35
Groundwater: Guelph, Ontario	Phthalate	4.4	35
Groundwater: Norman, OK (near landfill site)	DEP	4.1	36
Groundwater: Norman, OK (near landfill site)	DiBP	0.1	36
Groundwater: Norman, OK (near landfill site)	DCHP	0.2	36
Groundwater: Norman, OK (near landfill site)	DOP	2.4	36
Raw drinking water: 10 U.S. cities	DBP	tr—5	37
Raw drinking water: 10 U.S. cities	DEP	tr—1.0	37
Raw drinking water: 39 public water system wells, New York State	DEHP	nd—170	38
Raw drinking water: 39 public water system wells, New York State	DnBP	nd—470	38
Raw drinking water: 39 public water system wells, New York State	DEP	nd—4.6	38
Raw drinking water: 39 public water system wells, New York State	BBP	nd—38	38
Brackish Water			
Brackish water: Mississippi Delta	DEHP	0.07	39
Brackish water: Mississippi Delta	DnBP	0.10	39

Table 6 (continued)
ENVIRONMENTAL LEVELS OF PAEs IN AQUATIC MILIEUX[2,22]

Type of water and location	Compound	Concentration (µg/l)	Ref.
Brackish Water			
Brackish water: Gulf Coast (Gulf of Mexico)	DEHP	0.13	39
Brackish water: Gulf Coast (Gulf of Mexico)	DnBP	0.07	39
Saltwater			
Saltwater: open Gulf of Mexico	DEHP	0.08	39
Saltwater: open Gulf of Mexico	DnBP	0.09	39
Freshwater and Marine Sediments µg/kg			
Lake sediments: Lake Ontario near mouth of Niagara River; 0 to 3 cm depth	Phthalate	65	40
Lake sediments: Black Bay, Lake Superior	DEHP	200	27
Lake sediments: Black Bay, Lake Superior	DnBP	100	27
Lake sediments: Lake Superior, 28 sites	DEHP	nd—1.5	41
Lake sediments: Lake St. Clair	DEHP	3.8—5.3	31
Lake sediments: Nipigon Bay, Lake Superior	DOP	0.7	30
Lake sediments: Lake Erie, Detroit River	DBP	3—6	31
Lake sediments: Lake Erie, Detroit River	DEHP	1—5	31
Lake sediments: Munising Harbor, MI	Phthalate	4,100	42
Lake sediments: Lake Michigan	DEHP	nd—218,000	43
Lake sediments: Lake Michigan	DBP	25,000—120,000	43
River sediments: settleable solids from 3 Michigan rivers	DEHP	≤1,000—25,000	44
River sediments: Upper Saginaw River, MI	BBP	567 (ave.)	13
River sediments: Rhine River, Netherlands	DEHP	7,000—71,000	26
River sediments: Rhine River, Netherlands	DnBP	nd—16,000	26
River sediments: Ijssel River, Netherlands	DEHP	3,000—36,000	26
River sediments: Ijssel River, Netherlands	DnBP	nd—8,000	26
River sediments: Meuse River, Netherlands	DEHP	1,000—17,000	26
River sediments: Meuse River, Netherlands	DnBP	nd—2,000	26
River and estuarine sediments: River Mersey and Mersey Estuary, U.K.	Phthalate	48—11,400	45
Estuarine sediments: Mississippi Delta	DEHP	69	39
Estuarine sediments: Mississippi Delta	DnBP	13	39
Coastal sediments: Gulf Coast, Gulf of Mexico	DEHP	6.6	39
Coastal sediments: Gulf Coast, Gulf of Mexico	DnBP	7.6	39
Ocean sediments: open Gulf, Gulf of Mexico	DEHP	2.0	39
Ocean sediments: open Gulf, Gulf of Mexico	DnBP	3.4	39
Biota			
Fish: walleye, Black Bay, Lake Superior	DEHP	800	27
Fish: burbot, Straits of Mackinac and Goderich, Ontario	Phthalate	10—100	46
Fish: lake trout, Lake Superior	DEP	250	47
Fish: lake trout, Lake Superior	DBP	1,700	47
Fish: lake trout, Lake Superior	DEHP	150	47
Fish: lake trout, Lake Superior	DEP	1,320	47
Fish: lake trout, Lake Superior	DBP	40	47
Fish: lake trout, Lake Superior	DEHP	440	47
Fish: channel catfish, MS and AR	DEHP	1,000—7,500 (3,200 mean)	27

Table 6 (continued)
ENVIRONMENTAL LEVELS OF PAEs IN AQUATIC MILIEUX[2,22]

Type of water and location	Compound	Concentration (µg/l)	Ref.
Biota			
Fish: channel catfish, Fairport National Fish Hatchery, IA	DEHP	400	27
Fish: channel catfish, Fairport National Fish Hatchery, IA	DnBP	200	27
Fish: fish available to Canadian consumers: Fairport National Fish Hatchery, IA	DEHP	0—160	48
Fish: fish available to Canadian consumer: Fairport National Fish Hatchery, IA	DnBP	0—78	48
Fish: yellow perch, Spirit Lake, IA	DnBP, DEHP	nd	49
Fish: brook trout, Clover Leaf Lake, CA	DnBP, DEHP	nd	49
Marine biota: mostly fish, Gulf of Mexico	DEHP	4.5	39
Marine biota: mostly fish, Gulf of Mexico	DnBP	≤0.1	39
Freshwater invertebrates Dragonfly naiads: Fairport National Fish Hatchery, IA	DEHP	200	27
Freshwater invertebrates Dragonfly naiads: Fairport National Fish Hatchery, IA	DnBP	200	27
Fish: hatchery-reared juvenile Atlantic salmon (lipid)	DEHP	13,000—16,000	50
Oceanic invertebrates: deep sea Jellyfish, North Atlantic (lipid)	Phthalate	13% of Lipid content	51
Freshwater vertebrates: tadpoles, Fairport National Fish Hatchery, IA	DEHP	300	27
Freshwater vertebrates: tadpoles, Fairport National Fish Hatchery, IA	DnBP	500	27
Freshwater vertebrates; double-crested cormorant and herring gull egg yolks (lipid basis)	DnBP	11,000—19,000	50
Oceanic vertebrates; Common seal pup blubber (lipid basis)	DEHP	11,000	50
Commercial fish food: (lipid basis)	DEHP	8,000—9,000	50
Commercial fish food: (lipid basis)	DnBP	5,000—8,000	50

NOTE: nd = non-detectable; tr = trace.

A number of reports have focused on acute and chronic toxicity as well as chronic lethality and acute lethality of specific PAEs to a variety of aquatic species. However, as Pierce et al.[2] have pointed out, the lack of initial or sustained solubility of PAEs in water has forced researchers to resort to use of emulsifiers or solvent aids to achieve aqueous concentrations sufficiently high to elicit toxicological responses. Depending on the agents used to this end (some of which may be toxic themselves) unrealistic results may be obtained. While there is some justification for assuming that PAEs can attain elevated aqueous concentrations in natural waters as a result of complexation with soluble fulvic acid,[27] there appears to have been little work done to quantify ancillary ramifications of this as far as synergistic/antagonistic effects, types of fulvic acid, other chemical factors and actual PAE/fulvic acid concentrations are concerned.

Few investigators have described their toxicological methodologies and confusion exists as to the actual concentrations used. Contamination during testing cannot always be ruled out when dealing with sensitive organisms, given that certain aquarium apparatus may include PAE plasticizers. The number of detailed chronic and subchronic studies relating to specific PAEs is still not large.[2] Controlling factors such as pH, temperature, water hardness, and the physicochemical states of the dissolved or solubilized PAEs have not been studied

Table 7
ACUTE TOXICITY OF PHTHALATES TO
MACROINVERTEBRATES AND FISH

Species	Chemical	96-h LC_{50}/EC_{50} (µg/l)	Ref.
Cladocera, *Daphnia magna*	BBP	92,300	62
	BBP	3,700	13
	DEP	52,100	62
	DMP	33,000	62
	DEHP	11,100	62
Scud, *Gammarus pseudolimnaeus*	DnBP	2,100	53
	DEHP	≥32,000	59
Midge, *Chironomus plumosus*	DnBP	4,000	63
Rainbow trout, *Salmo gairdneri*	BBP	3,300	13
	DnBP	1,200	64
	DnBP	6,470	53
	DEHP	540,000	64
Fathead minnow, *Pimephales promelas*	BBP	5,300	13
	BBP	2,100	13
Bluegill, *Lepomis macrochirus*	DnBP	1,300	53
	BBP	43,300	62
	BBP	1,700	13
	BEHP	≥77,000	62
	DEP	98,200	62
	DMP	49,500	62
	DnBP	730	53
	DnBP	1,200	62
Channel Catfish, *Ictalurus punctatus*	DnBP	2,910	53
Crayfish, *Orconectes nais*	DnBP	≥10,000	53

From Lupp, M. and McCarty, L. S., Guidelines for Surface Water Quality, Vol. 2, Organic Chemical Substances, Phthalic Acid Esters, Environment Canada, Inland Waters Directorate, Water Quality Branch, Ottawa, 1983. With permission.

extensively, even though some indications of their importance in PAE toxicology have been intimated.[2,13] Nevertheless, the range of lethal toxicity in aquatic organisms based on DnBP is about 1.0 to 10. mg/l over 96 h. Lethal thresholds have not frequently been cited, though for rainbow trout *(Salmo gairdneri)* a lethal threshold of 0.5 mg/l for DnBP has been given.[2] Acute toxicity levels for DMP, DEP, and DnBP appear to have similar values while the higher molecular weight esters, DEHP and DnPP, appear to be similar and much less toxic.

Table 7 provides literature data on chronic toxicity levels. Reported effects include reduced hatchability of larvae of brine shrimp,[54] reduced growth rates in certain microbes,[55-57] heartbeat depression in goldfish,[58] reproduction depression (reduced egg laying) in water flea *(Daphnia magna)*,[59] alteration of vertebral collagen content of bone and hydroxyproline content of collagen in rainbow trout fry, adult brook trout *(Salvelinus fontinalis)*, and fathead minnow *(Pimephales promelas)* fry,[60] interference of steroid hormone biosynthesis in male Atlantic cod *(Gadus morhua)*,[61] decreased serum potassium levels in coho slamon *(Onchorhynchus kisutch)*,[27] tetanus in zebra fish *(Brachydemia revio)* (possibly related to a modified calcium metabolism), and reduced fry survival in zebra fish.[53]

2. Aquatic Flora

Few data are available on toxicity levels for aquatic plants. The range in EC_{50} (concentration at which an effect is noted in 50% of the test subjects) is wide, as shown in Table 8.

Table 8
TOXICITY OF PAEs ON AQUATIC FLORA

Species	Chemical	Effect	96-h $EC_{50}(\mu g/l)$	Ref.
Selenastrum capricornutum	BBP	Chlorophyll *a*	110	62
	BBP	Cell number	130	62
	BBP	Cell number	400	13
	DEP	Chlorophyll *a*	90,300	62
	DEP	Cell number	85,600	62
	DMP	Chlorophyll *a*	42,700	62
	DMP	Cell number	39,800	62
Microcystis aeruginosa	BBP	Cell number	1,000,000	13
Navicula peliculosa	BBP	Cell number	600	13

From Lupp, M. and McCarty, L. S., Guidelines for Surface Water Quality, Vol. 2, Organic Chemical Substances, Phthalic Acid Esters, Environment Canada, Inland Waters Directorate, Water Quality Branch, Ottawa, 1983. With permission.

3. Other Fauna Utilizing Aquatic Ecosystems

Higher vertebrates making extensive use of aquatic habitat may also be exposed to PAEs from this source. Very few data are available on the toxicity of aqueous PAEs to waterfowl, semiaquatic furbearers, herpetofauna, or livestock which either utilize or dwell in water subject to PAE contamination. Belisle et al.[65] fed mallard ducks a diet containing 10 mg/kg DBP or DEHP but found no accumulation during a five-month period. No residues of DBP were detected while occasional low levels of DEHP (about 0.15 mg/kg in lung and breast) were determined.[65]

While nontrivial concentrations of certain phthalate esters have been reported from certain livestock tissues, most studies have not focused on waterborne PAEs but rather dietary sources. Some influences of the following nature (related to dietary sources) were found in domestic fowl: food consumption decrease;[66] lowered egg hatchability;[66] adult plasma lipid and cholesterol level declines;[67] liver weight, liver lipid, liver cholesterol increases;[67] and specified muscle fat content declines.[67]

E. Biodegradation, Accumulation, and Elimination Considerations
1. Biodegradation

Pierce et al.[2] have summarized data on PAE biodegradation; a general similarity of catabolism and excretion is apparent in a wide range of biological system. The following four basic reaction sequences were described:

- Hydrolysis of the ester
- Benzene ring cleavage
- Oxidation of the liberated alcohol moiety
- Oxidation of the alkyl side chain while bound to the phthalate ester

The catabolic pathways are shown in Figure 2.

Only a few fungi and bacteria are apparently able to degrade PAEs completely, including benzene ring cleavage.[2] Natural populations of some of these organisms occur in water and sediments, as well as in soil and waste disposal situations. In most higher vertebrates an initial hydrolysis of the diester (effected by nonspecific esterases[68]) is the only catabolic reaction, after which glucuronate conjugation and complex excretion follow.[2] A second hydrolysis yielding phthalic acid may occur, followed by the same conjugation/excretion process.[69,70] Some of the alcohols released during hydrolysis may be toxic and may or may

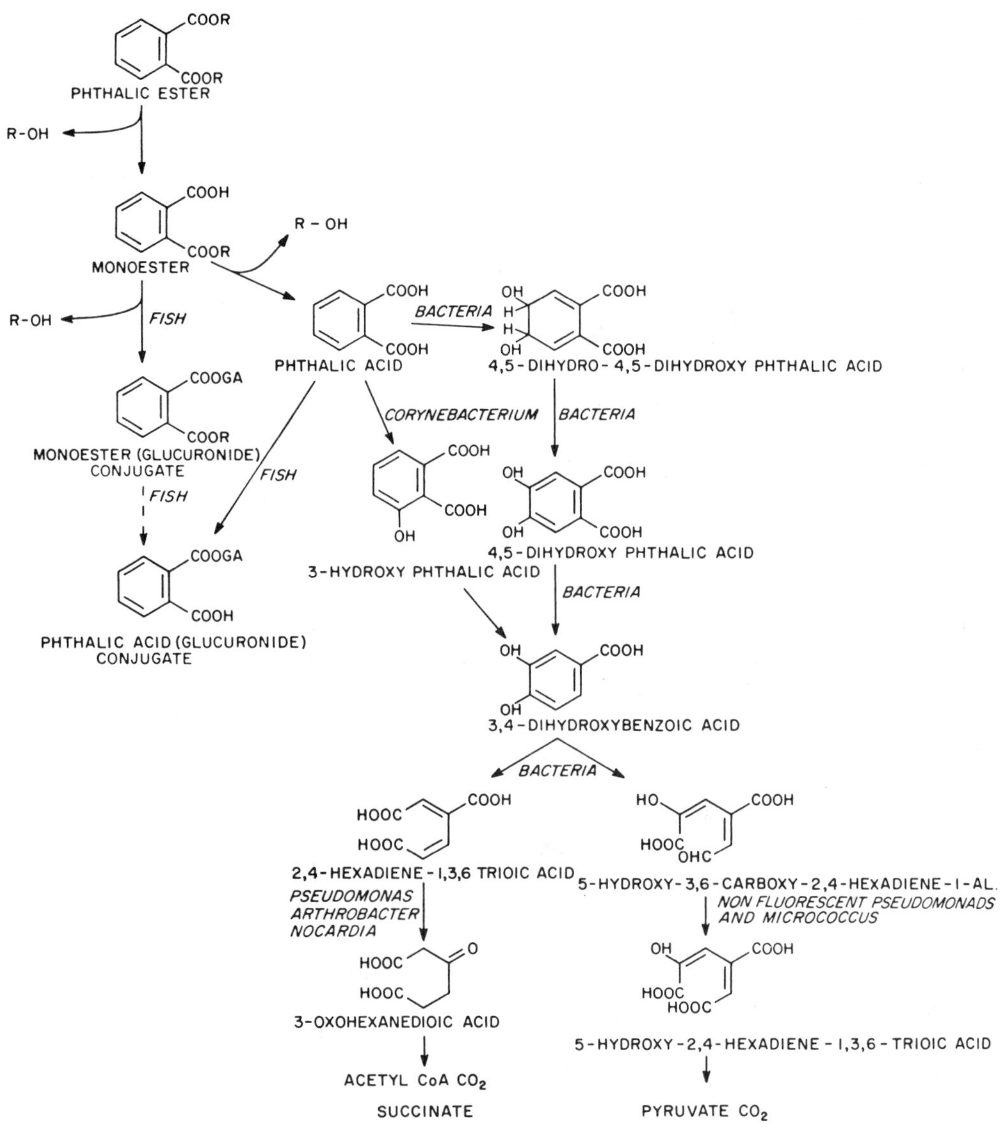

FIGURE 2. The catabolic pathways of phthalate esters. (From Pierce, R. C., et al., NRCC Publication No. 17583, 1980.

not be further metabolized by the host. Rates of PAE biodegradation appear to be rapid. Many bacteria and fungi require a lag phase prior to induction of the required biodegradation enzyme systems when newly exposed to significant levels of PAEs. A few genera and species of bacteria (e.g., *Serratia*, *Marcescens*, *Pseudomonas testosteroni*, and *Micrococcus* spp.)[71] apparently have the necessary cellular chemistry to directly biodegrade PAEs. Not all PAEs induce the same enzymes in a given strain of organism. Pierce et al.[2] indicated that over 100 different bacteria and fungi have been isolated from plastics exposed to the environment and tested for PAE biodegradation capabilities. Only a few of these partially or completely catabolize PAEs.[2] Phthalate esters are apparently less readily metabolized by microbes than are other components of certain plastics. The effect of specific PAE and its structure on biodegradability has yet to be clarified in most species.

Although data are limited, PAEs with shorter and partially oxidized side chains, are more readily degraded than those with long, branched side chains. Environmental factors of temperature, nutrient availability, and pH, influence degradation rates in microbes but their influence in vertebrate biodegradation is unknown.

The bacterium *Pseudomonas acidovorans* was found to degrade a 0.3% DEHP culture solution in 52 h and a 0.5% one in 72 h.[72,73] Stalling et al.[49] and Melancon and Lech[69] found an 85% reduction of DEHP during 24-h continuous exposure (static test) studies using, respectively, channel catfish *(Ictalurus punctatus)* and rainbow trout. Tests of the rates of disappearance of PAEs in aquatic systems led Johnson and others[74-76] to suggest that aquatic and sediment flora hydrolyze the ester linkage and decarboxylate the phthalic acid moiety. With ^{14}C ring-labeled DEHP, 20% of radioactivity (as [14]CO_2) was released in 28 days in a hydrosoil medium, indicating a substantial degree of complete mineralization. Hattori et al.[77] found that DnBP, DMP, DEP, and DiBP (each at 25 mg/l) were completely decomposed upon incubation with river water for 4 to 10 days while DEHP was only partially decomposed even after 2 weeks. They reported similar findings using seawater. Biological degradation of DEHP and BBP has been reported[1] using a semicontinuous activated sludge system where 91 and 99% of the two receptive esters were degraded in 48 h at a charge rate of 5 mg/48 h.

2. Bioaccumulation and Biomagnification

Bioaccumulation occurs when uptake by an organism is faster than the combined rate of degradation and elimination. Biomagnification occurs when consumers achieve higher concentrations of a substance than their prey, resulting from the feeding process. Continuous exposure to PAEs lead to bioaccumulation in the few species of aquatic organisms and for the limited number PAEs thus far examined.

Aquatic invertebrates appear to lack the biodegradation ability demonstrated (for DEHP) in several fish species.[2] However, it is not certain the extent to which this uptake represents cellular tissue vs. exoskeleton accumulation. Invertebrates, as common prey organisms, may thus afford an important means of PAE transfer in the food chain. However, biomagnification in fish, the most prominent aquatic invertebrate consumers, appears unlikely insofar as the bulk of certain PAEs is rapidly metabolized in fish. Within fish, a two-compartment model has been invoked to account for rapid and slow catabolism/excretion. Most of the PAEs taken up are readily eliminated, while under continuous exposure, smaller quantities are less readily eliminated and may accumulate. At a certain time (56 days for fathead minnows in one study)[78] an equilibrium concentration is apparently reached in fish, but the relationship between PAE water concentration and fish body concentration may not always be a simple or linear one. For aquatic invertebrates a similar equilibrium is apparently reached in only a few days, suggesting either a very slow or negligible biodegradation rate. After exposure to DEHP a linear rate of elimination has been observed in fathead minnows; a biological half-life of 12 days for DEHP was computed from this, while that of the prime metabolite MEHP, was 5.7 days.[78]

Accumulation of PAEs may be specific to certain organs or other body components in some species or in some stages of species life cycles. For example, 96% of DEHP in trout eggs was associated with the chorion while only 4% was in the ovum.[79] Accumulation appears also to be specific to the ester in question. The magnitude of bioaccumulation appears to correlate with the octanol/water partition coefficient (Log P_{ow}) (Table 1).[78] Short-chain PAEs with Log P_{ow} values of 1.5 to 2.2 appear to have little bioaccumulation potential while longer side-chain esters with Log P_{ow} of 3 to 4 may bioaccumulate 100- to 1000-fold compared to water concentrations.[2]

F. Human Health Implications

Because of the low solubility of PAEs in water and the comparatively high quantities of PAEs required to elicit responses in human subjects or in other mammals (e.g., mice, rats, guinea pigs) regularly used to infer human toxic response levels, there is virtually no reason to presume that there are acute effects resulting from human contact or consumption of water saturated with PAEs. Chronic or subacute effects may occur but data are inconclusive.

Toxic responses have been reported for various PAEs but rarely has the environmental medium been aqueous. Indirectly, several potentially toxic routes may be traced whereby PAEs in water could conceivably form the original source of toxicity. Examples include consumption of foods such as fish or aquatic plants which have accumulated PAEs from the water. Alternatively, toxic responses may be due to intake of PAEs from several sources which may in part include aqueous sources. Other sources include inhalation, dermal uptake, medical uptake (via leaching of PAEs from blood bags or during hemodialysis, etc.), and uptake from foods.

Drinking water may have PAE contamination at its source or during transport, handling, or storage (e.g., via PVC water pipes and plastic containers).[4] Placed in perspective, Leah[18] estimated that the average Canadian receives a daily dose of 85 μg of PAEs — 80 μg from food intake, 3 μg from inhalation and 2 μg from water consumption. Health and Welfare Canada[35] estimated a worst case daily intake of 230 μg — 192 μg from food, 18 μg from air, and 20 μg from water. These values (average daily and worst case daily) translate to intake rates of 1.2 and 3.3 μg/kg body weight for a 70 kg man.[3] In comparison, rat LD_{50} values ranges from 1.7 to over 68 g/kg body weight for various PAEs — i.e., about 6 orders of magnitude higher.

In laboratory animals, acute oral toxicity (LD_{50}) has been shown to be related in a statistically significant linear regression analysis, to PAE molecular weight, with lowest molecular weight esters being more toxic. Converting these regression data using the USEPA method,[81] estimated human LD_{50} values (70 kg man basis) are 48 g for DMP, 421 g for DEHP, and 872 g for DTDP.

Long term accumulation of PAEs does not appear to be a problem in mammals. Phthalates are apparently metabolized rapidly by humans and laboratory animals and are excreted in the urine, feces, and bile.[4,83,84] The liver, pancreas, and gut have substantial phthalate hydrolase activity and there does not appear to be any singly high specific target organs or sites for accumulation.

Factors influencing toxicity include ester water solubility and measures of solubility in organic solvents (Log P_{ow}). Water solubility is generally inversely related to the length of the alcohol side chains. In addition, however, toxicity and solubility are related to the specific structure of these side chains, and solubility data have not been rigorously determined for all PAEs.

Subacute and chronic toxicity studies reported in the literature, including those on teratogenicity, carcinogenicity, and mutagencity, in addition to variously reported physiological effects, have been largely based on tests with small mammals. Methods of PAE application include oral and nonoral means, the latter of which may have limited application to human effects derivation. Other factors such as use of carrier compounds and variables related to dose and time period for exposure, preclude a comprehensive summary of probable human implications. Frequently reported symptoms in laboratory animals include liver enlargement with changes in activity of certain enzymes, changes in the microstructure of hepatocytes, inhibition of lipid metabolism, enlargement of the kidney, hematuria, kidney morphology changes, testicular atrophy and accompanying tissue damage, reduction in growth rate, occasional loss of body weight, swelling of heart and brain, increased gamma globulin levels, and decreased food consumption rates.[3] Limited data reveal that many symptoms disappear after cessation of treatment.[16,83-88]

Environment Canada[3] reviewed a number of studies and observed that for various esters, effects were noted with administration levels of 0.26 to 11.2 mmol per kilogram body weight per day, but that no-effect dosages reported in the most recent literature were being revised downward, possibly due to improved effect detection procedures. There appears to be no consistent ester-specific concentration related differences in subacute/chronic toxicity, unlike the situation for acute toxicity. Finally, they observed that the lowest no-effect dose estimate (for DEHP) is 0.13 mmol per kilogram body weight per day, so that a value of 0.1 mmol per kilogram body weight per day is probably a conservative estimate for a threshold value for all esters.

Data listed in Table 6 for PAE levels reported in biota could be of importance relative to human toxicology, should contaminated organisms (notably fish) form a significant portion of human diet. Fish intake by reference or standard man is relatively small but in certain geographic areas freshwater fish form a significant part of the diet. Few or no computations of potential doses have been given in the recent literature.

III. CRITICAL REVIEW OF ANALYTICAL METHODS

A. Review of Literature Methods
1. Sampling and Sample Storage

There has been little information available on sampling and storage of samples containing PAEs. However, it is well known that PAEs undergo vaporization, adsorption, hydrolysis, and biodegradation. Several PAEs also occur as background contamination given the widespread use of PAEs in everyday articles, in the ecosystem, and laboratory environment.

Bowers et al.[109] compared the loss of PAEs and other organic compounds, during evaporation-reconstitution, in organic environmental sample extracts and standard solutions. DMP was entirely lost and large losses occurred for DEP during the evaporation-reconstitution. Shouten et al.[106] observed that the PAE levels, especially DEHP and DBP, in river waters rapidly decreased with time and the resultant loss was attributed to biodegradation. The authors performed extraction at the sampling station to prevent losses of PAEs. Sullivan et al.[102,104] reported significant losses of PAEs from seawater during storage, as a result of adsorption; the hydrolysis of PAEs was investigated by Wolfe et al.[103] Fishbein and Albro[110] and Hartung[111] reported sample contamination by PAEs in a variety of matrices. The contamination is believed to occur during sample processing, due to ubiquitous presence of PAEs with special reference to impurities originating from commercial reagents and equipment commonly used in laboratories.[138] Sample contamination can also occur during sample collection by unskilled or untrained personnel involved in sampling due to the ease by which PAEs can be transferred from fingers to glassware.[92] The use of plastic containers, plastic tubing and connections can also contribute to the potential contamination during sampling.

Standard sampling procedures outlined by ASTM,[89] Garrison,[90] and Wilson[91] should be used in collecting samples for PAE analysis. The chapter of this book dealing with PCBs also covers the subject of sampling and storage for organic contaminants. The procedures for sampling various matrices are briefly outlined below:

a. Water

Care must be taken to collect a representative water sample. Glass or Teflon containers are recommended for the collection of water samples and the containers should be filled to the top to minimize headspace so as to minimize loss due to volatization. Teflon or aluminum foil (dull side faced up) capliners are recommended. Ideally, water samples should be extracted on site to minimize loss of PAEs. Otherwise, water samples placed in properly sealed containers should be kept at low temperature (approximately 4°C) in the dark and extracted as soon as the samples arrive at the laboratory. The inside of the container should be rinsed with the appropriate extraction solvent in order to recover adsorbed phthalates and the resultant rinse added to the extract prior to analysis.

b. Sediments

Wet representative surface samples or sediment cores should be collected in precleaned mason jars with Teflon or aluminum capliners. The samples should be frozen until analysis.

c. Fish

Fish or other biota should be wrapped in solvent-cleaned aluminum foil or placed in precleaned glass jars and frozen until analysis. The skin or shells of fish should be removed, when possible, and discarded before analysis.

2. Extraction

Pierce et al.[2] have reviewed the isolation and extraction of water, biota, and sediment samples prior to PAE analysis. The type of sample and analytical technique generally dictates the method used for isolation and concentration of PAEs. Procedures for extraction and concentration of PAEs from water samples employ both liquid-liquid extraction (LLE) with organic solvents and solid absorbents such as XAD-resins, activated carbon, polyurethane foams, and n-octadecyl trichlorsilane treated beads (ODS-columns). Soxhlet extraction, ultrasonic extraction, and use of various blending, tumbling, and LLE have been employed for sediments. Biota and fish are first homogenized in a blender followed by blending with anhydrous sodium sulfate prior to soxhlet extraction using organic solvents. Direct solvent extraction is also employed for isolation of PAEs from biota and fish samples.

a. Water

Liquid-liquid extractions are carried out in a separatory funnel or by stirring sample and organic solvent with a Teflon® stir bar. Custom designed water extractors, employing large volume water samples — up to 200 l, have been used for lowering the detection limits of PAEs and other organic contaminants in natural waters.

The solvents used for the extraction of water include n-hexane,[46,107] n-heptane,[108] petroleum ether,[99] dichloromethane,[32] chloroform,[28,30] and mixture of solvents.[112] Dichloromethane was found to be superior to other nonpolar solvents for extraction of PAEs from water.[108]

Solid adsorbents are also used as alternatives for isolation of PAEs from water samples, XAD-resins having been used in the analysis of PAEs and other organic contaminants.[93-96] The recovery efficiencies of XAD-resins (-2, -4, -7, -8,) and resin mixtures were determined using water containing PAEs and other organic pollutants. The results indicated that these resins were not effective for quantitative extraction of PAEs.[97] Activated carbon is reported to be effective for extraction of PAEs; however, the recovery of the adsorbed material is not quantitative.[100] Porous polyurethane foams have been successfully used to remove some phthalate esters from water at the part-per-million level; however, the foams were not able to retain the larger molecular weight PAEs such as DEHP and DnOP.[100] The ODS-Columns were also employed for trace enrichment of PAEs which were easily recovered from the columns with organic solvents.[118]

b. Sediments

Soxhlet extraction and Polytron® homogenization, using various solvents and/or solvent systems, have been successfully employed for extraction of sediments. Giam[99] utilized Polytron® homogenization and acetonitrile for initial extraction. The resultant acetonitrile was then mixed with 5% aqueous sodium chloride solution and the PAEs were partitioned using a methylene chloride-petroleum ether (1:10 V/V) mixture prior to cleanup and analysis. Schwartz et al.[26] extracted freeze dried sediment samples using Soxhlet extraction with a n-hexane-acetonitrile-methanol (8:1:1 V/V) mixture. The Soxhlet extractor, using appropriate solvent and/or solvent mixture, is considered as a preferable technique since it is known to be the most exhaustive procedure for extracting PAEs and other trace organic contaminants from sediments.

In Soxhlet extraction, a sufficient amount of precleaned celite is added to the sintered glass thimble in order to prevent the sintered disc from becoming plugged with fine particles of sediment samples and to prevent the disc from being contaminated. A known amount of freeze-dried or wet sediment is then added and the resultant mixture is extracted with suitable solvent and/or solvent mixture for a period of 2 to 6 h. In some cases wet sediments are mixed with anhydrous sodium sulfate prior to extraction since excess of water is known to affect the extraction causing low and varying recoveries.

c. Biota Samples

Various tumbling, blending, and liquid-liquid extractions have been utilized to extract PAEs and their metabolites in fish and other biota samples.[48,49,122,123] Williams[48] employed hot *n*-hexane with occasional stirring to extract PAEs from frozen and canned fish samples. Takeshita et al.[123] mixed biological samples with anhydrous sodium sulfate prior to extraction with *n*-hexane using Waring blender. Giam et al.[39,99,105] extracted PAEs from macerated biota samples by blending with acetonitrile. The acetonitrile was mixed with 5% aqueous sodium chloride solution and extracted with methylene chloride-petroleum ether (1:5 V/V) to isolate PAEs.

Stalling et al.[49] extracted PAEs and their metabolites by grinding a fish sample in a blender with anhydrous sodium sulfate (1:4 W/W). The mixture was packed into a glass column (1 to 2 — cm I.D.) and extracted with 1% phosphoric acid in acetone. Kaiser[115] extracted organic residues including PAEs with *n*-hexane: ether (2:1 V/V) mixture using a Polytron® homogenizer. The homogenization with Polytron® and extraction with suitable organic solvent is considered a preferred technique for extracting PAEs and their metabolites from biological samples. Phthalate ester metabolites are extracted under acidic conditions.

3. Cleanup Procedures

Several procedures are used for cleanup of the extracts from various environmental matrices. The cleanup procedure depends upon the nature of the sample and the method of quantitation employed. The data indicates that the use of column chromatography using adsorbents such as Florisil, alumina, and silica gel are the procedures of choice over liquid/liquid partition.

Water extracts usually have low concentrations of interfering substances thus requiring single step cleanup procedure.[96,99] However, sediments and biological extracts require multistep cleanup procedures. These may include use of appropriate techniques to remove large concentrations of major extractants such as lipids, fulvic/humic acids and sulfur containing compounds followed by column chromatography to fractionate PAEs prior to quantitative analysis.

a. Removal of Bulk Coextractants

Biota and sediment extracts contain large concentrations of lipids, fulvic/humic acids, and sulfur-containing compounds. The majority of cleanup columns, using solid adsorbents, have limited capacity to retain high concentration of the above coextractants. In many cases, a certain concentration of the coextractants pass through the cleanup columns causing interference with gas chromatographic determination of PAEs. Therefore, a partial cleanup is required to remove the bulk coextracts before column chromatography for the cleanup of sediment and biota extracts.

Lipid removal can be achieved by the use of commercially available adsorbents such as Micro Cel-E (a synthetic calcium silicate) possessing high adsorptive capacity for lipids.[116,117] Liquid-liquid partitioning has also been employed using acetonitrile, dimethyl formamide, and concentrated sulfuric acid to remove lipids prior to cleanup and analysis of trace organics in biota extracts. These techniques were not found suitable to remove lipids and obtain a selective fractionation of PAEs. Stalling et al.[49] and Giam et al.[105] found that partitioning of organic extracts, containing lipids and PAEs, with acetonitrile and dimethyl formamide did not selectively remove lipids. DEHP coextracted with lipids when the *n*-hexane extract was partitioned with acetonitrile. Similarly, certain PAEs coextracted with lipids using the petroleum extract and dimethyl formamide.

In addition, the partitioning of diethyl ether and dimethyl formamide yielded difficult emulsions resulting in varying recoveries of PAEs. Zitko[122] found that PAEs undergo condensation in the presence of concentrated sulfuric acid producing fluorescent phthalic anhydride and, therefore, the sulfuric acid treatment cannot be used for removal of lipids as a partial cleanup step during the determination of PAEs by gas chromatography.

Gel permeation chromatographic technique for the separation of PAEs from lipids and other large molecular weight compounds is a preferred technique used for fish extracts. Baker[125] separated phthalate esters from lipids using gel filtration on a Bio-Bead SX-2 (copolystyrene, 2% divinyl benzene) resin. Stalling et al.[49] found gel permeation chromatography to be an efficient and reproducible technique for removing lipids from fish extracts using a Bio-Bead SX-2 column with cyclohexane as eluent. Burns et al.[119] also separated lipids and phthalate esters using a Bio-Bead SX-3 column with 1:1 methylene chloride-cyclohexane. The authors found that PAEs eluted earlier than the organochlorine pesticides and a small percentage of total fish lipid as carried into the phthalate fraction. However, subsequent cleanup of the extract on Florisil was effective in fractionation of PAEs prior to quantitative analysis provided the lipid content was maintained below a critical level.[105]

b. Removal of Sulfur Compounds

Metallic mercury has been found effective in removing sulfur-containing compounds from the organic extracts. Giam et al.[105] found that when the sample extracts were shaken with metallic mercury, better chromatograms were obtained because of the removal of EC active sulfur-containing compounds.

c. Adsorption Column Chromatography for Cleanup

Chlorinated hydrocarbons including the DDT family and polychlorinated biphenyls are present in almost all marine biota samples and interfere with the gas chromatographic analysis of phthalate esters using electron capture detection.[105] Organochlorine pesticides and polychlorinated biphenyls have a higher electron capture response than phthalates and, therefore, an efficient technique is required to separate the phthalate esters from interfering organochlorine residues.[45,105]

Deactivated alumina has been employed for the cleanup of water and sediment extracts. A majority of PAEs are eluted in the 15% diethylether-hexane fraction.[45] Alumina deactivated with 5% water has been used for the cleanup of fish extracts with the phthalate esters eluting in the 10% diethylether-hexane fraction.[122] However, these columns have limited capacity for lipids and may not be able to completely separate lipids from other organochlorine residues.[123]

Activated Florisil has been used for the cleanup of water, sediment, and biota extracts. All PAEs were eluted in 15% diethylether in petroleum ether fraction. Giam et al.[99,105] found that more reproducible and quantitative fractionation of PAEs was achieved when Florisil was deactivated with 3% water and the column eluted with 15% diethylether in petroleum ether.

Thin layer chromatography on silica gel[45] and high performance liquid chromatography on μ-Porasil[96] have also been employed for the isolation of phthalate esters from environmental extracts.

4. Quantitative Analysis

Spectrophotometry,[126] fluorescence,[122] thin layer chromatography,[127] gas liquid chromatography,[124] and high performance liquid chromatography[32] have been employed for quantitative analysis.

a. Gas Chromatography

Gas Chromatography is commonly used for the quantitative analysis of PAEs. Isothermal column conditions are normally employed during the gas chromatographic analysis. Both packed columns and capillary columns, with nonpolar stationary phases such as SE-30, OV-101, OV-17, are employed for obtaining the desired resolution. Quantitative analysis is carried out using a number of detectors. These have included thermal conductivity,[128] flame ionization,[129] photoionization,[130] and electron capture.[131] Although both flame ionization[34,48,122] and electron capture[105,45,124,27,119] have also been employed for the gas chromatographic analysis of trace quantities of phthalates in environmental samples, electron capture detection is preferred over flame ionization detection because of its increased sensitivity[105,132] as well as its specificity.[105,133] Gas chromatography-mass spectrometry is also reported to be a very sensitive and selective method for analysis and confirmation of PAEs. Amounts as low as 10 pg per injection may easily be detected using single ion monitoring whereas the typical detection limit of flame ionization detection is near 1 ng per phthalate ester injection.[118]

b. High Performance Liquid Chromatography

High performance liquid chromatography offers a great potential for environmental analysis and is reported to have an advantage over gas chromatography, for several reasons. A wide variety of compounds can be analyzed ranging from volatile to nonvolatile with molecular weights differing by as much as 3 to 4 orders of magnitude. It is nondestructive, requires less time for analysis and does not normally employ a precleanup of the extract prior to quantitative analysis.[136] Furthermore, high performance liquid chromatography also enables large volumes to be injected into the system, thereby achieving detection limits at least equivalent to those obtained using gas chromatography with flame ionization detection.[106]

The normal phase high performance liquid chromatography for PAEs is simple and rapid. However, other modes with ultraviolet (UV) absorption detectors are also employed for the analysis of PAEs. Schouten et al.[106] and Schwartz et al.[26] employed a normal phase system using silica gel/*n*-hexane-dichloromethane (1:2, V/V) containing 0.1 to 0.2% of ethanol for the analysis of PAEs in river water and sediments. Mori[107] utilized all three separation modes (normal and reversed phase chromatography and gel chromatography) with UV-detector to determine PAEs in river water. Reversed phase liquid chromatography was also successfully employed for the analysis of polar metabolites of phthalate esters using ionic suppression.[121] On-line trace enrichment, separation and quantitative analysis can simultaneously be achieved using the above technique;[134] however, care must be taken to avoid recurrent clogging of the precolumn during on-line trace enrichment. In some cases, coextractants during on-line trace enrichment, may cause interferences which can easily be removed by the use of a cleanup step prior to analysis.

5. Confirmation

Confirmation of the identity of PAEs is achieved by the use of the following techniques:

- Dual column GC-ECD[105]
- Monitoring the disappearance of phthalate esters peaks with GC-ECD following hydrolysis[119]
- Alkaline hydrolysis and derivatization with 2-chloroethylamine hydrochloride GC-ECD[137]

- GC-MS[49,124]
- Infrared spectroscopy of gas-liquid chromatography fractions[34]
- Monitoring the spectrophotofluorescence of phthalic anhydride following reaction of PAEs with concentrated sulfuric acid[122]

B. Factors Affecting Reliability of Current Methodologies

The ubiquitous nature of phthalate esters is one of the major problems in obtaining precise and accurate results for these plasticisers. Phthalate contamination of laboratory reagents and materials is recognized as a major problem in analysis of PAEs.[45,92,105,138] Airborne PAE contaminants from the out gassing of plastic floor tiles, tubings, air conditioning filters, etc., can significantly contaminate clean glassware in a relatively short period of time.[52,92,105] Minimizing this background as well as the use of distilled-in-glass solvents is essential in order to minimize the cross-contamination. Some procedures reported in the literature minimize the contamination during the sample preparation and analysis. These have included the decontamination of glassware and reagents by washing with clean solvents and/or heating in a muffle furnace.[45,105,138] Rinsing the glassware with high quality solvent just prior to its use is the most practical method to minimize cross-contamination. Muffling can be used, provided the temperature of muffling is strictly controlled, because it affects the surface properties of chromatographic adsorbents and glass surfaces.

Losses of PAEs can occur during storage of aqueous samples. The loss of phthalate esters, particularly the less polar phthalates such as BEHP, is reported from aqueous solutions during storage. The loss is attributed to the adsorption of the phthalates on the walls of glass vessels.[102]

PAEs are known to undergo biodegradation under aerobic conditions. Biodegradation of low molecular weight PAEs is reported in natural waters.[106] The addition of known microbial inhibitors did not completely eliminate the biodegradation of PAEs.[106]

Certain PAEs such as DMP and DEP are relatively volatile and can be lost during sample pretreatment. For example, the analysis of PAEs involves the use of large volumes of organic solvents during concentration and cleanup. Subsequently these extracts are evaporated to nearly dryness or reconstituted to a small volume prior to analysis and can result in loss of certain PAEs of high volatility. Some workers have reported low recoveries of PAEs, particularly of low boiling esters such as DMP and DEP, during the concentration of organic extracts.[109]

PAEs are also known to exhibit a wide range of polarities and are difficult to separate as a group using column chromatography in relatively small volumes of eluents. Similarly it is difficult to use the same isothermal GC conditions to analyze both low- and high-boiling esters due to their wide range of volatilities.[45]

C. Critical Evaluation of Literature Methods and Optimization of Methodology for PAE Analysis

In view of the problems associated with the analysis of PAEs, Ryan and Scott[113] undertook a detailed investigation to critically evaluate the current extraction, cleanup, and quantitative analysis procedures reported in the literature. Methods for minimizing sample contamination were also investigated. The authors selected six PAEs: DMP, DEP, DBP, BBP, BEHP, and DOP to represent wide range of volatilities and polarities of PAEs. Both high performance liquid chromatography-UV detection (HPLC-UV) and high resolution gas chromatography with electron capture detection (GC-ECD) were critically evaluated. An optimum method was developed for the analysis of PAEs and organochlorine pesticides including PCBs in a wide variety of environmental samples. The results of the study are summarized in the following.

1. Contamination

PAEs, particularly BEHP and DBP, were found to be contaminants in disposable glassware, solvents, adsorbents, and other materials used in cleanup. The contaminants of disposable glassware such as GC vials and Pasteur pipets were readily removed by rinsing with acetone prior to use. Solvent contamination (distilled-in-glass grade) was not found to be a problem except in the case of the BioBead SX-3 GPC eluate, which required about a 100-fold concentration. The most practical solution to the latter problem was to decrease the overall concentration factor to about 20. By far the most serious contamination problem was with the materials used for the adsorption chromatography column. The silanized glass wool, which was found to be the most heavily PAE-contaminated material, was readily cleaned with acetone rinsing prior to use. The anhydrous sodium sulfate was also readily decontaminated by washing with acetone prior to muffling overnight at 600°C.

Attempts to wash the silica with acetone were unsuccessful because it appeared to react with the acetone, particularly on oven drying at 130°C, producing a yellowed silica, laden with extractable compounds. Although the removal of PAE contaminants was accomplished using other polar solvents, such as ethyl acetate and acetonitrile, the elution profile of the PAEs from the treated silica gel column was altered using these solvents. Solvent washing appeared to decrease the activity of the silica gel as manifested by the premature elution of the least polar PAEs in the other silica fraction which was to contain the organochlorine pesticides.

The alternate approach was the prewashing of the complete silica gel cleanup column with a polar solvent just prior to its use. This was generally successful in removing PAE contaminants. However, the polar solvents that were capable of removing PAE contaminants, also caused the elution profile of the PAEs to be altered despite the rinsing of the polar solvents from the column material prior to use. Methylene chloride was found to be one solvent that did not alter the PAE elution pattern as a result of prewashing of silica gel column. However, the methylene chloride washing was capable of removing only the most nonpolar of the PAEs such as BEHP and DOP. Therefore, for the analysis of more polar PAEs such as BBP and DBP, a source of silica gel free of BBP and DBP must be used. Woelm, active silica (100 to 200 μm) packed in 500 gm quantities in aluminum cans was found to be free of BBP and DBP contamination.

The PAE contamination of neutral silica and neutral alumina, in various forms of packing and from various sources, was also evaluated. The Woelm neutral silica in aluminum cans was found to be the least contaminated of the adsorbents tested. The silica contained only traces of BEHP, easily removed by washing the cleanup column prior to use. The following procedure was found to be the most suitable. A silica column, consisting of 8 cm of 3% deactivated Woelm silica gel (from aluminum cans) was layered between the acetone washed anhydrous sodium sulfate, in a 1 cm I.D. column. Acetone rinsed silanized glass wool was used as a plug at the bottom of the column. The silica gel column was washed with 20 ml of methylene chloride and 20 ml of hexane just prior to use in order to remove background contamination of BEHP from the silica gel.

Evaporating solvent extracts or chromatography fractions to dryness under vacuum using a rotary evaporator (Buchi Rotavapor) resulted in a significant loss of the low boiling esters such as DMP and DEP. However, this loss was greatly diminished using a Buchler Vortex evaporator for the removal of solvent under reduced pressure. Therefore, for the analysis of dimethyl phthalate and diethyl phthalate, special care must be taken in removing the solvent from extracts or eluates. Solvent should not be taken to dryness using large rotary evaporators. A small volume of keeper (i.e., iso-octane) should be used where possible and/or a vortex evaporator should be employed if available.

2. HPLC-UV Methods

HPLC was evaluated for the analysis of PAEs using reverse phase, normal phase, and size-exclusion chromatography with various mobile phases and HPLC columns.

Reverse phase HPLC columns with C-18 (5 μm) and C-8 (10 μm) bonded packings were found unsuitable for PAE analysis. The large difference in the partition coefficients of the PAEs, compounded by the fact that the PAEs were eluted in descending order of their sensitivity to UV adsorbance, resulted in detection limits (based on peak height adsorbance measurements) being grossly different for the 6 PAEs of interest. The reverse phase HPLC separation of BBP and DBP was difficult and the isochratic separation of all 6 PAEs using reverse phase chromatography was impractical. Both BEHP and DOP were highly retained under the conditions required for the critical separation of BBP and DBP. Gradient elution was not pursued because of sensitivity limitations resulting from the trace enrichment and elution of solvent impurities and the change in the baseline due to the changing composition of the mobile phase.

Size exclusion HPLC was investigated using 5-μm and 10-μm styrene-divinylbenzene GPC columns with exclusion limits of 100 and 2000, and 20 and 700 Da, respectively. Results showed that this method of separation was unsuitable for the HPLC analysis of the 6 PAEs of interest because similarities in molecular size prevented the separation of DOP and BEHP as well as BBP and DBP.

Normal phase HPLC was found to be the most suitable HPLC method for the separation of the PAEs. The elution order of the PAEs was generally the inverse of that obtained using reverse phase HPLC. Normal phase HPLC out-performed reversed phase HPLC by providing fast, effective isochratic separations of the phthalates with the peak heights of all PAEs of similar magnitude. Normal phase HPLC was optimized using a Water's μ-Porasil (10 μm) column with a mobile phase of 2 ml/min methylene chloride. All 6 PAEs were well resolved with retention time variances of less than 5%.

The μ-Porasil normal phase HPLC system showed potential as a cleanup procedure. Similarly, μ-Styragel (100 Å) GPC showed potential as a size exclusion cleanup step using 1 ml/min. methylene chloride as mobile phase. The major advantage of these HPLC systems for cleanup purposes is the relatively small eluate volumes in which the PAEs are collected, such that any significant concentration of solvent PAE contaminants is avoided. The μ-Porasil HPLC system has the added advantage of overcoming the PAE contamination problem of LC adsorbents which is encountered using conventional column chromatography.

Both UV absorbance and molecular fluorescence were investigated for PAE detection following HPLC separation. An evaluation of the spectrophotometric and spectrofluorometric characteristics of the PAEs revealed that the PAEs had absorbance maxima at 224 nm and 276 nm but were nonfluorescent. The absorbance maxima at 276 nm was found to be only about 14% of that at 224 nm.

Phthalic anhydride, a condensation product of the PAEs, is known to be fluorescent. An evaluation of the spectrofluorometric characteristics of the concentrated sulfuric acid condensation product of a mixture of all 6 PAEs revealed that the emission wavelength was 352 nm and the excitation wavelengths were 268 nm and 306 nm. Excitation at 268 nm produced about twice the emission of that at 306 nm. However, this procedure did not lend itself to automation for the monitoring of HPLC effluents and therefore was not pursued for the HPLC detection of PAEs.

Both a fixed wavelength UV detector and a variable wavelength UV detector were evaluated for the HPLC detection of PAEs using the optimized μ-Porasil/methylene chloride HPLC conditions. It was found that the absorbance maxima at 224 nm cannot be used with methylene chloride as a mobile phase due to the absorbance of methylene chloride in this region resulting in high background noise. Therefore, a fixed wavelength detector using a 254 nm filter or variable wavelength detector at 235 nm was employed. The detection limits of 6 PAEs, based on a signal to noise ratio of 5:1 with a methylene chloride mobile phase and fixed wavelength detector at 254 nm and a variable wavelength detector at 235 nm, ranged from 20 to 30 ng and 6 to 8 ng, respectively.

The detection limit of the μ-Porasil/methylene chloride HPLC system using UV absorbance at 235 nm is about 3 orders of magnitude less than that of the GC-ECD method. However, the practical detection limit of the optimized HPLC system can be increased substantially to the ppb level by using 1 ml injection volumes. In order to use large injection volumes the sample must be dissolved in iso-octane or similar solvent which is sufficiently less polar than the methylene chloride mobile phase in order to facilitate the concentration of the PAEs at the head of the HPLC column. Using 1 ml methylene chloride injections resulted in a loss of resolution and the incomplete separation of the 4 least polar PAEs.

The potential interference from polychlorinated biphenyls (PCBs) organochlorine pesticides (OCs) polynuclear aromatic hydrocarbons (PAHs), phenols, carbamate pesticides, and carbamate pesticide degradation products, was investigated using the μ-Porasil/methylene chloride HPLC system. The majority of PAEs such as DEP, DBP, BEHP, and DOP were subject to interference from various phenols, carbamate pesticides, and carbamate pesticide degradation products. These compounds also interfered with a reversed-phase HPLC separation of PAEs, as did organochlorines and PAHs.

3. GC-ECD Methods

A variety of methods for the separation and analysis of PAEs using high resolution gas chromatography were evaluated. The use of nonpolar wall coated open tubular capillary columns such as DB-5 with a temperature programming was found to be the most effective technique for separation of PAEs. All 6 PAEs of interest were well resolved and readily detectable at pg levels using the conditions recommended for the analysis of PAEs as described in the later part of this chapter. The run time could be shortened in the absence of interferences without loss of resolution by increasing the rate of the second temperature ramp.

All samples require a suitable cleanup step and/or procedures prior to analysis of PAEs. Even simple matrices such as extracts from distilled water and tap water require single step silica gel column chromatography or the μ-Porasil HPLC cleanup prior to gas chromatographic analysis of PAEs. A precleanup step using μ-Styragel (100 Å) HPLC system and methylene chloride as the mobile phase was found essential for sediment extracts prior to fractionation of PAEs and subsequent analysis of the extracts by gas chromatography. The precleanup step removed the highly colored material from sediment extracts and iso-octane insoluble coextractants which could otherwise occlude PAEs during the processing of the extracts. The technique also removed a majority of the ECD-active coextractants from sediment extracts and heavily polluted water samples. An alternate precleanup technique utilizing Bio-Beads SX-3 (copolystyrene, 3% divinyl benzene) resin was found satisfactory for the removal of bulk lipids from fish extracts prior to further cleanup using the silica gel column. The GPC system employing Bio-Beads, can be operated using 1:1 mixture of cyclohexane and dichloromethane as a mobile phase. The examination of the elution profile, using Bio-Beads SX-3 and 1:1 cyclohexane/dichloromethane, indicated that the majority of lipids and high molecular weight compounds eluted in the first 100 ml of GPC effluent at the rate of 5 ml/min using a 48 cm × 2.5 cm I.D. column. This fraction was discarded prior to collection of the phthalate esters in the following 90 ml of eluent which was subjected to further cleanup.

The study of the elution profile of the PAEs on the silica column revealed that all 6 PAEs of interest readily separated from other coextractants by the following elution scheme. The silica gel column was first washed with 20 ml of methylene chloride followed by a 20 ml rinse with *n*-hexane. The reconstituted iso-octane extract containing PAEs was applied to the column and the column was eluted with 25 ml of *n*-hexane (Fraction A). This fraction contained chlorobenzenes, PCBs and selected organochlorine pesticides. The column was then eluted with 20 ml of benzene (Fraction B) to isolate remaining organochlorine pesticides and toxaphene. A further elution of the column with 10 ml ethylacetate recovered all PAEs of interest.

4. Comparison of HPLC-UV and GC-ECD Methods

A comparison of both methods indicated that the GC-ECD method was far superior for the analysis of PAEs in water, sediment, and fish samples. There was no correlation between the data obtained for the majority of PAEs using both methods except for BEHP. Furthermore, UV-detection was too nonspecific and lacks sensitivity for the quantitative analysis of ultratrace quantities of PAEs in environmental samples. The normal phase HPLC system was found to be effective as a cleanup procedure to remove interfering coextracts prior to quantitative analysis of PAEs by the GC-ECD method.

D. Recommended Method for Analysis of PAEs (DBP, BBP, DEHP, DOP) and Organochlorines in Environmental Samples

The following method is based on the work of Ryan and Scott.[113] The method is presented in details for those who wish to apply it to obtain reliable data or for those analysts who wish to build the skill to carry out this type of analysis.

1. Water

Natural water samples are extracted with dichloromethane, either on-site or in the laboratory using solvent extraction in separatory funnels or a large volume sample extractor. The dichloromethane is evaporated under reduced pressure and the extract is reconstituted in a small volume of iso-octane. The resultant iso-octane solution is subjected to silica column cleanup and the appropriate fraction containing the PAEs is analyzed by gas chromatography using a non-polar WCOT capillary column with splitless injection, temperature programming and electron capture detection.

a. Extraction and concentration
  (1) Natural water samples should be extracted with methylene chloride. The extraction technique employed will depend upon the nature of the sample. For the extraction of ultratrace levels, a large-sample extractor can be employed using volumes up to 200 l. Otherwise conventional solvent extraction of 1 to 2 l of aqueous sample can be employed.
  (2) Water samples, ideally, should be extracted as the sampling site without delay in order to prevent any loss of PAEs as a result of biodegradation. Alternatively, dichloromethane can be added to the sample container, either in the laboratory or at the sampling site, and the sample thoroughly mixed with the dichloromethane at the time of collection, before transporting the sample back to the laboratory.
  (3) If water samples are not extracted at the time of collection, then the inside of glass storage vessels should be extracted with 50 ml of dichloromethane. This is necessary because PAEs in aqueous solutions are rapidly adsorbed onto glass surfaces.
  (4) The dichloromethane extracts are dried by passing the extract through acetone washed sodium sulfate. The extracts should not be concentrated by subjecting them to evaporation under a stream of nitrogen because of the possible introduction of PAE contaminants as a result of the outgassing of PAEs from plastic tubing, filters, or molecular sieves.
  (5) The extract should be reconstituted into 1 ml of iso-octane.

b.   Cleanup and fractionation

(1)   Prepare a 35 cm × 1 cm I.D. glass column using a plug of acetone rinsed, silanized glass wool, followed by 2.5 cm of anhydrous sodium sulfate (prewashed with acetone and sintered), 8 cm of 3% deactivated silica gel and a second 2.5 cm of anhydrous sodium sulfate (prewashed with acetone and sintered).

(2)   Prewash the column with 20 ml of dichloromethane followed by 20 ml of hexane. Ensure that the column once wetted with solvent, is not allowed to dry past the top of the packing material. Discard the wash eluates and place a clean 100 ml round bottom flask under the column to collect the first fraction.

(3)   Using an acetone rinsed disposable 9 in. Pasteur pipette, transfer the iso-octane concentrate of the extract a.(5) to the head of the column.

(4)   Allow the meniscus of the extract concentrate to just disappear below the top of the upper layer of anhydrous sodium sulfate.

(5)   Rinse the flask that contained the extract concentrate with 5 ml of hexane and transfer the rinsings to the column.

(6)   Allow the hexane rinsings to drain down to the top layer of anhydrous sodium sulfate in the column.

(7)   Repeat steps b.(5) and b.(6).

(8)   Add 15 ml hexane eluant to the column and allow the hexane to drain down to the top layer of anhydrous sodium sulfate. Collect and label the hexane eluate along with the flask rinsings and the extract's iso-octane — Fraction A.

(9)   Place a clean 50 ml round bottom flask under the column and elute the column with 20 ml of benzene. Allow the benzene to drain down to the top layer of anhydrous sodium sulfate and collect and label the eluate — Fraction B.

(10)   Place a clean 15 ml screw cap centrifuge tube under the column and elute the column with 10 ml of ethyl acetate (see e.(2) under Remarks). Allow the column to totally drain dry with this eluant and collect and label the eluate — Fraction C.

(11)   Add 10 ml of iso-octane to Fraction A and Fraction B and concentrate by vacuum distillation on a rotary evaporator to 3 ml. Do not concentrate to less than 3 ml.

(12)   Quantitatively transfer the 3 ml concentrates of Fraction A and Fraction B to graduated centrifuge tubes using an additional three rinsings of 2 to 3 ml iso-octane.

(13)   Place both fractions under a stream of nitrogen and evaporate to 1.0 ml using a water bath temperature of 30 to 35°C.

(14)   Evaporate Fraction C just to dryness under reduced pressure using a vortex evaporator or a rotary evaporator using a temperature of 40°C.

(15)   Dissolve the residue of Fraction C in 1.0 ml of iso-octane.

(16)   Fraction A contains chlorobenzenes, PCBs Heptachlor, Aldrin, Mirex, and DDT metabolites (p,p'-DDE, o,p'-DDT, p,p'-TDE, and p,p'-DDT).

(17)   Fraction B contains α-BHC, Lindane, Heptachlor Epoxide, α-Chlordane, γ-chlordane, α-Endosulfan, β-Endosulfan, Dieldrin, Endrin, Methoxychlor, Toxaphene, Polyaromatic hydrocarbons, and DDT metabolites (o,p'-TDE, p,p'-DDT, and possibly o,p'-DDT and p,p'-DDT).

(18)   Fraction C contains the phthalate esters.

c. Quantitative analysis

(1) The PAEs in Fraction C are quantitatively analyzed for the PAEs using the gas chromatographic condition described below:

Injection volume:	1 µl
Injection port:	Splitless mode, 60 s purge activation time
Infection port temperature:	200°C
Detector:	Nickel-63 electron capture
Detector temperature:	300°C
Detector make-up gas:	Argon/methane (95/5) at 40 ml/min
Carrier gas:	Hydrogen @ 1.0 ml/min (260°C), linear velocity 41.5 to 33.5 cm/s, (80 to 260°C)
Column head pressure:	10 psi/g
Initial column temperature:	80°C, hold for 3 min
Temperature program rate 1:	20°C/min to 150°C
Temperature program rate 2:	2°C/min to 260°C
Final column temperature:	260°C, hold for 10 min
Column:	30 m × 0.25 mm I.D., 0.25 µm film DB-5 capillary column (J&W Scientific, Inc.). The column is conditioned with repeated injections of standard solution.

(2) Calibrate the GC/ECD using a standard solution containing 200 (µg/l) of the PAEs of interest in iso-octane. Be sure to correct for the blank values obtained for the analysis of iso-octane solvent.

(3) Establish the reagent blank values for the PAEs of interest by concentrating the same volume of dichloromethane solvent used for the extraction of the water samples and subjecting it to the same cleanup procedure.

(4) Analyze the PAE fraction (Fraction C) of the water sample extracts, analyzing a standard before and after every five water sample extracts and running an iso-octane blank after every standard and extract.

(5) Measure the peak height and/or peak area of the sample extract peaks whose retention times correspond to those of the PAE standards within ±0.03 min.

(6) Correct the sample extract peak values for the reagent blank values obtained in step c.(3).

d. Calculations

(1) Determine the correction factor (C_f) for each of the PAEs of interest, for the extraction technique employed, i.e., solvent extraction in separatory funnels or in a large-volume extractor. The correction factor is obtained by spiking a known amount of standard PAEs into the same volume of samples used in the analysis and running the spiked sample through the extraction, concentration, cleanup and GC-ECD analyses. The level of the PAE spike should be at least double the indigenous concentration of that PAE in the sample. The correction factor is calculated by dividing the GC-ECD value (area or height) obtained for the spiked PAE in the sample extract by the value obtained for the same amount of PAE standard injected directly into the GC-ECD.

(2) Calculate the concentration of individual PAEs in the aqueous sample using the following formula:

$$\text{Aqueous concentration ng/l} = \frac{A_{ext}}{A_{std}} \times \frac{C_{std}}{C_f} \times \frac{V_{ext}}{V_{aqu}} \times 1000$$

Where A_{ext} = peak area or height of a response in the sample extract; A_{std} = peak area or height of the corresponding PAE standard; C_{std} = ng/ml concentration of the PAE standard injected to give response A_{std}; C_f = correction factor to compensate for the loss of PAE during the processing of the sample (concentration and cleanup); V_{ext} = final volume (ml) of the iso-octane solution of Fraction C; V_{aqu} = total volume in (ml) of the aqueous sample extracted.

(3)   Confirmation of PAEs can be achieved by the GC-ECD analysis of sample extracts on two dissimilar nonpolar capillary columns.

(4)   Final confirmation of the PAEs presence in sample extracts is achieved using GC-MS in the electron impact mode and monitoring the common mass ion of phthalate esters at m/e 149 for all phthalates esters other than dimethyl phthalate in which case the base peak is at m/e 163. Alternatively, chemical-ionization mass spectrometry can be employed to confirm the phthalate esters identity.

e.   Remarks

(1)   The elution pattern of the organochlorines and PAEs must be determined for each batch of silica gel prepared.

(2)   If only BEHP and DOP are to be analyzed, then the 10 ml of ethyl acetate eluant for Fraction C may be replaced with 20 ml of dichlormethane and the eluate collected in a 50 ml round bottom flask instead of a 15 ml centrifuge tube.

(3)   For the GC-ECD analysis of PAEs, organochlorine pesticides, and PCBs are separated from the PAEs using adsorption chromatography. Either conventional column chromatography on high quality silica gel or normal phase HPLC on a μ-Porasil analytical size column can be used. The factor limiting the use of conventional column chromatography is the inability to remove the more polar PAE contaminants from the silica adsorbent without affecting its chromatographic behaviour. For the analysis of all priority pollutant PAEs, particularly in the absence of a suitable quality of silica gel, cleanup on a 30 cm × 3.9 mm I.D. μ-Porasil HPLC column using 2 ml/min methylene chloride containing approximately 0.1% isopropanol is recommended. To optimize this normal phase HPLC system as a screening and cleanup tool, the PAEs should be eluted between 3 and 9.5 min in a 12.5 + 0.5 ml fraction. This permits the resolution of the 6 PAEs of interest and provides adequate separation of the OCs from the PAEs. The isopropanol concentration in the mobile phase should be adjusted to optimize the elution pattern. (Isopropanol deactivates the silica packing of the μ-Porasil column by hydrogen bonding with the Si-OH functionality of the packing). A 1 ml aliquot of the iso-octane concentrate of a water extract or the GPC eluate of a fish or sediment extract is injected into the normal phase HPLC system and the PAE fraction collected in a 15 ml pyrex centrifuge tube. The PAE eluate is carefully evaporated to dryness using a vortex evaporator and redissolved in 1 ml of iso-octane for GC-ECD analysis. The initial volume of eluate preceeding the PAE fraction contains the OC pesticides, PCBs, and PAHs and can be collected and processed for their analysis. Periodic injections of 2 ml aliquots of tetrahydrofuran between runs regenerates the μ-Porasil column and elutes the highly retained coextractants.

2. Sediment and Fish

Sediment and fish samples can be extracted using soxhlet extraction and/or Polytron® homogenization. Ethyl acetate and methylene chloride were the preferred solvents for sediments and fish respectively. The resultant extracts are reconstituted into 5 ml of iso-octane. A precleanup column employing μ-Styragel or Bio-Beads SX3 is used to remove large molecular weight coextractants and lipids, respectively. The sulfur-containing coextracts are removed by using mercury. The resultant extract is then carried through the same procedure as described for the water matrix to determine PAEs.

a. Fish lipid is removed by size exclusion chromatography on a Bio-Bead SX-3 system using 5 ml/min 1:1 methylene chloride/cyclohexane. The solvent is carefully removed from the fish extract under reduced pressure, using a rotary evaporator and the remaining fish lipid is made up to 10 ml with 1:1 methylene chloride/cyclo-hexane. A 5 ml aliquot of the lipid solution, containing a maximum of 1 g of fish lipid is run through the Bio-Bead SX-3 system. The elution pattern of the PAEs will depend upon the size of the Bio-Bead SX-3 column; but for a 48 cm × 2.5 cm I.D. column the 6 priority pollutant PAEs are collected in a 90 ml fraction after discarding the first 100 ml of eluate. Add a small volume of iso-octane to the PAE fraction and remove the methylene chloride solvent and concentrate the fraction to a final volume of 5 ml using a rotary evaporator.

b. Large molecular size coextractants are removed from sediment extracts by size ex-clusion chromatography on a 30 cm × 7.8 mm I.D. μ-Styragel HPLC column using 1 ml/min methylene chloride. The sediment extract is concentrated to about 5 ml using a rotary evaporator and the resultant concentrate quantitatively transferred to a 15 ml centrifuge tube for the complete removal of solvent using a vortex evaporator. The dried extract is redissolved in 500 μl of methylene chloride and a 100 μl aliquot is run through the μ-Styragel HPLC system.

c. The 6 priority pollutant PAEs the collected in a 2 ml fraction after discarding the first 6.2 ml of eluate — depending on the length of transmission tubing and the value of the flow cell. The PAE fraction is carefully evaporated to dryness using a vortex evaporator and redissolved in 1 to 2 ml of iso-octane.

IV. APPENDIX

ABBREVIATIONS USED IN TEXT FOR SPECIFIC PHTHALIC ACID ESTERS

DMP	dimethylphthalate
DEP	diethylphthalate
DnPP	di-*n*-propylphthalate
DiPP	di-*iso*-propylphthalate
DnBP (DBP)	di-*n*-butylphthalate
DiBP	di-*iso*-butylphthalate
DAMP	diamylphthalate
DnHP	di-*n*-hexylphthalate
DCHP	dicyclohexylphthalate
DnHiP	di-*n*-heptylisophthalate
DnHtP	di-*n*-heptylterephthalate
DnOP (DOP)	di-*n*-octylphthalate
DiOP	di-*iso*-octylphthalate
DEHP, BEHP	di-(2-ethylhexyl)phthalate, bis(2-ethylhexyl)phthalate
DnNP	di-*n*-nonylphthalate
DnDP	di-*n*-decylphthalate
BBP	butylbenzylphthalate

nBnDP	*n*-butyl-*n*-decylphthalate
BDP	benzyldecylphthalate
DHnNDP	diheptyl-*n*-nonylundecylphthalate
nBBP	*n*-butylbenzylphthalate
nOnDP	*n*-octyl-*n*-decylphthalate
DiDP	di-isodecylphthalate
DTDP	ditridecylphthalate
nBnOP	*n*-butyl-*n*-octylphthalate
BOP	butyloctylphthalate
DMTP	dimethylterephthalate

ANALYTICAL ABBREVIATIONS

GC-FID	Gas chromatography; flame ionization detection
GC-ECD	Gas chromatography; electron capture detection
GC-MS	Gas chromatography; mass spectrometry
GPC	Gel permeation chromatography
HPLC-UV	High performance liquid chromatography-ultra violet detection
TLC	Thin layer chromatography

REFERENCES

1. **Graham, P. R.,** Phthalate ester plasticizers — why and how they are used, *Environ. Health Perspect.,* 3, 3, 1973.
2. **Pierce, R. C., Mathur, S. P., Williams, T. D., and Boddington, M. J.,** Phthalate Esters in the Aquatic Environment, National Research Council Canada Publication NRCC No. 17583, Associate Committee on Scientific Criteria for Environmental Quality, 1980.
3. **Wolkober, Z.,** The influence of inorganic fillers and pigments on volatility of phthalic acid plasticizers, in *The Stability of Ester Type Plasticizers,* National Aeronautics and Space Administration, Washington, D.C., February 1972, chap. 2.
4. **Peakall, D. B.,** Phthalate esters: occurrence and biological effects, *Residue Rev.,* 54, 1, 1975.
5. **Darby, J. R. and Sears, J. K.,** Plasticizers, in *Encyclopedia of Polymer Science and Technology,* Vol. 10, Interscience, New York, 1969.
6. **Penn, W. S.,** *PVC Technology,* 3rd ed., Applied Science Publications, London, 1971.
7. **Graham, P. R.,** Environmental aspects of plasticizers, *Soc. Plast. Eng. Tech. Paper,* 21, 275, 1975.
8. **Ogner, G. and Schnitzer, M.,** Humic substances — fulvic acid — dialkyl phthalate complexes and their role in pollution, *Science,* 170, 317, 1970.
9. **Matsuda, K. and Schnitzer, M.,** Reactions between fulvic acid, a soil humic material, and dialkyl phthalates, *Bull. Environ. Contam. Toxicol.,* 6, 200, 1971.
10. **Boehm, P. D. and Quinn, J. G.,** Solubilization of hydrocarbons by the dissolved organic matter of sea water, *Geochim. Cosmochim. Acta,* 37, 2459, 1973.
11. **Rao, G. V. and Venkatasubramanian, N.,** Alkaline hydrolysis of aromatic dicarboxylic esters in aqueous DMSO, *Aust. J. Chem.,* 24, 201, 1971.
12. **Klopman, G. and Calderazzo, F.,** Synthesis and reactivity of chromium tricarbonyl complexes of substituted benzoic esters, *Inorgan. Chem.,* 6, 977, 1967.
13. **Gledhill, W. E., Kaley, R. G., Adams, W. J., et al.,** An environmental safety assessment of butylbenzyl phthalate, *Environ. Sci. Technol.,* 14, 301, 1980.
14. **Paciorek, K. L., Kratzer, R. H., Kaufman, J., Nakahara, J., and Hartstein, A. M.,** Oxidative thermal decomposition of polyvinyl chloride compositions, *J. Appl. Polym. Sci.,* 18, 3723, 1974.
15. **DeCoste, J. B.,** Soil burial resistance of vinyl chloride plastics, *Indust. Eng. Chem. Prod. Res. Develop.,* 7, 238, 1968.

16. Ambient Water Quality Criteria for Phthalate Esters, Report EPA 440/5-80-067, U.S. Environmental Protection Agency, Washington, D.C., 1980.
17. Synthetic organic chemicals, U.S. production and sales, International Trace Commission, Washington, D.C., 1978.
18. **Leah, T. D.,** The production, use and distribution of phthalic acid esters in Canada, Environmental Contaminants Inventory Study No. 4, Report Series 47, Inland Waters Directorate, Ontario Region, Water Planning and Management Branch, Burlington, Ontario, Fisheries and Environment Canada, Ottawa, 1977.
19. **Mathur, S. P.,** Phthalate esters in the environment: pollutants or natural products, *J. Environ. Qual.,* 3, 189, 1974.
20. Quality Criteria for Water, Report EPA 440/9-76-023, U.S. Environmental Protection Agency, 1976.
21. National Academy of Sciences, National Academy of Engineering, Water Quality Criteria 1972, Report EPA-R3-73-033, U.S. Environmental Protection Agency, Washington, D.C., 1973.
22. **Lupp, M. and McCarty, L. S.,** Guidelines for Surface Water Quality, Vol. 2, Organic Chemical Substances, Phthalic Acid Esters, Environment Canada, Inland Waters Directorate, Water Quality Branch, Ottawa, 1983.
23. **Kingsbury, G. L., Sims, R. C., and White, J. B.,** Multimedia Environmental Goals for Environmental Assessment, Report EPA-600/7-79-176a,b, U.S. Environmental Protection Agency, Washington, D.C., 1979.
24. **Autian, J.,** Toxicity and Health Threats of Phthalate Esters: Review of the Literature, Report ORNL-TIRC-72-2, Oak Ridge National Laboratory, Toxicology Information Response Center, 1972.
25. **Shackleford, W. M. and Keith, L. H.,** Frequency of Organic Compounds Identified in Water, Report EPA-600/4-76-062, Environmental Research Laboratory, Office of Research and Development, U.S. Environmental Protection Agency, Athens, GA, 1976.
26. **Schwartz, H. E., Anzion, C. J. M., Van Vliet, H. P. M., Peerebooms, J. W. C., and Brinkman, U. A. T.,** Analysis of phthalate esters in sediments from Dutch rivers by means of high performance liquid chromatography, *Int. J. Environ. Anal. Chem.,* 6, 133, 1979.
27. **Mayer, F. L., Stalling, D. L., and Johnson, J. L.,** Phthalate esters as environmental contaminants, *Nature,* 238, 411, 1972.
28. **Strachan, W. M. J.,** Chloroform-extractable organic compounds in the International Great Lakes, in *Identification and Analysis of Organic Pollutants in Water,* Keith, L. M., Ed., Ann Arbor, Science Publishers, Ann Arbor, MI, 1976, 479.
29. **Ewing, B. B., Chian, E. S. K., Cook, J. C., Evans, C. A., and Hopke, P. K.,** Monitoring to Detect Previously Unrecognized Pollutants in Surface Waters — Appendix, Organic Analysis Data, U.S. NTIS PB-273-350, 1977.
30. **Brownlee, B. and Strachan, W. M. J.,** Distribution of some organic compounds in the receiving waters of a Kraft pulp and paper mill, *J. Fish Res. Board Can.,* 34, 830, 1977.
31. Michigan Department of Natural Resources, STORET retrieval, cited by Pierce et al., 1978.
32. **Hites, R. A.,** Phthalates in the Charles and the Merrimack rivers, *Environ. Health Perspect.,* 3, 17, 1973.
33. **Sheldon, L. S. and Hites, R. A.,** Organic compounds in the Delaware River, *Environ. Sci. Technol.,* 12, 1188, 1978.
34. **Morita, M., Nakamura, H., and Mimura, S.,** Phthalic acid esters in water, *Water Res.,* 8, 781, 1974.
35. Guidelines for Canadian Drinking Water Quality 1978: Supporting Documentation, Health and Welfare Canada, 1980.
36. **Dunlap, W. J., Shew, D. C., Scalf, M. R., et al.,** Isolation and identification of organic contaminants in groundwater, in *Identification and Analysis of Organic Pollutants in Water,* Keith, L., Ed., Ann Arbor Science, Ann Arbor, MI, 1976, 453.
37. Preliminary Assessment of Suspected Carcinogens in Drinking Water, December 1975 report to Congress, National Reconnaissance Survey, U.S. Environmental Protection Agency, Washington, D.C., 1975.
38. **Kim, N. K. and Stone, D. W.,** Organic Chemicals and Drinking Water, New York State Dept of Health, Albany, 1980.
39. **Giam, C. S., Atlas, E., Chan, H., and Neff, G.,** Estimation of fluxes of organic pollutants to the marine environment: phthalate ester concentration and fluxes, *Rev. Int. Oceanogr. Med.,* 47, 79, 1977.
40. Murdoch, A., personal communication, cited by Lupp and McCarty in Reference 22.
41. **Kinkead, J. D. and Chatterjee, R. M.,** A limnological survey of nearshore waters of Lake Superior, in *Proc. 17th Conf. Great Lakes Res., Int. Assoc. Great Lakes Res.,* 1974, 549.
42. Upper Lakes Reference Group, The Waters of Lake Huron and Lake Superior, Report to the International Joint Commission, Windsor, Ontario, 1977.
43. **Schacht, T.,** Pesticides in the Illinois water of Lake Michigan, Illinois State EPA-660/3/74/002, PB-245 150, 1974.
44. Michigan Water Resources Commission, Study on transport of trace metals and organics to Lake Michigan by tributary rivers, draft report, 1973.

45. **Webster, R. D. J. and Nickless, G.,** Problems in the environmental analysis of phthalate esters, *Proc. Anal. Div. Chem. Soc.,* 13, 333, 1976.

46. **Glass, G. E., Strachan, W. M. I., Willford, W. A., Armstrong, F. A. I., Kaiser, K. L. E., and Lutz, A.,** Organic Contaminants — Lake Huron, Tech. Report No. PB-277-149, U.S. Environmental Protection Agency, Washington, D.C., 1977.

47. **Swain, W. R.,** Chlorinated organic residues in fish, water and precipitation from the vicinity of Isle Royale, Lake Superior, *J. Great Lakes Res.,* 4, 398, 1978.

48. **Williams, D. T.,** Dibutyl- and di-(2-ethylhexyl) phthalate in fish, *J. Agr. Food Chem.,* 21, 1128, 1973.

49. **Stalling, D. L., Hogan, J. W., and Johnson, J. L.,** Phthalate ester residues — their metabolism and analysis in fish, *Environ. Health Perspect.,* 3, 159, 1973.

50. **Zitko, V.,** Determination toxicity and environmental levels of phthalate plasticizers, *J. Fish Res. Board Can.,* Tech. Report No. 344, 1972.

51. **Morris, R. J.,** Phthalic acid in deep sea jellyfish, *Atolla, Nature,* 227, 1264, 1970.

52. **Tepper, L. B.,** Phthalic acid esters — an overview, *Environ. Health Perspect.,* 3, 179, 1973.

53. **Mayer, F. L. and Sanders, H. O.,** Toxicology of phthalic acid esters in aquatic organisms, *Environ. Health Perspect.,* 3, 153, 1973.

54. **Sugawara, N.,** Effect of phthalate esters on shrimp, *Bull. Environ. Contam. Toxicol.,* 12, 421, 1974.

55. **Perez, J. D., Downs, J. E., and Brown, P. J.,** The effects of dimethylphthalate on the growth of *Pseudomonas aeruginosa, Bull. Environ. Contam. Toxicol.,* 16, 486, 1976.

56. **Wilson, W. B., Giam, C. S., Goodwin, T. C., Aldrich, A., Carpenter, V., and Hrung, T. C.,** The toxicity of phthalates to the marine dinoflagellate *Gymnodinium breve, Bull. Environ. Contam. Toxicol.,* 20, 149, 1978.

57. **Bringmann, G. and Kuhn, R.,** Comparison of the toxicity thresholds of water pollutants to bacteria, algae and protozoa in the cell multiplication inhibition test, *Water Res.,* 14, 231, 1980.

58. **Pfudener, P. and Francis, A. A.,** Phthalate esters: heart rate depressors in the goldfish, *Bull. Environ. Contam. Toxicol.,* 13, 275, 1975.

59. **Sanders, H. O., Mayer, F. L., Jr., and Walsh, D. F.,** Toxicity, residue dynamics and reproductive effects of phthalate esters in aquatic organisms, *Environ Res.,* 6, 84, 1973.

60. **Mayer, F. L., Mehrle, P. M., and Schoettger, R. A.,** Collagen metabolism in fish exposed to organic chemicals, in *Recent Advances in Fish Toxicology: A Symposium,* Tubb, R. A., Ed., EPA Ecological Research Ser. EPA-600/3-77-085, U.S. Environmental Protection Agency, Corvallis, OR, 31, 1977.

61. **Freeman, H. C. and Sangalang, G.,** personal communication, cited by Pierce et al. in Reference 2.

62. In-depth Studies on Health and Environmental Impact of Selected Water Pollutants, Contract No. 68-01-4646, U.S. Environmental Protection Agency, Washington, D.C., 1978.

63. **Streufert, J. M.,** Some effects of two phthalic acid esters on the life cycle of the midge *(Chironomus plumosus)* , M.S. thesis, University of Missouri, Columbia, MO, 1977.

64. **Hrudey, S. E., Sergy, G. A., and Thackeray, T.,** Toxicity of oil sands plant wastewaters and associated organic contaminants, in *Proc. 11th Can. Symp. Water Pollut. Res.,* 34, 1976.

65. **Belisle, A. A., Reichel, W. L., and Spann, J. W.,** Analysis of tissues of mallard ducks fed two phthalate esters, *Bull. Environ. Contam. Toxicol.,* 13, 129, 1975.

66. **Jacobs, R. D. and Ringer, R. K.,** The effects of di-(2-ethylhexyl) phthalate in rats, dogs, and miniature pigs, *Food Cosmet. Toxicol.,* 18, 637, 1976.

67. **Wood, D. L. and Bitman, J.,** The effect of feeding di(2-ethylhexyl) phthalate (DEHP) on the lipid metabolism of laying hens, *Lipids,* 15, 151, 1980.

68. **Lake, B. G., Phillips, J. C., Linnell, J. C., and Gangolli, S. D.,** The *in vitro* hydrolysis of some phthalate diesters by hepatic and intestinal preparations from various species, *Toxicol. Appl. Pharmacol.,* 39, 239, 1977.

69. **Melancon, J. J. and Lech, J. J.,** Distribution and biliary excretion products of di-(2-ethylhexyl) phthalate in rainbow trout, *Drug Metab. Disp.,* 4, 112, 1976.

70. **Daniel, J. W.,** Toxicity and metabolism of phthalate esters, *Toxicol. Ann.,* 3, 257, 1979.

71. **Keyser, P., Pujar, B. G., Eaton, R. W., and Ribbons, D. W.,** Biodegradation of the phthalates and their esters by bacteria, *Environ. Health Perspect.,* 18, 159, 1976.

72. **Kurane, R., Suzuki, T., Takahara, Y., and Komagata, K.,** Identification of phthalate ester-assimilating bacteria, *Agric. Biol. Chem.,* 41, 1031, 1977.

73. **Kurane, R., Suzuki, T., Takahara, Y., and Komagata, K.,** Isolation of microorganisms growing on phthalate esters and degradation of phthalate esters by *Pseudomonas acidovorans, Agric. Biol. Chem.,* 41, 2119, 1977.

74. **Johnson, B. T. and Lulves, W.,** Biodegradation of di-n-butyl phthalate and di-(2-ethylehexyl) phthalate in freshwater hydrosoil, *J. Fish Res. Board Can.,* 32, 333, 1975.

75. **Johnson, B. T., Stalling, D. L., Hogan, J. W., and Schoettger, R. A.,** Dynamics of phthalic acid esters in aquatic organisms, *Adv. Environ. Sci. Technol.,* 8, 283, 1977.

76. **Fish, T. D. and Johnson, B. T.,** Various chemical and physical factors which influence the biodegradation of di-(2-ethylhexyl) phthalate (DEHP) in hydrosoil, *Trans. Mo. Acad. Sci.,* 10, 296, 1977.
77. **Hattori, Y., Kuge, Y., and Nakagawa, S.,** Microbial decomposition of phthalate esters in environmental water, *Mizu Shori Gijutsu,* 16, 951, (Chem. Abst. 85, 83043v), 1975.
78. **Mayer, F. L.,** Residue dynamics of di-(2-ethylhexyl) phthalate in fathead minnows *(Pimephales promelas), J. Fish Res. Board Can.,* 33, 2610, 1976.
79. **Mehrle, P. M. and Mayer, F. L.,** Di-(2-ethylhexyl) phthalate: residue dynamics and biological effects in rainbow trout and fathead minnows, *Trace Sub. Environ. Health,* 10, 519, 1976.
80. **Veith, G. D., Macek, K. J., Petrocelli, S. R., et al.,** An evaluation of using partition coefficients and water solubility to estimate bioconcentration factors for organic chemicals in fish, in *Aquatic Toxicology,* Eaton, J. G. et al., Eds., ASTM STP 707, 1980.
81. Water Quality Criteria, Availability: Request for Comments, Federal Register, 44, Part 3, U.S. Environmental Protection Agency, Washington, D.C., July 25, 1970.
82. **Thomas, J. A., Darby, T. D., Wallin, R. F., et al.,** A review of the biological effects of di-(2-ethylhexyl) phthalate, *Toxicol. Appl. Pharmacol.,* 45, 1, 1978.
83. **Morton, S. V.,** The hepatic effects of dietary di-2-ethylhexyl phthalate, *Dissert. Abstr. Int.,* 40B, 4236, 1980.
84. **Foster, P. M. D., Thomas, L. V., Cook, M. W., et al.,** Study of the testicular effects and changes in zinc excretion produced by some *n*-alkyl phthalates in the rat, *Toxicol. Appl. Pharmacol.,* 54, 392, 1980.
85. **Kwano, M.,** Toxicological studies on phthalate esters. II. Metabolism accumulation and excretion of phthalate esters in rats, *Jpn. J. Hyg.,* B5, 693, 1980.
86. **Mushtag, M., Srivastava, S. P., and Seth, P. K.,** Effects of di-2-ethylhexyl phthalate (DEHP) on glycogen metabolism in rat liver, *Toxicology,* 16, 153, 1980.
87. **Oishi, S. and Hiraga, K.,** Testicular atrophy induced by phthalic acid esters: effect on testosterone and zinc concentrations, *Toxicol. Appl. Pharmacol.,* 53, 35, 1980.
88. **Mangham, B. A., Foster, J. R., and Lake, B. G.,** Comparison of the hepatic and testicular effects of orally administered di(2-ethylhexyl) phthalate and dialkyl 7,9 phthalate in the rat, *Toxicol. Appl. Pharmacol.,* 61, 205, 1981.
89. *Annual Book of ASTM Standards, Part 31, Water,* American Society for Testing and Materials, 1978.
90. **Garrison, A. W.,** Analysis of organic compounds in water to support health effect studies, *Ann. N.Y. Acad. Sci.,* 298, 2, 1977.
91. **Wilson, A. L.,** The chemical analysis of water, *Analytical Sciences Monograph No. 2,* Society of Analytical Chemistry, London, 1974.
92. **Singmaster, J. A. and Crosby, D. G.,** Plasticizers as intereferences in pollutant analysis, *Bull. Environ. Cont. Toxicol.,* 16, 291, 1976.
93. **Junk, G. A., Richard, J. J., Grieser, M. D., Witiak, D., Witiak, J. L., Arguello, R. V., Svek, H. J., Fritz, J. S., and Calder, G. V.,** Use of macroreticular resins in the analysis of water for trace organic contaminants, *J. Chromatogr.,* 99, 745, 1974.
94. **Shinohara, R., Koga, M., Shinohara, K., and Hori, T.,** Extraction of traces of organic compounds from water with Amberlite XAD-2 resin, *Bunseki Kagaku,* 26, 856, 1977.
95. **Tateda, A. and Fritz, J. S.,** Mini-column procedure for concentrating organic contaminants from water, *J. Chromatogr.,* 152, 329, 1978.
96. **Thruston, A. D.,** High pressure liquid chromatography techniques for the isolation and identification of organics in drinking water extracts, *J. Chromatogr. Sci.,* 16, 254, 1978.
97. **Van Rossum, P. and Webb, R. G.,** Isolation of organic water pollutants by XAD resins and carbon, *J. Chromatogr.,* 150, 381, 1978.
98. **Buelow, R. W., Carswell, K. J., and Symons, J. M.,** An improved method for determining organics by activated carbon adsorption and solvent extract, *J. Am. Water Works Assoc.,* 65, 195, 1973.
99. **Giam, C. S.,** Trace analysis of phthalates and chlorinated hydrocarbons in marine samples, in *Strategies for Marine Pollution Monitoring,* Goldberg, E. D., Ed., John Wiley & Sons, New York, 61, 1976.
100. **Gough, K. M. and Gesser, H. D.,** The extraction and recovery of phthalate esters from water using porous polyurethane foam, *J. Chromatogr.,* 115, 838, 1975.
101. **Carmignani, G. M. and Bennett, J.,** Filter media for the removal of phthalate esters in water of closed aquaculture systems, *Aquaculture,* 8, 291, 1976.
102. **Sullivan, K. F., Atlas, E. L., and Giam, C. S.,** Loss of phthalic acid esters and polychlorinated biphenyls from seawater samples during storage, *Anal. Chem.,* 53, 1718, 1981.
103. **Wolfe, N. L., Steen, W. C., and Burns, L. A.,** Phthalate ester hydrolysis: linear free energy relationships, *Chemosphere,* 9, 403, 1980.
104. **Sullivan, K. F., Atlas, E. L., and Giam, C. S.,** Adsorption of phthalic acid esters from seawater, *Environ. Sci. Technol.,* 16, 428, 1982.
105. **Giam, C. S., Chan, H. S., and Neff, G. S.,** Sensitive method for determination of phthalate ester plasticizers in open-ocean biota samples, *Anal. Chem.,* 47, 2225, 1975.

106. **Schouten, M. J., Peereboom, J. W. C., and Briukman, U. A. T.,** liquid chromatographic analysis of phthalate esters in dutch river water. *Intl. J. Environ. Anal. Chem.,* 7, 13, 1979.

107. **Mori, S.,** Identification and determination of phthalate esters in river water by high performance liquid chromatography, *J. Chromatogr.,* 129, 53, 1976.

108. **Arbin, A. and Ostelius, J.,** Determination of electron capture gas chromatography of mono and di(2-ethylhexyl) phthalate in intravenous solutions stored in poly(vinylchloride) bags, *J. Chromatogr.,* 193, 405, 1980.

109. **Bowers, W. D., Parsons, M. L., Clement, R. E., and Karasek, F. W.,** Component loss during evaporation-reconstitution of organic environmental samples for gas chromatographic analysis, *J. Chromatogr.,* 207, 203, 1981.

110. **Fisherbein, L. and Albro, P. W.,** Chromatographic and biological aspects of the phthalate esters, *J. Chromatogr.,* 70, 365, 1972.

111. **Hartung, R.,** An evaluation of toxicological and environmental issues associated with phthalic acid esters, Research report prepared for the Manufacturing Chemists Association, Washington, D.C., May 28, 1974.

112. **Rhoades, J. W., Nulton, C. P.,** Priority pollutants analysis of individual wastewaters using a microextraction approach, *J. Environ. Sci. Health,* A15, 467, 1980.

113. **Ryan, J. F. and Scott, B. F.,** Gas chromatographic determination of bis-(2-ethylhexyl) phthalate and di-*n*-octylphthalate in water samples, NWRI Contribution number 86-214, Canada Centre for Inland Waters, Burlington, Ontario, Canada.

114. **Yohe, T. L., Suffet, I. H., and Cairo, P. R.,** Specific organic removals by granular activated carbon pilot contractors, *J. Am. Water Works Assoc.,* p.402, August, 1981.

115. **Kaiser, K. L.,** Organic contaminant residues in fishes from Nipigon Bay, Lake Superior, *J. Fish. Res. Board Can.,* 34, 850, 1977.

116. **Veith, G. D., Kuehl, J. W., and Rosenthal, J.,** Preparative method for gas chromatographic/mass spectral analysis of trace quantities of pesticides in fish tissues, *J. Assoc. Off. Anal. Chem.,* 58, 1, 1975.

117. **Rogers, W. M.,** The use of a solid support for the extraction of chlorinated pesticides from large quantities of lipids and oils, *J. Assoc. Off. Anal. Chem.,* 55, 1053, 1972.

118. **Ehrhardt, M. and Derenbach, J.,** Phthalate esters in the Kiel Bight, *Marine Chemistry,* 8, 339, 1980.

119. **Burns, G. B., Musial, C. J., and Uthe, J. F.,** Novel clean-up method for quantitative gas chromatographic determination of trace amounts of di-2-ethylhexyl phthalate in fish lipid, *J. Assoc. Off. Anal. Chem.,* 64, 282, 1981.

120. **Musial, C. G., Uthe, J. F., Sirota, G. R., and Burns, B. G.,** Di-n-hexyl phthalate (DHP), a newly identified contaminant in Atlantic Herring *(Clupea harengus harengus)* and Atlantic Mackerel *(Scombes scombrus), Can. J. Fish Aquat. Sci.,* 38, 856, 1981.

121. **Draviam, E. J., Kerkay, J., and Pearson, K. H.,** Separation and quantitation of urinary phthalates by HPLC, *Anal. Let.,* 13(B13), 1137, 1980.

122. **Zitko, V.,** Determination of phthalates in biological samples, *Intern. J. Environ. Anal. Chem.,* 2, 241, 1973.

123. **Takeshita, R., Takabatake, E., Minagawa, K., and Takizawa, Y.,** Micro-determination of total phthalate esters in biological samples by gas-liquid chromatography, *J. Chromatogr.,* 133, 303, 1977.

124. **Thomas, G. H.,** Quantitative determination and confirmation of identity of trace amounts of dialkyl phthalates in environmental samples, *Environ. Health Perspect.,* 3, 23, 1973.

125. **Baker, R. W. R.,** Gel filtration of phthalates esters, *J. Chromatogr.,* 154, 3, 1978.

126. **Wildbrett, G.,** Diffusion of phthalic acid esters from PVC milk tubing, *Environ. Health Perspect.,* 3, 29, 1973.

127. **Schultz, C. O. and Rubin, R. J.,** Distribution, metabolism and excretion of di-2-ethylhexyl phthalate in the rate, *Environ. Health Perspect.,* 3, 129, 1973.

128. **Zulaica, J. and Guiochon, G.,** Fast qualitative and quantitative microanalysis of plasticizers in plastics by gas liquid chromatography, *Anal. Chem.,* 35, 1724, 1963.

129. **Krishen, A.,** Programmed temperature gas chromatography for identification of ester plasticizers, *Anal. Chem.,* 43, 1130, 1971.

130. **Langhorst, M. L.,** Photoionization detector sensitivity of organic compounds, *J. Chromatogr. Sci.,* 19, 98, 1981.

131. **Bunting, W. and Walker, E. A.,** Quantitative determination of trace amounts of some dialkyl phthalates by gas liquid chromatography, *Analyst,* 92, 575, 1967.

132. **Weisenberg, E., Schoenberg, Y., and Ayalon, N.,** A rapid method for monitoring low levels of di-(2-ethylhexyl) phthalate in solutions, *Analyst,* 100, 857, 1975.

133. **Vessman, J. and Rietz, G.,** Determination of di(ethylhexyl) phthalate in human plasma and plasma proteins by electron capture gas chromatography, *J. Chromatogr.,* 100, 153, 1974.

134. **Van Vliet, H. P. M., Bootsman, T. C., Frei, R. W., and Brinkman, U. A. T.,** On-line trace enrichment in high-performance liquid chromatography using a pre-column, *J. Chromatogr.,* 185, 483, 1979.

135. **Persiani, C. and Cukor, P.,** Liquid chromatographic method for the determination of phthalate esters, *J. Chromatogr.,* 109, 413, 1975.

136. **Amundson, S. C.,** Determination of di(2-ethylhexyl) phthalate, mono (2-ethylhexyl) phthalate and phthalic acid by high pressure liquid chromatography, *J. Chromatogr. Sci.,* 16, 170, 1978.

137. **Giam, C. S., Chan, H. S., Hammargren, T. F., Neff, G. S., and Stalling, D. L.,** Confirmation of phthalate esters from environmental samples by derivatization, *Anal. Chem.,* 48, 78, 1976.

138. **Ishida, M., Suyama, K., and Adachi, S.,** Background contamination by phthalates commonly encountered in the chromatographic analysis of liquid samples, *J. Chromatogr.,* 189, 421, 1980.

139. **Bowers, W. D., Parsons, M. L., Clement, R. E., Eiceman, G. A., and Karasek, F. W.,** Trace impurities in solvents commonly used for gas chromatographic analysis of environmental samples, *J. Chromatogr.,* 206, 279, 1981.

140. **Weschler, C. J.,** Identification of selected organics in the arctic aerosol, *Atmos. Environ.,* 15, 1365, 1981.

141. **Giam, C. S., Atlas, E., Chau, H. S., and Neff, G. S.,** Phthalate esters, PCB and DDT residues in the Gulf of Mexico, *Atmos. Environ.,* 14, 65, 1980.

Chapter 8

ORGANOMETALLIC COMPOUNDS IN THE AQUATIC ENVIRONMENT

Y. K. Chau and P. T. S. Wong

TABLE OF CONTENTS

I. INTRODUCTION

Organometallic compounds designate a class of compounds formed from the direct union of metal and carbon of the organic group. Thus carbonates of metals, complexes of metals with organic ligands or metal salts or organic acids are not classified as organometals. This discussion in this chapter also includes organometalloids such as organoarsenic and organoselenium compounds.

Variations in the organic groups in the molecules can change drastically the properties of the compounds, therefore, organometals can be tailor-synthesized to suit a specific purpose. Such synthesis is fully demonstrated in the specific biocidal actions of alkyltin compounds towards bacteria, molds, and a variety of organisms by varying the nature of the organic groups. This is what makes organotin compounds versatile biocides. Other widely used organometals are lead alkyls which have been effectively used in internal combustion engines as antiknocking agents. Synthesis of organometals represents one of man's greatest achievements in chemical technology.

Production of organometals has been steadily increased during the last four decades to meet their ever increasing demands. They are used in almost any kind of commodities, industrial products, and farm produce. Consequently, organometals are intimately related to our daily life and are widespread in the environment.

The analysis of organometallic compounds has conventionally been carried out by spectrophotometric or fluorimetric determination of the metal elements after destruction of the compounds. This is normally done after effective chemical separation of the compound. In recent years, determination of the metal has been carried out by atomic absorption or electrochemical methods. All these methods, however, determined the total concentration of an organometal as a class without knowledge of the organic moiety. As many environmental studies involve the integrated knowledge of several disciplines in order to understand the biogeochemical processes of organometals in the ecosystems, most chemical analyses are now required to give the information of the form of the element present in the sample. Such information is particularly important in the study of mechanisms of mobilization and transport, and is essential in toxicological study. It is only during the last few years that species analysis of metals and organometals in the environment has become utterly important, and has opened up a new horizon of research and knowledge.[1]

II. OCCURRENCE OF ORGANOMETALS IN THE ENVIRONMENT

It has long been recognized that organometallic compounds in the environment are all derived from anthropogenic sources because it is unlikely any mechanism for their formation in the natural system exists. Recently environmental formation of organometals has been discovered and related to the process of biological methylation which described mechanisms of transfer of methyl groups to metal and metalloid in biological systems. The first organometalloid bearing environmental significance is the production of Goico gas in a damp room by the action of molds growing on the wallpaper containing arsenic pigments. The gas was later identified by Challenger[2] as trimethylarsine.

Studies on the environmental aspects of organometallic compounds began to gain momentum after the human tragedies of mercury poisonings at Minimate Bay and Niigata in Japan in the late fifties. These poisonings were caused by ingestion of fish containing methylmercuric compounds derived partly through biological methylation of inorganic mercury by microorganisms.[3] A new area of research has been developed in the environmental aspects of organometallic compounds including their occurrence, formation, and fate. The input of organometallic compounds to the environment due to biological transformation has now been widely recognized in addition to their anthropogenic inputs. Possibilities of methylation of elements have been predicted by the relative ease of formation of the metal-carbon bonds[4] and by the redox potentials of the elements.[5]

Environmental occurrence of organometals and organometalloids continues to appear in the literature. Several reviews on the methylation of metals are available.[6-10] Different forms of methylarsenic acids have been identified in natural waters and in other environmental samples.[11-14] Arsenobetaine has been identified by Edmonds and Francesconi[15] as the major organoarsenic component in lobster. Methyltin and butyltin species have also been found in the natural waters and in a variety of environmental materials especially in harbors and marinas where butylin compounds are widely used in antifouling paints.[16-23] Organolead compounds are also widespread in the environment. Tetraalkyllead compounds have been identified in the street air[24-26] and in fish.[27] Recently, the dialkyllead (R_2Pb^{2+}) and trialkyllead (R_3Pb^+) compounds have also been found for the first time, in fish, aquatic weeds, sediment, and water.[28-29]

Other organometals and organometalloids formed in the environment include alkylselenides[30-32] and dimethylsulfide.[33] Table 1 summarizes the occurrence of selected organometallic compounds reported in environmental samples.

III. APPROACHES IN THE ANALYSIS OF ORGANOMETALS

Industrial organometallic compounds are mostly synthesized for specific purposes. These compounds exhibit characteristic properties as a result of the combination of the metal atoms and the organic groups. The field of organometallic chemistry is extremely vast. The variety and numbers of organometallic compounds produced and known are numerous and are still increasing every year. It is for this reason that only organometals of current environmental interest, such as organotin and organolead, are selected in the discussion. Of the numerous analytical methods only those suitable for environmental samples are reviewed.

There are generally two types of organometals dispersed in the aquatic environment. One is molecular in nature, hydrophobic and volatile, such as tetraalkyllead (R_4Pb) and tetraalkyltin (R_4Sn). Their presence in the water column is only transient before they will finally be adsorbed onto particulates, partitioned into living organisms with lipid content, or volatilized into the atmosphere. The other type is ionic, hence polar and hydrophilic, such as dialkyllead (R_2Pb^{2+}), trialkyllead (R_3Pb^+), and trialkyltin (R_3Sn^+) which behave as solvated metal ions in solution either in the form of salts or coordinated with organic ligands such

Table 1
SELECTED ORGANOMETALS FOUND IN
ENVIRONMENTAL SAMPLES

Methylarsenic Acids

Sample	Location	MMA[a]	DMA[b]	Ref.
Water (μg/l)				
River water	Moira River, Ontario	4.0	5.2	34
Lake water	Bend Bay, Ontario	9.0	93.0	34
Seawater	Scripps, CA	0.02	0.12	12
River water	Owen River, CA	0.06	0.22	12
River water	Colorado River, AZ	0.18	0.25.	12
Lake water	Lake Carrol, FL	0.18	0.25	11
Weeds (mg/kg, wet wt.)				
S. fluitans	S. Bermuda	0.01	0.18	35
S. filipendula	E.Gulf Mexico	0.01	0.06	35
Fish (mg/kg wet wt.)				
Yellow perch	Bay of Quinte, Ontario	0.74	2.45	34
Yellow sucker	Bay of Quinte, Ontario	0.64	0.24	34
White perch	Bay of Quinte, Ontario	0.84	nd[c]	34

Alkyltin

Sample	Location	$MeSn^{3+}$	Me_2Sn^{2+}	Me_3Sn^+	Ref.
Water (ng/l)					
Seawater	San Diego Bay, CA	2—8	15—45	nd	17
Lake water	Grand Haven, MI	6—18	nd—63	nd	17
Rainwater	FL	5.9	7.4	0.2	16
Tap water	FL	4.3	1.3	1.5	16
Lakes, rivers	Ont.	60—1200	nd—400	nd—50	19

Sample	Location	$BuSn^{3+}$	Bu_2Sn^{2+}	Bu_3Sn^+	Ref.
Lake water	Grand Haven, MI	22—1200	10—1600	nd	17
Lake, rivers	Ontario	nd—8000	nd—7300	nd—2900	19
Algae (μg/kg, wet wt.)					
mixed algae	Mission Bay, CA	nd—0.4	nd—0.2	nd	37
	San Diego Bay, CA	nd—1.2	0.4—1.1	nd—0.3	37
Fish (μg/kg dry wt.)					
U. moluccensis	Mediterranean	27	2.6	1.2	36

Table 1. (continued)
SELECTED ORGANOMETALS FOUND IN ENVIRONMENTAL SAMPLES

Alkyllead

Sample	Location	I^d	IIe	III^f	IV^g	V^h	Ref.
Fish (ng/g, wet st.)							
Coho salmon	Vineland Creek, Ontario	nd	4.3	2.3	2.8	nd	27
Yellow perch	Stocco Lake, Ontario	1.7	nd	7.1	4.4	2.7	27

Sample[i]	Location	Me_4-	Et_4-	Me_3-	Me_2-	Et_3-	Et_2-	Pb(II)
Carp	St.Lawrence River	137	780	2735	362	906	707	1282
Pike		nd	1018	215	nd	nd	nd	1040
Sediment (ng/g, dry wt.)		nd	1152	nd	nd	187	22	10,000
Macrophytes, mixed (ng/g, wet wt.)		nd	68	nd	nd	132	nd	4327

Methylmercury

Sample	Location	CH_3Hg	Ref.
Sediment (ng/g)	Monte Amiata, Italy	20—40	41
	San Francisco Bay, CA	0.4—1.9	42
	River Mersey, U.K.	1.6—60.6	43
Lake water (ng/l)	Canada	0.5—1.7	44
Seawater (ng/l; n = 5)	Japan	2.0	39
Fish (µg/wet wt.)			
Swordfish steak (n = 20)	Washington, D.C.	0.49—2.44	45
Canned tuna (n = 11)	Washington, D.C.	0.07—0.53	45
Northern pike (n = 3)	Sweden	1.17—2.19	45

Alkylselenides[j]

Sample	Location	Me_2Se	Me_2Se_2	Me_2SeO_2
Air (ng/m³)				
Campus Lake	Antwerp, Belgium	0.47	0.35	0.10
River (Scheldt)	Antwerp, Belgium	0.85	0.30	0.10
Sewage treatment	Antwerp, Belgium	2.40	0.30	18.8
North Sea shore	Ostend, Belgium	0.15	0.30	0.10

Dimethysulfide

Sample	Location	Me_2S	Ref.
Water (ng/l)			
Sea water	English Channel	116.4	33
	Gulf of Mexico	28.0	33
	S. Atlantic	37.1	33

[a] MMA = monomethylarsonic acid.
[b] DMA = dimethylarsinic acid.
[c] nd = not detectible.
[d] I = Me_4Pb.
[e] II = Me_3EtPb.
[f] III = Me_2Et_2Pb.
[g] IV = $MeEt_3Pb$.
[h] V = Et_4Pb.
[i] Reference 29.
[j] Reference 40.

as humates and fulvates. Analytical methods must deal with specific types of organometals in question through sampling techniques and specific methods of separation and determination.

A. Determination of the Metal Element

As the metal atom is the indestructible part of an organometal molecule, analyses of the compound have frequently been carried out by specifically determining the metal element after decomposition or isolation of the organometal. The chemical separation procedure for this method must be specific and effective. The determination part normally presents little problems. Kojima[46] determined tributyltin and dibutyltin compounds by graphite furnace atomic absorption spectrometry after partition extraction. Farnsworth and Pekola[47] studied the various methods for destroying the organotin compounds. After breaking down of the organometallic compounds, many techniques have been used to determine the metals such as polarography,[48-49] amperometry,[50] atomic absorption and atomic emission spectrometry.[51] Using an extraction method, Bolanowska[52] extracted triethyllead salts from aqueous solutions in the presence of saturated sodium chloride and determined as lead dithizonate complex after mineralization in acids. Of all the techniques used for total tin or lead analyses, atomic absorption spectrometry (AAS) appears to be the most convenient and rapid technique which requires very little skill of the operator. Separation is the key step for this type of analysis. The most common methods used are extraction,[46,51-52] distillation,[48] or total sample digestion.[49] These methods as a whole are simple and rapid for the analysis of organometals, but they lack specificity.

B. Determination of the Organometallic Compounds

1. Spectrophotometric Methods

Specific methods for the determination of organometals are abundant among which spectrophotometric methods are the most widely used. They are selective, relatively specific and also sensitive, provided that the sample matrices are simple and interferences due to cations and anions are properly removed or masked. Many complexing agents form with organometallic compounds, complexes with characteristic absorption spectra suitable for spectrophotometric measurements. Diorganotin, triorganotin, and trioganolead have been individually determined by Aldridge and Street[53] in biological materials spectrophotometrically with dithizone and fluorimetrically with 3-hydroxyflavone at nanomole concentration. Similarly, Blunden and Chapman[54] determined triphenyltin compounds in water by fluorimetric method with 3-hydroxyflavone with detection limit of 0.004 ppm. In this method, trimethyltin compounds react in a similar manner with 3-hydroxyflavone and, therefore, should be absent in the sample. Other organotin methods include the use of diphenylcarbazone[55] for spectrophotometric determination of dibutyltin dichloride which reacts specifically with the reagent. The only interfering compound, butyltin trichloride, can be removed by extraction with EDTA. Fluorimetric determination of monoalkyl-, dialkyl-, trialkyl-, and triphenyltin compounds with morin with detection limits in the submicromolar level have been recently reported by Arakawa and Wada.[56]

Dithizone reacts with a number of metals and has also been used extensively in spectrophotometric determination of organolead. Cremer[57] used dithizone to determine triethyllead compounds in tissues and blood. Making use of the dithizone reactions, Moss and Browett[58] determined tetraalkyllead by converting them to their corresponding dialkyllead with iondine monochloride, which was determined as dithizonates spectrophotometrically. Similarly, Hancock and Slater[59] extracted the coverted dialkyllead with dithizone from solution in the presence of EDTA which complexed any inorganic lead present. The lead is determined by AAS after decomposition of the dialkyllead complexes with acids. An alternate reagent, 4-(2-pyridylazo)-resorcinol (PAR) was used by Pilloni and Plazzogna[60] to determine the dialkyllead species specifically. The dithizone reactions with organolead were further expanded in a method to determine simultaneously triethyllead, diethyllead, and inorganic lead in solutions by measuring the absorbance at three different wavelengths based on their characteristic absorption spectra of their dithizonates.[61]

Based on all the above work, Noden[62] developed a separation scheme to determine the components in a mixture containing tetraalkyl-, trialkyl-, dialkyllead, and lead(II) in water. Tetraalkyllead was first quantitatively extracted in hexane. The trialkyllead compounds were then separated by toluene extraction. The dialkyllead species were determined by dithizone or PAR, and the trialkyllead was also determined by the same method after conversion to dialkyllead with iodine monochloride. This method however, distinguishes only the dialkyllead, trialkyllead, and tetraalkyllead as a class without identification of the alkyl groups. It can be used, for example, in following the fate of tetramethyllead or tetraethyllead in water, but not suitable for species determination of the individual alkyllead compounds if they are all present in a sample.

Spectrophotometric techniques are sensitive, convenient, and generally within reach of most laboratories. They are suitable for determination of a specific compound in a known matrix or in a complex mixture after isolation. The dithizone reagent, although widely used for organolead and organotin analyses, forms color complexes with many metals and organometals which constitute the main source of interferences.

2. Gas Chromatography

Gas chromatography has been used extensively in the separation of volatile alkyllead compounds in gasoline. Parker et al.[63] separated the methyl-ethyl-mixed tetraalkyllead compounds in gasoline and collected the fractions in methanolic iodine solution for measurement of the lead content with dithizone. It is an elaborate and time-consuming technique, although the method is quite specific. The method was subsequently improved by Lovelock and Zlatkis[64] who used an electron capture detector, and Dawson[65] who used temperature programming to achieve a better separation. Further refinements were made by Bonelli and Hartmann[66] by using a scrubber containing silver nitrate to remove the halogenated scavengers used in gasoline. The use of a flame ionization detector (FID) was reported by Soulages[67] who analyzed the mixed alkyl lead and lead scavengers in gasonline after decomposition of all these components to methane and ethane by a catatylic column. Using the same detector, De Jonghe and Adams[68] were able to determine the trialkyllead halides in aqueous solutions. Detection limits were not particularly impressive, however, it demonstrated the feasibility of flame ionization detection for direct alkyllead analysis. Although electron capture is more sensitive to alkyllead compounds, the lack of specificity is its major drawback.

Hill and Aue[69-70] described an element-specific detector for use in the analysis of volatile organometallics. It consists of a spectrophotometric channel added to a regular flame ionization detector to identify and quantify the element of interest by its emission spectro lines. When the hydrogen-rich flame was used, lower nanogram ranges of iron, tin and lead compounds could be measured. Other element-specific detectors used with GC include helium plasma emission spectrometry[71] and a cold vapor mercury analyzer[72] for organomercury compounds. Major instrumentation has been interfaced to a GC as detector, forming a specific analytical system which will be discussed in detail under the speciation section.

3. Electrochemical Techniques

Electrochemical techniques are in principle simple. They make use of the electrochemical characteristics, such as redox potentials, to differentiate the individual substances. These include organic and inorganic, ionic, or molecular substances that can be dissolved in a dielectric solvent, and can either be oxidized or reduced at an electrode. Voltammetry is a major branch of the electrochemical techniques which deals with the effect of potential of an electrode in an electrolytic system on the current which flows through it. Differential pulse polarography and stripping voltammetry are the improved and more sensitive techniques that have been applied to the determination of organometals. Details of their applications will be discussed under the speciation section.

C. Combination Techniques

Tandem combination of a separation technique and an element specific detection has been the most up-to-date and state-of-the-art approach to speciation of organometallic compounds. There are many possible combinations depending on the nature of the analytes in the sample and the specificity requirement. Chromatography separation including gas chromatography (GC), high pressure liquid chromatography (HPLC), and thin layer chromatography (TLC) have been used in the separation part. Detection systems include mass spectrometry, various forms of atomic spectrometry, and electrochemical techniques. Developments in this area during the last decade have been extensive and rapid. New combination analytical systems continue to appear. All these techniques and features will be discussed in the following section under speciation analysis.

IV. SPECIATION OF ORGANOMETALS

In environmental analysis especially in areas related to biological and toxicological aspects, the form of the compounds present in the sample is critically important information. The most commonly used combination analytical systems are discussed in the following sections.

A. Gas Chromatography-Atomic Absorption Spectrometry (GC-AAS)

The use of an atomic absorption detector for metals has been most extensive in species analysis. Reviews on this topic and general applications of the chromatography and atomic spectrometry techniques have been provided by Van Loon[73] and Ebdon et al.[74] The GC-AAS system provides a sensitive technique for speciation of volatile organometallics in the molecular forms such as tetraalkyllead compounds, and volatile metal and organometal hydrides. In the analysis of complex samples containing organometals and other organic compounds, the AAS detector will respond only to those analyte species containing the selected metal. Thus a very simple chromatogram can be obtained. The interfacing of the two instruments is extremely simple without any major modification to either instrument. The component parts of the two instruments that are essential to organometal work are discussed below.

1. GC Column

The choice of stationary phase in the GC column is important in obtaining clear separation of components without tailing of the individual peaks. For organometals such as alkyltin and alkyllead compounds, nonpolar silicone phases are suitable. OV-1 phase has been used by Chau et al.[75] and Reamer et al.,[76] and OV-101 by De Jonghe et al.[77] in the separation of mixtures of tetraalkyllead compounds. Light phase loading (3 to 5%) on a neutral Chromosorb W support is generally preferred to obtain well-defined chromatographic peaks. For extremely volatile compounds such as the hydride-derivatized alkyltins, the liquid nitrogen-cooled trap containing glass beads or glass wool can serve as a chromatographic column to allow separation of the different hydrides.[16-17] Capillary columns have been used in the separation of alkyllead compounds.[78-79] In the analysis of biological samples, some high boiling greasy residues are often found to accumulate at the upper part of the packed column after a number of determinations. When this happens, the upper part of the packing material can be removed and replaced, after the column interior has been cleaned with acetone. This is an advantage of using packed columns over capillary columns.

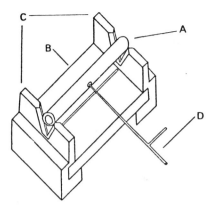

FIGURE 1. Atomization tube with fixed-height tube support. (A) Ceramic tube; (B) air-acetylene burner head; (C) stainless steel knife-edge ceramic tube support; (D) glass-lined T-piece. (From Ebdon, L., Ward, R. W., and Leathard, D. A., *Analyst,* 107, 129, 1982. With permission.)

2. GC-Flame AAS Interface

Both the flame and furnace modes of AAS can be used as atomization source depending on the sensitivity required. Normally three orders of magnitude in sensitivity enhancement can be gained with a furnace. In either case the atomization unit has to be continuously heated to the desired temperature to decompose the organometallic compounds in the effluent flow through a transfer-line from the GC. The simplest interface was achieved by feeding the GC effluent through a transfer tube directly into the nebulizer of the burner.[80-82] In this interfacing, Chau et al.[81] suggested that the spoiler of the nebulizer was removed and a glass tube lining was installed inside the nebulizer to reduce the absorption of alkyllead on the Penton surface. Air-acetylene flame was used with detection limit about 10 ng as Pb. Improvements were made by feeding the effluent directly into the flame[83-84] to minimize peak broadening due to dilution of analytes with large volumes of gases in the nebulizer. The flame is continuously operated during analysis. Sensitivity of flame AAS is generally in the μg range. Recently, Ebdon et al.[85] were able to increase the residence time of the atoms in the tube and reach a sensitivity of 17 pg for tetramethyllead, by feeding the GC effluent into a flame-heated ceramic tube. Figure 1 shows the design of the atomization tube. The atomization does not take place in the flame, but rather in the flame-heated ceramic tube. In essence this is not a genuine flame mode AAS. However, the device is simple, robust, sensitive, and low cost. The air-acetylene flame can reach a temperature as high as 2300°C which is equivalent to the atomization mode in the graphite furnace operation. Therefore, this simple device is similar to a graphite furnace with a long atomizing tube which provides an optical path four times longer than that of a graphite tube. The detection limit for tetraalkyllead is certainly the lowest achieved for a GC-AAS system including those employing graphite furnace atomization.

3. GC-Furnace AAS Interface

As the AA furnace for atomization has to be continuously heated to the desired temperature (~ 900°C for organolead and organotin) during the analysis, a special furnace must be designed for such a purpose. Chau et al.[75] used an open-ended electrothermal silica furnace (7 mm I.D., 4 cm long), for the determination of tetraalkyllead compounds. The silica tube is wrapped with 26-gauge chromel wire to give a resistance of ~ 5 Ω. A layer of asbestos tape is used to cover the wire and to insulate the tube. The whole assembly is housed in a

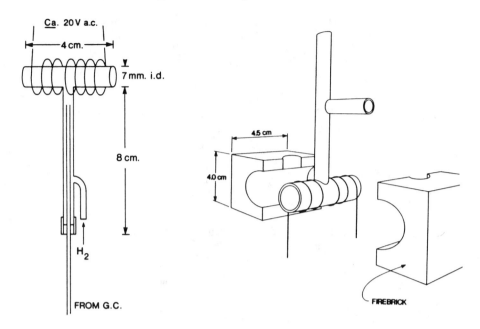

FIGURE 2.   Silica furnace and assembly. (From Chau, Y. K. and Wong, P. T. S., in *Environmental Analysis,* Ewing, G. W., Ed., Academic Press, New York, 1977, 215, With permission,)

block of preshaped fire brick which can be conveniently mounted on top of the burner unit. The burner alignment controls can be used to position the furnace to the optical path. Hydrogen and air can be separately introduced to the furnace through a side arm to enhance atomization and to elevate the furnace temperature. Voltage is regulated by a variable transformer. When the applied voltage is ~ 20 V AC the furnace temperature, with hydrogen burning, can reach ~ 850 to 900°C, which is sufficient for many of the organometals studied. The construction of the furnace unit assembly is shown in Figure 2.

In the analysis of certain elements such as As and Se, with absorption lines at 194 and 196 nm, respectively, in the far UV region, organic solvents, particularly chlorinated hydrocarbons, may absorb and cause interferences which the deuterium background corrector may not be able to compensate effectively. A precombustion section made from a silica tubing can be installed before the entrance of the main furnace to eliminate such interferences. Air is introduced to this section to burn off the contaminants.[86] Both parts of the furnace are wrapped with chromel wire and heated to ~ 850 to 900°C through a variable transformer (Figure 3.). The "home-made" furnace is robust, economical, and can be continuously operated daily over a period of 1 month without deterioration.

Interfacing of the GC and the silica furnace can be achieved by a transfer line connected to the column outlet at the original GC detector base after the detector has been removed. A reduction Swagelock joint is suitable for this connection. A stainless steel tubing (2 mm, O.D.) can be used as a transfer line.[75] The other end of the transfer line is positioned at the junction of the side tube and the furnace, care being taken not to block the light beam. Several types of tubing can be used as transfer lines depending on the properties of the organometals analyzed. For example, methylmercury compounds are decomposed at heated stainless steel surfaces and Teflon tubing should be used instead.[87] We found that stainless steel tubings are suitable for organotin and organolead compounds. The transfer line should be taken out and cleaned with acetone after heavy use with biological samples as indicated by decreasing sensitivity and increasing peak tailings. Other materials such as nickel, glass-lined stainless steel, and glass tubings of similar diameter have also been used. The transfer

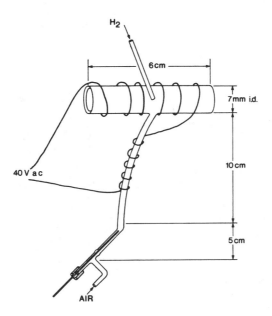

FIGURE 3. Silica furnace with precombustion section. (From Chau, Y. K. and Wong, P. T. S., in *Environmental Analysis,* Ewing, G. W., Ed., Academic Press, New York, 1977, 215. With permission.)

line is wrapped with heating tape with a thermister so that its temperature can be monitored. Figure 4 shows the interfacing of a GC-AAS system. The separation of four *n*-butyl derivatized methyltin compounds by the GC-AAS system is illustrated in Figure 5.

4. GC-Graphite Furnace Interface (GC-GFAAS)

Commercial graphite furnaces have been interfaced to gas chromatograph by a transfer line through different designs. Segar[88] used a tungsten tube to deliver the GC effluent into a heated graphite tube at *circa* 1700°C to separate the five methyl/ethyl tetraalkyllead mixtures in gasoline, with detection limit of ca. 10 ng. Robinson et al.[89] designed a carbon rod atomizer operated at a constant temperature of 2000°C and achieved a sensitivity in the order of 0.1 ng for tetramethyllead. Parris et al.[90] interfaced a commercial furnace to a GC for the analysis of several organometallic compounds generated in biological systems. The transfer tube was connected to the inert gas purge passageways of the furnace by means of a stainless steel T-fitting, so that the effluent gas was symetrically introduced into the furnace from both ends. The system was optimized by assessing the effects of varying the atomizing temperature, liner surface of the furnace, and composition of carrier gases. Detection limits for As, Se and Sn were found to be 5 ng, 7 ng, and 12 ng, respectively. A similar approach of utilizing the inert gas flow of the graphite furnace for introduction of the GC effluent was taken by Radziuk et al.[25] and De Jonghe et al.[77] in the analysis of tetraalkyllead. The former group used a friction fit tantalum connector sitting in the central injection opening of the furnace tube to couple the stainless steel transfer tube from the GC. The latter group used a glass tubing to deliver the GC effluent to the furnace from both inert gas entraces, similar to the design of Parris et al.[90] Both designs reached absolute detection limits of about 0.04 ng. Similar modification was made by Andreae and Froelich[91] on a commercial furnace, by introducing the gaseous germanium hydrides to the left inert gas inlet port using the center sample port and the right inert gas port as exits.

In all this work, the graphite furnace was operated continuously at 1500 to 2000°C during

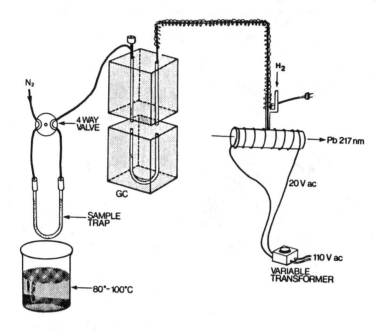

FIGURE 4.   GC-AAS interface. (From Chau, Y. K. and Wong, P. T. S., in *Trace Element Speciation in Surface Waters and Its Ecological Implications*, Leppard, G. G., Ed., Plenum Publishing, New York, 1983, 87. With permission.)

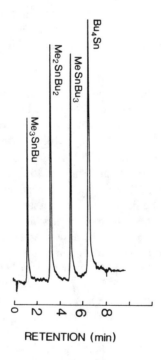

RETENTION (min)

FIGURE 5.   GC-AAS chromatograms of methyltin and Sn(IV) species after butylation. (Reprinted with permission from Chau, Y. K., Wong, P. T. S., and Bengert, G. A., *Anal. Chem.*, 54, 246, 1982. Copyright 1982 American Chemical Society.)

chromatography. Under these conditions we experienced a slow deterioration of the graphite tube, which was indicated by loss of sensitivity. Such a phenomenon began to appear after about 20 separate chromatographic runs, each lasting about 15 min. Radziuk et al.[25] reported that the lifetime of a furnace tube ranged between 10 to 15 h. Thus analysis with graphite furnace tubes could be quite expensive. In this aspect, the electrically-heated silica furnace which lasts about one month on daily operation is simpler and more economical.

5. Sampling
a. Volatile Organometals

Tetraalkylated metals such tetraalkyllead, tetraalkytin (R = Me, Et), and methylarsines are volatile and can be trapped cryogenically in a cold trap containing glass beads, glass wool, or common chromatographic column packing materials such as OV-1, Apiezon L, SE-52, etc. Mixtures of liquid nitrogen and methanol are commonly used to make up a cold bath from −70 to −170°C for collecting volatile organometals in the atmosphere, such as tetraalkyllead,[26] alkylselenide,[32] or organometals metabolic gases generated in culture systems. Chau and co-workers[1,92] trapped the atmosphere of culture flasks incubated with inorganic metals with and without sediment, to study biolobical generations of volatile organometallic compounds. The trap containing the samples was mounted to a 4-way valve installed between the carrier gas and the injection port of the GC, so that the sample could be swept into the column by the carrier gas (Figure 4). A hot water bath was used to warm up the trap to speed up the desorption. Alternatively, Reamer et al.[76] used a small volume of solvent to absorb tetraalkyllead compounds in the atmosphere for direct injection into the GC system through the normal injection port. The cold trap technique utilizes the whole sample and is, therefore, more sensitive.

b. Ionic Organometals

This type of organometallic compound is polar, hydrated, nonvolatile, and behaves like salts in solution. Extraction of these compounds from an aqueous medium is normally difficult. They can be speciated in the GC-AAS system after derivitization to the more volatile forms, or directly separated by liquid chromatography, followed by AAS determination. The techniques for analysis are seperately discussed under speciation analysis.

6. Other Element-Specific Detectors

Many element-specific detectors have been used in tandem combination with the GC. The use of atomic fluorescence spectrometry was reported by Van Loon[73] for tetraalkyllead compounds with sensitivity of 0.1 ng for Pb, which was similar to that obtained with AAS with an electrothermally heated silica tube.[75] Braham et al.,[11] with DC arc discharge atomic emission detector, have detected methylarsines in the nanogram range. Later in their investigations of methyltin speciation, higher sensitivity was achieved in a hydrogen-rich flame emission spectrometric detector.[16] Further improvements have been made by Talmi and Bostick[94], and Reamer et al.,[76] by using microwave plasma exited emission which enhanced the sensitivity to the pg level for alkylarsenic and alkyllead compounds. Other plasma such as microwave helium plasma emission has been used by Estes et al.[93] It is obvious that plasma discharge is a more efficient exitation source and, therefore, is more sensitive.

Emission spectrometric detectors have the capability of multielement detection which can be a useful feature for environmental analysis. Another detection system includes a spectrophotometric technique utilizing dithizone to measure the concentration of alkyllead collected individually in an iodine scrubber after the GC separation.[63] This method is cumbersome and lacks the sensitivity for environmental samples.

Gas chromatography-mass spectrometry (GC-MS) systems have not been extensively used in the analysis of organometallic compounds, apparently because of the following reasons:

(1) high initial cost of installation, (2) sensitivity is not particularly superior to atomic spectrometry, (3) deposits of metals on the interface are detrimental to the system, and (4) a specially trained operator is required. All these disadvantages discourage the use of GC-MS systems.

Laveskog[95] used a GC-MS system to determine alkyllead species in the air and reported sensitivity of ca. 10 pg. A similar system was used by Meinema et al.[96] to analyze butyltin species in aqueous systems after conversion to their methyl derivatives. The reported detection limit was 0.01 ppm Sn, when 500 ml of water was used.

B. Liquid Chromatography-Element Specific Detector

Many organometallic compounds of high molecular weights, or of polar nature are not volatile nor are readily converted to volatile derivatives for GC-AAS analysis. Liquid chromatography (LC) in combination with an element-specific detector can provide a suitable system for these types of compounds. With the advent of efficient high pressure liquid chromatography systems (HPLC), the separation of ionic organometals of biological origin has become a simple task. The HPLC separates the compounds directly without chemical modification of the molecules. The elution is much faster, and the choice of isocratic or gradient elution mode further widens its applicability. A comprehensive review on the use of HPLC in organic analysis has been prepared by Willeford and Veening.[97]

1. Liquid Chromatography

Columns used in HPLC for separation of organometals are generally of the reversed-phase type. Brinckman et al.[98] used "Lichrosorb", a C2 reversed-bonded phase column to separate triphenyltin, tributyltin, and tripropyltin species followed by GFAAS determination. The system was semiautomated. Criteria on solvent, column, analyte polarity, and complexation equilibria on peak resolutions were discussed. In another study of tetraalkyllead, Messman and Rains[99] separated the alkyls with a reversed-phase C_{18} column interfaced to a flame AAS, and achieved a detection limit of 10 ng for each compound. A similar column has been applied by Stockton and Irgolic[100] to separate organoarsenic compounds containing arsenobetaine, arsenocholine, and inorganic arsenic.

Columns of ionic exchange type have also been used. In a study of alkylarsenic acids, arsenite, and arsenate in soil extracts and drinking water, Brinckman et al.[101] employed strong anion exchange (SAX), strong cation exchange (SCX), and C_{18} reversed-phased columns coupled to a GFAAS to identify and quantify the arsenic compounds at 10-20 μg/l level. Alternately, Woolson and Aharonson[102] separated the same arsenic compounds on a low capacity anion exchange column that did not affect the GFAAS signal response of the arsenicals. All methylated arsenic acids and As(III), As(V) gave similar calibration curves with detection limit of 5 ng As injected in aliquots of 20 μl.

Simple gravity flow ion exchange chromatography also finds application in speciation systems. For example, Chau and Wong[103] used a basic cation exchange column to separate a mixture of methylarsenic acids. The column effluent was coupled to an on-line hydride generation system, and the hydride derivatives were determined by AAS using a heated silica furnace.

The solvents used in HPLC such as acetonitrile, methanol, and in the case of ion exchange columns, a dilute buffer solution of ammonium acetate or carbonate, did not cause interferences in the AAS furnace atomization.

2. LC-AAS Interface

Interface of simple gravity flow ion exchange chromatography and HPLC to a flame AAS unit can be easily achieved by directly feeding the LC effluent into the burner nebulizer. It is advisable to match approximately the flow rate of the column to that of the nebulizer.

Koropchak and Coleman,[104] in a study of the effect of the nebulizer flow rate on the sensitivity of the flame AAS, indicated that substantial signal improvements might be obtained in the LC-AAS system through adjustments of the nebulizer to apply back pressure to the column. Similar observations were reported by Messman and Rains[99] in their determination of tetraalkyllead in gasoline.

The sensitivity of flame AAS is normally in the μg range. A further loss of sensitivity can arise as a result of peak spreading as the analyte passes through the chromatographic column.[105] Furnace atomizers can generally give detection limits three orders of magnitude lower than that of flame, and are therefore preferable in environmental analysis.

The inherent difficulty of interfacing a graphite furnace to a LC is the incapability of the furnace to handle a continuous flow of liquid while the furnace is heated at high temperatures. There are several approaches to tackle this problem. One approach is to install a sampling device between the AA furnace and the LC to sample the effluent and to inject it into the furnace. Cantillo and Segar[106] suggested the use of a multiport sampling and injection valve controlled by a sequencer. Thus an aliquot of the LC effluent can be sampled and injected, while the main effluent stream is stopped during the heating cycles of the furnace. The main drawback of this technique is the lengthy analysis time and the poor resolution of separation. A commercial autosampler for AA furnace was used by Brinckman et al.[98] to sample the effluent from an over-flowing well, while Stockton and Irgolic[100] used a sequencer-controlled slider injector valve to sample the effluent on-line. All these techniques give pulse signals, the frequency of which depend on the cycle times of the furnace operation ~ 30 to 50 s per sample). The precision of these techniques is governed by the number of samples analyzed for a chromatographic peak. For a narrow peak which lasts only 1 to 2 min, the AA analysis, used to define a component, can vary considerably depending on how the autosampler operation is synchronized to the emergence of the peak. Another limitation of this technique is the small fraction of the effluent being used in the analysis, about 3%, assuming 40 μl sample is injected into the furnace while the main effluent flows 2 ml min^{-1}.

An improvement in interfacing suggested by Vickery et al.[107] is the storage of the peak-containing effluent stream in a Teflon capillary tube during the chromatographic run, and to analyze the peak-containing fraction off-line by injecting it incrementally into the AA furnace. Thus a greater number of analyses can be performed on an analyte peak to achieve a more accurate calculation of its area. A detection limit of 480 pg for a solution of Ph_4Pb was obtained in the LC-GFAAS analysis. The procedure is lengthy, but it is obviously a much improved technique to quantify a chromatographic peak. In a similar manner, Koizumi et al.[108] collected the sample in fractions off-stream, and injected aliquots from these fractions into the furnace for the determination of Co and vitamin B-12 complex mixtures.

The second approach for furnace interfacing does not involve sampling of the effluent, but coverts the analytes on-line to volatile derivatives, for continuous feeding to a silicone furnace. For elements that form covalent hydrides, such as As, Se, Ge, Sn, Pb, Sb, etc., the on-line hydride generation can isolate these elements in a gas separator. Using this idea, a procedure has been developed by Ricci et al.[109] to determine As(III), As(V), mono-methyl arsenic acid, dimethylarsinic acid, and p-aminophenylarsonate after separation by ion chromatography with detection limits of 10 ng/ml for each species. At the same time, Chau and Wong[103] also developed a method for methylarsenic acids and As(III) and As(V) by coupling a conventional cation exchange column to an automatic hydride system. The post-column hydride generation has definite advantages over the hydride generation in a batch. The possible rearrangements of the alkyl groups[94] during the batch hydride reactions can be avoided; furthermore, the on-line automated hydride generation is more uniform and reproducible, compared to the manual introduction of the borohydride reagent to a gas bubbler.[13]

Recently, Burns et al.[110] interfaced an HPLC to an electrothermal silica furnace through a post-column automatic hydride generator for the species analysis of methyltin compounds

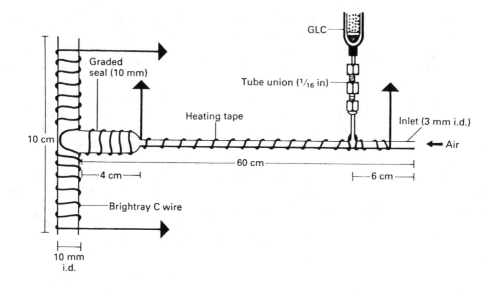

A

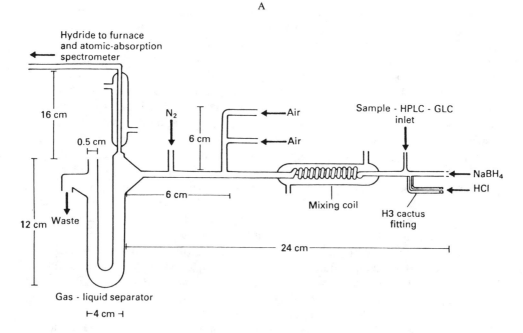

B

FIGURE 6.   HPLC-furnace AAS interface. (A) Using a quartz furnace; (B) using an automatic hydride generator. (From Burns, D. T., Glockling, F., and Harriott, M., *Analyst*, 106, 921, 1981. With permission.)

with excellent sensitivity (2 to 20 pg). The hydride generation is currently the state-of-the-art in LC furnace AAS interfacing. Figure 6 shows the interfacing of the HPLC and silica furnace through an on-line hydride system in the species analysis of methyltin compounds. A review of the HPLC-AAS techniques applied to environmental analysis is available.[111]

3. Sampling

Sampling for the LC-AAS is simple. Liquid samples of water, extracts or digests of sediment and biological samples can be directly loaded to the LC column through micro-sampling valves, with sample capacities from a fraction of a microliter to several milliliters. The LC column can also be used to concentrate analytes in case of very dilute samples, such as natural waters. A large sample can often be charged to the LC column if the initial mobile phase solvent has a low eluting power. In general, sampling preparations are minimal for the LC in comparison to the GC techniques. There is no need to prepare volatile and stable derivatives as required for GC analysis.

4. Other Systems

Many other atom spectrometric detections have been used with LC. Uden et al.[112] directly interfaced a DC argon-phase emission detector to an HPLC in the determination of a number of transition metal complexes. Inductively coupled plasma emission spectrometry (ICP-AES) has been coupled to LC by Hausler and Taylor[113] through a spray chamber as an interface in the analysis of complex organometallic compounds in solvent-refined coal fractions. With the recent advancements of microbore and capillary columns in HPLC, flame photometric detectors can be used to accept a slow flowing effluent to achieve specific detection.[114]

Electrochemical detectors have been widely used with LC in organic analysis.[115-116] They are based on the detection of the oxidation or reduction of the analytes in solution. Their applications in orgnometallic compounds have only recently been developed. A cyclic voltammetric detection system has been interfaced to an HPLC by McCrehan et al.[117] for the analysis of alkyl- and phenyl-derivatives of Hg, Sn, Sb, and Pb with submicrogram sensitivity. Further improvements in selectivity and sensitivity were achieved by the above workers[118] by using the differential pulse mode of amperometry to discriminate against other reducible metals. The method has been applied to determine methylmercury in tuna fish and sharks, with detection limits of 40 pg. A new flow-through polarographic detector has recently been designed by Kutner et al.[119] for use with the HPLC. The system was used in the analysis of cations and in the separation of a mixture of testeroids. It has potential applications in the analysis of organometals. Reviews[115-120] on the use of electrochemical detectors in liquid chromatography are available.

C. Interferences

Atomic spectrometers used as detectors suffer similar spectro interferences, as they are used individually in metal analysis. Nonspecific interferences, including light scatter and molecular absorption with atomic absorption, and light scatter and molecular fluorescence in atomic fluorescence are the main courses of interference. Effluents from a GC contain a solvent peak which normally causes nonspecific absorption at the ultraviolet region, especially for halogenated solvents. This type of interference is minimal in the flame mode of AAS. In furnace AAS, however, the background corrector often fails to compensate for the large amounts of solvent used in carrying the sample. The remedial methods that are available are either to have the solvent peak emerge well before the analyte peaks or to use a furnace provided with a precombustion chamber to burn off the solvent, as designed by Chau et al.[86] in the analysis of volatile methyl selenides.

Light scatter can be a serious interference in atomic fluorescence, and its correction is not simple. Van Loon[73] measured the scatter of radiation from a lamp of another element not present in the sample (e.g., Au) to obtain background compensation in atomic fluorescence. However, a method is still not yet available for background compensation for interferences in molecular fluorescence.

Interferences in electrochemical detection are mainly from solution matrix, organic coating of electrodes, which will be discussed under electrochemical methods.

D. Electrochemical Methods

Electrochemical techniques have been used in the analysis of organometals with good sensitivity and a certain degree of specificity. These techniques have been used extensively for the determination of a single compound. Anodic stripping voltammetry and potentiometric titration were used by Booth and Fleet[121] and Litan et al.,[122] respectively, in the determination of triphenyltin compounds at submicrogram levels. Both butyltin chloride and dibutyltin dichloride in an ethanol-water mixture were determined by Hasebe et al.[123] at a 10^{-8} – 10^{-7} M level by differential pulse polarography after elution of these compounds from fishing nets.

In a mixture of organometals, Plazzogna and Pilloni[124] determined dialkyllead and trialkyllead species in the presence of each other at μg/l levels by amperometric titrations with ferrocynide and sodium tetraphenylboron, respectively. Dialkyltin and trialkyltin compounds were determined by titrating their total amounts potentiometrically with alkali and then determining the diakyltin species in another aliquot by amperometric titration with 8-hydroxyquinoline solution. Greater sensitivity was achieved for dialkyllead and trialkyllead species by Hodges and Noden,[125] using differential pulse anodic stripping voltammetric techniques. Unfortunately all the techniques mentioned above can only differentiate the dialkyllead and trialkyllead as a class without identification of the alkyl groups. These techniques are therefore not considered as genuine speciation techniques.

Recently, a comprehensive scheme was published by Colombini et al.[126] for consecutive determination of the following alkyllead species in natural waters: Me_4Pb, Et_4Pb, Me_3Pb^+, Et_3Pb^+, Me_2Pb^{2+}, Et_2Pb^{2+}, Pb^{2+}. The method is based on a selective solvent extraction scheme coupled with differential pulse anodic stripping voltammetry. The method involves many steps of calculation by difference. The technique is sensitive and selective but the scheme may be too elaborate for practical use in the analysis of natural waters.

Electrochemical methods are sensitive, and have adequate specificity in the determination of a single organometal in relatively simple matrices. Difficulties have been experienced in the analysis of environmental samples such as polluted water samples and complex biological extracts. The electrodes used in most of the electrochemical methods are vulnerable to coating by organic matter and thus subject to interference from complex matrices. This is especially the case for mercury electrodes. The electrochemical methods are also sensitive to changes in sample matrices. These methods are still not widely used in spite of their many merits.

V. SPECIATION OF IONIC ORGANOMETALS BY GC-AAS

Ionic organometals, e.g. dialkyllead and trialkyllead salts, are polar, aquated, and behave like metal cations in solution. There are two major difficulties to overcome in their analysis. First is the difficulty in quantitive extraction from an aqueous medium: and second, the preparation of stable and volatile derivatives for gas chromatographic separation. All the previous methods were based on extraction of the trialkyllead compounds in a saturated sodium chloride medium, followed by spectrometric determination with dithizone[52,62] and by gas chromatography with ECD[127] and FID[68] detectors. The authors have evaluated the GC technique using GC-AAS analysis,[103] but could not reproduce the high extraction recovery which one of the workers claimed.[127] The sensitivity of the GC-AAS measurement for trialkyllead was much lower compared to that for tetraalkyllead. It is possible that thermal decomposition of the trialkyllead takes place during the GC separation,[68] or in the heated transfer line before it reaches the furnace.[103]

Several other attempts have been made using GC-microwave helium plasma,[93] GC-FID,[68] and GFAAS,[128] to determine the trialkyllead compounds in aqueous solution without satisfactory results. Recently, significant achievements have been made by two independent research groups, namely Chau et al.[129] and Forsyth, and Marshall[79] in quantitative extraction and analytical aspects for simultaneous speciation of the dialkyl- and trialkyllead compounds. There are three steps involved in the analysis, namely, extraction, derivatization, and GC-AAS determination.

A. Extraction of Samples

Extraction of the dialkyl- and trialkyllead salts from an aqueous medium with solvents is difficult because of the highly polar nature of these compounds. Bolanowska[52] extracted triethyllead from a saturated salt medium. Noden[62] used sodium chloride and sodium iodide to improve the recovery. Birnie and Hodges[28] found that sodium benzoate aided the recovery of the dialkyllead species. In spite of all these efforts, recovery of the dialkyllead species is still not satisfactory. Recently, a complexing agent, sodium diethyldithiocarbamate (NaDDTC), has been used by Chau et al.[129] to extract all the dialkyl- and trialkyllead quantitatively into benzene from an aqueous solution. At the same time, Forsyth and Marshall[79] reported the use of dithizone in chloroform to recover the same species at quantitative levels from water, phosphate buffer, and from the enzyme-hydrolyzed whole egg homogenate. It is obvious that the use of a complexing agent is essential in achieving quantitative recovery of the ionic alkyllead.

The mono-, di-, and tributyltin compounds were extracted from water by Meinema et al.[96] using tropolone in benzene. Chau et al.[23] improved the extraction efficiency for the methyltin species by using sodium chloride in the medium. The extraction of ionic organometals from sediment and biological matrices requires further modification and treatment, and will be discussed in detail separately under the sections on Sampling, Sample Preservation, and Sample Treatment (see Section VI).

B. Hydride Derivatization

Many alkyl metals or alkyl metalloids, like their inorganic ions, react with borohydride to form covalent hydrides. For example, monomethylarsonic acid and dimethylarsinic acid form monomethylarsonic hydride and dimethylarsinic hydride respectively. These reactions have been made use of by Braman et al.[11] in the determination of inorganic arsenic and methylarsenic compounds in water. In this method, the volatile hydrides were generated from a water sample, trapped cryogenically on glass beads in a column, and allowed to distil into a DC arc emission detector. Andreae,[12] in a similar manner, analyzed these compounds in a variety of environmental samples, except that he used atomic absorption in a heated quartz furnace for the final determination. The hydride derivatization technique has been extended to determine methyltin, and tin(IV) species by Braman and Tompkins[16] and Hodge et al.[17] The hydride reaction can be applied to other alkylmetals of the $R_nM^{(4-n)+}$ type, such as that of Pb(IV), Ge(IV), and other hydride-forming elements. A similar approach has been adopted by Braman and Tompkins[130] and Hambrick et al.[131] to determine methylgermanium derivatives in sea water.

C. Alkylation Derivatization

Another derivatization method involves the use of a selected R′ Grignard reagent to convert the polar organometal species, $R_nM^{(4-n)+}$ type, to the tetra-substituted forms $R_nMR'_{(4-n)}$, which are stable, more volatile, and are suitable for GC separation. Meinema et al.[96] methylated the butyltin species after extraction into a tropolone/benzene solution. This technique was modified by Maguire and Huneault[132] by using a pentyl Grignard reagent to speciate the butyltin compounds. The pentylated butyltin derivatives have slightly higher boiling points thus facilitating sample concentration for subsequent GC-AAS analysis. Similarly, butyl Grignard reagent was used by Chau et al.[23] to butylate the methyltin species. The reactions involved in the butylation of the methyltin and tin(IV) species are:

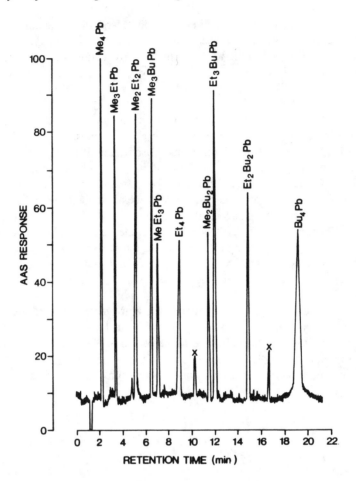

FIGURE 7.   GC-AAS chromatograms of 5 tetraalkyllead (10 ng each); 4
butyl derivatives of dialkyl- and trialkyllead (8 ng each) and Pb(II) (15 ng).
(x) Unidentified lead compounds. (From Chau, Y. K., Wong, P. T. S., and
Kramar, O., *Anal. Chim. Acta,* 146, 211, 1983. With permission.)

$$MeSn^{3+} + BuMgCl \rightarrow MeSnBu_3 + Mg^{2+} + Cl^- \text{ (bp 122 to 124°C)}$$

$$Me_2Sn^{2+} + BuMgCl \rightarrow Me_2SnBu_2 + Mg^{2+} + Cl^- \text{ (bp 70°C)}$$

$$Me_3SN^+ + BuMgCl \rightarrow Me_3SnBu + Mg^{2+} + Cl^- \text{ (bp 41 to 42°C)}$$

$$Sn^{4+} + BuMgCl \rightarrow Bu_4Sn + Mg^{2+} + Cl^- \text{ (bp 145°C)}$$

the equations are not balanced for simplicity.

Butyl Grignard reagent was also used by Estes et al.[78] in the derivatization of trialkyllead,
and later on by Chau et al.[129] to include the dialkyllead species for GC-AAS analysis. The
derivatization products, $R_nPbBu_{(4-n)}$ (R = Me, Et), not only have lower boiling points,
but are thermally more stable than the original dialkyl- and trialkyllead. Thus the thermal
decomposition of the organolead was circumvented. The GC-AAS chromatograms showing
the separation of ten organolead compounds are shown in Figure 7.

Based on the same principle, further improvements in extraction procedure and sensitivity
have been achieved.[133,134] Other alkylating methods, such as ethylation[135] and propylation,[136]
have been suggested for derivatization of the ionic alkyllead for GC separation. They are

excellent alternative techniques for verification of the alkyllead species should uncertainties occur.

Aryl Grignard reagent, phenylmagnessium bromide, has been used by Forsyth and Marshall[79] for derivatization of the dialkyl- and trialkyllead extracted in dithizone solution. Similar advantages have been gained in the GC separation. The weakness of this method is the use of ECD as detector which is sensitive but nonspecific.

D. Comparison of Derivatization Methods

There are advantages and disadvantages of these two derivatization methods. In the hydride method, the analytes are isolated directly from the sample matrix as volatile derivatives without laborious chemical manipulation, thus free from interference of the matrix. It is a total sampling technique where the total amount of the alkylmetal in the sample is converted to hydride for determination. In the alkylation method, the derivatization products are in a solvent medium and only a fraction of it is injected into the analytical system. Thus the overall sensitivity of the hydride method is superior based on the same sample size taken for analysis. On the other hand, the alkylation method has an extraction step to isolate the alkylmetals from the sample. A larger sample size can be extracted to achieve a concentration effect.

There is a possibility of alkyl group rearrangement during the hydride reaction which was observed earlier by Talmi and Bostick[94] and Wong et al.[13] in the analysis of organoarsenic compounds, and also in the analysis of methyltin.[142] The cause of the alkyl rearrangement in the hydride method is not fully understood although Talmi and Bostick[94] suggested the use of borohydride tablets to minimize it.

From the experience in the authors' laboratories, the hydride method is easily subject to contamination even if the most rigorous precautions are practiced. The hydride generation is affected by the way the borohydride is added to the sample, for example, in one addition, or in multiple additions. An automatic system is recommended for uniform results.[13]

VI. SAMPLING AND SAMPLE PRESERVATION AND TREATMENT

A. Sampling

Contamination in sampling for organometals analysis is not as serious a problem as it is for heavy metal analysis. However, certain precaution must be taken in obtaining representative samples. For example, plastic samplers such as Van Dorn bottles and plastic containers should not be used for taking samples for organotin analysis, because of the alkyltin stabilizers used in plasticware. Boat paints also contain butyltin derivatives as antifouling agents, which are continuously leached out from the paints. Similarly, tetraalkyllead in gasoline used in boat motors is a source of contamination of alkyllead.

B. Sample Preservation

Alkyltin compounds are stable in water and no preservatives are required in natural water samples. It is however, recommended by Chau et al.[23] to add a saturated amount of sodium chloride to the sample at the time of sampling so as to reduce the risk of adsorption of Sn(II) and Sn(IV) on the container walls, and to facilitate subsequent extraction.

Ionic alkyllead compounds slowly degrade in water in the presence of light. However, lake water samples enriched with dimethyllead and trimethyllead chloride are stable over a period of at least one month if stored in the dark and refrigerated.[129] Alternately, the samples can be extracted, butylated, and dried over anhydrous sodium sulfate until analysis. To avoid adsorption of the ionic alkyllead species on the container walls, De Jonghe et al.[128] suggested the addition of 0.5 ml of hydrochloric acid to 1 l of water sample. Maguire[137] also found it desirable to acidify the water sample to pH 1 with hydrochloric acid for butyltin and inorganic tin sample for short storage.

Tetraalkyllead compounds are volatile and are not stable in water on storage. Water samples should not be filtered by suction but should be extracted with hexane immediately after collection. It was found convenient by Chau et al.[138] to add 5 ml of hexane to the water sample on collection to seal off the surface to prevent volatilization loss. The hexane phase can be used for the determination of tetraalkyllead.

C. Sample Preparation.
1. Sediment Extraction
Sediment samples can be analyzed dry or wet. In normal practice, freeze-dried and ground sediment samples are more homogeneous and reproducible in replicate analysis than samples taken from a wet batch. Dry sediment samples should be stored in tightly capped amber-glass bottles. For determination of volatile organometals such as tetraalkyllead, the sediment sample should be extracted wet with a solvent after collection.[138]

Organometals are mostly derived from anthropogenic sources or formed as a result of biological and chemical alkylation and dealkylation. These types of compounds are generally not involved in mineralogical processes. Complete dissolution of sediment samples as often required in total sediment analysis is, therefore, not considered necessary. An efficient extraction method would be appropriate for organometal analysis.

Potter et al.[139] extracted tetraalkyllead from sediment with petroleum ether, and trialkyllead with benzene, after saturation of the suspension with sodium chloride. Tetraalkyllead and trialkyllead were extracted from street dust by ammoniacal methanol by Harrison et al.[140] The tetraalkyllead compounds are hydrophobic and are readily extracted into nonpolar solvents such as benzene and hexane.

Chau et al.[29] extracted the dialkyl- and trialkyllead compounds from sediment with benzene after addition of a chelating agent, NaDDTC in the presence of sodium chloride, sodium iodide, and sodium benzoate. Similarly, methyltin species were extracted from sediment with a solution of tropolone in benzene in the presence of sodium chloride.[141] Maguire[20] found that for complete extraction of the butyltin species, the sediment had to be refluxed in a tropolone benzene solution for 2 h. In all these procedures, the organometallic compounds are extracted from sediment without changing their authentic forms.

2. Biological Samples
The major difficulties in the speciation of organometals in biological samples are (1) the lack of a suitable digestion method to dissolve the sample without changing the chemical forms of the organometal, and (2) the strong affinity of organometal with protein and lipids which cause incomplete recovery. Previous work included extraction of the tissue homogenate with benzene in the presence of sodium chloride[52,57] and sodium benzoate,[28] all of which did not adequately recover the dialkyllead species. No procedure was available to digest the tissues to release these compounds for speciation. Recently, two methods have been published on biological samples. Chau et al.[29] digested fish tissue and macrophyte samples in a tissue solubilizer, tetramethylammonium hydroxide (TMAH), followed by chelation extraction of the dialkyl- and trialkyllead with sodium diethyldithicarbamate (NaDDTC). The other method was enzyme hydrolysis developed by Forsyth and Marshall,[79] to digest the homogenate of whole eggs for dithizone/chloroform extraction of the alkyllead species. Both procedures recovered the alkyllead species, including the inorganic lead(II), satisfactorily. The extracts were correspondingly butylated[29] and phenylated[79] with appropriate Grignard reagents for final determination as described under Section V.C. TMAH digestion technique has also been applied to the analysis of methyltin compounds in fish and biological samples. In this case, the digest was extracted with a tropolone/benzene solution followed by butylation of the ionic alkyltin species for GC-AAS determination.[141]

VII. PROCEDURES FOR THE DETERMINATION OF ALKYLLEAD COMPOUNDS IN ENVIRONMENTAL SAMPLES

The following procedures are extracted from published documentation to illustrate the speciation of alkyllead compounds. Details in principles and discussion of the methods have been partially included in previous sections. Further details can be found in the original publications.

A. Water

Equipment and reagents: the details of the GC-AAS system have been described previously[75] (see Section IV.A). The sample extract is introduced directly into the gas chromatograph by a syringe. The following instrumental parameters are used.

GC: glass column, 1.8 m long, 6 mm diameter packed with 10% OV-1 on Chromosorb W (80 to 100 mesh); nitrogen carrier gas, 65 ml/min; injection port temperature 150°C, transfer line temperature, 160°C, oven temperature programs, 80 to 200°C at 5°C per min.

AAS: Pb spectral line, 217.0 nm; electrodeless discharge lamp, 10 Watts; silica furnace, electrically heated to *circa* 850 to 900°C, hydrogen flow, 85 ml/min; deuterium background corrector used.

Peak area is integrated with Autolab Minigrator (Spectra-Physics, CA). Tetramethyllead and tetraethyllead (Alfa Chemicals, Danver, MA); Standard trialkyllead solution (100 μg/ml as Pb). Dissolve 0.0152 and 0.0171 g, respectively, of trimethyllead acetate and triethyllead acetate (Alfa Chemicals, Danver, MA) in distilled water and make up to 100 ml in amber volumetric flasks.

Standard dialkyllead solution (100 μg/l as Pb). Add 3 drops of iodine monochloride to 3 to 4 ml of the above trialkyllead standard solutions. The trialkyllead compounds are spontaneously converted to the corresponding dialkyllead species. The addition of a small amount of ICI does not significantly change the concentration of trialkyllead solutions. A dialkyllead solution prepared by this method should not be mixed with the trialkyllead standards in case the residual ICI should act on the trialkyllead. Commercially supplied dialkyllead compounds generally contain trialkyllead and lead(II) impurities because of their instability in storage.

Iodine monochloride solution. Add 11 g of potassium iodide to a mixture of 40 ml water and 44.5 ml of concentrated hydrochloric acid. Slowly add, while stirring, 7.5 g potassium iodate until the iodine so formed gradually redissolves to give a light brown solution. This solution can be kept for months if stored in an amber bottle at room temperature.

n-Butyl Grignard reagent (Alfa Chemicals, Danver, MA). 1.9 *M* in tetrahydrofuran.

Sodium diethydithiocarbamate (NaDDTC). 0.5 *M*. Dissolve 113 g of NaDDTC·3H$_2$0 in 100 ml of solution.

Procedure: to one l of water (pH 6-8) add 50 ml of 0.5 *M* NaDDTC, 50 g of sodium chloride and 50 ml of benzene. Shake the mixture for 30 min. Separate the benzene phase into a 200 ml round bottom flask and reduce its column to 0.8 ml first in a rotary evaporator followed by a vortex evaporator in a 10 ml graduated centrifuge tube. Add 0.2 ml of *n*-butyl Grignard reagent, mix, and let stand for ~ 10 min. Wash the mixture with 2 ml of 0.5 *M* sulfuric acid to destroy the excess Grignard reagent. Pipet the organic phase into a small vial containing anhydrous sodium sulfate. Inject 10 to 20 μl of this solution into the GC-AAS system. The worked up sample can be stored for at least 2 weeks without deterioration. The GC-AAS chromatograms of 5 mixed tetraalkyllead, 4 ionic alkyllead and lead(II) are illustrated in Figure 7.

1. Calibration of Method

Peak areas for equal quantities of all the alkyllead species expressed as Pb are identical. Lead(II) ion after butylation gives a broader peak which is mainly due to chromatographic behaviors. A component peak emerging at a higher temperature inevitably suffers from some degree of tailing and broadening.

The butylated trialkyllead compounds are more stable and can be used as a standard for calibration. Standardize the method by adding 10 μg of each of trimethyllead and triethyllead and 20 μg of lead(II) in one liter of distilled water, and process with the samples. Use the peak area of trimethyllead for the calculation of other species in the sample. The peaks of triethyllead and of lead(II) are used as markers for retention times. Calibration graphs for all the dialkyllead and trialkyllead compounds are linear from 1 to 100 ng above which some overlapping of component peaks occur. If all the alkyllead species are not simultaneously present, and those present are not adjacent to each other, the linear range can be extended much further (to 1000 ng) until the furnace is saturated. The absolute detection limit of lead is 0.1 ng.

The above procedure has a detection limit for water (1 l) of 5 ng/l with a relative standard deviation ranging from 5.4% for Me_3Pb^+ to 9.5% for lead(II) at the 100 μg/l level.

2. Sample Storage

Tetraalkyllead compounds are volatile, they do not form true solutions but unstable suspensions. Their presence in water is unstable and can only be taken as transient concentration. They may decompose to other alkyllead species or evaporate into the atmosphere.

All ionic alkyllead compounds (Rb_2Pb^{2+}, R_3Pb^+) slowly degrade in the presence of light. However, lake water samples enriched with dimethyllead chloride and trimethyllead chloride at 100 μg/l level are stable over a period of at least one month in the laboratory when stored in the dark and refrigerated (5°C). There is no need to add any preservatives to the samples.

B. Fish and Biological Materials[29]

Equipment and chemicals: the GC-AAS system, operation parameters and standard solutions are identical to that for water analysis (Section VII.A). Tetramethylammonium hydroxide (TMAH) (Fisher Chemicals), 20% in water.

Procedure: homogenize fish samples for a minimum of five times in a commercial meat grinder. Weigh about 2 g of the homogenate in a test tube with cap and digest with 5 ml of TMAH solution in a water bath at 60°C for 2 h or until the tissues have completely dissolved into a pale yellow solution. After cooling, neutralize the mixture with 50% hydrochloric acid to pH 6-8. Extract the mixture with 3 ml of benzene for 2 h in a mechanical shaker after the addition of 2g NaCl, 3 ml of 0.5 NaDDTC. Centrifuge, and remove a measured amount (~ 1 ml) of the benzene phase into a glass vial with a stopper. Follow the same procedure as described for water (Section VII.A). Inject 10 to 20 μl of the benzene extract into the GC-AAS system.

This procedure has a detection limit of 8 ng/g for fish sample, with a relative standard deviation of 6.5% for triethyllead to 20% for diethyllead at 2.5 μg/g level.

Clam and fish intestines can be treated in a similar manner. Macrophyte samples are shredded to thin pieces before TMAH digestion. It takes longer to dissolve macrophytes than fish tissue. It is convenient to leave the mixture in a capped tube in a water bath (60°C) overnight. The dark green color pigment in macrophyte extraction does not interfere with the subsequent butylation reaction.

C. Sediment[29]

Procedure: extract a dried (2 g) or wet (5 g) sediment sample in a capped vial with 3 ml of benzene after additions of 10 ml of water, 6 g of NaCl, 1 g of KI, 2 g of sodium benzoate, 3 ml of 0.5 *M* NaDDTC and 2 g of coarse glass beads (20 to 40 mesh) for 2 h in a mechanical shaker. Centrifuge and transfer a measured amount (1 ml) of the benzene phase into a glass vial with a stopper. Follow the same procedure as described for water (Section VII.A). Inject 10 to 20 μl of the benzene extract into the GC-AAS system.

The procedure has a detection limit of 8 ng/g and a relative standard deviation of 4% for trimethyllead and triethyllead, to 15% for the dialkyllead species at 2.5 μg/g level.

VIII. CONCLUSIONS

The combination system consisting of a separation technique and an element-specific detector is currently the best analytical system for the species analysis of organometallic compounds. The separation techniques that are used, are mostly chromatographic, either in gas or liquid phase, and the element-specific detectors are atomic spectrometric in absorption, emission or fluorescence mode. Atomic absorption spectrometry has been the most widely used detection technique because of its simplicity, sensitivity, and ease of operation.

Liquid chromatography has great potential in the direct separation of organometals in solution without further derivatization, but there are still limitations in its interfacing with the furnace atomic absorption spectrometer. Plasma emission spectrometry, in spite of its inferior sensitivity, has the unique advantage of multielement detection capability, for the simultaneous analyses of organometals in one sample. Such an advantage may be preferred over the high sensitivity detection of one element. The lack of development in this technique may be due to the high cost of plasma emission instrumentation that is not easily accessible in ordinary laboratories. Another area of development is the electrochemical detection system for use with the GC or HPLC. Once the separation has been achieved, the detection of the analytes can become specific.

REFERENCES

1. **Chau, Y. K. and Wong, P. T. S.,** Occurrence of biological methylation of elements in the environment, in *Organometals and Organometalloids — Occurrence and Fate in the Environment,* ACS Symposium Series No 82, Brinckman, F. E., and Bellama, J. M., Eds., American Chemical Society, 1978, chap. 3.
2. **Challenger, F.,** Biological methylation, *Chem. Rev.,* 36, 315, 1945.
3. **Jensen, S. and Jernelov, A.,** Biological methylation of mercury in aquatic organisms, *Nature,* 223, 753, 1969.
4. **Wood, J. M.,** Biological cycles for toxic elements in the environment, *Science,* 183, 1049, 1974.
5. **Ridley, W. P., Dizikes, L. J., and Wood, J. M.,** Biomethylation of toxic elements in the environment, *Science,* 197, 329, 1977.
6. **Jernelov, A. and Martin, A.-L.,** Ecological implications of metal metabolism by microorganisms, *Ann. Rev. Microb.,* 29, 62, 1975.
7. **Saxena, J. and Howard, P. H.,** Environmental transformation of alkylated and inorganic forms of certain metals, in *Advances in Applied Microbiology,* Vol. 21, Academic Press, New York, 1977, 185.
8. **Iverson, W. P. and Brinckman, F. E.,** Microbial metabolism of heavy metals, in *Water Pollution Microbiology,* Vol. 2, Mitchell, R., Ed., John Wiley & Sons, New York, 1978, 201.
9. **Craig, P. J.,** Metal cycles and biological methylation, in *Handbook of Environmental Chemistry,* Vol. 1, Part A, Hutzinger, O., Ed., Springer-Verlag, Berlin, 1980, 169.
10. **Thayer, J. S. and Brinckman, F. E.** The biological methylation of metals and metalloids, in *Advances in Organometallic Chemistry,* Vol. 20, Stone, F. G. and West, R., Eds., Academic Press, New York, 1982, 313.
11. **Braman, R. S., Johnson, D. L., Foreback, C. C., Ammons, J. M., and Bricker, J. L.,** Separation and determination of nanogram amounts of inorganic arsenic and methylarsenic compounds, *Anal. Chem.,* 49, 621, 1977.
12. **Andreae, M. O.,** Determination of arsenic species in natural waters, *Anal. Chem.,* 49, 820, 1977.
13. **Wong, P. T. S., Chau, Y. K., Luxon, L., and Bengert, G. A.,** Methylation of arsenic in the aquatic environment, in *Trace Substances in Environmental Health,* Vol. 11, Hemphill, D. D., Ed., University of Missouri Press, Columbia, MO, 1977, 100.
14. **Andreae, M. O.,** Distribution of arsenic in natural waters and some marine algae, *Deep Sea Res.,* 25, 391, 1978.

15. **Edmonds, J. C. and Francesconi, K. A.,** Isolation, crystal structure and synthesis of arsenobetaine, the arsenical constituent of the western rock lobster *Panulirus longpipes cygnus George, Tetrahedron Lett.,* 18, 1543, 1977.

16. **Braman, R. S. and Tompkins, M. A.,** Separation and determination of nanogram amounts of inorganic tin and methyltin compounds in the environment, *Anal. Chem.,* 51, 12, 1979.

17. **Hodge, V. F., Seidel, S. L., and Goldberg, E. D.,** Determination of tin(IV) and organotin compounds in natural waters, coastal sediments and macro algae by atomic absorption spectrometry, *Anal. Chem.,* 51, 1256, 1979.

18. **Jackson, J. A., Blair, W. R., Brinckman, F. E., and Iverson, W. P.,** Gas chromatographic separation of methylstannanes in the Chesapeake Bay using purge and trap sampling with a tin selective detector, *Environ. Sci. Tech.,* 16, 110, 1982.

19. **Maguire, R. J., Chau, Y. K., Bengert, G. A., Hale, E. J., Wong, P. T. S., and Kramar, O.,** Occurrence of organotin compounds in Ontario lakes and rivers, *Environ. Sci. Tech.,* 16, 698, 1982.

20. **Maguire, R. J.,** Butyltin compounds and inorganic tin in sediment in Ontario, *Environ. Sci. Tech.,* 18, 291, 1984.

21. **Byrd, J. T. and Andreae, M. O.,** Tin and methyltin species in seawater: concentration and fluxes, *Science,* 218, 565, 1982.

22. **Brinckman, F. E., Jackson, J. A., Blair, W. R., Olsen, G. J., and Iverson, W. P.,** Ultratrace speciation and biogenesis of methyltin transport species in estuarine waters, in *Trace Metals in Seawater,* Wong, C. S., Boyle, E., Bruland, K. W., Burton, J. D., Goldberg, E. D., Eds., Plenum Publishing, New York, 1983, 39.

23. **Chau, Y. K., Wong, P. T. S., and Bengert, G. A.,** Determination of methyltin(IV) and tin(IV) species in water by gas chromatography-atomic absorption spectrometry, *Anal. Chem.,* 54, 246, 1982.

24. **Harrison, R. M.,** The analysis of tetraalkyllead compounds and their significance as urban air pollutants, *Atmos. Environ.,* 11, 87, 1977.

25. **Radziuk, B., Thomassen, Y., Van Loon, J. C., and Chau, Y. K.,** Determination of alkylated lead compound in air, *Anal. Chim. Acta.,* 105, 255, 1979.

26. **De Jonghe, W. R. A., Chakraborti, D., and Adams, F. C.,** Sampling of tetraalkyllead compounds in air for determination by gas chromatography-atomic absorption spectrometry, *Anal. Chem.,* 52, 1974, 1980.

27. **Chau, Y. K., Wong, P. T. S., Kramer, O., Bengert, G. A., Cruz, R. B., Kinrade, J. O., Lye, J., and Van Loon, J. C.,** Occurrence of tetraalkyllead compounds in the aquatic environment, *Bull. Environ. Contam. Toxicol.,* 24, 265, 1980.

28. **Birnie, S. E., Hodges, D. J.,** Determination of ionic alkyl lead species in marine fauna, *Environ. Tech. Lett.,* 2, 433, 1981.

29. **Chau, Y. K., Wong, P. T. S., Bengert, G. A., and Dunn, J. L.,** Determination of dialkyllead, trialkyllead, tetraalkyllead and lead(II) compounds in sediment and biological samples, *Anal. Chem.,* 56, 271, 1984.

30. **Chau, Y. K., Wong, P. T. S., Silverberg, B. A., Luxon, L., and Bengert, G. A.,** Methylation of selenium in the aquatic environment, *Science,* 192, 1130, 1976.

31. **Reamer, D. C. and Zoller, W. H.,** Selenium biomethylation products from soil and sewage sludge, *Science,* 208, 500, 1980.

32. **Jiang, S., De Jonghe, W., and Adams, F.,** Determination of alkylselenide compounds in air by gas chromatography-atomic absorption spectrometry, *Anal. Chim. Acta,* 136, 183, 1980.

33. **Andreae, M. O. and Barnard, W. R.,** Determination of trace quantities of dimethyl sulfide in aqueous solutions, *Anal. Chem.,* 55, 608, 1983.

34. Chau, Y. K., unpublished data, 1977.

35. **Johnson, D. L. and Braman, R. S.,** The speciation of arsenic and the content of germanium and mercury in members of the pelagic Sargassum community, *Deep Sea Res.,* 22, 503, 1975.

36. **Tugrul, S., Balkas, T. I., and Goldberg, E. D.,** Methyltins in the marine environment, *Mar. Poll. Bull.,* 14, 297, 1983.

37. **Ishii, T.,** Tin in marine algae, *Bull. Jpn. Soc. Sci. Fish.,* 18, 1609, 1982.

38. **Seidel, S. L., Hodge, V. F., and Goldberg, E. D.,** Tin as an environmental pollutant, *Thallassia Jugoslavica,* 16, 209, 1980.

39. **Chiba, K., Yoshida, K., Tanabe, K., Haragushi, H., and Fuwa, K.,** Determination of alkylmercury in seawater at the nanogram per litre level by gas chromatography/atmospheric pressure helium microwave-induced plasma emission spectrometry, *Anal. Chem.,* 55, 450, 1983.

40. **Jiang, S., Robberecht, H., and Adams, F.,** Identification and determination of alkylselenide compounds in environmental air, *Atmos. Environ.,* 17, 111, 1983.

41. **Batti, R., Magnaval, R., and Lnazola, E.,** Methylmercury in river sediments, *Chemosphere,* 1, 13, 1975.

42. **Olsen, B. H. and Cooper, N. C.,** Comparison of aerobic and anaerobic methylation of mercuric chloride by San Francisco Bay sediments, *Water Res.,* 10, 113, 1976.

43. **Bartlett, P. D., Craig, P. J., and Morton, S. F.,** Total mercury and methyl mercury levels in British estuarine and marine sediments, *Sci. Total Environ.,* 10, 245, 1978.

44. **Chau, Y. K. and Saitoh, H.**, Determination of methylmercury in lake water, *Int. J. Environ. Anal. Chem.*, 3, 133, 1973.

45. **Kamps, L. R., Carr, R., and Miller, H.**, Total mercury and monomethylmercury content of several species of fish, *Bull. Environ. Contam. Toxicol.*, 8, 273, 1972.

46. **Kojima, S.**, Separation of organotin compounds by using difference in partition behaviour between hexane and methanolic solution, *Analyst*, 104, 660, 1979.

47. **Farnsworth, M. and Pekola, J.**, Determination of tin in inorganic and organic compounds and mixtures, *Anal. Chem.*, 31, 410, 1959.

48. **Beccaria, A. M., Mor, E. D., and Poggi, G.**, A method for the analysis of traces of inorganic and organic lead compounds in marine sediments, *Ann Chim.*, 68, 607, 1978.

49. **Nembrini, P. G., Dogan, S., and Haerdi, W.**, Simultaneous determination of tin, lead and copper by AC polarography, *Anal. Lett.*, 13(A11), 947, 1980.

50. **Nangniot, P. and Martens, P. H.**, Application de la chroneampèrométrique par redissolution anodique au dosage de traces d'acétate de triphenyl-etain, *Anal. Chim. Acta.*, 24, 276, 1961.

51. **Marr, I. L. and Anwar, J.**, Micro-determination of tin in organotin compounds by flame emission and atomic absorption spectrophotometry, *Analyst*, 107, 260, 1982.

52. **Bolanowska, W.**, A method for the determination of triethyllead in blood and urine, *Chem. Anal., (Warsaw)*, 12, 121, 1967.

53. **Aldridge, W. N. and Street, B. W.**, Spectrophotometric and fluorimetric determination of tri- and di-organotin and -organolead compounds using dithizone and 3-hydroxyflavone, *Analyst*, 106, 60, 1981.

54. **Blunden, S. J. and Chapman, A. H.**, Fluorimetric determination of triphenyltin compounds in water, *Analyst*, 103, 1266, 1978.

55. **Skeel, R. T. and Bricker, C. E.**, Spectrophotometric determination of di-butyltin dichloride, *Anal. Chem.*, 33, 428, 1961.

56. **Arakawa, Y. and Wada, O.**, Extraction and fluorometric determination of organic compounds with morin, *Anal. Chem.*, 55, 1901, 1983.

57. **Cremer, J. E.**, Biochemical studies on the toxicity of tetraethyllead and other organo-lead compounds, *Brit. J. Ind. Med.*, 16, 191, 1959.

58. **Moss, R. and Browett, E. V.**, Determination of tetraalkyllead vapour and inorganic lead dust in air, *Analyst*, 91, 428, 1966.

59. **Hancock, S. and Slater, A.**, A specific method for the determination of trace concentration of tetramethyl and tetraethyllead vapours in air, *Analyst*, 100, 422, 1975.

60. **Pilloni, P. and Plazzogna, G.**, Spectrophotometric determination of dialkyllead and dialkyltin ions with 4-(2-pyridylazo)-resorcinol, *Anal. Chim. Acta*, 35, 325, 1966.

61. **Henderson, S. R. and Snyder, L. J.**, Rapid spectrophotometric determination of triethyllead, diethyllead and inorganic lead ions, and application to determination of tetraorganolead compounds, *Anal. Chem.*, 33, 1172, 1961.

62. **Noden, G. F.**, The determination of tetraalkyllead compounds and their degradation products in natural water, in *Lead in the Marine Environment*, Branica, M. and Konrad, Z., Eds., Pergamon Press, Oxford, 1980, 83.

63. **Parker, W. W., Smith, G. Z., and Hudson, R. L.**, Determination of mixed lead alkyls in gasoline by combined gas chromatographic and spectrophotometric techniques, *Anal. Chem.*, 33, 1170, 1961.

64. **Lovelock, J. E. and Zlatkis, A.**, A new approach to lead alkyl analysis: gas phase electron absorption for selective detection, *Anal. Chem.*, 33, 1958, 1961.

65. **Dawson, H. J., Jr.**, Determination of methy-ethyl lead alkyls in gasoline by gas chromatography with an electron capture detector, *Anal. Chem.*, 35, 542, 1963.

66. **Bonelli, E. J. and Hartmann, H.**, Determination of lead alkyls by gas chromatography with electron capture detector, *Anal. Chem.*, 35, 1980, 1963.

67. **Soulages, N. L.**, Simultaneous determination of lead alkyls and halide scavengers in gasoline by gas chromatography with flame ionization detection, *Anal. Chem.*, 38, 28, 1966.

68. **De Jonghe, W. and Adams, F.**, Gas chromatography with flame ionization detection for the speciation of trialkyllead halides, *Fresenius Z. Anal. Chem.*, 314, 552, 1983.

69. **Hill, H. H., Jr. and Aue, W. A.**, Selective detection of organometallics in gas chromatographic effluents by flame photometry, *J. Chromatogr.*, 74, 311, 1972.

70. **Aue, W. A. and Hill, H. H., Jr.**, A hydrogen-rich flame ionization detector sensitive to metals, *J. Chromatogr.*, 74, 319, 1972.

71. **Bache, C. A. and Lisk, D. J.**, Gas chromatographic determination of organic mercury compounds by emission spectrometry in a helium plasma, *Anal. Chem.*, 43, 950, 1971.

72. **Dressman, R. C.**, A new method for the gas chromatographic separation and detection of dialkylmercury compounds — application to river water analysis, *J. Chrom. Sci.*, 10, 472, 1972.

73. **Van Loon, J. C.**, Metal speciation by chromatography/atomic absorption spectrometry, *Anal. Chem.*, 51, 1139A, 1979.

74. **Ebdon, L., Hill, S., and Ward, R. W.,** Directly coupled chromatography-atomic spectroscopy. I. Directly coupled gas chromatography-atomic spectroscopy — a review, *Analyst,* 11, 1113, 1986.
75. **Chau, Y. K., Wong, P. T. S., and Goulden, P. D.,** Gas chromatography-atomic absorption spectrometry for the determination of tetraalkyllead compounds, *Anal. Chim. Acta,* 85, 421, 1976.
76. **Reamer, D. C., Zoller, W. H., and O'Haver, T. C.,** Gas chromatograph microwave plasma detector for the determination of tetraalkyllead species in the atmosphere, *Anal. Chem.,* 50, 1449, 1978.
77. **De Jonghe, W., Chakraborti, D., and Adams, F.,** Graphite furnace atomic absorption spectrometry as a metal specific detection system for tetraalkyllead compounds separated by gas-liquid chromatography, *Anal. Chim. Acta,* 115, 89, 1980.
78. **Estes, S. A., Uden, P. C., and Barnes, R. M.,** Determination of n-butylated trialkyllead compounds by gas chromatography with microwave plasma detection, *Anal. Chem.,* 54, 2402, 1982.
79. **Forsyth, D. S., and Marshall, W. D.,** Determination of alkyllead salts in water and whole eggs by capillary gas chromatography with electron capture detection, *Anal. Chem.,* 55, 2132, 1983.
80. **Kolb, B., Kemner, G., Schlesser, F. H., and Wiedeking, E.,** Elementspezifische Anzeige gaschromatographisch getrennter Metallverbindungen mittels Atom- absorptions- Spektroskopie (AAS), *Fresenius Z. Anal. Chem.,* 221, 166, 1966.
81. **Chau, Y. K., Wong, P. T. S., and Saitoh, H.,** Determination of tetraalkyllead compounds in the atmosphere, *J. Chrom. Sci.,* 14, 162, 1976.
82. **Harrison, R. M., Perry, R., and Slater, D. H.,** An adsorption technique for the determination of organic lead in street air, *Atmos. Environ.,* 8, 1187, 1974.
83. **Coker, D. T.,** Determination of individual and total lead alkyls in gasoline by a single rapid gas chromatography/atomic absorption spectrometry technique, *Anal. Chem.,* 47, 386, 1975.
84. **Wolf, W. R.,** Coupled gas chromatography-atomic absorption spectrometry, *J. Chromatogr.,* 134, 159, 1977.
85. **Ebdon, L., Ward, R. W., and Leathard, D. A.,** Development and optimization of atom cells for sensitive coupled gas chromatography flame absorption spectrometry, *Analyst,* 107, 129, 1982.
86. **Chau, Y. K., Wong, P. T. S., and Goulden, P. D.,** Gas chromatography-atomic absorption method for determination of dimethyl selenide and dimethyl diselenide, *Anal. Chem.,* 47, 2279, 1975.
87. **Chau, Y. K. and Wong, P. T. S.,** An element- and speciation-specific technique for the determination of organometallic compounds, in *Environmental Analysis,* Ewing, G. W., Ed., Academic Press, New York, 1977, 215.
88. **Segar, D. A.,** Flameless atomic absorption gas chromatography, *Anal. Lett.,* 7, 89, 1974.
89. **Robinson, J. W., Kiesel, E. L., Goodbread, J. P., Bliss, R., and Marshall, R.,** The development of a gas chromatography-furnace absorption combination for the determination of organic lead compounds. Atomization processes in furnace atomizers, *Anal. Chim. Acta,* 92, 321, 1977.
90. **Parris, G. E., Blair, W. R., and Brinckman, F. E.,** Chemical and physical considerations in the use of atomic absorption detectors coupled with a gas chromatograph for determination of trace organometallic gases, *Anal. Chem.,* 49, 378, 1977.
91. **Andreae, O. M. and Froelich, P. N., Jr.,** Determination of germanium in natural waters by graphite furnace atomic absorption spectrometry with hydride generation, *Anal. Chem.,* 53, 287, 1981.
92. **Ahmad, I., Chau, Y. K., Wong, P. T. S., Carty, A. J., and Taylor, L.,** Chemical alkylation of lead(II) salts to tetraalkyllead(IV) in aqueous solution, *Nature,* 287, 716, 1980.
93. **Estes, S. A., Uden, P. C., and Barnes, R. M.,** High resolution gas chromatography of trialkyllead chloride with an inert solvent venting interface for micro-wave excited helium plasma detection, *Anal. Chem.,* 53, 1336, 1981.
94. **Talmi, Y. and Bostick, D. T.,** Determination of alkylarsenic acids in pesticide and environmental samples by gas chromatography with a microwave emission spectrometric detection system, *Anal. Chem.,* 47, 2145, 1975.
95. **Laveskog, A.,** A method for determination of tetramethyl lead (TML) and tetraethyl lead (TEL) in air, *Proc. 2nd Intl. Clean Air Congr.,* 1970, 549.
96. **Meinema, H. A., Burger-Wiersma, T., Versluis-de Haan, G., and Gevers, E. Ch.,** Determination of trace amounts of butyltin compounds in an aqueous system by gas chromatography/mass spectrometry, *Environ. Sci. Tech.,* 12, 288, 1978.
97. **Willeford, B. R. and Veening, H.,** High-performance liquid chromatography: applications to organometallic and metal coordination compounds, *J. Chromatogr.,* 251, 61, 1982.
98. **Brinckman, F. E., Blair, W. R., Jewett, K. L., and Iverson, W. P.,** Application of a liquid chromatograph coupled with a flameless atomic absorption detector for speciation of trace organometallic compounds, *J. Chrom. Sci.,* 15, 493, 1977.
99. **Messman, J. D. and Rains, T. C.,** Determination of tetraalkyllead compounds in gasoline by liquid chromatography-atomic absorption spectrometry, *Anal. Chem.,* 53, 1632, 1981.

100. **Stockton, R. A. and Irgolic, K. L.,** The Hitachi graphite furnace-Zeeman atomic absorption spectrometer as an automated element-specific detector for high pressure liquid chromatography, *Int. J. Environ. Chem.,* 6, 313, 1979.

101. **Brinckman, F. E., Jewett, K. L., Iverson, W. P., Irgolic, K. J., Ehrhardt, K. C., and Stockton, R. A.,** Graphite furnace atomic absorption spectrophotometers as automated element-specific detectors for high pressure liquid chromatography. The determination of arsenite, arsenate, methylarsonic acid and dimethylarsenic acid, *J. Chromatogr.,* 191, 31, 1980.

102. **Woolson, E. A. and Aharonson, N.,** Separation and detection of arsenical pesticide residues and some of their metabolites by HPLC-graphite furnace-atomic absorption spectrometry, *J. Assoc. Off. Anal. Chem.,* 63, 523, 1980.

103. **Chau, Y. K. and Wong, P. T. S.,** Direct speciation analysis of molecular and ionic organometals, in *Trace Element Speciation in Surface Waters and Its Ecological Implications,* Leppard, G. G. Ed., Plenum Publishing, New York, 1983, 87.

104. **Koropchak, J. A. and Coleman, G. N.,** Investigations of nebulizer parameters for on-line flame atomic absorption detector of liquid chromatographic effluents, *Anal. Chem.,* 52, 1252, 1980.

105. **Jones, D. R., IV and Manahan, S. E.,** Detection limits for flame spectrophotometric monitoring of high speed liquid chromatographic effluents, *Anal. Chem.,* 48, 1897, 1976.

106. **Cantillo, A. Y. and Segar, D. A.,** Metal species identification in the environment, a major challenge for the analyst, *Proceedings of the International Conference on Heavy Metals in the Environment,* Toronto, 1975, 183.

107. **Vickrey, T. M., Howell, H. E., and Paradise, M. T.,** Liquid chromatogram peak storage and analysis by atomic absorption spectrometry, *Anal. Chem.,* 51, 1880, 1979.

108. **Koizumi, H., Hadeishi, T., and McLaughlin, R.,** Speciation of organometallic compounds by Zeeman atomic absorption spectrometry with liquid chromatography, *Anal. Chem.,* 50, 1770, 1978.

109. **Ricci, G. R., Colovos, G., and Hester, N. E.,** Ion chromatography with atomic absorption spectrometric detector for determination of organic and inorganic arsenic species, *Anal. Chem.,* 53, 610, 1981.

110. **Burns, D. T., Glockling, F., and Harriott, M.,** Investigation of the determination of tin tetraalkyls and alkyltin chloride by atomic absorption spectrometry after separation by gas-liquid chromatography or high pressure liquid-liquid chromatography, *Analyst,* 106, 921, 1981.

111. **Chau, Y. K.,** Occurrence and speciation of organometallic compounds in freshwater systems, *Sci. Total Environ.,* 49, 305, 1986.

112. **Uden, P. C., Quimby, B. D., Barnes, R. M., and Elliott, W. G.,** Interfaced D.C. argon-plasma emission spectroscopic detection for high-pressure liquid chromatography of metal compounds, *Anal. Chim. Acta,* 101, 99, 1978.

113. **Hausler, D. W. and Taylor, L. T.,** Non-aqueous on-line simultaneous determination of metals by size exclusion chromatography with inductively coupled atomic emission spectrometric detection, *Anal. Chem.,* 53, 1223, 1981.

114. **McGuffin, V. L. and Novotny, M.,** Micro-column high-performance liquid chromatography and flame-based detection principles, *J. Chromatogr.,* 218, 179, 1981.

115. **Buchta, R. C. and Papa, L. J.,** Electrochemical detector for liquid chromatography, *J. Chrom. Sci.,* 14, 213, 1976.

116. **Kissinger, P. T.,** Amperometric and coulometric detectors for high-performance liquid chromatography, *Anal. Chem.,* 49, 447A, 1977.

117. **McCrehan, W. A., Durst, R. A., and Bellama, J. M.,** Electrochemical detection in liquid chromatography: application to organometallic speciation, *Anal. Lett.* 10, 1175, 1977.

118. **McCrehan, W. A. and Durst, R. A.,** Measurement of organomercury species in biological samples by liquid chromatography with differential pulse electrochemical detection, *Anal. Chem.,* 50, 2108, 1978.

119. **Kutner, W., Debowski, J., and Kemula, W.,** Polarographic detection for high-performance liquid chromatography using a flow-through detector, *J. Chrom.,* 191, 1980.

120. **Pungor, E., Toth, K., Feher, Zs, Nagy, G., and Varadi, M.,** Application of electroanalytical detectors in chromatography, *Anal. Lett.,* 8, ix, 1975.

121. **Booth, M. D. and Fleet, B.,** Electrochemical behaviours of triphenyltin compounds and their determination of submicrogram levels by anodic stripping voltammetry, *Anal. Chem.,* 42, 825, 1970.

122. **Litan, R., Basters, J., Martijn, A., Van der Molen, T., Pasma, T., Rabenort, B., and Smink, J.,** Potentiometric method for determining triphenyltin compounds in formulations — collaborative study, *J. Assoc. Off. Anal. Chem.,* 61, 1504, 1978.

123. **Hasebe, K., Yamamoto, Y., and Kambara, K.,** Differential pulse polarography determination of organotin compounds coated on fishing nets, *Fresenius Z. Anal. Chem.,* 310, 234, 1982.

124. **Plazzogna, G. and Pilloni, G.,** Amperometric titration of organolead and organotin ions, *Anal. Chim. Acta,* 37, 260, 1967.

125. **Hodges, D. J. and Noden, F. G.,** The determination of alkyl lead species in natural waters by polarographic techniques, *Proceedings of the Management and Control of Heavy Metals in the Environment,* CEP Consultants, Edinburgh, 1979, 408.

126. **Colombini, M. P., Corbini, G., Fuoco, R., and Papoff, P.,** Speciation of tetra-, tri-, dialkyllead compounds and inorganic lead at nanomolar levels (sub-ppb) in water samples by differential pulsed electrochemical techniques, *Ann. Chim.,* 71, 609, 1981.

127. **Hayakawa, K.,** Microdetermination and dynamic aspects of *in vivo* alkyllead compounds, *Jpn. J. Hyg.,* 26, 377, 1971.

128. **De Jonghe, W. R. A., Van Mol, W. E., and Adams, F. C.,** Determination of trialkyllead compounds in water by extraction and graphite atomic absorption spectrometry, *Anal. Chem.,* 55, 1050, 1983.

129. **Chau, Y. K., Wong, P. T. S., and Kramar, O.,** The determination of dialkyllead, trialkyllead, tetraalkyllead and lead(II) ions in water by chelation/extraction and gas chromatography/atomic absorption spectrometry, *Anal. Chim. Acta,* 146, 211, 1983.

130. **Braman, R. S. and Tompkins, M. A.,** Atomic emission spectrometric determination of antimony, germanium, and methylgermanium compounds in the environment, *Anal. Chem.* 50, 1088, 1978.

131. **Hambrick, G. A., III, Froelich, P. N., Jr., Andreae, M. O., and Lewis, B. L.,** Determination of methylgermanium species in natural waters by graphite furnace atomic absorption spectrometry with hydride generation, *Anal. Chem.,* 56, 421, 1984.

132. **Maguire, R. J. and Huneault, H.,** Determination of butyltin species in water by gas chromatography with flame photometric detection, *J. Chromatogr.,* 209, 458, 1981.

133. **Chakraborti, D., DeJonghe, W. R. A., Van Mol, W. E., Van Cleuvenberg, R. J. A., and Adams, F. C.,** Determination of ionic alkyllead compounds in water by gas chromatography/atomic absorption spectrometry, *Anal. Chem.,* 56, 2692, 1984.

134. **Harrison, R. M. and Radojevic, M.,** Determination of tetraalkyllead and ionic alkyllead compounds in environmental samples by butylation and gas chromatography — atomic absorption, *Environ. Technol. Lett.,* 6, 129, 1985.

135. **Rapsomonikis, S., Donard, O. F. X., and Weber, J. H.,** Speciation of lead and methyllead ions in water by chromatography/atomic absorption spectrometry after ethylation with sodium tetraethylborate, *Anal. Chem.,* 58, 35, 1986.

136. **Radojevic, M., Allen, A., Rapsomanikis, S., and Harrison, R. M.,** Propylation technique for the simultaneous determination of tetraalkyllead and ionic alkyllead species by gas chromatography/atomic absorption spectrometry, *Anal. Chem.,* 58, 658, 1986.

137. **Maguire, R. J.,** personal communication.

138. **Chau, Y. K., Wong, P. T. S., Bengert, G. A., and Kramar, O.,** Determination of tetraalkyllead compounds in water, sediment and fish samples, *Anal. Chem.,* 51, 186, 1979.

139. **Potter, H. R., Jarvie, A. W. P., and Markall, R. N.,** Detection and determination of alkyl lead compounds in natural waters, Water Pollution Control, England, 1977, 123.

140. **Harrison, R. M.,** Organic lead in street dust, *J. Environ. Sci. Health,* A11(6), 417, 1976.

141. **Chau, Y. K.,** unpublished data.

142. **Chau, Y. K.,** personal communication.

Chapter 9

HUMIC ACID AND RELATED SUBSTANCES IN THE ENVIRONMENT

J. Lawrence

TABLE OF CONTENTS

I. INTRODUCTION

Natural organic compounds, including humic substances (humic and fulvic acids) tannins, lignins, amino acids, phenols, carbohydrates, and hydrocarbons are widely distributed throughout the environment. The plethora of information available on these natural products precludes the possibility of a detailed presentation of the subject in one chapter. Many excellent reviews and chapters have been written on natural organic compounds[1-10] and the reader is referred to these for a more complete coverage of the subject. In addition the chemistry of tannins and lignins has been discussed at length in numerous plant biochemistry texts.[11-14] Rather than present the entire subject superficially, the present chapter will concentrate on the extraction, characterization, and analysis of the dissolved form of the four major classes: humic acid, fulvic acid, tannins, and lignins. Brief mention of the other compounds will only be made when it is pertinent to the main discussion. Although much knowledge has been gained about these natural products over the last 100 years, there is still much to be learned about their origin, structure, and role in the aquatic environment.

Humic acid, fulvic acid, tannins, and lignins occur abundantly throughout the aquatic environment[9,14,15] in marshes, lakes, streams, rivers, and to a lesser extent the sea.[9,17] They are complex mixtures of plant and animal origin in various stages of biological and chemical decomposition. These compounds are mainly responsible for the yellow-brown coloration of many waters[18] and account for the major portion of dissolved organic matter in natural waters, often totalling 50 to 80% of the dissolved organic carbon.[9]

Unlike most chemical constituents in water, these compounds are not well defined with specific chemical structure. Berzelius[19] first isolated dissolved humic material in 1833 from a mineral spring in Sweden. He noted similarities between this material and the humic material extracted from soil. Humic substances are a series of acidic, dark colored, predominantly aromatic, chemically complex, high molecular weight polyelectrolytes. They are highly resistant to microbial decomposition and represent one of the most stable fractions of natural organic matter. Dissolved humic material can be divided into two main fractions, humic acid and fulvic acid, based on their solubility in acid and alkali.[1] Humic acid is that fraction that is soluble at pH 9 but insoluble at pH 2, whereas fulvic acid is that fraction which is soluble at pH 2 and at pH 9.

Tannins and lignins are complex polycyclic aromatic compounds sythesized by plants. They are, in fact, the major constituents of most plants and according to Brauns and Brauns[12] in the case of pine needles they represent 25 to 30% of the organic matter. Lignins consist of a series of cross linked phenyl propane monomers with some phenolic and methoxyl groups attached to the phenyl rings.[11] The name tannin was first used by Seguin in 1796 to describe the substances present in vegetative extracts which were responsible for converting putrescible animal skins into a stable leather product. Tannins are either polymers of gallic acid linked to glucose residues (hydrolysable tannins) or polymers containing flavonoid nuclei[14] (condensed tannins). Both tannins and lignins are highly resistant to biodegradation, however, certain microorganisms, notably the white wood rot fungi, are known to decompose lignin oxidatively via phenolase enzyme.[20] Although tannins and lignins have always been present in the aquatic environment as a result of vegetative degradation, they are now frequently found in the effluents from paper pulping and tanning industries.[21]

By comparison with the humics, tannins, and lignins, the other naturally occurring organic compounds in aquatic systems are of relatively low molecular weight and are more easily degraded by microorganisms. Hence, they are relatively short lived. The main compounds in this group are the amino acids, carbohydrates, hydrocarbons, and phenols.[22] Amino acids are the building blocks of proteins and as such occupy a central position in the structure of all living systems.[23] They are simple organic molecules with at least one carboxylic and one amino group per molecule. Of the numerous structures that are theoretically possible, only

20 occur in proteins of organisms and find their way into the aquatic environment. Carbohydrates make up the basic units of plant cells. The most common monosaccharides found in biotic sources are the pentoses and hexoses.[24] Monosaccharides occur as straight chain or ring compounds and polymerise with loss of water to yield di-, tri-, tetra-, and polysaccharides. The presence of hydrocarbons in the aquatic environment is derived from terrigenous and aquatic biotic processes. Terrigenous plants tend to yield n-alkanes with chain lengths of from 23 to 35 carbon atoms, while aquatic plants yield n-alkhanes with from 14 to 23 carbon atoms per chain. Phenolic compounds are widely distributed throughout the aquatic environment. They range from the high molecular weight polycyclic compounds such as humics tannins and lignins to low molecular weight monomeric compounds containing from 6 to 40 carbon atoms such as simple phenols and flavonoids.

Most dissolved natural organic matter originates from the biodegradation of the flora and fauna. Living systems consist of high molecular weight polymeric materials which on degradation break down into yellow colored substances that we frequently find in our surface waters. Several mechanisims have been proposed for the formation of humic substances, tannins and lignin.[13,14,25] The plant alteration hypothesis proposed by Felbeck[25] suggests that plant tissue, which is relatively resistant to microbial decay, undergoes superficial decomposition to a high molecular weight humic type substance which bears strong resemblance to the original plant material. Further microbial attack then reduces this humic material to fulvic acid and eventually to carbon dioxide and water. Another mechanism proposed by Felbeck involves microbial degradation of plant and animal matter to small molecular weight components which then undergo chemical oxidation and polymerization to form humic like substances which bear little resemblance to the original organic material. Two other mechanisms also proposed by Felbeck suggest that humic substances are produced by autolysis of plant and microbial cells after their death or synthesized intracellularly and released upon death of the organism. In reality it is likely that all four mechanisms probably occur simultaneously.

A study carried out by Nord[26] attempted to examine the chemical properties of isolated pine and spruce lignins after decay by several white rot fungi. Nord was able to identify some 15 compounds after degradation, including vanillic, p-hydroxybenzoic, feralic and p-hydroxycinnamic acids, 4-hydroxy-3-methoxy-phenylpyruvic acid, vanillin, dehydrodivanillin, coniferaldehyde, p-hydroxycinnamaldehyde, guaiacylglycerol, and its β-coniferyl ether. It appears from the work of Henderson[27] that the first step in the degradation of softwood lignin by white rot fungi consists of loss of methoxyl groups and the cleavage of certain ether linkages. Vanillic acid and vanillin appear to be the major degradation products. The dissolved lignin component of surface waters is likely a complex of monomeric and polymeric degradation products. Tannins are sometimes found in high concentrations in lakes and rivers as a result of leaching of wood bark or leaf litter.[14] In addition to vegetative degradation, tannins also enter the aquatic environment in the discharge from the tanning industry and spent boiler water where it is used to reduce scale formation.

II. ENVIRONMENTAL CONCENTRATIONS

The concentration of natural organic compounds varies widely in different waters. Ground water[28] and sea water[17,29,30] usually contain the lowest concentrations of around 0.02 to 0.5 mg/l (carbon) while relatively colorless streams rivers and lakes contain from about 0.5 to 4 mg/l.[9,15,16] Concentrations of 25 to 30 mg/l[9,15,31] have been reported for some highly colored lake, river, and bog waters. Comparison of these levels with total pesticide concentrations which are typically 0.01 μg/l or lower[32] shows that the natural organics account for by far the major fraction of organic compounds, especially in the areas removed from direct industrial influence. Table 1 shows the concentration of the predominant classes of

Table 1

PREDOMINANT ORGANIC COMPOUNDS IN SOME CANADIAN RIVERS

Sample location	DOC (mg/l)	Rel. color (col. units)	Humic acid (mg/l)	Fulvic acid (mg/l)	Tannins and lignins (mg/l)	Phenols (mg/l)	Carbohydrate (mg/l)	Amino acids (µg/l)	Total pesticides (µg/l)
Thompson River @ Savona (B.C.)	6	17	N.D.	2.1	0.4	N.D.	0.03	172	0.006
Saskatchewan River above Carrot River	13	10	1.7	8.0	1.3	N.D.	0.1	high	0.006
Yukon River @ Dawson (Yukon)	5		N.D.	2.7	0.6	N.D.	0.2	676	0.002
Qu'Appelle River @ Welby (Sask.)	12	20	0.8	12.0	1.4	N.D.	0.07	low	0.004
Moose River @ Mouth (Ont.)	18		7.1	11.3	4.0		N.D.	291	0.017
Richelieu River @ St.Helaire (Que.)	8	10	N.D.	7.7	0.3	N.D.	0.09	189	0.006
St. Lawrence River @ Levis (Que.)		10	0.3	6.8	0.7	0.003	0.08	low	0.004
Annapolis River @ Wilmot (NS)	7		0.6	5.2	1.2	N.D.		253	N.D.
St. John River @ Woodstock (NB)	14	30	N.D.	12.6	4.0	N.D.	0.1	203	0.001
Exploits River @ Millertown (Newf.)	7		0.6	5.5	1.4	N.D.	0.11	100	0.004

Table 2
CONCENTRATION OF NATURAL ORGANICS
IN STREAMS IN THE MARMOT BASIN[16]

	Cabin Creek	Twin Forks	Middle Creek
Humic and fulvic acid	3.3	7.3	5.5
Tannins and lignins	1.0	1.0	1.1
Phenols	0.05	0.03	0.04
Amino acids	0.03	0.11	0.12
Fatty acids	0.002	0.01	0.07
Carbohydrates	0.17	0.65	0.34

organic compounds for a selection of Canadian rivers.[15] Each site was chosen to represent typical conditions for that area. Inspection of the data indicates that humic fulvic acid, tannins, and lignins (the refractory organics) account for the major part of the soluble organic content of the water and in fact, the sum of these parameters agrees reasonably well with the measured DOC. From this one can infer that no major organic components have been neglected, unless they are of such a nature that they have been included in the humic- and fulvic acid fractions or excluded from the measured DOC. In most cases the concentration of fulvic acid is at least a factor of 10 greater than that of humic acid. This ratio of 10 has been observed by other workers[33] but one might have expected it to have been less in waters where the pH was relatively high (~pH 8). Most of the rivers with relatively high tannins and lignins were fed from forested drainage basins or had known local sources such as a pulp mill upstream of the sampling location. The concentrations are of the same order as those reported by other researchers.[34]

Telang et al.[16,33] have reported the level of organics in several pristine mountain streams in the Canadian Rocky Mountains about 100 km southwest of Calgary. Table 2 shows values for three streams in the Marmot drainage basin. These levels are very comparable to the concentrations found in the ground water in the Marmot basin. This is perhaps not surprising since the authors showed that with the exception of during the spring run-off, the stream flow of the Marmot is derived mainly from groundwater sources. Compared with the larger rivers in Table 1, the Marmot streams are very low in natural organics, and the refractory organics only account for about 25% of the DOC. Wetlands such as marshes, bogs, and swamps contain very high levels of aquatic humus due to the decomposition of vegetative material.[9,15] Since the primary productivity of plant material is high in wetlands, the amount of decomposition product is correspondingly high. One sample collected from the Hudson Bay Lowlands (fen drainage water) had a DOC of 25 mg/l. The DOC was principally fulvic acid, 23 mg/l while humic acid was only 1.7 mg/l.

In lakes the concentration of natural organics varies with the trophic status of the lake[9] with olegotrophic lakes having the lowest concentration (< 1 mg/l) through mesotrophic (1.0 to 1.5 mg/l carbon) to eutrophic lakes with levels in excess of 1.5 mg/l. The origin of humic substances in lakes can be aquatic or terrestrial depending on whether the inputs are from marshes, streams, or algae from within the lake itself. Marshes and streams tend to contribute more to small lakes whereas in larger lakes the algae inputs are of greater significance.

The concentrations of amino acids in Tables 1 and 2 are within the range of 0.01 to 0.9 mg/l with the major fractions being in the combined form. The values in Table 1 are higher than those reported by Telang et al.[16] in Table 2 but the latter were measured on filtered water samples while the former were measured on whole water samples. Of the 11 amino acids measured in Table 1,* proline, alanine, glycine, and ornithine were the most abundant. Proline, alanine, and glycine are three of the more stable amino acids in aquaous solution which would account for their higher concentrations.

Very little information is available on the seasonal variations of organic compounds in aquatic systems. Telang et al.[16] report seasonal concentrations for winter (January to April), spring/summer (May to August), and fall (September to December) for the Marmot stream system. These authors reported the concentration of humic and fulvic acids to be relatively stable throughout the fall and winter but to rise by about 60% during the spring. Since humic substances are not likely to be found in the stream so close to the source, the high organic loading in spring is probably due to run-off from the soil and litter leachout. The peak concentration in spring corresponds to the period of peak flow of the stream. Oliver et al.[35] have shown that intense ultraviolet radiation can photochemically degrade humic substances so this could account for the decreased concentration levels observed in the Marmot during the fall and winter when the suspended particle load is likely to be much reduced. A similar peak in concentration of fulvic acid during the spring has been reported by Thurman for an alpine stream in Colorado.[9]

In contrast to the humics and fulvics, the concentrations of tannins and lignins are lowest in the spring/summer, rising gradually throughout the fall to reach a maximum during the winter months.[16] The authors were unable to explain this variation other than to predict that it is due to some combination of microbial oxidation and photochemical degradation. The concentration of carbohydrates showed a peak in the spring months possibly due to biomass releasing carbohydrates throughout the winter freeze-up period with a resulting increase in loading during the spring. For the amino acids the seasonal fluctuations differed for various individual acids.

Klotz and Matson[36] studied the seasonal and diurnal fluctuations of dissolved organic carbon in the Shetuchet River of eastern Connecticut. They reported low concentrations at all stations during the high-flow winter months and higher concentrations during the summer months. Maximum seasonal concentrations of up to 2200 μmol/l occurred in late February. Maximum diurnal fluctuations accounted for up to 53% of the annual range at each station. No attempt was made by the authors to determine the components of the DOC but they were not all of natural origin since an activated sludge, sewage treatment plant discharged into the upper reaches of the study area. The impact of organic carbon has been measured for many water sheds in an attempt to understand the functioning of ecosystems[37,38,39] but most of the studies have not gone beyond defining the carbon budget in terms of dissolved, fine particulate, coarse particulate, and total organic carbon.

III. SIGNIFICANCE OF NATURAL ORGANIC COMPOUNDS IN AQUATIC SYSTEMS

For a long time it has been known that humic substances form complexes and chelates with metals and some organic compounds. Many books on soil organic matter have dealt with the ion exchange and chelation properties of humic substances in great detail and the reader is referred to one of these texts for further details.[1,7,40] Methods are available for determining stability constants of metal-fulvic acid complexes and the values provide some indication of the availability of metals to biota.[1,41] This is particularly important in the case of essential trace elements and also for toxic metals.[9,42] Humic-clay interactions play an important role in soil structure but this is beyond the scope of the present text.[1,43] Humic substances are known to combine with many organic compounds including pesti-

---

* Proline, alanine, glycine, threonine, serine, leucine, ornithine, hydroxyproline, phenylalanine, glutamic acid, and lysine.

cides.[1,44,45,46,47] Very little is known about the mechanism of the adsorbtion process although many studies have been carried out because of the economic implications to the agricultural chemical industry. In addition to adsorption of organic compounds by soil humics, aqueous fulvic, and humic acids are known to solubilize some organic compounds. A study carried out by Wershaw et al.[46] showed that a humic solution was capable of increasing the solubility of DDT by a factor of 20 probably due to a lowering of the surface tension. Similar studies with herbicides[48,139] have shown that humic and fulvic acid can increase the mobility of some pesticide products in the aquatic environment and hence presumably change the physiological characteristics of the compounds.

The direct physiological properties of fulvic and humic acids on plants have been studied by a number of researchers.[1] Low concentrations of approximately 60 ppm have been shown by Kononova[6] to enhance root growth, while some Russian researchers[49] hypothesized that humic compounds may enter young plants and serve as an additional source of polyphenols and act as respiratory catalysts.

One of the major concerns of aqueous humic and fulvic acids in recent years has been in the water treatment field.[50-52] These compounds are known to form trichloromethanes during chlorination of potable water for disinfection.[53-57] Many studies have been carried out on the conditions and kinetics of trichloromethane production but at present there seems to be no way of preventing their formation other than by first removing the fulvic precursors or by employing an alternative disinfectant such as ozone or chlorine dioxide.[58] Oliver and Lawrence[55] have shown that for a given set of reaction conditions, the production of trichloromethanes is almost directly proportional to the dissolved organic carbon in the raw water. Some municipalities which are forced to use water with high fulvic concentrations have had to take steps to either reduce the organic concentrations by alum flocculation prior to chlorination or treat the disinfected water with activated carbon to remove most of the trichloromethanes produced. Tannins, lignins, and amino-acids will also form trichloromethanes during chlorination[57] but because their concentrations in most surface waters are much lower they rarely give rise to concern.

IV. ISOLATION

Most of the early work on humic and fulvic acid was carried out on material extracted from soil by dilute aqueous NaOH.[1,6,25] There has been some controversy in the literature as to whether auto-oxidation of humic compounds occurs in the presence of oxygen and dilute base but many researchers claim that there is no difference in terms of functional group analysis and in spectrophotometry between materials extracted with dilute aqueous NaOH, NaF, water, or dilute acid.[1]

Various methods have been proposed for extracting dissolved organic compounds from water. These include resin adsorption,[59,60,61] carbon adsorption,[62,63] ion exchange,[64] electrodialysis, liquid-liquid extraction,[18] freeze concentration,[65] ultrafiltration,[64] and coprecipitation. Of these eight procedures, adsorption and ultrafiltration have received the most attention. The others are either very slow and tedious or not specific to the organic fraction (freeze concentration and coprecipitation). The solid adsorbent methods lend themselves to processing large volumes of water (in the field if necessary) and hence are more suitable for waters with relatively low organic content. The isolation of humic and fulvic acids by adsorption onto activated carbon, macroreticular resins or a combination of the two has been studied by numerous researchers.[59-64,66] Prior to this Jeffrey and Wood[62] and other workers used inorganic adsorbents such as alumina, silica gel, magnesia, and calcium carbonate[67] but the yields were low due to inefficiency of adsorption and/or difficulty of eluting the adsorbed acids. The essential requirements for a useful adsorbent system are (1) high adsorption and elution efficiencies, (2) a stable inert adsorbent, (3) ability to achieve reasonably high flow rate, and (4) an assurance that the extracted material is chemically identical to that in the original sample.

Mantoura and Riley[61] studied the thermodynamics of adsorption of humic and fulvic acids on Amberlite XAD-2 (Rohm and Haas), a polystyrene based, hydrophobic adsorbent which adsorbs neutral molecular solutes by Van der Waals forces. The adsorption of these previously purified acids followed a Langmuir isotherm which can be represented by:

$$\frac{1}{\Gamma_A} = \left(\frac{1}{K\Gamma_\infty}\right)\frac{1}{a_A} + \frac{1}{\Gamma_\infty} \tag{1}$$

where Γ_A is the adsorption density of the acid on the resin in mol/g, Γ_∞ is the theoretical adsorptive capacity, K the equilibrium adsorption constant (mole per liter) and a_A the activity of the acid solution. At low concentrations a_A can be equated to the concentration C_A. The standard enthalpy of adsorption ΔH_A° was calculated in the usual way from a plot of Log K as a function of 1/T. ΔH_A° for humic acid was -5.4 kJ/mol while the standard free energy of adsorption ΔG_A° was -36.4 kJ/mol at 21°C. The relatively low value of ΔH_A° is indicative of predominantly hydrophobic bonding between the adsorbate and the adsorbent.

The authors also studied the influence of pH and flow rate on the efficiency of humic acid isolation with Amberlite XAD-2. As would be expected with a hydrophobic adsorbent, the adsorption increased with decreasing pH since the degree of dissociation would be decreased. Uptakes of >92% for humic acid and >75% for fulvic acid were achieved at pH 2.2 and a flow rate of 35 bed volumes per hour. It was also observed that only about 20% of the theoretical adsorptive capacity of the resin could be used before bleeding or leakage of the column started to occur. Increasing the ionic strength of the solution increased the adsorbtion efficiency, although the pH was the principal controlling factor. Good recoveries of the adsorbed acids were obtained by eluting the columns with 0.2 M NaOH (95% desorption) or aqueous methanolic ammonia (91%). The latter reagent has the advantage that it can easily be removed by evaporation: overall recoveries of humic products from streams varied from 82 to 87%. Stuermer and Harvey[60] employed a similar procedure for extracting organic acids from seawater except that they used aqueous ammonia (pH 11.6) as eluent. Later Harvey and co-workers[68] scaled-up and partly automated the procedure for the extraction of humics from very large quantities of sea water.

Weber and Wilson[66] extracted fulvic and humic acids from Jewell Pond and the Oyster River by passing 100 to 150 gal of prefiltered water through two columns packed with Rohm and Haas IRA-458 or XE-279, anion exchange resins (OH form) and eluting with 2 *M* aqueous sodium chloride. The excess NaCl was then removed by evaporation and chrystallization or by passing the eluate through an XAD-2 column after acidification to pH-1 and eluting with 1% aqueous sodium hydroxide. It is doubtful that anything is achieved by using the double column procedure and direct adsorbtion at pH ~2 with XAD-2 is preferable. Some attempts have been made to elute adsorbent columns with organic solvents.[25] However, the low solubility of humic compounds in most organic solvents results in very low recoveries.

Since the adsorbate molecules are held to the adsorbent by hydrophobic bonding it follows that many of the polar constituents and lower molecular weight compounds will not be completely adsorbed. To achieve complete recovery of organic compounds from natural waters, Fu and Pocklington[63] employed a mixed column of XAD-2 and granular activated carbon. Whereas the efficiency of organic removal from seawater with XAD-2 or XAD-8 was only about 40%, the mixed bed column yielded almost quantitative recovery for the first 500 bed volumes of sample. After that the efficiency decreased, presumably as a result of saturation of the most active sites on the adsorbent. The organic material was eluted by a sequence of aqueous ammonia (7 *M*), methanol and methanolic ammonia. For all the seawater sampled, about 50% of the adsorbed organic carbon was removed by the aqueous ammonia while the other two fractions contained about 25% in each.

Thurman and Malcolm[59] used macroreticular resins (XAD-8) in a multiple cycle adsorbtion sequence to extract and fractionate organic acids from ground water. After filtration and acidification the sample was passed through an XAD-8 column, eluted with 0.1 N sodium hydroxide and the eluate, after reacidification, recycled through a smaller XAD-8 column. The eluate from this column was then chromotographed on Enzacryl Gel and reconcentrated on an XAD-8 column. The eluate was then ready for the humic/fulvic fractionation. Another organic adsorbent which has been examined for organic removal from colored water is nylon[69] (bleached stockings). Although about 70% of the humic acid was adsorbed from the water at pH 3.5, the overall efficiency was only 54% due to low recovery (0.2 M aqueous sodium hydroxide eluent). This low recovery has been explained by Snyder[70] as due to chemical bonding of the quinoid groups on the humic acid to the free amino groups present on the nylon.

Chemical precipitation of metal humates and fulvates from water has been described in the literature. Weber and Wilson[66] precipitated lead(II) and iron(III) humates or fulvates from water according to the following equations:

$$Pb^{2+} + H_x Hum \rightarrow Pb H_{x-2}Hu_{(S)} + 2H^+ \tag{2}$$

$$Fe^{3+} + H_x FuL \rightarrow Fe H_{x-3}Ful_{(S)} + 3H^+ \tag{3}$$

These equations are gross simplifications because the metal-humic/fulvic ratios are unknown. Humic acid was precipiated from 450 l Jewell Pond water by precipitation with 1 l 0.2 M lead(II) nitrate. The resulting lead humate/fulvate was dissolved in 1 l 0.4 M ammonia to which had been added 5 g dithiazone. Lead dithiazonate was precipitated leaving ammonium humate/fulvate in solution. After purification, the addition of acetic acid and diethyl ether precipitated ammonium humate/fulvate which was then separated by filtration. Although this is an alternative method for separating colored organic acids from water it is relatively complicated and is not recommended for routine use. Aluminum and other metal humates have also been precipitated.[1]

Ultrafiltration[71] and gel chromatography[72] have been used for the simultaneous extraction and fractionation of dissolved organic material from waters. Oliver and Visser[71] employed a series of Amicon ultrafilters to extract stream and lake fulvic and humic acid. Eight molecular weight fractions of fulvic acid (500 to 300,000 MW) and seven fractions of humic acid (1000 to more than 300,000 MW) were obtained. This is a very tedious method of extraction but if specific fractions are required it has advantages. However, if the water contains appreciable quantities of nonhumic material it may be faster to separate the humic/fulvic acid by resin adsorbtion and then fractionate by ultrafiltration.

Liquid extraction has been used with some success for the extraction of colored organic acids from water. Shapiro[18] has extracted coloured acids with ethyl acetate and butanol while Martin and Pierce[73] isolated only the humic acid fraction using isoamyl alchohol and acetic acid. Humic substances have been isolated as ion-pairs using tetrabutyl ammonium salt and chloroform.[74] Greater than 90% of the color was removed by this method.

V. FRACTIONATION AND PURIFICATION

From the previous discussion it is apparent that some fractionation occurs during many of the extraction procedures. These will be discussed in more detail below. The traditional method of fractionating extracted humic material into fulvic and humic components was based on solubility at different pH values.[1,33] The fraction which was soluble at pH 9 and pH 2 was referred to as fulvic acid while the fraction which was soluble at pH 9 but insoluble at pH 2 was considered to be humic acid. Other fractions of soil humic materials (humin, hymatomelonic acid) have also been fractionated on the basis of pH solubility[1] but will not be considered here.

More recently Salfield[75] and Martin et al.[76] have fractionated samples of soil humic and fulvic acids, respectively, on the basis of differential solubility in aqueous tetrahydrofuran containing 5, 10, and 20% water. Fractions produced by such a scheme probably differ in overall polarity rather than specific chemical speciation or molecular weight ranges. Similarly, fractional salting out with ammonium sulfate[77] produced varying fractions depending on the degree of saturation. At relatively low saturation the material was less highly charged and contained more aliphatic material per unit weight than fractions produced at higher degrees of saturation.

The availability of gels and ultrafine membrane filters with a broad range of molecular weight exclusion properties has resulted in renewed attempts to fractionate humic and fulvic acids on the basis of molecular weight. Ferrari and Dell'Agnola[78] used Sephadex® G-50 to obtain three fractions, two of which resembled the humic and fulvic fractions obtained by conventional pH precipitation except that differences were observed in the electrophoretic properties. Similarly Gjessing and Lee[79] used Sephadex® columns to produce 10 discrete fractions which differed in amounts of oxydizable material, color and organic nitrogen. By using a number of gels of different molecular weight cut-offs it is possible to separate both humic and fulvic acids into numerous molecular weight ranges. Ultrafine membrane filters can also be used to produce varying molecular weight fractions although care must be taken to ensure that the filters do not become partly blocked or the effective size exclusion will be reduced.[80] Oliver and Vesser[71] used a range of Diaflo (Amicon Corp.) filters to produce 8 fractions of fulvic acid with molecular weights of 500 to 1,000; 1,000 to 10,000; 10,000 to 20,000; 20,000 to 30,000; 30,000 to 50,000; 50,000 to 100,000; 100,000 to 300,000 and 300,000.

Column adsorption chromatography has been used to separate various humic and fulvic fractions.[63,81-83] As many as 13 fractions with differing chemical and spectroscopic properties have been reported using activated carbon, activated alumina, or activated carbon and alumina.[84] As mentioned earlier, adsorption of the hydrophobic part of organic molecules by solid adsorbents is primarily by Van der Waals forces with the hydrophilic functional groups being oriented towards the aqueous phase. The adsorption interaction is relatively weak (5 to 10 k cal/mol) and hence is easily reversible. It has been observed[9] that the acrylic ester resins (Rohm and Haas XAD-7 and XAD-8) have superior adsorption capacities than the styrene-divinyl benzene resins such as the XAD-1, 2, and 4. Also the ease of desorption is better for the acrylic ester resins. The adsorption properties vary with the ratio of the number of carbon atoms to the number of carboxyl groups in the organic molecule. The ionic character of the molecule when the ratio reaches 1 to 12 is approximately equal to the hydrophobic adsorption energy. At this ratio the adsorption energy is approximately equal to the desorption energy or the ion-dipole interaction between the ionic carboxyl and the water molecules. Consequently fractional desorption in buffer solutions can be used to separate the organics on the basis of chain length to functional group ratios.

Fractional freezing of humic acid has been reported by two Russian research teams.[85,86] Two discrete fractions were obtained when humic acid solutions were subjected to a very slow freezing process. The fractions appeared to be similar to the conventional humic/fulvic reactions produced by acid/alkali separation. Due to the difficulty of the method it is not likely to receive much attention as a practical method of fractionation; pH gradient desorption can be used to separate the predominant phenol functional groups.[9,87] Phenolic groups are not ionic until pH − 10 so they remain adsorbed on the resins for longer than the carboxylic functional groups as the pH of the eluting solvent is gradually raised. At a low pH (≈2),

Table 3
ELEMENTAL COMPOSITION OF SOME FULVIC AND HUMIC ACIDS

	Carbon	Hydrogen	Nitrogen	Oxygen	Sulfur	Ash	Ref.
Groundwater (fulvic)	60	5.9	0.9	32	0.3	1.2	9
(humic)	62	4.9	3.2	24	0.5	5.1	
Willow brook (fulvic)	42	6.4	1.3	46	1.7	2.6	80
Suwannee River (fulvic)	51	4.3	0.6	43	<0.2	<0.05	88
Marsh (fulvic) (humic)	51	4.3	0.74	40	0.4	2.0	9
	51	4.4	0.56	41	0.6	2.0	
Soil (fulvic) (humic)	51	3.3	0.7	45	0.3		
	57	5.2	2.3	35.4	0.4		1
Lake sediment (humic)	54	5.8	5.4	35			1

all the organic material is adsorbed on the column. As the pH of the eluting solution is raised, the first eluting fraction will contain polycarboxylic acid fractions with less than 12 carbon atoms per functional group. This will be followed by polycarboxylic acids with more than 12 carbon atoms per functional group and finally by the polyphenolic fraction. This phenolic fraction will contain the tannin component as well as any phenolic fraction of humic or fulvic acids. In most river, lake, and bog waters this fraction is much smaller than those containing polycarboxylic acids.

Thurman and Malcolm[88] have employed weak-base ion-exchange chromatography to fractionate humic and fulvic acids. These authors obtained three fractions: the first at pH 8 (0.1 N NaHCO$_3$) containing mostly carboxylic acids with some phenols; the second at pH 13 (0.1 N NaOH) containing mostly phenols with some carboxylic acids and the third also at pH 13 (50% methanol and 50% 0.1 N NaOH) containing the weakly ionized functional groups. With a combination of column chromatographic fractionation techniques (gel, adsorption, weak base) it is possible to generate a variety of fractions based on molecular size, functional group density and functional group type.

Several researchers have used paper chromatography to fractionate components of fulvic acid. In most cases, 5 to 10 yellow fluorescing zones were noted. Siebarth and Jensen[69] employed two-dimensional paper chromatography (developed with water in one direction and acetic acid-ethyl acetate-water in the other) and obtained a number of fluorescent spots.

VI. CHARACTERIZATION

The characterization of natural organic compounds whether of aquatic or terrestrial origin, has generally been achieved by a combination of chemical and physical methods including elemental analysis, functional group analysis, spectroscopy, molecular weight determination, acidity, viscosity, and thermogravemetric analysis. The first four of these techniques are the most informative and will be discussed in more detail than the remainder.

A. Elemental Analysis
Elemental analysis of natural organic compounds is a relatively simple method of characterization. The main components are carbon (~50%), oxygen (~40%), hydrogen (~5%), nitrogen (~4%), and sulfur (~1%).[9] Table 3 shows the chemical composition of a number of aqueous humic and fulvic acids together with one or two from terrestrial sources for comparison.[1,9] In general terms soil humic acids contain more carbon and less oxygen than soil fulvic acids while the elemental composition of the aqueous counterparts are about the same and are closer to the soil fulvic acid than soil humic acid. There seems to be very little difference between the aqueous organic acids from various origins except those from groundwater contain about 10% more carbon and 10% less oxygen than those from streams,

Table 4
MAJOR FUNCTIONAL GROUP ANALYSIS OF SOME FULVIC AND HUMIC ACIDS

	Carboxyl (meq/g)	Phenolic OH (meq/g)	Carbonyl (meq/g)	Hydroxyl (meq/g)	Ref.
Groundwater (fulvic)	5.3	1.6			9
Oyster River (fulvic) (humic)	6.8	4.3	4.3		66
	4.5	3.7	4.3		
Jewell Pond (humic)	4.9	2.2	5.1		66
Suwannee River (fulvic)	6.1	2.5	1.7	7.0	88
Marshes (fulvic) (humic)[9]	5.3	2.5			9
	4.3	2.5			9
Soils (fulvic) (humic)	9.1	3.3	3.1	3.6	1
	1.5	4.2	0.9	2.8	

lakes, seawater, and marshes. This may be a reflection of the anaerobic conditions under which ground water exists which would tend to deplete the oxygen content.

Alder, in 1957,[89] proposed a structure for spruce wood lignin which corresponded to a carbon content of 65% and oxygen 30% with the remaining 5% composed of hydrogen. How closely this lignin resembles the lignin-like compounds found in water is not known. Similarly, vegetative tannins[14] have a composition of about 62% carbon, 35% oxygen, and 3% hydrogen, but again there is no information available as to the composition of tannins in natural waters.

The hydrogen content of fulvic acids is normally in the range of from 3 to 6% with fulvic acid being slightly higher than humic acid from the same source. However, in many instances the values are so close as to be statistically insignificant. According to Thurman,[9] the average hydrogen content of fulvic acid extracted from ground water is 5.9%, from streams and rivers 5.0%, from lakes 5.2%, and from marshes 4.3%, with the corresponding values for humic acid being slightly lower. He attributes the increased hydrogen content of fulvic acid as one goes from marshes to streams to groundwater to an increase in the aliphatic/aromatic carbon ratio which is also borne out by ^{13}C-NMR measurements. This trend is also suggested by the small increase of the H/C ratio on going from marshes to groundwater.

The nitrogen content of humic and fulvic acids varies from about 1 to 6% with marine and terrestrial organics having a much higher content than streams and lake water. In general the nitrogen content of humic acid is about twice that of corresponding fulvic acid. Khan and Sowden[90,91] have shown that acid hydrolysis of soil humic and fulvic acids results in about 18 easily identifiable amino acids with aspartic acid and glycine being the two most abundant. It has been suggested that the total nitrogen content of humic extracts and the distribution after acid hydrolysis depends on the method of extraction used. However, this may not be the case for acids extracted from water since in most cases much milder extraction conditions are employed, although care must be taken not to use XAD extraction with ammonia desorption if one subsequently intends to study the nitrogen content on the extract. The sulfur content of the fulvic and humic acids is usually less than 1% except in the cases where they are extracted from coal or soil with high sulfur content.[9] Groundwater fulvic and humic acids sometimes have a higher S content than those from streams or lakes due to dissolved H_2S, some of which may become incorporated into the organic structure. The structural significance of sulfur in organic acids is not known.

B. Functional Group Analysis

The predominant oxygen containing functional groups in aquatic organic acids are carboxyl, hydroxyl, carbonyl, and phenolic. Table 4 shows typical values for the functional groups for several aquatic and terrestrial humic and fulvic acids.[1,9] Most aquatic fulvic acids contain about 4 to 7 meq/g carboxyl and 1 to 5 meq/g carbonyl. In general, the total acidity (sum of carboxyl and phenolic) of fulvic acid is higher than that for humic acid from the same source by about 25%. The precipitation of humic acid from fulvic-humic mixtures at pH-2 removes the less soluble compounds containing few carboxyl groups. The average carboxyl content of fulvic acid of 5.5 meq/g correspond[5] to one carboxyl group for every 7.6 carbon atoms whereas that for humic acid is about 4 meq/g corresponding to one carboxyl group for every 12 carbon atoms. Thurman and Malcolm[92] discuss the significance of this ratio in terms of it being the maximum ratio for which the hydrophilic carboxyl group is able to solubilize the hydrophobic molecule.

The phenolic content of aquatic humic materials from lakes and streams is usually much less than the carboxyl content (around 0.5 to 2 meq/g)[9] For marsh acids it is somewhat higher, about 2.5 meq/g which is approaching the value for soil and sediment extracts.

The hydroxyl content of aqueous fulvic acid is usually in the range of 6.5 to 8 meq/g,[9] whereas for fulvic and humic acids extracted from marine sediments or soils, typical values have been reported as 5 meq/g or less and in some cases as nondetectable.[1] However, there are still some uncertainties with the methodologies employed for hydroxyl determinations so some of the values reported may not be completely reliable. The development of novel NMR techniques may soon enable unequivocal determination of carboxyl, carbonyl, and hydroxyl groups.[88]

While several authors have attempted to determine the carbonyl content of aquatic humic and fulvic acids, major discrepancies exist depending on the methods used. Weber and Wilson[66] using the method developed by Schnitzer and Khan[1] found 4.3 mmol/g for fulvic acid extracted from the Oyster River. However, Thurman[9] using ^{13}C-NMR found values of only 1 to 2 mmol/g.

C. Physical Characterization

Many articles have appeared in the literature on the physical characterization of natural organic compounds[1,9,93] but their contribution to elucidating the structure of the compounds is really quite limited. Since this chapter is not intended to be an exhaustive treatment of natural organics, only those physical methods of characterization which shed some light on the molecular structure will be discussed. Such methods include spectroscopy, electrometric titrations, and molecular weight determinations. For a more detailed discussion on the physical characterization the reader is referred to treatices by Schnitzer and Khan,[1] or Thurman,[9] or Thurman and Malcolm.[88]

In the early part of the twentieth century, color intensities were used as a criterion for characterizing humic and fulvic acids. The dark color in neutral or basic solutions led to visual color comparisons being used to compare acids from different origins. However, for a given humic or fulvic acid solution the color intensity is strongly dependent on the pH and the amount of chelated metal complex present. Visible and ultraviolet spectroscopy soon replaced visual characterization but the uncharacteristic spectra of these acids (Figure 1) are not, in themselves, very informative. While the extinction coefficients at 400 nm of aquatic fulvic and humic acids do depend to a small extent on the origin of the sample (increasing from groundwater through streams and rivers to marshes with the humic component usually higher than the fulvic component) the ratio of optical densities or extinctions at two wavelengths such as 465 and 665 nm is more useful for characterization. As shown by Kononova,[6] for soil humic acids, these ratios are independent of concentration but dependent on the origin of the sample. Thurman[9] has shown that there is similar dependence for aquatic humic and fulvic material. Ratios of other wavelengths could be also used but the E_4/E_6 (465 and 665 nm) is generally used. Several researchers have attempted to correlate E_4/E_6 to molecular weight and degree of humification. The lower ratios in humic acid have been attributed to increased humification while fulvic acid, which is less aged has higher ratios.[1,6]

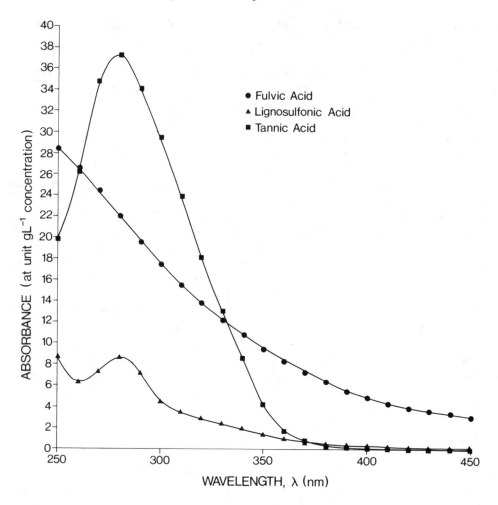

FIGURE 1.   Adsorption spectra of fulvic acid, tannic acid, and liquosulfonic acid in phosphate buffer.[131]

Table 5
PREDOMINANT IR ABSORBTION BANDS
OF HUMIC MATERIALS

Frequency (cm⁻¹)	Band assignment
3400	Hydrogen bonded OH
2900	Aliphatic C–H stretch
1725	C=O stretch
1630	Aromatic C–C
1450	Aliphatic C–H
1400	COO⁻
1200	C–O or O–H deformation

By comparison, the spectra of tannins and lignins are more characteristic, especially at the UV end (Figure 1). Again ratios of extinction coefficients could be used for characterization but the author is unaware of anyone having done so.

Infrared spectroscopy has also been used for the characterization of humic substances,[81] but it gives us little additional information that cannot be obtained from chemical functional group analysis. Table 5 shows some of the main IR adsorption bands of humic substances.

Comparison of IR spectra of humic and fulvic acids shows that humic acid contains more aliphatic C-H groups than fulvic acid.[1] This is in agreement with the chemical data which shows fulvic acid containing significantly more carboxyl groups than humic acid.

The last 10 years has seen a rapid growth in the use of liquid and solid ^{13}C-NMR spectroscopy applied to the structure of aquatic and terrestrial humic substances.[9,88,94-98] Aromatic, aliphatic, and carboxylic resonances are very easily distinguishable. In a typical aquatic fulvic acid spectrum well defined peaks are obtained for aliphatic carbon, carbon-oxygen (hydroxyl or ester), aromatic carbon, carbon-phenolic hydroxyl, carboxyl, or ester and carboxyl. Thurman[9] found significant differences in the carboxyl and phenolic contents of humics and fulvics from different origins. In general the composition determined by ^{13}C-NMR agrees well with data generated from functional group analysis, titration, or elemental analysis.

Electron spin resonance, which measures unpaired electrons in paramagnetic substances, is another tool for examining humic substances. Humic and fulvic acids contain relatively large concentrations of free radicals which can be detected by this technique. For soil humic acids, tannins and lignins the subject has been reviewed by Steelink and Tollin.[99] Nagar et al.[100] found that the free radical content of lignin was much greater than soil humic acid which in turn was greater than a microbial humic acid. From the data it was apparent that the free radicals in lignin were significantly different from the soil or microbial humic acids whereas those from the two humic acids were similar.

Many natural organic compounds fluoresce under UV and visible light due to the aromatic, cyclic, or closed ring structure. According to Rockwell and Larson[101] there is considerable similarity in the fluorescence spectra of humic and fulvic acids from different origins but they are quite different from the spectra of lignins.[102] They suggest that the fluorescence of humic materials may be partly due to coumarins and propose a possible mechanism for their formation in the aquatic environment. In a study of stream ecosystems in southeastern Pennsylvania the authors observed that the total fluorescence generally increased in a downstream direction as did the fluorescence per unit mass of dissolved organic carbon. All stream samples had identical fluorescent characteristics but differed somewhat from those taken from the forest seeps (predominantly leaf litter leachate) where the excitation and emmission were shifted to shorter wavelengths. The spectra of these samples resembled that of degraded lignin and oxygenated cinnamic acid derivatives known to be found during microbial lignin decomposition. Fluorescence and UV-visible spectroscopic techniques have also been used by Carey[103] to characterize and quantitate natural organic compounds in small streams in several drainage basins in southern Ontario with differing vegetative covering.

The low pH of colored marshes, swamps and dystrophic lakes suggests that organic acids contribute significantly to water acidity.[104-106] Potentiometric and conductometric titrations have been used to characterize acidic functional groups of organic acids. To assess the organic acid contribution to acidity, Oliver et al.[107] used the change in slope of Gran titrations to calculate the concentration and pK of the weak acids in various rivers, streams, lakes and groundwaters from Canada and the U.S. The carboxyl content was operationally defined as those acidic groups titrated to pH 7 (after the work of Perdue et al.[108]). Despite the diversity of samples the authors found only small variations in the humic carboxyl content.

Gamble[109] proposes two general types of carboxyl groups: type 1 which is ortho to a phenolic hydroxyl group and type II which is not adjacent. The two types can be distinguished by conductometric titration but not by potentiometric titration which only provides information on the total carboxyl content.

Characterization of aqueous organic matter by molecular weight determination has been reviewed in the literature.[1] Molecular weights varying from 300 to >100,000 have been reported for aquatic humic material with most being 10,000 or less. Methods used include small angle X-ray scattering,[110] gel chromatography,[111,138] ultrafiltration,[112] vapor pressure

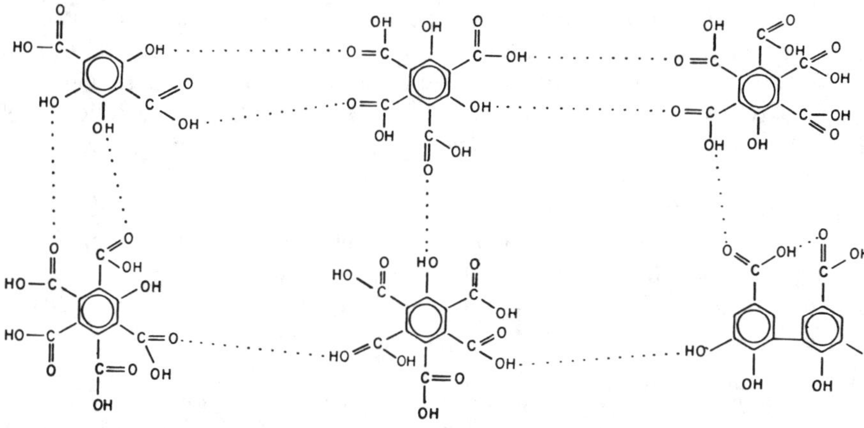

FIGURE 2.   Proposed structure of fulvic acid.[1]

osmometry,[113] ultracentrifugation,[114] freezing point depression,[115] and viscosity.[116] Schnitzer and Khan[1] have observed that in general there is disagreement between the methods based on number average, weight average and z-average (sedimentation) with the order being $\overline{Mn} > \overline{Mw} > \overline{Mz}$. Gel chromatography and ultrafiltration can be used to separate humic and fulvic acids into fractions with relatively narrow molecular weight ranges.[71,80]

VII. STRUCTURE

Although much structural information on humic and fulvic acids can be learned from the chemical and physical characterization discussed earlier and from chemical and microbial degradation studies,[1,117] the detailed structure of the compounds have eluded chemists. In recent years we have learned a great deal about the type and relative proportions of functional groups and the degree of aromaticity but we still lack sufficient structural information to propose an unequivocal polymeric structure. Undoubtedly, part of this difficulty is because these complex substances probably exist as a mixture of compounds in various stages of humification from complex lignins to relatively low molecular weight fulvic acids. The nature of the originating biological material would also be expected to influence the structure especially in the early stages of humification. According to Burges et al.[118] the overall similarity in properties of humic acids from widely different sources arises from the fact that they are large aromatic polymers with characteristics mainly governed by physical properties and surface functional groups.

Haworth[119] proposed that soil humic acid consists of an undefined protein core attached to peptides, carbohydrates, metals, and phenolic acids. The nature of the bonding was not established other than the suggestion that the peptides may be bound by hydrogen bonding which would explain their removal by hot water.

A more definitive structure of soil fulvic acid based on nondegrative studies has been proposed by Schnitzer.[1] He suggests that the molecule consists of phenolic and benzene-carboxylic acids (all of which have been isolated) joined by hydrogen bonds to form a stable polymeric structure. This proposed structure is shown diagramatically in Figure 2. The structure is punctured by voids of varying dimensions which can trap organic molecules such as alkanes, fatty acids, dialkyl phthalates, carbohydrates, and peptides. Some inorganic compounds such as metals could also be trapped in this way. While this structure does account for many of the observed properties, it is uncertain as to the amount of humic structure it really represents.

Thurman[9] reports that many of the properties of aqueous fulvic acid cannot be explained

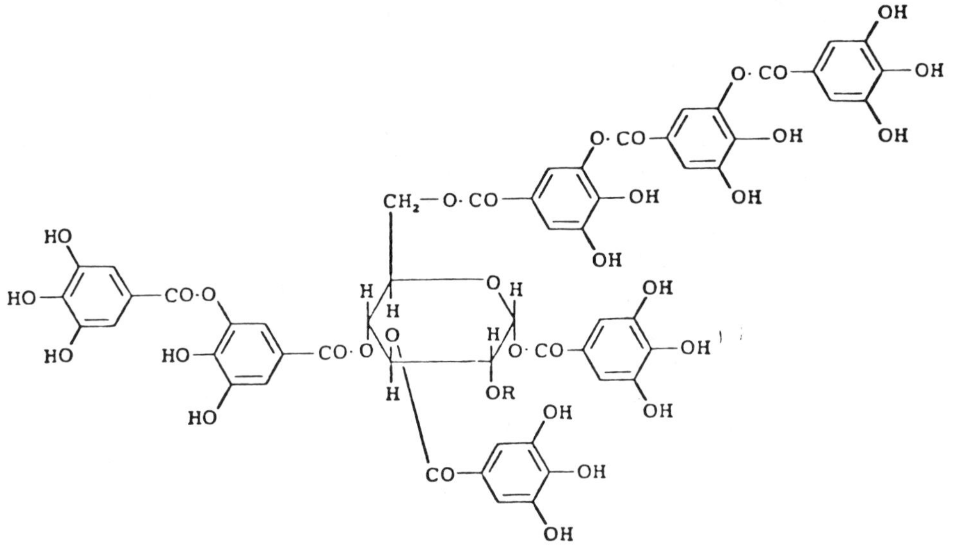

FIGURE 3.   General structure of Chinese and Turkish hydrolysable tannin.[14]

by Schnitzer's structure. These include the ratios of carboxyl and phenolic groups to the number of carbon atoms, the aromatic to aliphatic ratio (from ^{13}C-NMR spectra) and the abundance of nitrogen hydroxyl and carboxyl groups. Thurman has not yet proposed a definitive structure incorporating all of the above properties, but he suggests that in contrast to Schnitzer's it would have a complex aliphatic/aromatic basis.

The structure of tannins and lignins has received considerable attention from plant bio chemists. How closely the aqueous tannins and lignins resemble those extracted from vegetative material is not known, but nothing has yet been reported on the aqueous materials. The hydrolyzable tannins (abundant in hardwood) appear to be a mixture of compounds in which glucose is linked to an average of 9 to 10 gallic acid residues.[14] A simplified structure for tannin, as proposed by Swain,[14] is shown in Figure 3. However, although this may represent the basic structure, many alternatives are known to exist for tannins of varying origins. In the presence of oxygen in aqueous solutions tannins are oxidized to quinones and unidentified polymer material so aqueous tannins probably bear only slight resemblance to the structure shown in Figure 3. The condensed tannins are less well defined than the hydrolyzable class and specific structures have not been determined. For a more detailed discussion of the structure of tannins the reader is referred to the review by Swain.[14] A similar situation exists for the structure of lignin: it is again assumed to be a complex mixture of polymeric material which would readily oxidize in aqueous solution as part of the initial humification process. However, some success has been accomplished regarding the structure of lignin. Figure 4 shows the structural elements as proposed by Freudenberg[120] who extended earlier studies by Alder[89] and Brauns[12] (see Reference 121 for a summary of these structures).

The chemical structures of the low molecular weight natural organic compounds, amino acids, carbohydrates, hydrocarbons, and phenols are all well defined. Details can be found in many publications on organic chemistry[16,23,122] and they will not be discussed further here.

VIII. ANALYSIS

Due to the complex structure of aquatic organic matter, there is no known specific chemical reaction that can be used for analysis and with the wide variation in functional groups, molecular weight, and general composition, it is unlikely that such a reaction will ever be

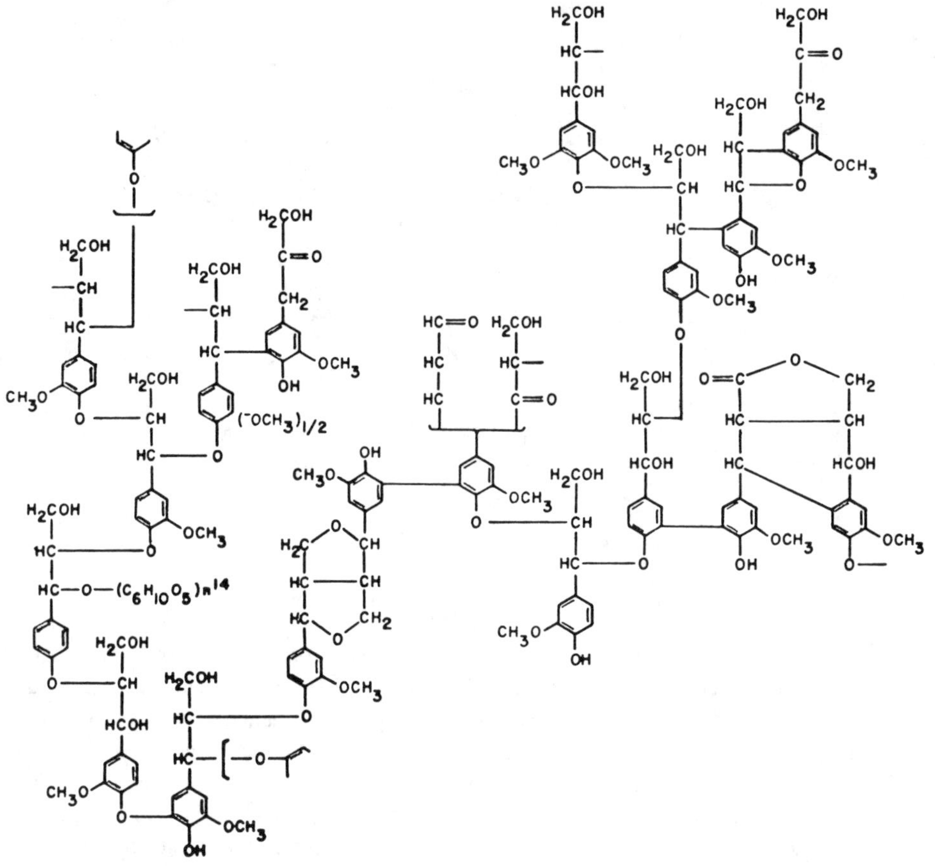

FIGURE 4.   Proposed structure of liquin.[120]

developed. In cases where humic and fulvic acids are present either alone or as the principal components, it is possible to determine gross concentrations by colorimetric,[123] spectrophotometric,[124] or thermogravimetric[1] means. Ever since Hazen introduced the scale of platinum standard solutions, attempts have been made to equate the yellow/brown-colored solutions of organic acids to this scale for quantitation. According to Shapiro[18], 1 mg/l of humic acid corresponds to between 3 and 5 Pt-Co color units depending on the pH of the solution. More recently, Martin and Pierce[73] described a method based on a standard set of humic acid solutions extracted from a bog soil. Although colorimetry has been used fairly extensively for humic and fulvic acids, Shapiro has shown that the color of a lake water sample may bear little resemblance to its concentration, since waters of the same color may differ in chemical composition. The converse is also true in that solutions of the same humic content but at different pH or in the presence of varying concentrations of metals such as iron, have distinctly different color intensities. The dynamic nature of humic substances in ecosystems probably precludes the use of color as a useful method of determination.

Ultraviolet and visible spectophotometry have also been frequently used for the quantitation of aqueous humic substances.[1,9,124-126] However the uncharacteristic nature of the UV spectrum (Figure 1) makes quatitation by this method very insensitive and the absorbence is strongly dependent on pH and metal-ion content. Carpenter and Smith[137] have described a spectrophotometric method for the simultaneous determination of humic acid and iron by measuring absorbance on two samples, one untreated and one treated to enhance iron absorptivity.

Fluorescence spectrophotometry has been used with some success for the analysis of aquatic humic solutions.[103,127] Considerable similarities have been noted in the fluorescence spectra of organic acids from a wide variety of origins[107] but several researchers have noted that fluorescence decreases with increasing molecular weight of the solute and decreasing pH of the solution[127-128] However, Stewart and Wetzel[129] and Hall and Lee[130] found only minor changes in emission intensity (3 to 5%) over the pH range from 4.5 to 10. Most researchers report excitation wavelengths in the range 365 to 375 nm with 2 emissions maxima at approximately 440 and 730 nm.

Most data on humic and fulvic acids are reported in the form of total organic carbon on the assumption that these substances account for the majority of organic compounds present in a natural water sample. Solutions are normally concentrated and the DOC measured at pH 9 (humic and fulvic acid) and pH 2 (fulvic acid). Independent determination of the fulvic acid content, the humic acid content, and the combined humic and fulvic acids provides a useful check on this method of analysis (usually 5% agreement is observed). Gravimetric methods for aquatic humics and fulvics have also been reported but like the DOC method they measure the total dissolved organic concentration. de Haan[136] attempted to analyze humic acid solutions in terms of milligrams equivalents of tyrosine generated by reaction with Folin-Ciocalteu phenol reagent. Although linear calibration curves were obtained it was concluded that the results did not represent the true humic acid content.

In a novel approach, Eberle and Schweer[74] proposed a semiquantitative UV spectrophotometric method for the simultaneous determination of humic acid and lignin. In this method, the humic and lignosulfonic acids were extracted with trioctylamine/chloroform and regenerated with aqueous sodium hydroxide. The humic acid (HS) and lignosulfonic acid (LS) were then determined spectrophotometrically using the following equations:

$$\text{g HS/l} = \left(\frac{E_{400}\epsilon_{280},\text{LS} - E_{280} \cdot \epsilon_{400},\text{LS}}{\epsilon_{400},\text{HS} \cdot \epsilon_{280},\text{LS} - \epsilon_{280},\text{HS} \cdot \epsilon_{400},\text{LS}} \right)/20 \tag{4}$$

$$\text{g LS/l} = \left(\frac{E_{280} \cdot \epsilon_{400},\text{HS} - E_{400} \cdot \epsilon_{280},\text{HS}}{\epsilon_{280},\text{LS} \cdot \epsilon_{400},\text{HS} - \epsilon_{400},\text{LS} \cdot \epsilon_{280},\text{HS}} \right)/20 \tag{5}$$

where E is the extinction at a cell thickness of 1 cm and ϵ the weight extinction coefficient at the wavelengths indicated. Because the spectra of humic acid and to a lesser extent ligninosulfonic acid are pH dependent, all measurements were made in buffered medium.

Lawrence[131] extended the above method to include a third component, tannic acid, and found that in most cases the extraction step was not necessary. Using Beer's Law for a three component system of fulvic acid (F), tannic acid (T) and ligninosulfonic acid (L), the absorbance could be expressed as:

$$E_\lambda = \epsilon_{\lambda,\text{F}}C_\text{F}d + \epsilon_{\lambda,\text{T}}C_\text{T}d + \epsilon_{\lambda,\text{L}}C_\text{L}d \tag{6}$$

where E is measured absorbance of wavelength λ. ϵ's are the extinction coefficients at that wavelength, C the concentrations, and d the optical path length. For simplicity, the path length d can be set to 1 cm and hence disappears from the Equation. Since the molecular weights of these compounds are not known, the gram extinction coefficients (calculated for unit g/l) are used and concentrations are expressed in g/l.

Combining Equation 1 for three different wavelengths and solving for C_F, C_T, and C_L resulted in the following:

$$C_\text{F} = \frac{D(E_{\lambda_3}\epsilon_{\lambda_1,\text{T}} - E_{\lambda_1}\epsilon_{\lambda_3,\text{T}}) - B(E_{\lambda_3}\epsilon_{\lambda_2,\text{T}} - E_{\lambda_2}\epsilon_{\lambda_3,\text{T}})}{(DA - BC)} \tag{7}$$

$$C_T = \frac{F(E_{\lambda_3}\epsilon_{\lambda_1,L} - E_{\lambda_1}\epsilon_{\lambda_3,L}) - E(E_{\lambda_3}\epsilon_{\lambda_2,L} - E_{\lambda_2}\epsilon_{\lambda_3,L})}{(ED - BF)} \tag{8}$$

$$C_L = \frac{-C(E_{\lambda_3}\epsilon_{\lambda_1,F} - E_{\lambda_1}\epsilon_{\lambda_3,F}) + A(E_{\lambda_3}\epsilon_{\lambda_2,F} - E_{\lambda_2}\epsilon_{\lambda_3,F})}{(CE - AF)} \tag{9}$$

where $A = (\epsilon_{\lambda_3,F}\epsilon_{\lambda_1,T} - \epsilon_{\lambda_1,F}\epsilon_{\lambda_3,T})$; $B = (\epsilon_{\lambda_3,L}\epsilon_{\lambda_1,T} - \epsilon_{\lambda_1,L}\lambda_{3,T})$; $C = (\epsilon_{\lambda_3,F}\epsilon_{\lambda_2,T} - \epsilon_{\lambda_2,F}\epsilon\lambda_{3,T})$; $D = (\epsilon_{\lambda_3,L}\epsilon_{\lambda_2,T} - \epsilon_{\lambda_2,L}\epsilon_{\lambda_3,T})$; $E = (\epsilon_{\lambda_3,F}\epsilon_{\lambda_1,L} - \epsilon_{\lambda_1,F}\epsilon_{\lambda_3,L})$; $F = (\epsilon_{\lambda_3,F}\epsilon_{\lambda_2,L} - \epsilon_{\lambda_2,F}\epsilon_{\lambda_3,L})$. Hence, measuring the absorbance at three wavelengths and knowning a total of nine extinction coeficients, C_F, C_L, and C_T could be calculated. The absorbance was measured at 10-nm intervals between 250 and 450 nm and a computer was used to select every combination of 3 wavelengths and calculate the concentrations of three components. This multiple wavelength approach had the advantage of detecting the prescence of other interfering compounds. If another compound with significant UV absorption was present, it quickly became apparent by the wide variation of concentrations obtained for different wavelength combinations. Data generated by this method agree well with those from other techniques (DOC, gravimetric, and molybdophosphoric acid[132] method for tanins and lignins) provided the samples are relatively free of metal ions. Alberts[133] has discussed this last aspect in more detail. It would appear that any colorimetric or spectophotometric method is doomed to failure unless careful control of the pH and metal ion content is maintained.

Tannins and lignins have most frequently been determined together by the molydophosphoric/tungstophosphoric acid method using lignosulfonic acid as standard.[132] In this well-documented method, the tannin and lignin like compounds reduce the molybdophosphoric acid and tungstophosphoric acid to produce a blue color which can be measured by light absorption in the 600 to 700 nm range.

Methods for analysis of low molecular weight, natural organic compounds such as amino acids,[134] carbohydrates,[135] and phenols[132] have been extensively covered in the literature and the reader is referred to these works for further details. The better coverage is undoubtedly due to better chemical characterization of these discrete compounds.

REFERENCES

1. **Schnitzer, M. and Khan, S. U., Eds.,** *Humic Substances in the Environment,* Marcel Dekker, New York, 1972.
2. **Poveldo, D. and Golterman, H. L, Eds.,** *Humic Substances: Their Structure and Function in the Biosphere,* Centre for Agricultural Publishing and Documentation, Wageningen, The Netherlands, 1973.
3. **Christman, R. F. and Gjessing, E. T.,** *Terrestrial and Aquatic Humic Substances,* Ann Arbor Science, Ann Arbor, MI, 1983.
4. **Visser, S. A.,** A review of the distribution of organic compounds in freshwater lakes and rivers, *Afr. J. Trop. Hydrobiol. Fish.,* 2(2), 91, 1972.
5. **Steelink, C.,** Humates and other natural organic substances in the aquatic environment, *J. Chem. Educ.,* 54(10), 599, 1977.
6. **Kononova, M. M.,** *Soil Organic Matter: Its Nature, Its Role in Soil Formation and in Soil Fertility,* 2nd ed., Pergamon Press, London, 1966.
7. **Gjessing, E. T.,** *Physical and Chemical Characteristics of Aquatic Humus,* Ann Arbor Science, Ann Arbor, MI, 1976.
8. **Christian, R. F. and Minear, R. A.,** Organics in lakes, in *Organic Compounds in Aquatic Environments,* Faust, H., Ed., Marcel Dekker, New York, 1971, 119.
9. **Thurman, E. M.,** *Organic Geochemistry of Natural Waters,* Nyhoff-Junk, Doerdrecht, 1985.
10. **Swain, F. M.,** Geochemistry of humus, in *Organic Geochemistry,* Breger, I. A., Ed., MacMillan, New York, 1963.
11. **Brown, S. A.,** Lignins, *Ann. Rev. Plant Physiol.,* 16, 223, 1966

12. **Brauns, F. E. and Brauns, D. A., Eds.,** *The Chemistry of Lignin,* Supplement Volume, Academic Press, New York, 1965.
13. **Freudenberg, K. and Neish, A. C.,** *Constitution and Biosynthesis of Lignin,* Springer-Verlag, Berlin, 1968.
14. **Swain, T.,** The tannins, in *Plant Biochemistry,* Bonner, J. and Varner, J. E., Eds., Academic Press, New York, 1965.
15. **Lawrence, J.,** National inventory of natural organic compounds — an interim report, National Water Research Institute, unpublished report, Burlington, Canada, 1978.
16. **Telang, S. A., Baker, B. L., Costerton, J. W., Ladd, T., Mutch, R., Wallis, P. M., and Hodgson, G. W.,** Biogeochemistry of mountain stream waters: the marmot system, Scientific Series No. 101, Inland Waters Directorate, Ottawa, Canada, 1982.
17. **Kalle, K.,** The problem of Gelbstaff in the sea, *Ocean Mar. Biol. Ann. Rev.,* 4, 91, 1966.
18. **Shapiro, J.,** Chemical and biological studies on the yellow organic acids of lake water, *Limnol. Oceanogr.,* 2, 161, 1957.
19. **Berzelius, J. J.** Les deux acides organiques qu'on trouve dans les eaux minérales, *Ann. Chim. Phys.,* 54, 219, 1833.
20. **Tabak, H., Chambers, C. W., and Kabler, P. W.,** Bacterial utilization of lignins. I. Metabolism of α-conidendrin, *J. Bacteriol.,* 78, 469, 1959.
21. **Baylis, P. E. T.,** Lignin, *Sci. Prog.,* 48(191), 409 1960.
22. **Larson, R. A.,** Dissolved organic matter of a low coloured stream, *Freshwater Biol.,* 8, 91, 1978.
23. **Meister, A.,** *Biochemistry of the Amino Acids,* Academic Press, New York, 1965.
24. **Lowe, L. E.,** Carbohydrates in soil, in *Soil Organic Matter,* Schnitzer, M. and Khan, S. U., Eds., Elsevier, Amsterdam, 1978.
25. **Felbeck, G. T.,** Chemical and biological chacterization of humic matter, in *Soil Biochemistry,* Vol. 2, McLaren and Skujins, Eds., Marcel Dekker, New York, 1971.
26. **Nord, F. F.,** The formation of lignin and its biochemical degredation, *Geochim. Cosmochim. Acta,* 28, 1507, 1964.
27. **Henderson, M. E. K.,** Metabolism of methonylated aromatic compounds by soil fungi, *J. Gen. Microbiol.,* 16, 686, 1957.
28. **Thurman, E. M.,** Isolation, Characterization and Geochemical Significance of Humic Substances from Groundwater, Ph.D. thesis, University of Colorado, Boulder, CO, 1979.
29. **Stuermer, D. H. and Harvey, G. R.,** Humic substances from seawater, *Nature,* 250, 480, 1974.
30. **Kerr, R. A. and Quinn, J. G.,** Chemical studies on the dissolved organic matter in seawater. Isolation and fractionation, *Deep Sea Res.,* 22, 107, 1975.
31. **Webber, J. H.,** Metal complexes of components of yellow organic acids in water, Report No. PB-237 519, National Technical Information Service, U.S. Department of Commerce, Washington, D.C.
32. **Glooschenko, W. A., Strachan, W. M. J., and Sampson, R. C. J.,** Distributiion of pesticides and PCB's in water, sediment and seston of the Upper Great Lakes, *Pest Monit. J.,* 10(2), 61, 1976.
33. **Telang, S. A., Baker, B. L., Costerton, J. W., and Hodgson, G. W.,** Water quality and forest management: the effects of clear-cutting on organic compounds in surface waters of the Marmot Creek drainage basin, Report of the Environmental Sciences Centre (Kananaskis), University of Calgary, Canada, 1976.
34. **Baker, B. L., Telang, S. A., and Hodgson, G. W.,** Organic water quality studies in the Red Deer basin: baseline data for effects of dam construction and muskeg leaching, a Report of the Environmental Sciences Centre (Kananaskis), University of Calgary, Canada, 1975.
35. **Oliver, B. G., Cosgrove, E. G., and Carey, J. H.,** Effect of suspended sediments on the photolysis of organics in water, *J. Am. Chem. Soc.,* 13, 1075, 1979.
36. **Klotz, R. L. and Matson, E. A.,** Dissolved organic carbon flumes in the Shetusket River of eastern Connecticut, *Freshwater Biol.,* 8, 47, 1978.
37. **Lewis, W. M. and Grant, M. C.,** Relationships between stream discharge and yield of dissolved substances from a Colorado mountain watershed, *Soil Sci.,* 128(6), 353, 1979.
38. **Mulholland, P. J. and Keunzler, E. J.,** Organic carbon export from upland and forested wetland watersheds, *Limnol Oceanogr.,* 24(5), 960, 1979.
39. **Hobbie, J. E. and Lichens, G. E.,** Output of phosphorous, dissolved organic carbon and fine particulate carbon from Hubbard Brook waterwheds, *Limnol Oceanogr.,* 18, 734, 1973.
40. **Schnitzer, M.,** Metal-organic matter interactions in soils and waters, in *Organic Compounds in Aquatic Envronments,* Faucst, J. S. and Hunter, J. V., Eds. Marcel Dekker, New York, 1971.
41. **Schnitzer, M. and Hansen, E. H.,** Organo-metallic interactions in soils: an evaluation of methods for the determination of stability constants of metal and fulvic complexes, *Soil Sci.,* 109, 333, 1970.
42. **Chau, Y. K.,** Complexing capacity of natural waters — its significance and measurement, *J. Chromatogr. Sci.,* 11, 579, 1973.
43. **Wershaw, R. L. and Pinckney, D. J.,** Isolation and characterization of clay-humic complexes, in *Contaminants in Sediments,* Vol. 1, Baker, R. A., Ed., Ann Arbor Science, Ann Arbor, MI, 1980.

44. **Perdue, E. M.,** Association of organic pollutants with humic substances partitioning equilibria and hydrolysis kinetics, in *Aquatic and Terrestial Humic Materials,* Christman, R. F. and Gjesing, E. T., Eds., Ann Arbor Science, Ann Arbor, MI, 1983.

45. **Hance, R. J.,** Influence of pH, exchangeable cation and the presence of organic matter on the adsorption of some herbicides by montmorillonite, *Can. J. Soil Sci.,* 49, 357, 1969.

46. **Wershaw, R. L., Burcar, P. J., and Goldberg, M. C.,** Interaction of pesticides with natural organic material, *Environ. Sci. Technol.,* 3, 271, 1969.

47. **Poapst, P. A. and Schnitzer, M.,** Fulvic acid and adventitious root formation, *Soil Biol. Biochem.,* 3, 215, 1971.

48. **Weber, J. B., Weed, S. B., and Ward, T. M.,** Adsorption of S-triazines by soil organic matter, *Weed Sci.,* 17, 417, 1969.

49. **Khristeva, L. A. and Luk'yanenko, N. V.,** Soil humic acids mobilization and their influence on crop yields (English trans.), *Sov. Soil Sci.,* 1137, 1962.

50. **Rook, J. J.,** Haloforms in drinking water, *J. Am. Water Works Assoc.,* 68, 168, 1976.

51. **Rook, J. J.,** Formation of haloforms during the chlorination of natural waters, *Water Treat. Exam.,* 23, 234, 1974.

52. **Stevens, A. A., Slocun, C. J., Seeger, D. R., and Robeck, G. G.,** Chlorination of organics in drinking water, *J. Am. Water Works Assoc.,* 68, 615, 1976.

53. **Christman, R. F., Johnson, J. D., Hass, J. R., Pfaender, F. K., Liao, W. T., Norwood, D. L., and Alexander, H. J.,** Natural and model aquatic humics: reactions with chlorine, in *Water Chlorination: Environmental Impact and Health Effects,* Jolly, R. L., Gorchev, H., and Hamilton, D. H., Eds., Ann Arbor Science, Ann Arbor, MI, 1978.

54. **Oliver, B. G.,** Chlorinated non-volatile organics produced by the reaction of chlorine with humic materials, *Can. Res.,* 11(6), 1978.

55. **Oliver, B. G. and Lawrence, J.,** Haloforms in drinking water: a study of precursors and precursor removal, *J. Am. Water Works Assoc.,* 71, 161, 1979.

56. **Rook, J. J.,** Chlorination reactions of fulvic acids in natural waters, *Environ. Sci. Technol.,* 11, 478, 1977.

57. **Youssefi, M., Zenchelsky, S. T., and Faust, S. D.,** Chlorination of naturally-occuring organic compounds in water, *J. Environ. Sci. Health,* A13, 629, 1978.

58. **Love, O. T., Coswell, J. K., Miltner, R. J., and Symons, J. M.,** Treatment for the prevention or removal of trihalomethanes in drinking water, Appendix 3, in *Interim Treatment Guide for the Control of Chloroform and Other Trihalomethanes,* Municipal Environmental Research Laboratory, Environmental Protection Agency, Cincinnati, OH, 1976.

59. **Thurman, E. M. and Malcolm, R. L.,** Preparative isolation of aquatic humic substances, *Environ. Sci. Technol.,* 15, 463, 1981.

60. **Stuermer, D. H. and Harvey, G. R.** The isolation of humic substances and alchohol-soluble organic matter from seawater, *Deep Sea Res.,* 24, 303, 1977.

61. **Mantoura, R. F. C. and Riley, J. P.** The analytical concentration of humic substances from natural waters, *Anal. Chim. Acta,* 76, 97, 1975.

62. **Jeffrey, L. M. and Wood, D. W.,** Organic matter in seawater; and evaluation of various methods of isolation, *J. Mar. Res.,* 17, 247, 1958.

63. **Fu, T. and Pocklington, R.,** Quantitative adsorption of organic matter from seawater on solid matrices, *Mar. Chem.,* 13, 255, 1983.

64. **Milanovich, F. P., Ireland, R. R., and Wilson, D. W.,** Dissolved yellow organics: quantitative and qualitative apects of extraction by four common techniques, *Environ. Lett.,* 8(4), 337, 1975.

65. **Black, A. P. and Christman, R. F.,** Chemical characteristics of fulvic acids, *J. Am. Water Works Assoc.,* 55, 897, 1963.

66. **Weber, J. H. and Wilson, S. A.,** The isolation and characterization of fulvic acids and humic acid from river water, *Water Res.,* 9, 1079, 1975.

67. **Williams, P. M. and Zirion, A.,** Scavenging of dissolved organic matter from seawater with hydrated metal oxides, *Nature,* 204, 462, 1964.

68. **Harvey, G. R., Boran, D. A., Chesal, L. A., and Tokar, J. M.,** The structure of marine fulvic and humic acids, *Mar. Chem.,* 12, 119, 1983.

69. **Siebarth, J. and Jensen, A.,** Studies on algal substances in the sea. I. Gelstaff (humic material) in terrestial and marine waters, *J. Exp. Mar. Biol. Ecol.,* 2, 174, 1968.

70. **Snyder, L. R.,** *Principals of Adsorption Chromotography,* E. Arnold, London, 1968.

71. **Oliver, B. G. and Visser, S. A.,** Chloroform production from the chlorination of aquatic humic material: the effect of molecular weight, environment and season, *Water Res.,* 14, 1137, 1978.

72. **Swift, R. S. and Posner, A. M.,** Gel chromotography of humic acid, *J. Soil Sci.,* 22, 237, 1971.

73. **Martin, D. G. and Pierce, R. H.,** A convenient method of analysis of humic acid in fresh water, *Environ. Lett.,* 1, 49, 1971.

74. **Eberle, S. H. and Schweer, K. H.,** Bestimmung von huminsaure und ligninsulfonsaure in wasser durch flunig-flussig estraktion, *Vom Wasser*, 41, 27, 1974.

75. **Salfield, J. C.,** Fractionation of a humic substance preparation with water containing organic solvents, *Landbauforsch. Volkenrode,* 14, 131, 1964.

76. **Martin, F., Durbach, P., Mehta, N. C., and Deuel, K.,** Determination of functional groups of humic substances, *Z. Pflanzenernaehr. Dueng. Bodenkd.,* 103, 27, 1963.

77. **Theng, B. K. G., Wake, J. H. R., and Posner, A. M.,** Fractional precipitation of soil humic acid by ammonium sufate, *Plant Soil,* 29, 305, 1968.

78. **Ferrari, G. and Dell'Agnola, G.,** Fractionation of organic matter of soil by gel filtration through Sephadex, *Soil Sci.,* 96, 418, 1963.

79. **Gjessing, E. T. and Lee, G. F.,** Fractionation of organic matter in natural waters on Sephadex columns, *Env. Sci. Technol.,* 1, 631, 1967.

80. **Lawrence, J. and Carey, J. H.,** unpublished work.

81. **Schnitzer, M. and Skinner, S. I. M.,** *Isotopes and Radiation in Soil Organic Matter Studies,* International Atomic Energy Agency, Vienna, 1968.

82. **Forsyth, W. G. C.,** Studies on the more soluble complexes of soil organic matter. I. A method of fractionation, *Biochem. J.,* 41, 176, 1947.

83. **Khan, S. U. and Schnitzer, M.,** Sephodex gel filtration of fulvic acid: the identification of major components in two low molecular weight fractions, *Soil Sci.,* 112, 231, 1971.

84. **Dragunov, S. S. and Murzakov, B. G.,** Heterogeneity of ordinary chernozen fulvic acids (Engl. trans.), Sov. Soil Sci., 220 1970.

85. **Archegova, I. B.,** Experimental freezing of humic acid solutions, *Sov. Soil Sci.,* 6, 757, 1967.

86. **Karpenko, N. P. and Karavazev, N. M.,** Extracting humic acids by freezing, *Sov. Soil Sci.,* 10, 1154, 1966.

87. **MacCarthy, P., Peterson, M. J., Malcolm, R. L., and Thurman, E. M.,** Separation of humic substances by pH gradient desorption from a hydrophobic resin, *Anal. Chem.,* 51, 2041, 1979.

88. **Thurman, E. M. and Malcolm, R. L.,** Structural study of humic substances: new approaches and methods, in *Aquatic and Terrestrial Humic Materials,* Christman, R. F. and Gjessing, E. T., Eds., Ann Arbor Science, Ann Arbor, MI, 1983.

89. **Adler, E.,** Structural elements of lignin and solonetzic and black chernozenic soils of Alberta, *Ind. Eng. Chem.,* 49, 1377, 1957.

90. **Khan, S. U. and Sowden, F. J.,** Distribution of nitrogen in the black solonetzic and black chernozemic soils of Alberta, *Can. J. Soil Sci.,* 51, 185, 1971.

91. **Khan, S. U. and Sowden, F. J.,** Distribution of nitrogen in fulvic acid fraction extracted from the black solonetzic and black chernozemic soils of Alberta, *Can. J. Soil Sci.,* 52, 116, 1972.

92. **Thurman, E. M. and Malcolm, R. L.,** Concentration and fractionation of hydrophobic organic constituents from natural waters by liquid chromotography, Paper No. 1817-G, *U.S. Geological Survey Water Supply,* 1979.

93. **Ertel, J. R. and Hedges, J. L.,** Bulk chemical and spectroscopic properties of marine and terrestrial humic acids, melanoidins and catechol-based synthetic polymers, in *Aquatic and Terrestrial Humic Materials,* Christman, R. F. and Gjessing, E. T., Eds., Ann Arbor Science, Ann Arbor, MI, 1983.

94. **Stuermer, D. H. and Payne, J. R.,** Investigations of seawater and terrestrial humic substances with carbon-13 and proton nuclear magnetic resonance, *Geochim. Cosmochim. Acta,* 40, 1109, 1976.

95. **Hatcher, P. G.,** The Origin, Composition, Chemical Structure and Diagenisis of Humic Substances, Coals and Keragens as Studied by Nuclear Magnetic Resonance, Ph.D. thesis, University of Maryland, College Park, MD, 1980.

96. **Gillam, A. H. and Wilson, M. A.,** Application of C-NMR spectroscopy to the structural elucidation of dissolved marine humic substances and their phytoplanktonic precursors, in *Aquatic and Terrestrial Humic Materials,* Christman, R. F. and Gjessing, E. T., Eds., Ann Arbor Science, Ann Arbor, MI, 1983.

97. **Hatchev, P. G., Breger, I. A., Dennis, L. W., and Maciel, G. E.,** Solid state C-NMR of sedimentary humic substances: new revelations on their chemical composition, in *Aquatic and Terrestrial Humic Materials,* Christman, R. F. and Gjessing, E. T., Eds., Ann Arbor Science, Ann Arbor, MI, 1983.

98. **Steelink, C., Mikita, M. A., and Thorn, K. A.,** Magnetic resonance studies of humates and related compounds, in *Aquatic and Terrestrial Humic Materials,* Christman, R. F. and Gjessing, E. T., Eds., Ann Arbor Science, Ann Arbor, MI, 1983.

99. **Steelink, C. and Tollin, G.,** Free radicals in soil, in *Soil Biochemistry,* Vol. I., McLaren, A. D. and Peterson, G. H., Eds., Marcel Dekker, New York, 1967.

100. **Nagar, B. R., Datta, N. P., Das, M. R., and Krakhar, M. P.,** Origin and structure of soil humic acids by electron spin resonance, *Indian J. Chem.,* 5, 587, 1967.

101. **Rockwell, A. L. and Larson, R. A.,** Fluorescence and its relation to the structure of dissolved organic polymers, extended abstract, Division of Environmental Chemistry, American Chemical Society, April, 1979.

102. **Larson, R. A.,** Dissolved organic matter of a low coloured stream, *Freshwater Biol.,* 8, 91, 1978.
103. **Carey, J. H.,** unpublished data.
104. **Moore, P. D. and Bellamy, D. J.,** *Peatlands,* Springer Verlag, Berlin, 1974.
105. **Verry, E. S.,** Streamflow chemistry and nutrient yields from upland peatland watersheds in Minnesota, *Ecology,* 56, 1149, 1975.
106. **Patrick, R., Bineeti, V. P., and Holterman, S. F.,** Acid lakes from natural and anthropogenic causes, *Science,* 446, 1981.
107. **Oliver, B. G., Thurman, E. M., and Malcolm, R. L.,** The contribution of humic substances to the acidity of coloured natural waters, *Geochim. Cosmochim. Acta,* 47, 2031, 1983.
108. **Perdue, E. M., Reuter, J. H., and Ghosal, M.,** The operational nature of acidic functional group analyses and its impact on mathematical descriptions of acid-base equilibrium in humic substances, *Geochim. Cosmochim. Acta,* 44, 1841, 1950.
109. **Gamble, D. S.,** Titration curves of fulvic acid: the analytical chemistry of a weak acid polyelectrolyte, *Can. J. Chem.,* 48, 1970.
110. **Wershaw, R. L., Burcar, P. J., Sutula, C. L., and Wiginton, B. J.,** Sodium humate solution studied with small angle X-ray scattering, *Science,* 157, 1429, 1967.
111. **Kemp, A. L. W. and Wong, H. K. T.,** Molecular weight distribution of humic substances from Lake Ontario and Erie sediments, *Chem. Geol.,* 14, 15, 1974.
112. **Wilander, A.,** A study on the fractionation of organic matter in natural water by ultrafiltration techniques, *Schweiz. A. Hydrol.,* 34, 190, 1972.
113. **Hansen, E. H. and Schnitzer, M.,** Molecular weight measurements of polycarboxylic acids in water by vapour pressure osnometry, *Anal. Chim. Acta,* 46, 247, 1969.
114. **Flaig, W. A. J. and Beutelspacher, H.,** *Isotopes and Radiation in Soil Organic Matter Studies,* International Atomic Energy Agency, Vienna, 1968.
115. **Wilson, S. A. and Weber, J. H.,** A comparative study of number-average dissociation-corrected molecular weights of fulvic acids isolated from water and soil, *Chem. Geol.,* 19, 285, 1977.
116. **Visser, S. A.,** A physico-chemical study of the properties of humic acids and their changes during humification, *J. Soil. Sci.,* 15, 202, 1964.
117. **Matsuda, K. and Schnitzer, M.,** The permanganate oxidation of humic acids extracted from acid soils, *Soil Sci.,* 114, 185, 1972.
118. **Burges, N. A., Hurst, H. M., and Walkden, S. B.,** The phenolic constituents of humic acid and their relation to the lignin of the plant cover, *Geoch. Cosmochim. Acta,* 28, 1547, 1964.
119. **Haworth, R. D.,** The chemical nature of humic acid, *Soil Sci.,* 111, 71, 1971.
120. **Freudenberg, K.,** Lignin. Its constitution and formation from p-hydroxycinnamyl alcohols, *Science,* 148, 595, 1965.
121. **Pearl, I. A.,** *The Chemistry of Lignin,* Marcel Dekker, New York, 1967.
122. **Vickery, H. B. and Schmidt, C. L. A.,** History of the discovery of the amino acids, *Chem. Rev.,* 9, 169, 1931.
123. **Wilde, S. A.,** Rapid colorometric determination of soil organic matter, *Soil Sci. Soc. Am. Proc.,* 7, 393, 1942.
124. **Schnitzer, M.,** Characterization of humic constituents by spectroscopy, in *Soil Biochemistry,* McLaren, A. D. and Shujins, Eds., Marcel Dekker, New York, 1971.
125. **Ertel, J. R. and Hedges, J. I.,** Bulk chemical and spectroscopic properties of marine and terrestrial humic acids, melanoidins and catechol-based synthetic polymers, in *Aquatic and Terretrial Humic Materials,* Christman, R. F. and Gjessing, E. T., Eds., Ann Arbor Science, Ann Arbor, MI, 1983.
126. **Stevinson, F. J.,** *Humus Chemistry,* John Wiley & Sons, New York, 1982.
127. **Visser, S. A.,** Fluorescence phenomena of humic matter of aquatic origin and microbial cultures, in *Aquatic and Terrestrial Humic Materials,* Christman, R. F. and Gjessing, E. T., Eds., Ann Arbor Science, Ann Arbor, MI, 1983.
128. **Ghassemi, M. and Christman, R. F.,** Properties of the yellow organic acids of natural waters, *Limnol. Oceanogr.,* 13, 583, 1968.
129. **Stewart, A. J. and Wetzel, R. G.,** Fluorescence: adsorbance ratios — a molecular weight tracer of dissolved organic matter, *Limnol. Oceanogr.,* 25, 559, 1981.
130. **Hall, K. J. and Lee, G. F.,** Molecular size and spectral characterization of organic matter in a meromictic lake, *Water Res.,* 8, 239, 1974.
131. **Lawrence, J.,** Semiquantitative determination of fulvic acid, tannin and lignin in natural waters, *Water Res.,* 14, 373, 1980.
132. *Standard Methods for the Examination of Wter and Wastewater,* 14th ed., published by American Public Health Association, American Water Works Association, and Water Pollution Control Federation, Washington, D.C., 1976.
133. **Alberts, J.,** The effects of metal ions on the U.V. spectra of humic acid, tannic acid and lignosulfonic acid, *Water Res.,* 16(7), 1273, 1982.

134. **Blackburn, S.,** *Amino Acid Determination — Methods and Techniques,* Marcel Dekker, New York, 1968.

135. **Pierce, A. E.,** *Silation of Organic Compounds,* Pierce Chemical Co., Rockford, IL, 1968.

136. **de Haan, H.,** On the determination of soluble humic substances in fresh water, in *Humic Substances: Their Structure and Function in the Biosphere,* Povoledo, D. and Golterman, H. L., Eds., Centre for Agricultural Publishing and Documentation, Wageningen, The Netherlands, 1973, 53.

137. **Carpenter, P. D. and Smith, J. D.,** Simultaneous spectrophotometric determination of humic acid and iron in water, *Anol. Chim. Acta.,* 159, 299, 1984.

138. **Plechanov, N.,** Studies of molecular weight distributions of fulvic and humic acids by gel permeation chromatography, *Org. Geochem.,* 5(3), 143, 1983.

139. **Madhun, Y. A., Young, J. L., and Freed, V. H.,** Binding of herbicides by water-soluble organic materials from soils, *J. Environ. Qual.,* 15(1), 64, 1986.

INDEX

H

I

K

L